Reproductive Genomics in Domestic Animals

Reproductive Genomics in Domestic Animals

Edited by

Zhihua Jiang
Troy L. Ott

WILEY-BLACKWELL
A John Wiley & Sons, Inc., Publication

Edition first published 2010

Blackwell Publishing was acquired by John Wiley & Sons in February 2007. Blackwell's publishing program has been merged with Wiley's global Scientific, Technical, and Medical business to form Wiley-Blackwell.

Editorial Office
2121 State Avenue, Ames, Iowa 50014-8300, USA

For details of our global editorial offices, for customer services, and for information about how to apply for permission to reuse the copyright material in this book, please see our website at www.wiley.com/wiley-blackwell.

Library of Congress Cataloging-in-Publication Data

Reproductive genomics in domestic animals / editors, Zhihua Jiang, Troy L. Ott. – 1st ed.
p. cm.
Includes bibliographical references and index.
ISBN 978-0-8138-1784-2 (hardback : alk. paper) 1. Domestic animals–Genetics.
2. Domestic animals–Reproduction. 3. Livestock–Genetics. 4. Livestock–Reproduction.
5. Genomics–Research. I. Jiang, Zhihua, 1959– II. Ott, Troy L.
SF105.R45 2010
636.08'21–dc22
2009049307

A catalog record for this book is available from the U.S. Library of Congress.

Set in 10 on 13 pt Trump Mediaeval by Toppan Best-set Premedia Limited
Printed in Singapore

1 2010

Contents

Contributors

Steve Bischoff, Department of Molecular Biomedical Sciences and Center of Comparative Medicine and Translational Research, College of Veterinary Medicine, North Carolina State University, Raleigh, NC 27606

Kyle Caires, Department of Animal Sciences, Center for Reproductive Biology, Washington State University, Pullman, WA

Eduardo Casas, USDA, Agricultural Research Service, U.S. Meat Animal Research Center, Clay Center, NE 68933-0166

Jie Chen, Department of Animal Sciences, Center for Reproductive Biology, Washington State University, Pullman, WA 99164-6351; and College of Animal Sciences and Technology, Nanjing Agricultural University, Nanjing 210095, China

Noelle E. Cockett, Department of Animal, Dairy and Veterinary Sciences, Utah State University, Logan, UT 84322-4900

Valéria Conforti, Cincinnati Zoo & Botanical Garden, 3400 Vine Street, Cincinnati, OH 45220-1399

Corinne Cotinot, CNRS, FRE 2857, F-78350, Jouy-en-Josas, France

Lennart Dencker, Department of Pharmaceutical Sciences at Biomedical Centre, P.O. Box 594, Uppsala University, SE-756 45 Uppsala, Sweden

Peter Dovc, Department of Animal Science, University of Ljubljana, Groblje 3, SI-1230 Domzale, Slovenia

J. Joe Ford, USDA, Agricultural Research Service, U.S. Meat Animal Research Center, Clay Center, NE 68933-166

Zhihua Jiang, Department of Animal Sciences, Center for Reproductive Biology, Washington State University, Pullman, WA 99164-6351

Jiri Jonak, Laboratory of Diagnostic for Reproductive Medicine, Institute of Molecular Genetics, Academy of Sciences of the Czech Republic, Videnska 1083, 142 20 Prague 4, Czech Republic

Vera Jonakova, Laboratory of Diagnostics for Reproductive Medicine, Institute of Molecular Genetics, Academy of Sciences of the Czech Republic, Videnska 1083, 142 20 Prague 4, Czech Republic

Larry A. Kuehn, USDA, Agricultural Research Service, U.S. Meat Animal Research Center, Clay Center, NE 68933

Tanja Kunej, Department of Animal Science, University of Ljubljana, Groblje 3, SI-1230 Domzale, Slovenia

Wansheng Liu, Department of Dairy and Animal Science, Center for Reproductive Biology and Health, The Pennsylvania State University, University Park, PA 16802

Ulf Magnusson, Department of Clinical Sciences and Centre for Reproductive Biology in Uppsala, P.O. Box 7054, Swedish University of Agricultural Sciences, SE-750 07 Uppsala, Sweden

Derek McLean, Department of Animal Sciences, Center for Reproductive Biology, Washington State University, Pullman, WA

John P. McNamara, Department of Animal Sciences, Center for Reproductive Biology, Washington State University, Pullman, WA 99164-6351

Jennifer J. Michal, Department of Animal Sciences, Center for Reproductive Biology, Washington State University, Pullman, WA 99164-6351

Sukanta Mondal, Division of Animal Physiology, National Institute of Animal Nutrition and Physiology (Indian Council of Agricultural Research), Adugodi, Bangalore—560030, Karnataka, India

Holly Neibergs, Department of Animal Sciences, Center for Reproductive Biology, Washington State University, Pullman, WA 99164-6351

Dan J. Nonneman, USDA, Agricultural Research Service, U.S. Meat Animal Research Center, Clay Center, NE 68933

Jon Oatley, Department of Dairy and Animal Sciences, Center for Reproductive Biology and Health, The Pennsylvania State University, University Park, PA

Troy L. Ott, Department of Dairy and Animal Sciences, Center for Reproductive Biology and Health, The Pennsylvania State University, University Park, PA 16802

Eric Pailhoux, INRA, UMR 1198 Biologie du Développement et Reproduction, F-78350 Jouy-en-Josas, France

Luc J. Peelman, Department of Nutrition, Genetics and Ethology, Faculty of Veterinary Medicine, Ghent University, 9820 Merelbeke, Belgium

Jorge A. Piedrahita, Department of Molecular Biomedical Sciences and Center of Comparative Medicine and Translational Research, College of Veterinary Medicine, North Carolina State University, Raleigh, NC 27606

Gary A. Rohrer, USDA, Agricultural Research Service, U.S. Meat Animal Research Center, Clay Center, NE 68933-0166

Beau Schilling, Laboratory of Mammalian Reproductive Biology and Genomics, Department of Animal Science, Michigan State University, East Lansing, MI 48824-1225

George W. Smith, Laboratory of Mammalian Reproductive Biology and Genomics, Department of Physiology, Michigan State University, East Lansing, MI 48824-1225

Thomas E. Spencer, Department of Animal Science, Center for Animal Biotechnology and Genomics, Texas A&M University, College Station, TX 77843-2471

Marie Ticha, Laboratory of Diagnostic for Reproductive Medicine, Institute of Molecular Genetics, Academy of Sciences of the Czech Republic, Videnska 1083, 142 20 Prague 4, Czech Republic

Shengdar Tsai, Department of Molecular Biomedical Sciences and Center of Comparative Medicine and Translational Research, College of Veterinary Medicine, North Carolina State University, Raleigh, NC 27606

Jeffrey L. Vallet, USDA, Agricultural Research Service, U.S. Meat Animal Research Center, Clay Center, NE 68933

Galen A. Williams, Devers Eye Institute, 1225 NE 2nd Ave., Portland, OR 97232

Ricardo Zanella, Department of Animal Sciences, Center for Reproductive Biology, Washington State University, Pullman, WA 99164-6351

Preface

Reproductive efficiency has been considered one of the most critical factors affecting the productivity and profitability of the livestock industries. Unfortunately, in spite of a significant improvement in growth, feed efficiency, and carcass and meat quality due to genetic selection and management advances, reproductive efficiency has declined in most livestock species. Due to low heritabilities, and sex-limited complexity, it has been very difficult to improve reproductive traits using traditional selection methods. The rapid development of molecular genetics and genomics in recent years has, however, enabled the identification, characterization, and utilization of genes and pathways that contribute to the genetic complexity of reproduction in domestic animals. This book reviews the current status of reproductive genomics, transcriptomics, and proteomics and highlights the current and potential genomics tools and reagents for improving reproductive efficiency in domestic animals. It is our goal to have in the book a broad coverage on genome sciences and bio-technologies that can help address and understand various aspects of fertility and infertility in domestic animals.

The book consists of three main parts. Part I has seven chapters that focus on genome resources and quantitative genomics of reproduction. Chapter 1 demonstrates genome resources specifically available to livestock species, such as well-characterized genome maps, whole genome and cDNA sequences, expression arrays, and high-density genetic marker chips. Chapter 2 defines the female reproductive phenotypes and updates the genes/quantitative trait loci associated with the traits. Chapter 3 defines the male reproductive phenotypes and updates the genes/quantitative trait loci associated with these traits. Chapters 2 and 3 also include methods and technologies for the development and discovery of genomic markers as well as their genotyping formats. Chapter 4 covers genetic and genomic aspects of reproductive disorders associated with the ovary, vagina, and uterus; pregnancy and placenta; male reproductive organs; and embryos and fetuses. Chapter 5 deals with genetic and genomic aspects of reproductive diseases, such as paratuberculosis, respiratory disease, and brucellosis in cattle, and leptospirosis, Aujeszky's disease, and porcine reproductive and respiratory syndrome in swine. Chapter 6 focuses on the structure, function, and evolution of the Y chromosome and its effect on male fertility. Chapter 7 describes both mitochondrial genomes in the cytoplasm and nucleus and their involvements in male reproduction, female reproduction, embryo development, and reproductive aging.

Part II of this book possesses five chapters that target transcriptomics and physiological genomics of reproduction, which link genes to physiology and pathways critical for reproductive success. Chapter 8 deals with transcriptomics of ovarian tissues involved in follicular growth and development, luteinization of the dominant follicle, corpus luteum regression, oocyte maturation, and oocyte competence. Chapter 9 focuses on transcriptomics related to different preimplantation development stages and systems. Chapter 10 focuses on the genomics endometrial responses to conceptus signals mediating corpus luteum rescue. Chapter 11 reviews hormones and peptides, and transcriptomics involved in placental development. Chapter 12 targets cellular, molecular, and genomic mechanisms regulating testis function in livestock with emphasis on transcriptomics of the testis in bulls and reproductive genomics in boars.

Part III has six chapters that deal with genomics of reproductive biotechnology and their applications. Chapter 13 discusses the importance of nuclear reprogramming during somatic cell nuclear transfer and its implications for normal fetal and placental development. Chapter 14 describes how to use immunocontraception and immunosterilization as methods of fertility control in animals. Chapter 15 deals with the structure and properties of seminal plasma proteins and their potential roles in fertilization affecting the oviductal reservoir, and as capacitation modulators, gamete interaction enhancers, and enzyme inhibitors. Chapter 16 reports the state of knowledge on sex differentiation in domestic mammals and sex determination in nonmammal domestic species. Chapter 17 addresses the disruption of the reproductive endocrine systems and the mechanisms of action of endocrine disrupting chemicals that exert hormone-like activity in humans and animals. Chapter 18 focuses on the mechanistic connections between nutrient flux and reproductive processes with emphasis on nutritional physiology of pregnancy and lactation and demonstrates how nutrigenetics and nutrigenomics approaches can improve fertility, pregnancy, and lactation.

This book is for researchers, instructors, extension experts, and students in animal, veterinary, and biomedical sciences who are interested in quantitative genomics, physiological genomics, mitochondriomics, pathological genomics, epigenomics, nutrigenomics, evolutionary genomics, and proteomics of reproduction. The 37 contributors to the book are all internationally recognized experts in their field, and they represent 15 different institutions from seven different countries. We thank them for their contributions to this first book on reproductive genomics of domestic animals. Support from both families has been essential for us to finish the book project, and we are grateful for their patience. Thanks also to Justin Jeffryes, Susan Engelken, Shelby Allen, the Wiley-Blackwell publishing team, and the team at Toppan Best-set Premedia for their extra care and patience in publishing the book.

Zhihua Jiang
Troy L. Ott

Reproductive Genomics in Domestic Animals

Part I

Quantitative Genomics of Reproduction

1

Reproductive Genomics: Genome, Transcriptome, and Proteome Resources

Noelle E. Cockett

1.1 Introduction

Genomic resources, tools, and technologies that can be applied to studies in livestock species, including investigations related to reproduction, have been under development for the last decade. While many of the genomic approaches were originally developed for use in humans or laboratory model animals, they have been successfully applied to studies in livestock. There are now a myriad of resources specific to livestock species, such as well-characterized genome maps, high-resolution genome, and complementary DNA (cDNA) sequences, expression arrays, and high-density genetic marker chips. In addition, there is an explosion of high-throughput technology that will enhance these investigations, increasing the scope and accuracy of the results beyond anything that was imagined just 5 years ago. These technologies advance studies of single gene expression to full gene networks, from single gene sequences to whole genomes, and from hundreds of genetic markers to tens of thousands markers—all assayable in a few weeks to months as opposed to years.

These resources and technologies can be combined in innovative ways to advance two areas of research on reproductive traits, specifically the identification of genes or genetic regions influencing phenotypes and the characterization of expression of genes that are associated with traits.

1.2 Discovery of underlying genetic influences

The first area of interest for researchers studying reproductive traits is the characterization of genetic variation among animals or populations underlying a phenotypic trait, leading to the identification of the genetic cause of the phenotype. Two general approaches have been successfully used over the last 10–15 years, with a third approach now on the horizon. In the first approach, polymorphisms in a candidate gene likely to be involved in the phenotype are tested for associations with different manifestations or phenotypes of the trait. The candidate genes are selected for analysis based on an understanding of trait physiology and/or because of their involvement in similar traits in other species. In the second approach, genetic markers are analyzed for linkage with the phenotype using pedigrees of animals segregating for the trait and the markers. This analysis identifies genetic regions that contain associated genes. By testing additional markers through the families, the interval is narrowed so candidate genes can be selected. The third approach, referred to as whole genome associations, will soon be possible for livestock species now that the development of high-density single nucleotide polymorphism (SNP) arrays are readily available. However, the application of whole genome associations requires very large numbers of phenotyped animals, which is a limitation for most research projects.

1.2.1 Candidate gene associations

As mentioned, the candidate gene approach uses information of the trait to determine likely candidates for the underlying gene(s). The choice of the gene is strengthened by its involvement in comparable traits in other species or its location in a region previously identified as containing a quantitative trait loci (QTL) with similar attributes.

In the past, polymorphisms in a candidate gene were routinely detected by polymerase chain reaction—restriction fragment length polymorphisms (PCR-RFLP), which involves steps of amplifying the gene, digesting the amplicon with a restriction enzyme, and then using gel electrophoresis to separate the resulting fragments. In the PCR-RFLP technique, gene sequence differences among animals are detected by whether or not a restriction enzyme cuts, resulting in different-sized fragments. The genetic differences are usually due to an SNP within the restriction enzyme recognition site, although there might be genetic differences due to insertions/deletions (in/del) in the gene, which will also result in fragment size differences, although there is no variation in the restriction enzyme recognition site. Animals are expected to have two alleles for every gene except those on the X and Y chromosomes in males, so that the presence of one fragment on the electrophoresis gel would indicate that an animal is homozygous for the PCR-RFLP allele whereas the presence of two different-sized fragments would suggest that an animal is heterozygous. However, an animal might be misclassified as a homozygote if there is a polymorphism in the PCR primer sequence, which prevents that allele from being amplified and therefore, not detected on the electrophoresis gel—referred to as a "null" allele. A null allele will often be detected when misparentages are routinely found for a marker system. An animal might also be misclassified if another, nonallelic form of the gene is amplified with the PCR primers and digestion with the restriction enzyme results in a different-sized fragment. A nonallelic form is revealed by sequencing the fragments contained within the electrophoretic bands, which is a recommended step when establishing any marker system.

However, new technologies have significantly advanced our ability to identify SNPs and then explore multiple candidate genes at one time at a much lower cost/polymorphism than the PCR-RFLP method. The identification of SNPs within a gene or genetic region is now relatively easy. To do this, the genomic DNA of key animals within a population is sequenced using high-throughput automatic sequencing and then compared with other sequences within the population or to sequences in publically available databases. The later approach is referred to as *in silico* SNP detection. Regardless of the approach, confidence of the SNP is dependent on the quality of the sequence across the multiple sources of data.

Once an SNP is identified, the polymorphism can be detected by establishing a PCR-RFLP assay. However, allele-specific PCR using allele-specific oligonucleotides (ASOs) is an emerging technique for detecting genetic variation created by the SNP (Saiki et al. 1986). The 3′ ends of the primers used in the PCR amplification step of the ASO technique are designed to include the polymorphic site so that amplification of the animal's DNA is dependent on the absence or presence of the polymorphism within the primer sequence. Allele-specific primers can be combined into a single amplification reaction and the presence of the specific allele detected by the melting temperature of the alleles (Papp et al. 2003; Wang et al. 2005). Appropriate controls and design of the primers (e.g., Strerath et al. 2007) are critical in the allele-specific amplification assay so that absence of amplification is due to the polymorphism and not because of technical problems.

SNP arrays are an extension of the ASO method, but by spotting multiple ASOs onto a membrane or bead, multiple alleles or even multiple genetic markers can be assayed in a single run. Custom-built SNP chips specific to a trait are usually designed in a 92-, 384-, or 1534-SNP format. While the cost/SNP is lower for the SNP chip than with the PCR-RFLP or allele-specific amplification techniques, the initial setup for the chip is substantially higher. Thus, the number of SNPs that are tested and the number of animals included in the analysis will determine whether a custom-built SNP array is economical.

Emerging technology is now allowing the detection of differences in copy number variant (CNV) among animals. For some time, copy number variation has been associated with diseases (McCarroll 2008; Schaschi et al. 2009), while the ongoing analyses of livestock whole genome sequences has revealed the presence of CNV in multiple gene systems involved with innate immunity, including milk composition traits (Rijnkels et al. 2009; Tellam and Bovine Genome Sequencing and Analysis Consortium 2009). Detection of differences among animals for genes that are known to be present in the genome in multiple copies is now possible using microarray technology (Baumbusch et al. 2008), with higher copy number resulting in greater intensity for that spot on the array.

Once the polymorphism is detected within a population, the genotypes are usually analyzed for association with the trait by comparing the trait means among the marker genotypes (Rocha et al. 1992). Appropriate statistical models are needed in order to account for additive, dominant, and epistatic effects. In addition, the selection of animals used in the analysis must be sufficiently broad; otherwise, the marker alleles will merely serve as a trace of unique families, particularly when one of the alleles is at a very low frequency in the population and present only in one family in the analysis. This situation can result in a spurious significant association, simply because the family differs for the trait and not because the allele itself is associated.

The choice of the candidate gene(s) can be strengthened by its association with similar traits in the same or other species. Possible candidate genes can be found through literature searches using key words based on

Table 1.1 Websites containing genomics information in livestock species.

Species	Website	Information
Cattle	www.animalgenome.org/QTLdb/cattle.html	QTL
	www.vetsci.usyd.edu.au/reprogen/QTL_Map/	QTL
	www.hgsc.bcm.tmc.edu/projects/bovine/	Genome sequence
	bovinegenome.org	Genome project
Goat	dga.jouy.inra.fr/cgi-bin/lgbc/main.pl?BASE=goat	Genome project
Horse	www.uky.edu/Ag/Horsemap/welcome.html	Genome project
	www.broad.mit.edu/mammals/horse	Genome sequence
Sheep	rubens.its.unimelb.edu.au/~jillm/jill.htm	Primary Web source
	www.livestockgenomics.csiro.au/perl/gbrowse.cgi/vsheep2/	Virtual sheep genome
	www.ncbi.nlm.nih.gov/genome/guide/sheep/index.html	NCBI resources
	www.sheephapmap.org/	International Sheep Genome Consortium
Pig	www.animalgenome.org/QTLdb/pig.html	QTL
	www.sanger.ac.uk/Projects/S_scrofa/	Genome sequence
	www.piggenome.org/index.php	Genome project

the trait physiology or through searches of databases devoted to genetic abnormalities. One such database for livestock traits is called Online Mendelian Inheritance in Animals (OMIA; www.omia.angis.org.au/). The OMIA database contains details on genes, inherited disorders, and traits for a large range of animals species, similar to what is found within Online Mendelian in Man (OMIM; www.ncbi.nlm.nih.gov/sites/entrez?db=omim).

There are also databases that describe the location of QTLs for traits of interest in livestock species (Table 1.1). Additional candidate genes can be identified by searching genetic sequences that lie within QTL intervals and have involvement in the physiology of the trait, providing not only functional evidence but also positional evidence for inclusion in the candidate gene analysis. These genes are therefore referred to as "positional candidate genes" (see below).

1.2.2 *Analysis of genetic variation*

The second approach for detecting genes or, more commonly, genetic regions involved in traits is based on identifying and characterizing genetic variation that is found in pedigrees of animals. This approach has most commonly been done using linkage analysis, which examines the segregation of marker alleles through animal families with known phenotypes (Nejati-Javaremi and Smith 1995; Knott and Haley 2000; de Koning et al. 2003) and subsequent refinement of the genetic interval containing the trait locus (Riquet et al. 1999; Farnir et al. 2002). The data are analyzed to determine the coinheritance of marker alleles with the causative genetic mutation, presumably because they are closely located within the genome.

Linkage mapping requires pedigrees with specific family structures; these pedigrees are most commonly reciprocal backcrosses or F2 crosses developed from lines or breeds of animals that significantly differ for the trait. The analysis can include families within a single breed or line but the key parents must be heterozygous for both the markers and the trait in order for linkage to be detected. As with the association analyses, appropriate statistical models are needed to detect genetic mutations that are controlled by complex gene actions, such as the imprinted callipyge (Cockett et al. 1994, 1996) and IGF2 (Van Laere et al. 2003) loci. The effects of these loci would not have been detected without the appropriate statistical model (see Sandor and Georges 2008).

To perform a screen of markers across the complete genome (i.e., a genome scan), markers are typically selected about one every 10–20 centimorgans (cM). Because the typical mammalian genome is about 3000 cM, around 150–300 markers are needed for a genome scan. The availability of genome-wide maps in livestock species provides the information needed to select markers at appropriate intervals, which is dependent on the number of informative offspring in the families and the genetic variability in the trait. Several reviews on conducting a genome scan and subsequent analysis are available, including Schwerin (2001), Rocha et al. (2002), Andersson and Georges (2004), and Georges (2007).

1.2.3 Whole genome sequence

Genetic markers for a genome scan are usually selected from a genome map. The most complete genome map for a species is produced from a whole genome sequence that has been assembled and annotated. Assembled whole genome sequences are now publically available for cattle, swine, and horses (Table 1.2). Millions of bases of sequences can be accessed for the analysis of genes, SNPs, regulatory features, and so on. Comparisons across species, including non-livestock species, are now possible using "landmark" loci that anchor segments of the genome from species to species. International consortiums of experts have been organized for annotation of the sequences; for example, "the Horse Genome Project is a cooperative international effort by over 100 scientists in 20 countries to define the genome, the DNA sequence, of the domestic horse" (www.uky.edu/Ag/Horsemap/welcome.html). A wealth of knowledge from the analysis of these sequences is now being released.

In addition, assembled whole genome sequences can serve as the "reference" for comparison of individual animal sequences generated with state-of-the-art high-throughput platforms such as ABI's SOLiD™ (Carlsbad, CA), Roche 454 FLX Titanium™ (Branford, CT), and Illumina's Solexa™ (San Diego, CA) systems. These technologies produce millions of reads of short sequences (50–400 bases) in a single run at relatively

Table 1.2 Whole genome sequence assemblies in livestock species.[1]

Species	Reference or website	Sequenced animal	Method	Coverage
Cattle[2]	Baylor School of Medicine	Hereford male L1 Domino 99375	WGS	7.1X
Horse[3]	Broad Institute	Thoroughbred female Twilight	WGS	6.8X
Pig[4]	Sanger Institute	Duroc female	BAC by BAC tile path	4X

[1] As of February 1, 2009.
[2] www.hgsc.bcm.tmc.edu/projects/bovine/.
[3] www.broad.mit.edu/node/318.
[4] www.sanger.ac.uk/Projects/S_scrofa/.
WGS, whole genome shotgun.

Table 1.3 Most recent published linkage maps in livestock species that do not have a whole genome sequence.

Species	Population	No. of loci	Reference
Goat	INRA	307	Schibler et al. (1998a)
Deer	Red deer × Pere David's deer	621	Slate et al. (2002)
Sheep	IMF	1062	Maddox et al. (2001)

low cost ($10,000/10Gb) from either single or pooled DNA samples. The sequences for each run can then be compared back to the reference genome sequences, allowing detection of genetic differences across animals.

1.2.4 *Linkage or genetic maps*

For those species without a whole genome sequence, linkage and physical maps are critical for the orientation of loci as well as comparisons across species. Linkage maps usually contain a preponderance of highly polymorphic anonymous markers, primarily microsatellites, and relatively few expressed genes, which have very limited genetic variability. Also, multiple linkage maps may exist for a species because different reference families were used to create the linkage maps. The various maps are often combined into a consensus linkage map, which is anchored by common markers genotyped in the different reference families (Table 1.3). The distances between loci on linkage maps are given in centimorgans (cM), with 1 cM representing 1% recombination between two loci.

1.2.5 *Physical map assignments*

In addition to the linkage maps, physical maps exist for each species. These maps are created by direct assignment of a gene or marker to an intact chromosome or chromosomal fragment. Physical mapping is usually done by *in situ* hybridization, somatic cell hybrid analysis, or radiation hybrid (RH) mapping. Because a genetic variant within the locus is not necessary for physical mapping, these maps contain a relatively large number of expressed genes.

Table 1.4 Physical map in livestock species that do not have a whole genome sequence.

Species	No. of loci	Reference
River buffalo	388	Di Meo et al. (2008)
Goat	202	Schibler et al. (1998a)
Deer	59	Bonnet et al. (2001)
Sheep	452	Di Meo et al. (2007)

One of the first reports assigning genes to physical locations within the genome was performed by hybridizing a radioactively labeled probe to a spread of metaphase chromosomes in a technique referred to as *in situ* hybridization. A significant adaptation of this method entailed labeling the probe with fluorophores, leading to the moniker of fluorescent *in situ* hybridization (FISH). Hundreds of genes and genetic markers have now been assigned to specific chromosomes in livestock species using *in situ* hybridization techniques (Table 1.4).

Chromosome painting is an approach for evaluating the conservation of chromosomal segments across species. In this technique, chromosomes of one species are fluorescently labeled and hybridized to metaphase chromosomes of another species. Reciprocal chromosome painting has been performed between humans and farm animal species including pigs, cattle, sheep, and horses (Chowdhary et al. 1996; Chowdhary and Raudsepp 2001). These studies have defined the borders of conserved syntenies among the species, but because of insufficient resolution, they do not allow the study of gene order.

Somatic cell hybrid panels were used frequently in the 1970s–1990s to assign genes to specific chromosomes in livestock, but this method is now used much less frequently than other physical mapping approaches that have a much better resolution. A somatic cell hybrid panel is generated by fusing cells of the target species with cells of a rodent species, such as hamsters. The rodent cells randomly eject chromosomes of the target species, until at some point the cells are immortalized, leaving a complement of target chromosomes that is the signature of each somatic cell clone. DNA is harvested from each of the clones in the panel (usually around 30 clones) and then amplified with primers specific to a gene or genetic marker. Those clones that contain a piece of the chromosome harboring the gene amplify with the primers, as well as other "concordant" or linked genes. The somatic cell hybrid approach can be used to identify genes found within long segments of the chromosome, but the order of the genes along the chromosomal segment cannot be determined.

RH mapping provides a higher level of resolution of gene location and gene order than those produced by *in situ* hybridization and somatic cell hybrids. This technique is based on detecting the presence/absence of loci that are contained on fragments of DNA maintained in a panel of hybrid clones, similar to the somatic cell hybrid approach, but the rodent cells are fused with target cells that have been irradiated. Varying the radiation dose on the target species cells will create different-sized fragments and therefore, vary the resolution between two loci. The higher the radiation dose, the smaller the fragments and the better resolution between tightly linked loci. Thus, high rad panels are suitable when fine-mapping markers within a specific region, but large numbers of random markers must be screened in order to detect linkage of loci across the genome. Whole genome maps, which do not require a saturation of markers, are best done with a lower rad RH panel. RH panels have been generated for several livestock species (Table 1.5) and used for generating chromosome and whole genome RH maps.

Distances on an RH map are measured in centiRay (cR), with a distance of 1 cR_{rad} between two markers corresponding to a 1% frequency of breakage between these two markers after exposure to a specific radiation (rad) dose. Statistical programs have been developed to analyze the RH panel data to give the most likely order based on the least number of break points (e.g., Boehnke

Table 1.5 Radiation hybrid maps in livestock species that do not have a whole genome sequence.

Species	RH panel	Rad	No. of loci[1]	Reference
River buffalo	BBURH5000	5,000	3,990	Amaral et al. (2009)
Sheep	USUoRH5000	5,000	2,300	Wu et al. (2007, 2008, 2009)
	INRA	12,000	67	Laurent et al. (2007)

[1] As of February 1, 2009.

1992; Lange et al. 1995; Lunetta et al. 1996). A measure of relative likelihood of one order versus another is given for each map developed with the RH data.

Distance between loci on an RH map is directly proportional to physical distance, measured as the frequency of retention of a given pair of markers. The more times two loci are retained together, the closer they are found on a chromosome. Retention frequency is calculated as the percentage of clones that retain a given marker and is usually between 18% and 30% for whole genome RH panels.

1.2.6 *Positional candidate genes*

Once the location of a trait within the genome is determined because of linkage to previously mapped genetic markers, possible candidate genes controlling the trait can be inferred because of their proximity to the linked markers. A typical genome scan usually assigns the trait locus or QTL to a ~20-cM interval, which can contain hundreds of genes. However, it is not necessary to have map locations of all possible genes in a single livestock species. Rather, a subset of genes that are mapped in well-studied species, such as humans and mice, are also mapped in farm animals; these genes serve as "anchors" across the comparative maps and allow inference of the locations of other genes within a region, based on what is known within the well-mapped species (Burt 2002).

Positional candidate genes can be identified for traits mapped by linkage analysis once markers used in the linkage analysis are located on the comparative map, either by direct mapping or because a gene linked to the marker is placed on the comparative map. Several online comparative map databases have been established, which allow comparisons of genes contained within common genetic regions. These comparative maps can also be used to localize a single gene across multiple species.

Numerous causative mutations for single gene traits in livestock have been identified. In contrast, although numerous QTL have been identified for economically important traits in livestock (see Table 1.1), very few of the causative mutations for QTL (referred to as quantitative trait nucleotides or QTNs) have been characterized. There are numerous challenges in identifying the mutation for a quantitative trait, including a limitation on animals and/or families suitable for narrowing the QTL interval, an often unwieldy number of candidate genes and mutations within the genetic region, difficulty in estimating the interactions of other QTL on the trait, and technological and biological limitations when establishing the functionality of the candidate mutations. However, step-by-step approaches for establishing the causality of mutations involved in QTL have been proposed (Grisart et al. 2001, 2004; Andersson and Georges 2004; de Koning et al. 2007; Georges 2007; Ron and Weller 2007; Sellner et al. 2007).

1.2.7 *Analysis of genetic fragments*

Several large insert libraries, including bacterial artificial chromosome (BAC), yeast artificial chromosome (YAC), and fosmid vectors, exist for each livestock species, with the vast majority being BAC libraries (Table 1.6). Most of the BAC libraries for livestock species have been prepared by Pieter de Jong's group at BACPAC Resources Center (bacpac.chori.org/) and contain inserts with an average size of 90–200 Mb. These libraries can be screened by PCR amplification of plate, row, and column pools or by probe hybridization of high-density filters. The

Table 1.6 Large insert libraries in livestock species.

Species	Library	Genome coverage	Reference for BAC map
Cattle	CHORI-240	10.7X	Snelling et al. (2007)
	RPCI-42	10X	
Horse	CHORI-241	11.8X	Gustafson et al. (2003)
Sheep	CHORI-243	12X	Dalrymple et al. (2007)
Goat	6:15 translocation	3.3X	Schibler et al. (1998b)
Pig	CHORI-242	11.4X	Humphray et al. (2007)
	RPCI-44	10.2X	

Table 1.7 High-density SNP chips in livestock species.

Species	SNP chip	Reference
Cattle	Affymetrix 25K GeneChip	Khatkar et al. (2007)
	Illumina 50K BeadChip	Van Tassell et al. (2008)
Horse	Illumina 50K BeadChip	Chowdhary and Raudsepp (2008)
Sheep	Illumina 50K BeadChip	Kijas J. et al. (unpublished)
Pig	Illumina 60K BeadChip	Schook L. et al. (unpublished)

screening provides an exact clone address that contains the DNA sequence or gene of interest, and the clones can be purchased for about $20 from the BACPAC Resources Center.

Large overlapping segments of DNA called contigs can be generated by chromosome walking. To do this, the ends of isolated clones are sequenced, and then primers designed from the new sequence and used to screen the library in subsequent rounds. Sequential screenings will provide overlapping clones that can be pieced together into a single continuous fragment.

1.2.8 Whole genome association

In contrast to genetic linkage methods that use pedigrees segregating for the trait, genome-wide association (GWA) mapping is an approach that tests for allelic association at the population level through an analysis of linkage disequilibrium (LD), using large samples of unrelated individuals (Meuwissen et al. 2001; Amos 2007; McCarthy et al. 2008). Assuming that the trait allele of interest has descended from one or a few ancestral chromosomes, animals displaying the trait of interest will possess the ancestral haplotype that contains the allele of interest surrounded by closely linked marker alleles. These haplotypes may be fixed in a population or breed because of natural or artificial selection for the favorable allele, which is known as a "selective sweep" (Kim and Nielsen 2004; Pollinger et al. 2005; Voight et al. 2006; McVean 2007). The trait haplotype will be identified from all the wild-type haplotypes through the GWA analysis. Because very large numbers of markers are used in the analysis (around one marker every 100–500 kb), the resulting LD maps are typically of higher resolution than genetic linkage scans, which helps to limit the size of the interval that contains positional candidate genes.

High-density SNP chips are now under construction for all major livestock species (Table 1.7). These chips usually include between 30,000 and 60,000 SNPs, suitable for large-scale genotyping applications such as the GWA analysis. The availability of whole genome high-density SNP chips at a relatively low price per animal ($150–$300) means that GWA analyses in livestock become a more common approach for localizing a trait locus within the genome. An application of GWA has recently been illustrated by the fine mapping of five recessive disorders in cattle using the high-density bovine BeadChip (Charlier et al. 2008) with less than 20 affected animals and 20 controls. However, the number of unrelated animals needed for mapping quantitative traits is predicted to be greater than 1000 (McCarthy et al. 2008; Orr and Chanock 2008; Tian et al. 2008). Unfortunately, very few livestock populations of that size currently exist.

1.3 Characterization of gene expression

Researchers are often interested in characterizing the expression of genes in a specific tissue at a specific time or under a specific set of circumstances. The expression of these genes can be "captured" by examining the messenger RNA (mRNA) within the tissue sample. The abundance of a particular mRNA within a tissue can now be measured with relative ease using techniques developed within the last decade, such as Northern blots, and newly developed techniques, such as real-time PCR. It is also possible to determine a "profile" of mRNAs within a tissue using an analysis system such as expression microarrays or serial analysis of gene expression (SAGE).

1.3.1 Synthesis and analysis of cDNA

The array of mRNAs found within a tissue is often converted into a cDNA library, which can then be examined in a variety of ways. Because RNases, enzymes that break down RNA, are quite ubiquitous and difficult to degrade, care must be taken to preserve the tissue without delay after collection, such as snap freezing the tissue in liquid nitrogen or immersing the tissue in a preservative/RNase inhibitor such as RNALater™ (QIAGEN, Inc., Valencia, CA). After extracting the RNA from the tissue, another enzyme, reverse transcriptase, converts the RNA into the first strand of cDNA followed by second strand synthesis using DNA polymerase. The cDNA does not directly correspond to genomic DNA because intronic segments have been spliced out when the RNA molecule was produced and only the exonic sequences are contained with the cDNA strand. The cDNA mixture can be analyzed for transcript content using various techniques or used in the creation of a cDNA library by cloning into vectors, usually plasmids, which are then transformed into competent *Escherichia coli* cells. Replication of the host cells in the cDNA library results in the replication of the plasmid as well as the unique cDNA sequence contained within the plasmid.

Numerous kits for synthesizing cDNA from mRNA followed by analysis are now commercially available. Kits for the construction of cDNA libraries are also available, or library construction can be contracted for a relatively modest price or a library made from a particular tissue can be purchased from a commercial company. The bacterial library can be gridded onto filters and then screened for a particular gene by hybridization with a probe.

1.3.2 Analysis of gene expression

Quantitative real-time reverse transcription PCR (qRT-PCR; see Logan et al. 2009) is a method for measuring levels of specific mRNA transcripts within a sample. After the RNA sample is treated with reverse transcriptase, the resulting cDNA is amplified in a PCR reaction using primers specific to a transcript and the amount of transcript quantified in "real time" after each amplification cycle.

Detection of the transcripts is usually done with fluorescent dyes that intercalate with double-stranded DNA, although nonspecific binding can occur, which decreases the accuracy of the quantification. Another method of detection in qRT-PCR uses DNA oligonucleotide probes specific to the transcript that fluoresce when hybridized with a cDNA molecule. This method is more accurate than the double-stranded dyes, but the synthesis of fluorescent reporter probes is expensive.

Relative concentration of the transcript is determined in the qRT-PCR by plotting fluorescence (dependent on the number of copies of the transcript within the sample) against cycle number on a logarithmic scale. The quantity of a control, such as a housekeeping gene, is also measured on each sample so as to normalize for possible variation in the amount and quality of RNA between different samples, with the assumption that the expression of the control is similar across all samples.

"Global" expression of genes within a sample is commonly analyzed using expression microarrays, which allow simultaneous analysis of hundreds to thousands of genes. Probes spotted on the arrays were originally cDNAs but more recently, expression arrays contain oligonucleotides, usually in the range of 25–75 mers, designed from cDNA sequences. The longer the oligonucleotide, the more specific the detection, especially in cross-species experiments (Walker et al. 2006), but the shorter probes can be spotted on the array in higher density and are cheaper to synthesize. Oligonucleotide arrays are usually preferred to the cDNA arrays because of more uniform hybridization and ease of probe synthesis (Barrett and Kawasaki 2003; Hardiman 2004). Detection of a specific transcript within a sample is based on hybridization to the probes on the array. Annotation of the probes on the arrays is a key consideration for their usefulness. Statistical analysis of the data is challenging and requires appropriate controls, normalization of the signals, and adjustments for multiple comparisons. Significant results from an expression array experiment are often verified by qRT-PCR.

SAGE allows whole genome analysis of gene expression (i.e., mRNA) within a sample (Velculescu et al. 1995, 1997). Based on the concept that 10–14 bp of sequence provides "sufficient information to uniquely identify a transcript" within a sequence database, "quantification of the number of times a particular tag is observed provides the expression level of the corresponding transcript" (www.sagenet.org/findings/index.html). Previously unreported genes can also be detected through the generation of tags that are not contained within the databases. Subsequent adaptations of SAGE, such as SuperSAGE (Matsumura et al. 2005), allow precise annotation of existing and new genes because of an increased tag length of 25–27 bp. However, SAGE is relatively much more expensive than DNA microarrays, so large-scale projects are typically not performed with SAGE.

1.3.3 cDNA libraries and reproductive transcriptomes

To date, there are at least 270 publically available cDNA libraries that were derived from different reproductive tissues/organs in

Table 1.8 cDNA libraries and EST sequences for reproductive tissues/organs in livestock species.

Tissue/organ	Cattle		Swine		Sheep	
	No. of libraries	No. of ESTs	No. of libraries	No. of ESTs	No. of libraries	No. of ESTs
Embryo	27	62,951	14	89,916	—	—
Fetus	14	72,914	16	6,468	—	—
Mammary	56	65,227	3	16,656	1	2,309
Ovary	17	13,813	28	75,026	6	2,899
Oviduct	2	70	2	3,556	—	—
Pituitary	5	2,102	7	12,404	—	—
Placenta	10	23,665	6	21,307	—	—
Testes	7	15,033	11	42,494	—	—
Uterus	12	31,380	15	43,392	1	2,722
Total	150	287,155	102	311,219	8	7,930

cattle, swine, and sheep (Table 1.8). The library names, tissue/organ/cell line sources, physiological or reproductive stages, and contributors can be retrieved from either the GenBank database at NCBI (www.ncbi.nlm.nih.gov/) or the Gene Index database at Harvard University (compbio.dfci.harvard.edu/tgi/). As seen in Table 1.8, cattle have a slight edge over swine in the number of constructed libraries (105 and 102, respectively), but swine lead cattle in the number of expressed sequence tags (ESTs) that have been placed in the public databases (311,219 and 287,155, respectively). To date, only eight libraries have been established in sheep, and less than 8000 ovine ESTs for reproductive tissues/organs have been released.

These resources have been widely used in the survey of reproductive transcriptomes, identification of some breed- and developmental-stage-specific genes or gene clusters, and investigations of the genetic and physiological mechanisms underlying reproduction quantitative traits in livestock species. In addition, comparisons of livestock ESTs with sequences from other species have served as a valuable resource for comparative map development.

1.4 Resources for protein analysis

A unique complement of proteins is present in the cells of an organism at any one time under any one condition. This complement of proteins does not necessarily match to the complement of mRNA transcripts within the cells because of posttranslational modifications, splicing variants, and protein and RNA degradation. A recently defined area of research called proteomics encompasses large-scale studies of proteins, including their structure, function, and quantity (Anderson and Anderson 1998; Blackstock and Weir 1999). Two-dimensional (2D) gel electrophoresis is a well-established method commonly used to analyze proteins (Berth et al. 2007), although there are challenges in automatic analysis software. Technologies that allow high-throughput analysis of proteins within a tissue are now available, such as high-performance chromatography and mass spectrometry, but these approaches require highly specialized equipment.

Because of increasing emphasis on systems biology, databases have been created that present whole biological systems of interconnected proteins, with access to underlying genes, their sequences, and the background studies with a click of a mouse (see www.biochemweb.org/systems.shtml and www.semantic-systems-biology.org/biogateway/querying).

1.5 Future research directions

Genomic resources are now available for all major livestock species. These resources will allow researchers to identify regions within the genome that influence reproductive traits with relative ease. While the pursuit of the causative mutation controlling a quantitative trait may be complicated, combining knowledge from several lines of investigations should lead to the successful identification of the responsible gene. There are also multiple approaches for estimating gene expression in livestock species at both the single and whole genome levels. However, resources for the study of proteins are much less developed in livestock species, and therefore, researchers will need to exploit available information from humans and biomedical animal models.

References

Amaral, M.E., Grant, J.R., Riggs, P.K., Stafuzza, N.B., Filho, E.A., Goldammer, T., Weikard, R., Brunner, R.M., Kochan, K.J., Greco, A.J., Jeong, J., Cai, Z., Lin, G., Prasad, A., Kumar, S., Saradhi, G.P., Mathew, B., Kumar, M.A., Miziara, M.N., Mariani, P., Caetano, A.R., Galvão, S.R., Tantia, M.S., Vijh, R.K., Mishra, B., Kumar, S.T., Pelai, V.A., Santana, A.M., Fornitano, L.C., Jones, B.C., Tonhati, H., Moore, S., Stothard, P., and Womack, J.E. 2009. A first generation whole genome RH map of the river buffalo with comparison to domestic cattle. *BMC Genomics* **9**: 631.

Amos, C.I. 2007. Successful design and conduct of genome-wide association studies. *Human Molecular Genetics* **16**: R220–R225.

Anderson, N.L. and Anderson, N.G. 1998. Proteome and proteomics: New technologies, new concepts, and new words. *Electrophoresis* **19**: 1853–1861.

Andersson, L. and Georges, M. 2004. Domestic-animal genomics: Deciphering the genetics of complex traits. *Nature Reviews in Genetics* **5**: 202–212.

Barrett, J.C. and Kawasaki, E.S. 2003. Microarrays: The use of oligonucleotides and cDNA for the analysis of gene expression. *Drug Discovery Today* **8**: 134–141.

Baumbusch, L.O., Aarøe, J., Johansen, F.E., Hicks, J., Sun, H., Bruhn, L., Gunderson, K., Naume, B., Kristensen, V.N., Liestøl, K., Børresen-Dale, A.L., and Lingjaerde, O.C. 2008 Comparison of the Agilent, ROMA/NimbleGen and Illumina platforms for classification of copy number alterations in human breast tumors. *BMC Genomics* **9**: 379.

Berth, M., Moser, F.M., and Kolbe, M. 2007. The state of the art in the analysis of two-dimensional gel electrophoresis images. *Applied Microbiology and Biotechnology* **76**: 1223–1243.

Blackstock, W.P. and Weir, M.P. 1999. Proteomics: Quantitative and physical mapping of cellular proteins. *Trends in Biotechnology* **17**: 121–127.

Boehnke, M. 1992. Multipoint analysis for radiation hybrid mapping. *Annuals in Medicine* **24**: 383–386.

Bonnet, A., Thevenon, S., Claro, F., Gautier, M., and Hayes, H. 2001. Cytogenetic comparison between Vietnamese sika deer and cattle: R-banded karyotypes and FISH mapping. *Chromosome Research* **9**: 673–687.

Burt, D.W. 2002. Comparative mapping in farm animals. *Briefings in Functional Genomics and Proteomics* **1**: 159–168.

Charlier, C., Coppieters, W., Rollin, F., Desmecht, D., Agerholm, J.S., Cambisano, N., Carta, E., Dardano, S., Dive, M., Fasquelle, C., Frennet, J.C., Hanset, R., Hubin, X., Jorgensen, C., Karim, L., Kent, M., Harvey, K., Pearce, B.R., Simon, P., Tama, N., Nie, H., Vandeputte, S., Lien, S., Longeri, M., Fredholm, M., Harvey, R.J., and Georges, M. 2008. Highly effective SNP-based association mapping and management of recessive defects in livestock. *Nature Genetics* **40**: 449–454.

Chowdhary, B.P., Fronicke, L., Gustavsson, I., and Scherthan, H. 1996. Comparative analysis of the cattle and human genomes: detection of ZOO-FISH and gene mapping-based chromosomal homologies. *Mammalian Genome* **7**: 297–302.

Chowdhary, B.P. and Raudsepp, T. 2001. Chromosome painting in farm, pet and wild animal species. *Methods in Cell Science* **23**: 37–55.

Chowdhary, B.P. and Raudsepp, T. 2008. The horse genome derby: Racing from map to whole genome sequence. *Chromosome Research* **16**: 109–127.

Cockett, N.E., Jackson, S.P., Shay, T.L., Nielsen, D., Steele, M.R., Green, R.D., and Georges, M. 1994. Chromosomal localization of the *callipyge* gene in sheep (*Ovis aries*) using bovine DNA markers. *Proceedings of the National Academy of Sciences of the United States of America* **91**: 3019–3023.

Cockett, N.E, Jackson, S.P., Shay, T.L., Farnir, F., Berghmans, S., Snowder, G., Nielsen, D., and Georges, M. 1996. Polar overdominance at the ovine *callipyge* locus. *Science* **273**: 236–238.

Dalrymple, B.P., Kirkness, E.F., Nefodov, M., McWilliam, S., Ratnakumar, A., Barris, W., Zhao, S., Shetty, J., Maddox, J.F., O'Grady, M., Nicholas, F., Crawford, A.M., Smith, T., de Jong, P., McEwan, J.C., Oddy, V.H., and Cockett, N.E. 2007. Constructing the virtual sheep genome. *Genome Research* **8**: R152.

de Koning, D.J., Pong-Wong, R., Varona, L., Evans, G.J., Giuffra, E., Sanchez, A., Plastow, G., Noguera, J.L., Andersson, L., and Haley, C.S. 2003. Full pedigree quantitative trait locus analysis in commercial pigs using variance components. *Journal of Animal Science* **81**: 2155–2163.

de Koning, D.J., Archibald, A., and Haley, C.S. 2007. Livestock genomics: Bridging the gap between mice and men. *Trends in Biotechnology* **25**: 483–489.

Di Meo, G.P., Perucatti, A., Floriot, S., Hayes, H., Schibler, L., Incarnato, D., Di Berardino, D., Williams, J., Cribiu, E., Eggen, A., and Iannuzzi, L. 2008. An extended river buffalo (*Bubalus bubalis*, 2n = 50) cytogenetic map: Assignment of 68 autosomal loci by FISH-mapping and R-banding and comparison with human chromosomes. *Chromosome Research* **16**: 827–837.

Di Meo, G.P., Perucatti, A., Floriot, S., Hayes, H., Schibler, L., Rullo, R., Incarnato, D., Ferretti, L., Cockett, N., Cribuiu, E., Williams, J.L., Eggen, A., and Iannuzzi, L. 2007. An advanced sheep (*Ovis aries*, 2n = 54) cytogenetic map and assignment of 88 new autosomal loci by fluorescence in situ hybridization and R-banding. *Animal Genetics* **38**: 233–240.

Farnir, F., Grisart, B., Coppieters, W., Riquet, J., Berzi, P., Cambisano, N., Karim, L., Mni, M., Moisio, S., Simon, P., Wagenaar, D., Vilkki, J., and Georges, M. 2002. Simultaneous mining of linkage and linkage disequilibrium to fine map quantitative trait loci in outbred half-sib pedigrees: Revisiting the location of a quantitative trait locus with major effect on milk production on bovine chromosome 14. *Genetics* **161**: 275–287.

Georges, M. 2007. Mapping, fine mapping, and molecular dissection of quantitative trait loci in domestic animals. *Annual Reviews of Genomics and Human Genetics* **8**: 131–162.

Grisart, B., Coppieters, W., Farnir, F., Karim, L., Ford, C., Berzi, P., Cambisano, N., Mni, M., Reid, S., Simon, P., Spelman, R., Georges, M., and Snell, R. 2001. Positional candidate cloning of a QTL in dairy cattle: Identification of a missense mutation in the bovine *DGAT1* gene with major effect on milk yield and composition. *Genome Research* **12**: 222–231.

Grisart, B., Farnir, F., Karim, L., Cambisano, N., Kim, J.J., Kvasz, A., Mni, M., Simon, P., Frère, J.M., Coppieters, W., and Georges, M. 2004. Genetic and functional confirmation of the causality of the *DGAT1 K232A* quantitative trait nucleotide in affecting milk yield and composition. *Proceedings of the National Academy of Sciences in the United States of America* **101**: 2398–2403.

Gustafson, A.L., Tallmadge, R.L., Ramlachan, N., Miller, D., Bird, H., Antczak, D.F., Raudsepp, T., Chowdhary, B.P., and Skow, L.C. 2003. An ordered BAC contig map of the equine major histocompatibility complex. *Cytogenetics and Genome Research* **102**: 189–195.

Hardiman, G. 2004. Microarray platforms—Comparisons and contrasts. *Pharmacogenomics* **5**: 487–502.

Humphray, S.J., Scott, C.E., Clark, R., Marron, B., Bender, C., Camm, N., Davis, J., Jenks, A., Noon, A., Patel, M., Sehra, H., Yang, F., Rogatcheva, M.B., Milan, D., Chardon, P., Rohrer, G., Nonneman, D., de Jong, P., Meyers, S.N., Archibald, A., Beever, J.E., Schook, L.B., and Rogers, J. 2007. A high utility integrated map of the pig genome. *Genome Biology* **8**: R139.

Khatkar, M.S., Zenger, K.R., Hobbs, M., Hawken, R.J., Cavanagh, J.A., Barris, W., McClintock, A.E., McClintock, S., Thomson, P.C., Tier, B., Nicholas, F.W., and Raadsma, H.W. 2007. A primary assembly of a bovine haplotype block map based on a 15,036-single-nucleotide polymorphism panel genotyped in Holstein-Friesian cattle. *Genetics* **176**: 763–772.

Kim, Y. and Nielsen, R. 2004. Linkage disequilibrium as a signature of selective sweeps. *Genetics* **167**: 1513–1524.

Knott, S.A. and Haley, C.S. 2000. Multitrait least squares for quantitative trait loci detection. *Genetics* **156**: 899–911.

Lange, K., Boehnke, M., Cox, D.R., and Lunetta, K.L. 1995. Statistical methods for polyploid radiation hybrid mapping. *Genome Research* **5**: 136–150.

Laurent, P., Schibler, L., Vaiman, A., Laubier, J., Delcros, C., Cosseddu, G., Vaiman, D., Cribiu, E.P., and Yerle, M. 2007. A 12,000-rad whole-genome radiation hybrid panel in sheep: Application to the study of the ovine chromosome 18 region containing a QTL for scrapie susceptibility. *Animal Genetics* **38**: 358–363.

Logan, J., Edwards, K., and Saunders, N. (eds.). 2009. *Real-Time PCR: Current Technology and Applications*. Wymondham, UK: Caister Academic Press.

Lunetta, K.L., Boehnke, M., Lange, L., and Cox, D.R. 1996. Selected locus and multiple panel models for radiation hybrid mapping. *American Journal of Human Genetics* **59**: 717–725.

Maddox, J.F., Davies, K.P., Crawford, A.M., Hulme, D.J., Vaiman, D., Cribiu, E.P., Freking, B.A., Beh, K.J., Cockett, N.E., Kang, N., Riffkin, C.D., Drinkwater, R., Moore, S.S., Dodds, K., Lumsden, J.K., Adelson, D., Birkin, H., Broom, J.E.,

Buitkamp, J., Cambridge, E., Cushwa, W.T., Gerard, G., Galloway, S., Harrison, B., Hawken, R.J., Hiendleder, S., Henry, H., Medrano, J., Paterson, K., Phua, S.H., Schibler, L., Stone, R.T., and van Hest, B. 2001. An enhanced linkage map of the sheep genome comprising more than 1000 loci. *Genome Research* **11**: 1275–1289.

Matsumura, H., Ito, A., Saitoh, H., Winter, P., Kahl G., Reuter, M., Kruger, D.H., and Terauchi, R. 2005. SuperSAGE. *Cellular Microbiology* 7: 11–18.

McCarroll, S.A. 2008. Extending genome-wide associations studies to copy-number variation. *Human Molecular Genetics* **17**: R135–R142.

McCarthy, M.I., Abecasis, G.R., Cardon, L.R., Goldstein, D.B., Little, J., Ioannidis, J.P., and Hirschhorn, J.N. 2008. Genome-wide association studies for complex traits: Consensus, uncertainty and challenges. *Nature Reviews Genetics* **9**: 356–369.

McVean, G. 2007. The structure of linkage disequilibrium around a selective sweep. *Genetics* **175**: 1395–1406.

Meuwissen, T.H., Hayes, B.J., and Goddard, M.E. 2001. Prediction of total genetic value using genome-wide dense marker maps. *Genetics* **157**: 1819–1829.

Nejati-Javaremi, A. and Smith, C. 1995. Assigning linkage haplotypes from parent and progeny genotypes. *Genetics* **142**: 1363–1367.

Orr, N. and Chanock, S. 2008. Common genetic variation and human diseases. *Advances in Genetics* **62**: 1–32.

Papp, A.C., Pinsonneault, J.K., Cooke, G., and Sadée, W. 2003. Single nucleotide polymorphism genotyping using allele-specific PCR and fluorescence melting curves. *Biotechniques* **34**: 1068–1072.

Pollinger, J.P., Bustamante, C.D., Fiedel-Alon, A., Schmutz, S., Gray, M.M., and Wayne, R.K. 2005. Selective sweep mapping of genes with large phenotypic effects. *Genome Research* **15**: 1809–1819.

Rijnkels, M., Lemay, D.G., Barris, W.C., Casey, T.M., German, J.B., Hinrichs, A.S., Kriventseva, E.V., Lynn, D.J., Martin, W.F., Maqbool, N.J., Medrano, J.F., Molenaar, A.J., Neville, M.C., Pollard, K.S., Rincon, G., Zdobnov, E.M., Tellam, R.L., and Bovine Genome Sequencing and Analysis Consortium. 2009. Milking the bovine genome: insights in role of milk and variation of milk composition. *Plant and Animal Genomes XVII Conference*, W088, January 10–14, San Diego, CA.

Riquet, J., Coppieters, W., Cambisano, N., Arranz, J.J., Berzi, P., Davis, S.K., Grisart, B., Farnir, F., Karim, L., Mni, M., Simon, P., Taylor, J.F., Vanmanshoven, P., Wagenaar, D., Womack, J.E., and Georges, M. 1999. Fine-mapping of quantitative trait loci by identify by descent in outbred populations: Application to milk production in dairy cattle. *Proceedings of the National Academy of Sciences of the United States of America* **96**: 9252–9257.

Rocha, J.L., Baker, J.F., Womack, J.E., Sanders, J.O., and Taylor, J.F. 1992. Statistical associations between restriction fragment length polymorphisms and quantitative traits in beef cattle. *Journal of Animal Science* **70**: 3360–3370.

Rocha, J.L., Pomp, D., and Van Vleck, L.D. 2002. QTL analysis in livestock. *Methods in Molecular Biology* **195**: 311–346.

Ron, M. and Weller, J.I. 2007. From QTL to QTN identification in livestock—Winning by points rather than knockout: A review. *Animal Genetics* **38**: 429–439.

Saiki, R.K., Bugawan, T.L., Horn, G.T., Mullis, K.B., and Erlich, H.E. 1986. Analysis of enzymatically amplified beta-globin and HLA-DQ DNA with allele-specific oligonucleotide probes. *Nature* **324**: 163–166.

Sandor, C. and Georges, M. 2008. On the detection of imprinted quantitative trait loci in line crosses: Effect of linkage disequilibrium. *Genetics* **180**: 1167–1175.

Schaschi, H., Aitman, T.J., and Vyse, T.J. 2009. Copy number variation in the human genome and its implication in autoimmunity. *Clinical and Experimental Immunology*. February 11 [Epub ahead of print].

Schibler, L., Vaiman, D., Oustry, A., Giraud-Delville, C., and Cribiu, E.P. 1998a. Comparative gene mapping a fine-scale survey of chromosome rearrangements between ruminants and humans. *Genome Research* **8**: 901–915.

Schibler, L., Vaiman, D., Oustry, A., Guinec, N., Dangy-Caye, A.L., Billault, A., Cribiu, E.P. 1998b. Construction and extensive characterization of a goat bacterial artificial chromosome library with threefold genome coverage. *Mammalian Genome* **9**: 119–124.

Schwerin, M. 2001. Molecular genome analysis in livestock at the beginning of a new millennium. *Reproduction in Domestic Animals* **36**: 133–138.

Sellner, E.M., Kim, J.W., McClure, M.C., Taylor, K.H., Schnabel, R.D., and Taylor, J.F. 2007. Board-invited review: Applications of genomic information in livestock. *Journal of Animal Science* **85**: 3148–3158.

Slate, J., Van Stijn, T.C., Anderson, R.M., McEwan, K.M., Maqbool, N.J., Mathias, H.C., Bixley, J.J., Stevens, D.R., Molenaar, A.J., Beever, J.E., Galloway, S.M., and Tate, M.L. 2002. A deer (subfamily Cervinae) genetic linkage map and the evolution of ruminant genomes. *Genetics* **160**: 1587–1597.

Snelling, W.M., Chiu, R., Schein, J.E., Hobbs, M., Abbey, C.A., Adelson, D.L., Aerts, J., Bennett, G.L., Bosdet, I.E., Boussaha, M., Brauning, R., Caetano, A.R., Costa, M.M., Crawford, A.M., Dalrymple, B.P., Eggen, A., Everts-van der Wind, A., Floriot, S., Gautier, M., Gill, C.A., Green, R.D., Holt, R., Jann, O., Jones, S.J., Kappes, S.M., Keele, J.W., de Jong, P.J., Larkin, D.M., Lewin, H.A., McEwan, J.C., McKay, S., Marra, M.A., Mathewson, C.A., Matukumalli, L.K., Moore, S.S., Murdoch, B., Nicholas, F.W., Osoegawa, K., Roy, A., Salih, H., Schibler, L., Schnabel, R.D., Silveri, L., Skow L.C., Smith, T.P., Sonstegard, T.S., Taylor, J.F., Tellam, R., Van Tassell, C.P., Williams, J.L., Womack, J.E., Wye, N.H., Yang, G., Zhao, S., and the International Bovine BAC Mapping Consortium. 2007. A physical map of the bovine genome. *Genome Biology* **8**: R165.

Strerath, M., Detmer, I., Gaster, J., and Marx, A. 2007. Modified oligonucleotides as tools for allele-specific amplification. *Methods in Molecular Biology* **402**: 317–328.

Tellam, R.L. and the Bovine Genome Sequencing and Analysis Consortium. 2009. What does analysis of the bovine genome sequence say about innate immunity? *Plant and Animal Genome XVII Conference*, W087, January 10–14, San Diego, CA..

Tian, C., Gregersen, P.K., and Seldin, M.F. 2008. Accounting for ancestry: Population substructure and genome-wide association studies. *Human Molecular Genetics* **17**: R143–150.

Van Laere, A.S., Nguyen, M., Braunschweig, M., Nezer, C., Collette, C., Moreau, L., Archibald, A.L., Haley, C.S., Buys, N., Tally, M., Andersson, G., Georges, M., and Andersson, L. 2003. A regulatory mutation in IGF2 causes a major QTL effect on muscle growth in the pig. *Nature* **425**: 832–836.

Van Tassell, C.P., Smith, T.P., Matukumalli, L.K., Taylor, J.F., Schnabel, R.D., Lawley, C.T., Haudenschild, C.D., Moore, S.S., Warren, W.C., and Sonstegard, T.S. 2008. SNP discovery and allele frequency estimation by deep sequencing of reduced representation libraries. *Nature Methods* **5**: 247–252.

Velculescu, V.E., Zhang, L., Vogelstein, B., and Kinzler, K.W. 1995. Serial analysis of gene expression. *Science* **270**: 484–487.

Velculescu, V.E., Zhang, L., Zhou, W., Vogelstein, J., Basrai, M.A., Bassett, D.E., Hieter, P., Vogelstein, B., and Kinzler, K.W. 1997. Characterization of the yeast transcriptome. *Cell* **88**: 243–251.

Voight, B.F., Kudaravalli, S., Wen, X., and Pritchard, J.K. 2006. A map of recent positive selection in the human genome. *PLoS Biology* **4**: e72.

Walker, S.J., Wang, Y., Grant, K.A., Chan, F., and Hellmann, G.M. 2006. Long versus short oligonucleotide microarrays for the study of gene expression in nonhuman primates. *Journal of Neuroscience Methods* **152**: 179–189.

Wang, J., Chuang, K., Ahluwalia, M., Patel, S., Umblas, N., Mirel, D., Higuchi, R., Germer, S. 2005. High-throughput SNP genotyping by single-tube PCR with TM-shift primers. *Biotechniques* **39**: 885–893.

Wu, C.H., Nomura, K., Goldammer, T., Hadfield, T., Dalrymple, B.P., McWilliam, S., Maddox, J.F., Womack, J.E., and Cockett, N.E. 2008. A high-resolution comparative radiation hybrid map of ovine chromosomal regions that are homologous to human chromosome 6 (HSA6). *Animal Genetics* **39**: 459–467.

Wu, C.H., Nomura, K., Goldammer, T., Hadfield, T., Dalrymple, B.P., McWilliam, S., Maddox, J.F., Womack, J.E., and Cockett, N.E. 2009. A radiation hybrid comparative map of ovine chromosome 1 aligned to the virtual sheep genome. *Animal Genetics* **40**: 435–455.

Wu, C.H., Nomura, K., Goldammer, T., Hadfield, T., Womack, J.E., and Cockett, N.E. 2007. An ovine whole-genome radiation hybrid panel used to construct an RH map of ovine chromosome 9. *Animal Genetics* **38**: 534–536.

2

Quantitative Genomics of Female Reproduction

Jeffrey L. Vallet, Dan J. Nonneman, and Larry A. Kuehn

2.1 Introduction

The purpose of this chapter is to review quantitative trait loci (QTL) and the development and discovery of genomic markers for female reproductive traits of domestic livestock species. For this chapter, we define a quantitative trait as one for which the phenotypes of individual animals in a population form a continuum, and the trait itself is influenced by the function of numerous genes in concert. The utility of the QTL is that they can be assessed in individual animals using DNA markers to determine which alleles of the QTL are present. Individual alleles of a QTL are associated with genetic variation in the trait of interest. Discovery of QTL relies on collecting appropriate phenotypes, accurately determining genotypes within the area of the genome affecting the trait, and demonstrating a statistical association between genotypes and phenotypes. Making use of QTL requires one more step, the appropriate incorporation of the QTL in selection schemes, which is reliant on accurately predicting the effect of different alleles of one or more QTL affecting a trait. In practice, each step presents challenges in the successful implementation of this technology.

2.2 Female reproductive phenotypes

2.2.1 Complexity of reproduction

Females contribute most of the complexity of reproduction in livestock. In order to produce offspring, females must efficiently reach puberty; display estrus; shed one or more competent ova; create the appropriate oviductal environment for fertilization to take place; undergo the necessary systemic, ovarian, and uterine modifications to support pregnancy; deliver the offspring; lactate; and successfully return to estrus after offspring are weaned. Puberty, estrous cyclicity, oocyte competence, and the oviductal contribution to fertilization are all controlled by the female. The conceptus contributes to implantation, placental development and function, fetal survivability during pregnancy, susceptibility to stillbirth, and preweaning survival and growth rate. Thus, both the conceptus and mother contribute genetically to the success of pregnancy, parturition, lactation, and postweaning return to estrus. In litter-bearing livestock like the pig, the genetic contribution to success of traits like litter size is the combined effect of many interacting genotypes (sow and all piglets in the litter) and thus is very complex. Negative correlations between the mother's and piglet's genetic contributions for birth weight, stillbirth, and preweaning mortality have been reported (Roehe 1999; Arango et al. 2006). Therefore, the effect of genes of the sow on a trait may have an opposing effect to the same genes of the piglet. Despite this potentially complex antagonism, most female reproductive QTL analyses have focused on the influence of genetic differences between dams, and ignored the influence of genes in the offspring. Since each offspring inherits half its genome from the dam, genetic effects attributed to the dam in QTL analyses for pregnancy traits are in fact some combination of maternal and fetal genetic effects.

2.2.2 *Complex phenotypes*

Many reproductive traits are combinations of several traits. For example, litter size in pigs combines ovulation rate; oocyte competence; oviductal factors necessary for fertilization; systemic, ovarian, and uterine factors needed to maintain pregnancy; the fertilization rate of the sperm; implantation; placental formation and function; fetal survivability; and piglet susceptibility to stillbirth. In this case, genomes of the dam, sire, and fetuses contribute to the trait, and each subphenotype is typically controlled by numerous genes. This genetic complexity contributes to the low heritability of reproductive traits (Table 2.1). In addition for some traits such as litter size in pigs, the genetic influence contributing to later pregnancy success may not be manifested if ovulation or fertilization rates are low. These issues are relevant to the search for QTL. Traits that are the result of multiple interacting processes will be affected by many genes each with small effects on the overall trait. In traits where genes influence the outcome sequentially, poor performance of genes early in the process renders later gene effects undetectable. Accurate detection and estimation of these interdependent genetic effects require the collection of phenotypes of component traits to isolate gene effects on specific subcomponents of these complex traits.

2.2.3 *Genetic correlation and pleiotropic effects*

A further complexity is that many genes contribute to more than one phenotypic trait. This situation is known as pleiotropy and results in genetic correlations between traits. Numerous genetic correlations between reproductive traits and other traits have been reported. Genetic correlations can result in beneficial or detrimental changes in traits other than the trait of interest. For example in pigs, litter size is negatively genetically correlated with lean meat content (Holm et al. 2004), birth weight (Mesa et al. 2005), and average birth interval (Canario et al. 2006) and positively genetically correlated with percent stillborn (Canario et al. 2006). Given these correlations, selection for litter size results in detrimental effects on the piglets, unless other traits are also considered in the selection program (e.g., birth weight). In contrast, farrowing survival and preweaning survival are positively genetically correlated (Mesa et al. 2006). Thus, selecting for decreased stillbirth rate would also improve preweaning

Table 2.1 Heritabilities for reproductive traits in livestock.

Trait	Heritability	Species	Reference
Age at puberty	0.31–0.40	Swine	Holm et al. (2005); Sterning et al. (1998)
Weaning to estrus interval	0.02–0.24	Swine	Holm et al. (2005); Sterning et al. (1998)
Ovulation rate	0.33	Swine	Rosendo et al. (2007)
	0.10 (single obs.)	Cattle	Gregory et al. (1997)
	0.35 (six obs.)	Cattle	Gregory et al. (1997)
Pregnancy rate	0.07–0.13	Cattle	Bormann et al. (2006); MacNeil et al. (2006)
Litter size	0.10–0.20 (dam only)	Swine	Canario et al. (2006); Holm et al. (2005); van der Steen (1985)
	0.08 (direct)	Swine	Mesa et al. (2005)
	0.08 (maternal)	Swine	Mesa et al. (2005)
	0.06–0.17	Sheep	Janssens et al. (2004); Okut et al. (1999)
Twinning	0.01–0.10	Cattle	Gregory et al. (1997); Komisarek and Dorynek (2002)
Birth weight	0.03–0.09 (direct)	Swine	Arango et al. (2006); Roehe (1999)
	0.19–0.26 (maternal)	Swine	Arango et al. (2006); Roehe (1999)
	0.39–0.42 (direct)	Cattle	Gregory et al. (1997); Gutierrez et al. (2007)
	0.11–0.21 (maternal)	Cattle	Gregory et al. (1997); Gutierrez et al. (2007)
	0.27 (direct)	Sheep	Hanford et al. (2002)
	0.25 (maternal)	Sheep	Hanford et al. (2002)
Stillbirth	0.001–0.14 (direct)	Swine	Arango et al. (2006); Mesa et al. (2006); White et al. (2006)
	0.002–0.16 (maternal)	Swine	Arango et al. (2006); Mesa et al. (2006); White et al. (2006)
Calving difficulty	0.19–0.43 (direct)	Cattle	Bennett and Gregory (2001); Gutierrez et al. (2007)
	0.14–0.23 (maternal)	Cattle	Bennett and Gregory (2001); Gutierrez et al. (2007)
Preweaning mortality	0.05–0.18 (direct)	Swine	Arango et al. (2006); Mesa et al. (2005)
	0.08–0.10 (maternal)	Swine	Arango et al. (2006); Mesa et al. (2005)
Length of productive life	0.17	Swine	Serenius and Stalder (2006)
Stayability	0.09–0.30	Cattle	Martinez et al. (2005)

survival. Age at first service is positively correlated with weaning to first service interval after first parity (Sterning et al. 1998; Holm et al. 2005) and negatively genetically correlated with sow lifetime productivity (Serenius and Stalder 2004). Selection for early puberty would improve early return to estrus after weaning and would be associated with improved sow productive lifetime. Because of these genetic correlations, QTL effects on a broad variety of health and economic traits should be determined before they are used to manipulate any one particular trait of interest. Any QTL with pleiotropic effects could thus be determined, and this information is used in selection schemes. Hence, collecting a variety of economically relevant phenotypes on every animal in a QTL population, not just the phenotype(s) of interest, is important.

2.2.4 Trait measurement

Another consideration regarding collection of female reproductive phenotypes is the ease with which phenotypes can be obtained. Traits like ovulation rate (i.e., by laparoscopy or ultrasound) and speed of parturition (i.e., by video surveillance or constant monitoring) can be measured, but are time-consuming and therefore costly. Factors involving oocyte competence, oviductal influences on fertilization, and the systemic, ovarian, and uterine changes required to maintain pregnancy are still being defined, such that the appropriate measurements for these traits are the subject of ongoing research. Generally speaking, traits that are easily and externally measured currently form the basis of routine genetic selection for female reproductive traits. These include age at puberty, pregnancy rate, litter size and nipple number (for polytocous species), calving difficulty/stillbirth rate, preweaning mortality, return to estrus after parturition, and stayability. Although a single incidence of pregnancy or stillbirth of a particular animal is clearly not a continuously distributed trait, this trait can be converted to a rate or probability after multiple observations on the same individual or on multiple female progeny of a given parent (either by calculating a rate of occurrence [i.e., number of pregnancies divided by the number of services or the number of stillborns divided by the number of offspring] or using other methods of statistical modeling of categorical traits). Thus, the chance of a particular outcome forms a continuum among individuals and can be thought of as a continuously distributed trait (Gregory et al. 1997).

2.3 Genetic markers and genotyping methods

The goal of genotyping is to detect DNA sequence variation between individuals in a population. Methods differ according to the type of genetic variation to be detected. Two general types of variation are found: single nucleotide polymorphisms (SNPs) and insertions/deletions (indel). Detection of an SNP relies on strategies to detect individual nucleotides within a sequence. Detection of indels depends on the size of the indel. Small indels can be detected using methods useful for either SNP detection or detection of DNA fragment sizes. Larger indels are typically detected by differences in fragment sizes. Indels can range in size up to the insertion or deletion of the entire genes (Redon et al. 2006; Beckmann et al. 2007; McCarroll and Altshuler 2007), resulting in differences in copy number of specific genes. Detection of copy number differences is a special case and is carried out using strategies that differ from the detection of SNP and smaller fragments (see below).

2.3.1 SNPs and genotyping

SNPs are differences in one nucleotide base at a specific position in the DNA between members of a population. The rate of conversion or mutation of adenine (A), cytosine (C), guanine (G), and thymidine (T) nucleotides to the other nucleotides between generations is low ($\sim 2 \times 10^{-8}$ per nucleotide; Nachman and Crowell 2000). The low rate means that the incidence of reverse mutation is extremely low; thus, SNPs are thought to be permanent changes in the DNA, distinguishing them from microsatellites, whose mutation rate is much higher ($\sim 1 \times 10^{-4}$ per gamete × locus; Crawford and Cuthbertson 1996; see below). Detection of SNP typically relies on the ability to distinguish individual nucleotides in a sequence. Methods include restriction endonuclease susceptibility, hybridization differences, or detection of incorporation of different nucleotides. Restriction fragment length polymorphism or RFLP analysis relies on the introduction of a new restriction endonuclease site within the DNA caused by the SNP. The DNA is amplified with specific primers flanking a genetic polymorphism; the resulting amplified product is digested with restriction endonuclease and subjected to gel electrophoresis. The new restriction endonuclease site introduced by the SNP is detected as digestion of the amplified product, allowing visualization of genotypes by differences in DNA banding patterns after electrophoresis.

Although RFLP analysis of SNP is an effective detection method, analysis of many individual SNP using this method is time-consuming and costly. Emphasis has been placed on multiplexing of genotype collection, such that genotypes from multiple sites are collected simultaneously. These higher-throughput genotyping platforms are based on either hybridization or primer extension/nucleotide incorporation to differentiate alleles. Sequenom genotyping technology employs detection of incorporation of different nucleotides by mass spectrometry, since each incorporated nucleotide differs in molecular weight. Amplification and detection of numerous fragments can be performed simultaneously, allowing dozens of genotypes to be determined from a single genomic DNA sample (www.sequenom.com/). Affymetrix genotyping chips rely on differences in hybridization of genomic DNA to thousands of oligonucleotide probes immobilized on the chip, allowing for the simultaneous detection of thousands of SNPs from a single sample (www.affymetrix.com/index.affx). Illumina genotyping methods detect incorporation of specific labeled nucleotides into oligonucleotides linked to beads and also allow for simultaneous detection of thousands of genotypes from a single sample (www.illumina.com/). These platforms are provided as examples of currently available strategies; others are available from a variety of companies. These methods all have in common multiplexing strategies that allow simultaneous genotyping from a variety of loci. These strategies reduce the average cost of a single genotype and allow the possibility of whole genome association studies, which will be discussed later.

2.3.2 Indels/microsatellites and genotyping

Small indels can also be detected using the above strategies, or by direct detection of the size of amplified fragments of DNA. Microsatellites, which fall into this category, have been used extensively in QTL analyses and are typically detected by electrophoresis after polymerase chain reaction (PCR) amplification of specific DNA regions in the presence of isotope or fluorescent dye-labeled nucleotides. Microsatellites are regions of nucleotide repeats (e.g., $(CA)_n$) of varying length n within genomic DNA. They occur most often in noncoding regions of the genome, with some exceptions, and this results in a random distribution throughout the genome. The number of repeats is highly polymorphic (mutation rate $\sim 1 \times 10^{-4}$ per gamete $\times$ locus) and is greater for repeats of large n. Microsatellites have several advantages over SNP for genetic analysis. Because they cause differences in DNA fragment size, they are easy to detect. Rather than two alternative alleles at a given locus, microsatellites may have several alleles (e.g., CA_{10}, CA_{12}, CA_{14}) within the population. The larger number of alleles allows better tracking of genetic variation through a pedigree during linkage analysis (see below). This advantage, which is due to the high mutation rate between generations, is also a disadvantage, in that the rate of interconversion between alternate alleles is high.

While SNP alleles are assumed to be identical by descent at some point in the history of the species, the high mutation rate of microsatellites does not allow this assumption. The same microsatellite allele could have been generated numerous times in a population. Thus, although microsatellites have been useful in detection and selection of QTL in defined pedigrees by exploiting linkage, they are of limited value as genetic markers across unrelated populations.

2.3.3 Gene copy number and genotyping

A more recent development in genetic analysis has been the detection of differences in gene copy numbers between individuals within a population (Redon et al. 2006; Beckmann et al. 2007; McCarroll and Altshuler 2007). Although differences in reproductive traits associated with differences in gene copy numbers have not yet been described in livestock, in mice, it appears that the number of copies of Qa-2 genes in the major histocompatibility locus influences embryo development and survival (Byrne et al. 2007). Although this region is deleted in swine (Renard et al. 2006), there is evidence that major histocompatibility genes are associated with litter size (Conley et al. 1988). In addition, the number of major histocompatibility genes that are expressed appears to vary in cattle (Ellis et al. 1999), which could be the result of differences in copy number. Differences in gene copy numbers can be detected by analyzing relative abundance of a particular sequence in the genome. Strategies include hybridizing genomic DNA with arrays of probes tiling the genome (e.g., arrays of bacterial artificial chromosomes [BACs] sufficient to cover the whole genome, or cDNA arrays designed for transcription analysis). Differences in gene copy number are detected as differences in individual probe hybridization signals between genomic DNA from different individuals. An alternative approach is to use quantitative polymerase chain reaction (qPCR) using genomic DNA as template. Correction of the results with qPCR values of a known single copy gene yields an accurate assessment of copy number of target regions. Although allelic duplication of large regions of the genome is much rarer than the incidence of SNP, the chance of these polymorphisms having an effect on the animal is much greater than individual SNP, most of which have no functional effect on the animal. Because of the greater chance that these differences will result in phenotypic differences, this is a growing area of research.

2.4 Association of phenotypes with genotypes

Once phenotypes and a method of genotyping are available, it is possible to associate differences in genotype with differences in phenotype, and there are two broad approaches: candidate gene and genome scan. The candidate gene approach relies on prior knowledge of the role of a specific gene in some aspect of the phenotype measured. The genome scan approach assumes no prior knowledge of the genes involved in the trait of interest, and the whole genome or individual chromosomes are surveyed for regions associated with differences in the trait. The genome scan approach falls into two general categories: linkage analysis and linkage disequilibrium (LD) analysis. In livestock, linkage analysis has historically dominated for surveys of the whole genome, but technology is rapidly becoming available to make LD analysis of the whole genome more feasible. LD analysis has historically been used to fine-map regions associated with a trait that was discovered using linkage analysis.

2.4.1 Candidate genes

Available scientific literature describes the role of numerous genes in female reproductive traits, and it is not difficult to create a list of potential genes that could have significant effects on reproduction. Estrogen receptor was one of the first genes investigated in livestock for association between genetic variation and a female reproductive trait, specifically litter size (Rothschild et al. 1996). Since then, other genes known to play roles in female reproduction have been investigated for association with female reproductive traits. Genetic variation in the prolactin receptor (Drogemuller et al. 2001), retinol-binding protein (Rothschild et al. 2000), folate-binding protein (Vallet et al. 2005a), and erythropoietin receptor (Vallet et al. 2005b) genes have been associated with litter size; and genetic variation in the genes for insulin-like growth factor and its binding proteins and receptors have been associated with sow productive lifetime (Mote et al. 2006). Superficially, any gene with a role in female reproduction is a legitimate candidate. However, some genes represent better candidates than others. For example, genes that have broad effects on traits other than reproduction (i.e., pleiotropy), such as transcription factors, are likely to be poor candidate genes, because changes in the function of the gene would have broad effects beyond female reproductive performance, unless the effect of the genetic variation within the gene modified expression only in reproductive tissues. Related to this, genetic variation in genes at the top of metabolic pathways that affect reproduction would have broader effects than those nearer to the end result of the pathway, and thus those nearer the end result would have more predictable effects on reproduction. Genes whose effects are essential in a particular process would seem to make bad candidates, because significant changes in the function of these genes have a good chance of being lethal. Thus, although selection of candidate genes for female reproduction traits, or indeed any trait, would seem to be straightforward given current scientific knowledge, more knowledge of the regulation of specific female reproductive traits and their relation to other traits would improve the selection of candidate genes. Furthermore, this strategy is unlikely to discover all the genetic variation influencing reproductive traits, because despite our extensive knowledge of the role of various genes in reproduction, the roles of numerous other genes is not yet known.

Once the candidate gene is selected, the gene is searched for DNA sequence variation in a population of interest. A typical approach is to amplify a region of the gene in several animals by PCR and sequence the product. Animals to be sequenced should be as genetically diverse as possible but still represent the population of interest. Diversity increases the probability that QTL explaining portions of the genetic variation within the population will be detected, while maintaining representation of the population ensures that the variation found will be useful in the population of interest. How many animals to sequence will be determined by the lowest allele frequency of interest. In a population in equilibrium, if the frequency of a given SNP is 90% for one allele and 10% for the other, then assuming random mating under Hardy–Weinberg equilibrium, 81% of animals will be homozygous for the major allele, 18% heterozygous, and 1% homozygous for the minor allele. Analysis of 10 animals is expected to yield 2 heterozygous animals, and these may be difficult to distinguish from PCR or sequencing-induced errors. Additionally, if only 10 animals are genotyped, the probability of having no heterozygous animals in the group is relatively high (~0.12, the probability of a homozygous animal raised to the 10th power, 0.81^{10}). Even if the low-frequency polymorphism is discovered, association with differences in a trait requires sufficient observations to reach statistical significance; therefore, low frequency limits statistical power as well. These two problems make detection of associations in low-frequency alleles much more difficult.

It has been a common practice to sequence a gene among animals only within the exons, and after the examination of the polymorphisms detected concludes that no useful polymorphisms are present because differences observed did not alter the amino acid sequence of the resulting protein. Aside from altering the amino acid coding sequence, genetic polymorphisms can have a variety of other effects on gene function. The DNA sequence upstream, downstream, and within the exons and introns has been shown in numerous reports to be involved in the control of gene function (Fedorova and Fedorov 2003). Effects of DNA elements on gene transcription through promoter and enhancer elements (Roeder 1991; Maniatis and Reed 2002), on the efficiency of mRNA splicing (Maniatis and Reed 2002), on translation of the mRNA through transcription initiation (Iida and Masuda 1996) or blockade or through codon bias (Kurland 1991; Akashi 2001), and on mRNA degradation and/or storage through mRNA–protein (Ruiz-Echevarria et al. 2001) or mRNA–microRNA interactions (Wienholds and Plasterk 2005; Kiefer 2006; Zhao and Srivastava 2007) have been described. MicroRNAs are likely to be especially relevant to early embryo development (Schier 2007; Stitzel and Seydoux 2007). Thus, due to our current lack of ability to predict whether a specific polymorphism does or does not have an effect on gene function, it should be assumed that any genetic variation in the proximity of a gene could alter its function in some way, and a better approach to this problem is to perform a comprehensive search for genetic variation in and around the gene, and perform association analyses using all of the SNP discovered.

2.4.2 Genome scans

Genome scans can be done either by linkage analysis or by LD analysis, although currently all genome scans in livestock have been done by linkage analysis. Linkage analysis is performed on a population of animals in which the pedigree and relationships between animals is known. The number of generations between animals in the population is typically limited, creating artificially large regions of LD between parents and offspring. These large regions allow tracking of specific chromosomal regions from parents to offspring within the population using markers spaced every 10 or 20 million bases; thus, the inheritance of the entire genome of livestock can be examined with 150–300 genetic loci. Microsatellites have typically been used for these analyses because each locus may have numerous alleles, improving the information content of each locus and increasing the ability to track regions from parent to offspring in the analysis. This approach has several advantages. No prior knowledge of the genes responsible is needed. Relatively few markers are needed to assess the effects of the entire genome. Founder parent animals can be selected from breeds or lines that are divergent in the trait of interest, making the discovery of genomic regions having a major influence on the phenotype more likely. Most of the QTL for reproductive traits for livestock generated by genome scans currently reported in the literature have employed variations of this strategy.

Although linkage analysis is very good at finding QTL regions, it has several disadvantages. Because a pedigreed population is needed for this type of analysis, generation of a suitable population may take years to accomplish for livestock and is expensive. Dairy cattle have had a significant advantage over other livestock species in this regard, in that large pedigrees were already available within the dairy industry. Similar resources have been used in other countries for swine (Tribout et al. 2008) but have not been exploited in the United States. Instead, QTL analysis in swine has relied on the generation of specific populations. A related disadvantage is that the only genomic regions that will be found to be associated with the trait of interest will be those that differed among the original founder animals of the population used in the QTL analysis. Thus, a different population with different founder animals may yield different results. Therefore, selection of sufficient founder animals to represent the inference population is important, because sampling of small numbers of animals will limit the number of QTL regions that are identifiable within the population. On the other hand, selection of too many founder animals increases the number of total animals needed in the population to provide sufficient statistical power to be able to detect the effect of the genetic contribution of each founder animal. The extent of newly created LD among the animals within the artificially generated population limits the ability to reduce the size of any QTL region discovered, because chromosomal segments are inherited in relatively large segments between generations. Linkage analysis often results in associations with genomic regions that are millions of bases in length harboring hundreds of genes. Our knowledge of individual gene function is typically insufficient to allow the effective selection of candidate genes within regions of this size. Finally, because the linkage between markers and QTL regions within the experimental population has been artificially created by the limited number of generations in the pedigree, and results are reliant on the founder animals used, marker associations with QTL found for the experimental population are specific to the population and typically do not transfer to livestock populations at large without further research using LD analysis. Linkage analysis will detect regions associated with differences in traits, but strategies related to those discussed for candidate genes must then be used to obtain markers that are of use to livestock populations beyond that used to discover the QTL. The LD analysis is needed to reduce the size of the associated region. Combined linkage, LD analysis of specific genome regions has been used successfully to fine-map QTL regions that were previously identified by linkage analysis (Olsen et al. 2005; Schnabel et al. 2005b).

2.4.3 LD

Discovery of the actual genetic variation responsible for the difference in a trait is not necessary for the association to be of use, because of LD. LD is defined as the simultaneous inheritance of adjacent regions on the same chromosome. LD in a randomly breeding population is removed by genetic recombination, in which corresponding regions of an individual animal's two chromosomes switch places during gamete production.

Genetic recombination takes place during meiosis at relatively random positions throughout the genome (although recombination hotspots have been described) and therefore occurs more frequently between distant genetic loci than adjacent loci. On average within a population, genetic loci on the same chromosome that are distant from each other are in less LD than those close together. The degree of LD in a population depends on the number of generations its members have been free to mate at random, whether any reduction in the overall population size took place in its history, and when mutations occurred relative to one another. Genetic selection of individuals in a population will tend to preserve LD surrounding genes influencing the selected trait. A reduction in population size increases the relatedness of offspring of the remaining individuals, which increases LD until sufficient generations pass to allow recombination to mix up the linked chromosomal regions. In humans, regions of LD, called "haplotype blocks," are relatively small (50 kb [Dunning et al. 2000; Abecasis et al. 2001; Shifman et al. 2003; Tsunoda et al. 2004]) compared with livestock (pig 250–1000 kb [Nsengimana et al. 2004; Harmegnies et al. 2006; Du et al. 2007], sheep 5000–10,000 kb [McRae et al. 2002], and cattle 500 kb [McKay et al. 2007]). The larger regions of LD in livestock are likely a reflection of periodic reductions in population size as well as the cumulative effect of genetic selection for various traits. The existence of LD means that alleles of nearby SNP are often inherited together. If a functional polymorphism is sufficiently near other SNP, inheritance of the adjacent loci can be used to track inheritance of the functional polymorphism. To understand how this can work, suppose a functional mutation occurs on a chromosome within a region in a sire with nearby alleles. Inheritance of the functional mutation is now in LD with those nearby alleles in his descendants until recombination breaks up the relationship. Now suppose another mutation occurs in the same region in a sibling male, who shares the same original alleles in the region. The new mutation is also in LD with the shared alleles in his descendants, but not with the functional mutation in descendants of the original sire, even though they are located right next to each other in the genome. Finally, suppose a third nearby mutation occurs in a descendant of the original sire. This third mutation will also be in LD with the functional mutation. Over generations, recombination events may disrupt any of these relationships over time. In this way, varying degrees of LD can result between nearby loci, because mutations that are close in distance but separated in time may vary in LD with each other. To exploit LD, one needs to have sufficient markers in a region that are sufficiently close to the functional polymorphism. How close depends on the structure of LD for that region of the genome within the population. Within the region of substantial LD, dividing the genetic variation into three or four roughly equal portions using genetic markers will give a reasonable assessment of the effect of a chromosomal region on a trait. Figures 2.1 and 2.2 illustrate how this can work for a high-frequency and a low-frequency functional polymorphism and surrounding SNP. Adjacent SNP illustrated in Figures 2.1 and 2.2 are hypothetical situations meant to represent possible arrangements. In practice, one must examine sufficient adjacent SNP to provide a good division of the genetic variation within the locus, but not so many that the variation is divided up so much that individual effects of haplotypes are no longer detectable.

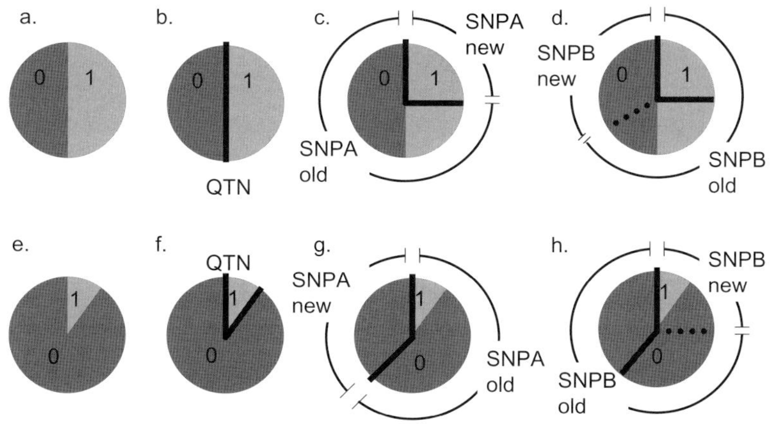

Figure 2.1 The pie charts illustrate an additive genetic effect in a population of animals in a hypothetical QTL region of the genome. In (a), two alleles of a functional SNP (quantitative trait nucleotide or QTN) with a frequency of 0.5 each exist at a locus. One allele does not change the phenotype (and is therefore 0); the other increases the phenotype by 1. This gives an allele substitution effect for this locus of 1; thus, in this population, this genomic region contributes from 0 (homozygous for the 0 allele) to 2 (homozygous for the 1 allele) units to the phenotypic trait. In (b), a genetic marker based on the QTN is illustrated. In (c), a nearby mutation occurs on a chromosome containing allele 1 of the QTN, creating an SNPA with alleles in linkage disequilibrium with the QTN. Overtime, half of the chromosomes with QTN allele 1 have the new allele of SNPA, and half have the old allele of SNPA, and all of the chromosomes with allele 0 of the functional SNP have the old allele of SNPA. For SNPA, the new allele is always present with allele 1 of the functional SNP; the old SNPA allele is a mixture of QTN alleles 0 and 1. In this arrangement, a commonly used measure of linkage disequilibrium (r^2) is 0.33 and is calculated as the squared correlation between the incidences of the alleles at each locus. The substitution effect of SNPA alleles would be 0.67, two-thirds that of the QTN. In (d), three adjacent SNPs are illustrated together, SNPA (from c) and SNPB, along with the QTN. SNPB in this case now occurred on a chromosome with allele 0 of the functional SNP, and overtime two-thirds of allele 0 has the new allele at SNPB, and one-third of allele 0 and all of allele 1 at the functional SNP have the old allele at SNPB. The r^2 = 0.5 for SNPB, and it can be calculated that SNPB will have a substitution effect of 0.75. This marker arrangement creates three haplotypes: SNPA old allele with SNPB new allele, SNPA old allele with SNPB old allele, and SNPA new allele with SNPB old allele. The r^2 for the haplotypes is 0.59, better than either SNP alone. The substitution effects of these three haplotypes are 0.6 for the substitution of SNPA old, SNPB old for SNPA old, SNPB new; 0.4 for the substitution of SNPA new, SNPB old for SNPA old, SNPB old; and 1 for the substitution of SNPA new, SNPB old for SNPA old, SNPB new. In this case, the haplotypes regenerate the substitution effect of the original QTN because the SNPA new, SNPB old haplotype is only present with the QTN 1 allele, and the SNPA old, SNPB new haplotype is only present with the QTN 0 allele. In (e), a different hypothetical QTN is illustrated that has an allele 1 with a frequency of 0.1 that increases the phenotype by 1 and an alternate allele 0 with a frequency of 0.9 that does not change the phenotype. In (f), a marker based on the new QTN is illustrated. In (g), a mutation generating a new allele for SNPA occurred sometime before the mutation resulting in the QTN and overtime results in a frequency of 40%, which will eventually be present only with QTN allele 0. The old SNPA allele (60%) will be present with a mixture of QTN alleles 0 and 1. In (h), another SNPB occurred sometime before the QTN on a chromosome with the old allele of SNPA, and overtime the frequency of the new allele is 25% and the old allele is 75%. This SNP arrangement results in three haplotypes as before: new allele SNPA with old allele SNPB (frequency 0.4), old allele SNPA with old allele SNPB (frequency 0.35), and old allele SNPA with new allele SNPB (frequency 0.25). Later in the history of the population, a mutation occurs on a chromosome with the old SNPA, new SNPB allele haplotype generating QTN allele 1, which rises in frequency to 0.1, representing 40% of chromosomes with the old SNPA, new SNPB haplotype. One can calculate r^2 and the substitution effects for SNPA, SNPB, and the three haplotypes. The r^2's are 0.07, 0.33, and 0.23, respectively. In this case, the haplotypes are in less linkage disequilibrium than SNPB. The substitution effects are 0.17 for SNPA, 0.4 for SNPB, and 0 for replacing SNPA old, SNPB old with SNPA new, SNPB old and 0.4 for replacing either alternative haplotype with SNPA old, SNPB new. In this example, the three haplotypes do not provide better detection of the additive effect of the genetic locus than SNPB alone. Both examples illustrate the potential advantage of exploiting linkage disequilibrium for the examination of genetic locus effects on a trait.

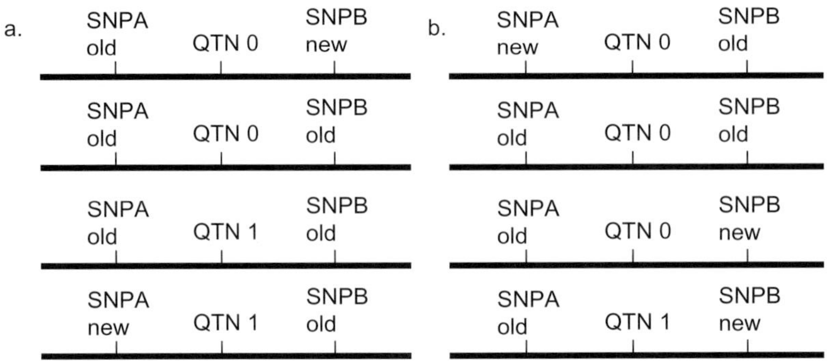

Figure 2.2 The arrangement of haplotypes described in Figure 2.1a through d is illustrated in (a); the arrangement of haplotypes described in Figure 2.1e through h is illustrated in (b). Without linkage disequilibrium, three loci, each with two alleles, would have 2^3 or 8 combinations of alleles. Linkage disequilibrium reduced the number of combinations to four in each case and broadened the range over which the effect of the QTL could be observed.

LD analysis is most useful when phenotyped animals represent a diverse sampling of the population of interest. Unrelated animals can be advantageous because linkage between adjacent loci is smaller and similar to that of the livestock population at large. Diverse and extensive sampling from the population of interest typically means that only those female reproductive traits that are routinely and easily obtained in livestock will be available for this type of analysis. Thus, the examination of difficult to collect reproductive phenotypes is at odds with the sampling necessary to reduce the size of regions associated with those traits. A compromise strategy that could be useful for these phenotypes is to perform continuous diverse sampling from the population of interest (e.g., sample males that are representative of the livestock population of interest by breeding them to a maintained research population of females such that over time sufficient diverse sampling of the livestock population occurs in the progeny of such matings).

One can skip linkage analysis and use the naturally occurring LD in a livestock population for a genome scan if genotyping using genetic markers that are sufficiently dense to assess genetic variation within the naturally occurring "haplotype blocks" alluded to earlier is available. Given that natural LD within livestock populations only occurs within 100 kb or so, a genomic scan capable of assessing the genetic variation across the entire genome (i.e., a genome scan using naturally occurring LD) requires the genotyping of thousands of SNP for each animal in a population. It is only recently that this technology has become available for livestock. Discovering the location of thousands of SNP in livestock is a monumental task. However, SNPs are a natural by-product of livestock genome sequencing efforts. The individual animals used as a source of DNA for genome sequencing efforts differ in DNA sequence between each other and are themselves heterozygous at numerous loci, resulting in a large number of SNP. Thus, availability of the sequenced genome for cattle has enhanced the effort to discover and map the needed SNP in this species, and similar efforts are being undertaken for swine and sheep. The availability of the needed SNP in cattle has resulted in the development of an SNP chip capable of simultaneously genotyping ~60,000 genetic loci (Illumina, Inc., San Diego, CA). LD analysis at the appropriate genotyping densities will have a significant advantage over linkage analysis, because if the markers used are sufficiently dense, they can be used on relatively unrelated animals and will be more predictive and of more utility to livestock populations at large. However, these advantages come with significant increases in the complexity of the statistical analysis required to reliably interpret the data.

2.4.4 Statistical analysis of genomic associations

Superficially, the association of SNP genotypes with phenotypes should be a simple task. Two alternate alleles at a locus generate three different genotypes in a population of animals. Analysis of variance can be used to analyze the phenotypes corresponding to the three genotypes and to arrive at a test of significance of the effect of genotype on the phenotype. One can use individual contrasts between genotypes to arrive at additive (comparison between homozygous animals) and dominance (comparison of the average of the two types of homozygous animals with the mean for heterozygous animals) effects. This approach can work for an unrelated population of animals, but such a population is rarely available for genomics research. Indeed, industry populations are often sampled under the assumption that the individuals sampled are unrelated and the results of these studies can produce inaccurate associations when the relationships between the animals sampled are ignored. Typically, genotype association studies are done on research or commercial herds in which the animals are related to each other in some degree. These relationships can originate from breed differences in animals used to generate the populations, or simply from parent–offspring (pedigree) relationships. Often, relationships are extremely high if the research herds are descended from populations used for linkage analysis. In addition, in many cases, analysis at more than a single locus is done. In these cases, association between the phenotype and genetic variation at many or all locations throughout the whole genome is sought. These two issues present some difficulties.

A founder animal within a population that is genetically superior at several independent loci, and is fixed for irrelevant loci on various chromosomes, would transmit both to its progeny. Separation of the superior loci from the irrelevant loci occurs in subsequent generations for loci on different chromosomes, but remains, at least in part, for loci on the same chromosome, because only recombination can break up the relationship. These associations can lead to erroneous conclusions about associations between genetic loci and phenotypic traits unless the polygenic effects between related animals are accounted for in the analysis (Calus and Veerkamp 2007). This is not trivial; relationships between animals must be known, and quantitative genetic analysis methods must be used to account for the relationships. Then, an individual genetic locus effect is fit simultaneously with the polygenic effects to distinguish one from the other. If this is not done, animals within the population are not independent from each other due to their interrelatedness, and incorrect associations can be obtained.

This type of analysis is adequate for a few SNP in a population, but when the goal is to determine the effects of genetic variation throughout the genome, as in a genome scan, the number of individual tests becomes a problem. Typical genome scan analyses test the effect of genotype at one centimorgan (~million base pair) intervals or less. Since the typical livestock genome is 3 billion bases, at least 3000 tests must be conducted for a complete genome scan. Use of new 50k SNP chips results in even greater numbers of tests. A statistical significance level of $P = 0.05$ for individual genetic loci on average results in one false-positive association by chance for every 20 tests performed. Typically, the problem of multiple tests is compensated by making analyses more stringent, such that the acceptable significance level is raised to a genome-wide error rate of 1 false positive in 20 genome scans, rather than 1 false positive in 20 tests. Although several QTL have met this criterion, this level of stringency risks missing true genotypic associations. In addition, the estimated effects of markers with associations at this significance level may be biased upward because of the stringent criteria required to accept them (Bogdan and Doerge 2005). In practice, a compromised statistical analysis strategy is typically adopted. The two or three QTL regions with the highest significance level for a particular phenotype become targets for further research, keeping in mind that one or more false positives may be present, depending on the actual significance level of the locus. In this way, for a given population, the two or three genetic loci with the greatest chance of being associated with a trait of interest are identified for further investigation using fine-mapping techniques to obtain more evidence of true association. This same issue is a problem for experiments that investigate a list of candidate genes for a particular phenotype. The more genes examined, the more false positives will be obtained. Acceptance of a candidate gene effect on a phenotype should have more evidence suggesting a real effect than just significance at $P < 0.05$ among multiple tests of candidate genes. This is one reason why the effect of a region of the genome on a particular trait should be examined in different populations to confirm the association. However, because results of genome scans depend on the founder animals chosen, failure to obtain a significant association in any given population does not mean the association is not present in another sampled population.

Appreciation of interrelatedness among individuals in a population and control of experiment-wide false-positive/false-negative rates is sufficient when the goal is to discover genetic loci having a true influence on a trait, and that has been the goal of much of genomics research to date. However, the real utility of genomics is in predicting individual animal performance given the genotype of the animal, and it is here that statistical analysis becomes even more complicated. With the availability of dense genome-wide genotyping in livestock, it will become possible to know with relative certainty how specific regions of the livestock genome or groups of haplotypes are passed from parent to offspring. Simulation studies have been reported on the utility of various methods to use this type of information to predict breeding values by examining the effects of multiple markers across the genome simultaneously. Methods that produce predicted breeding values from marker combinations densely sampling the entire genome have been broadly termed "whole genome selection" or WGS methods. It is clear from these studies that use of fixed least squares mean effects of statistically significant loci (selected using multiple regression model building methods) associated with a phenotype results in poor prediction of breeding values (Meuwissen et al. 2001). Statistical prediction of breeding values incorporating genotype information by fitting the markers as random effects with a common variance using a more quantitative genetics approach (best linear unbiased prediction [BLUP]) resulted in better prediction of breeding values. However, the best genotype-based prediction of breeding values was by using a Bayesian approach to fitting the effects of genotypes at multiple loci where most of the markers were allowed to have no effect. The best way to predict breeding values from genome-wide genotyping information is an area of ongoing intense research. Nevertheless, WGS shows great promise as a tool to use genomic data from field (e.g., dairy progeny tests) or research populations to predict breeding values of industry animals.

2.5 Some illustrative examples of reproductive QTL

Numerous QTL for female reproductive traits in livestock have been reported. A couple of recent reviews are available for swine (Buske et al. 2006; Rothschild et al. 2007). In addition, online resources are available. The AnimalQTLdb (Hu et al. 2005, 2007; Hu and Reecy 2007; www.animalgenome.org/QTLdb/) summarizes many QTL in swine and cattle, including their positions within the genome. The bovine QTL viewer (Polineni et al. 2006) is a website specific to cattle QTL (bovineqtlv2.tamu.edu/index.html). Genomic regions associated with most of the easily measured traits have been reported, along with some subcomponent traits that are more difficult to measure such as ovulation rate (Rathje et al. 1997; Rohrer et al. 1999; Wilkie et al. 1999; Kappes et al. 2000; Cassady et al. 2001; Sato et al. 2006) and uterine capacity (Rohrer et al. 1999; Vallet et al. 2005b). Associated genomic regions resulting from reported genome scans for QTL regions are typically very broad due to the small number of generations examined in these experiments. Numerous potential genes fall within the confidence interval of these reported QTL. Because the regions are so broad, it has been very difficult to identify causative genes and variation in those genes to convert many of the known QTL regions into genetic markers that are useful in a variety of populations. Candidate gene approaches have also yielded genetic loci associated with a variety of reproductive traits. Thus, associations with reproductive traits run the gamut from single SNP associations to huge genome regions from linkage analysis. Rather than attempt to describe them all, we thought it would be more useful to focus on a limited number of well-characterized associations with reproductive traits. These serve to illustrate the processes used and some of the limitations.

2.5.1 QTL mapping for ovulation rate in sheep

A good example of the application of genomics technology to reproductive traits is in the application of this technology to loci affecting ovulation rate in sheep. Although ovulation rate represents a quantitative trait in all livestock species, several gene loci with major effects on ovulation rate have been discovered in sheep. The first of these loci was reported to be present on the X chromosome (Davis et al. 1991) by virtue of its X-linked inheritance. This gene was named FecX by virtue of the fact that it was a gene affecting fecundity in sheep located on the X chromosome. This gene had the curious property that heterozygous animals displayed increased ovulation rate, while homozygous animals were sterile. A second Fec autosomal gene, FecB, was localized to an ovine chromosomal region similar to human chromosome 4 (Montgomery et al. 1993), because a genetic map for sheep at the time was not available. Subsequent studies localized the FecB gene to ovine chromosome 6 (Montgomery et al. 1994) based on detection of genes known to be within the human chromosomal region in sheep/hamster somatic cell hybrid clones of an ovine chromosome 6 translocation. Subsequent mapping confirmed that this region of the ovine genome is syntenic (similar) to human chromosome 4 (www.livestockgenomics.csiro.au/sheep/mapcreator). The effect of the FecB gene differed from that of FecX in that it appeared to be an additive; heterozygous sheep had ovulation rates midway between the ovulation rates of homozygous sheep. Galloway et al. (2000) reported that the FecX gene was explained by mutations in the bone morphogenic protein (BMP) 15 gene. Simultaneous reports (Souza et al. 2001; Wilson et al. 2001) indicated that the FecB gene was caused by a mutation in the BMP-1B receptor. Finally, other populations with phenotypic distributions (increased ovulation rate of heterozygous sheep, infertility of homozygous sheep) similar to FecX were explained by mutations in the growth differentiation factor (GDF) 9 gene on sheep chromosome 5 (Hanrahan et al. 2004). Both GDF9 and BMP15 are members of the transforming growth factor (TGF) β superfamily, and BMP-1B is a member of the TGFβ type 1 receptors (McNatty et al. 2004; Souza et al. 2004). Despite knowledge that various mutations in these genes are responsible for changes in ovulation rate, the mechanism whereby these changes result in differences in ovulation rate in sheep is still not clear. In all cases, it appears that an increase in ovulation rate is associated with partial loss of gene function, although complete loss of gene function is associated with sterility (Hanrahan et al. 2004). Curiously, in mice, knockout of the genes does not increase ovulation rate in heterozygous mice (Yan et al. 2001), pointing out that it is sometimes inappropriate to use results from other species to try to understand intricacies of control of reproductive traits in livestock. These mutations display several curious properties. Effects of BMP15 and GDF9 are additive in that ovulation rates for doubly heterozygous sheep are similar to the increase caused by each heterozygous genotype added together (McNatty et al. 2004). This result occurs despite the fact that increased ovulation rate for each gene is associated with partial but not complete loss of function. Failure of reproduction is likely associated with very low expression of this pathway. This result also illustrates the concept that redundancy in gene function may influence the success of a particular genetic association. The TGFβ family has many members with potential to have redundant functions among the individual genes. BMPR-1B is one of seven type 1 receptors and is expressed in bone during skeletal development. Knockout studies in the mouse result in subtle changes in skeletal development (Yi et al. 2000); however, changes in skeletal development have never been reported in FecB homozygous sheep. What has been described are reductions in live weight (Walling et al. 2000), although this may be caused by a separate adjacent locus in FecB sheep. It seems possible that subtle differences in weight could be the result of BMPR-1B effects on skeleton formation. Although these differences are not sufficient to outweigh the advantage in fecundity, it illustrates the point that gene alterations can have unintended effects, depending on the other pathways/functions in which that gene is involved.

2.5.2 QTL mapping for lactation in cattle

There have been several QTL studies in cattle for a variety of reproductive traits (Georges et al. 1995; Coppieters et al. 1998; Kappes et al. 2000; Ashwell et al. 2001, 2004; Boichard et al. 2003; Schrooten et al. 2004; Schnabel et al. 2005a; Muncie et al. 2006; Guillaume et al. 2007). Most of these studies examine lactation performance in dairy cattle, taking advantage of national dairy herd record programs in various countries (i.e., known pedigree information), along with the availability of sire performance testing for dairy bulls (resulting in the accurate prediction of breeding values for dairy traits), availability of semen for many of these bulls (to obtain DNA), and relatively widespread use of individual bulls by artificial insemination in the dairy industry. Thus, most cattle lactation QTL were discovered using high-accuracy, sire-predicted breeding values calculated from the evaluation of the lactation performance of resulting daughters. This once again reinforces the importance of pedigreed populations with phenotypes in QTL analysis. Two of the best characterized lactation QTL illustrate many of the issues involved in QTL analysis.

Georges et al. (1995) was the first to describe a QTL for protein yield on bovine chromosome 6, located midway along the chromosome, a finding subsequently confirmed by numerous reports (Ashwell et al. 2001, 2004; Boichard et al. 2003; Schrooten et al. 2004). Likewise, a QTL for fat percentage in milk was reported at the top of bovine chromosome 14 (Coppieters et al. 1998) and was subsequently confirmed by several other reports (Boichard et al. 2003; Ashwell et al. 2004; Schrooten et al. 2004).

Combined linkage–linkage disequilibrium analysis reduced the chromosome 6 QTL region to a small area of the chromosome containing just a few genes (Olsen et al. 2005; Schnabel et al. 2005b). Subsequently, polymorphisms in two different genes, osteopontin (SPP1; Leonard et al. 2005; Schnabel et al. 2005b) and the ATP-binding cassette transporter G2 (ABCG2; Cohen-Zinder et al. 2005), were suggested to be responsible for this QTL. This controversy (de Koning 2006) points out the difficulty in identifying a polymorphism in a gene as the causative polymorphism responsible for a QTL. The difficulty lies in the LD between polymorphisms in nearby genes. In order to distinguish the effects of one from the other, a population must be found where the LD is disrupted. In addition, to be convincing, a gene should have biological evidence suggesting that the gene is responsible for the differences in a trait of interest (Ron and Weller 2007). However, in this case, the discovered polymorphisms in both genes could reasonably be expected to have an effect on milk traits. A more recent report seems to suggest a resolution to the controversy, in favor of the ABCG2 gene polymorphism (Olsen et al. 2007), since the previously reported SPP1 polymorphism could be excluded in their study. Similar to the Fec genes in sheep, identification of the polymorphism and gene responsible did not immediately suggest the physiological mechanism. The ABCG2 gene affects the secretion of xenobiotics into milk (van Herwaarden and Schinkel 2006) and excludes xenobiotics from uptake from the gastrointestinal (GI) tract and from cells elsewhere in the body. It is difficult to understand how this function translates into changes in milk protein concentrations, and research to answer this question is ongoing.

Turning to the fat percentage QTL on chromosome 14, combined linkage–linkage disequilibrium analysis was again used to narrow this QTL to a small region near the centromere (Farnir et al. 2002). A subsequent polymorphism that alters the coding sequence of the acyl-CoA:diacylglycerol acyl transferase 1 (DGAT1) gene (substitutes a lysine K for alanine A at amino acid 232 of the coding sequence) within the QTL region was reported to be the cause of the QTL (Grisart et al. 2002; Winter et al. 2002; Thaller et al. 2003). This gene codes for the enzyme responsible for the final stage of triglyceride synthesis. It has been shown that the K allele of DGAT1 has greater triglyceride synthesizing activity than the A allele (Grisart et al. 2004); thus, it makes sense that a polymorphism that increases fat synthesis in the mammary gland would be associated with an increase in milk fat percentage. Although there is a solid case for the effect of this polymorphism on fat percentage, it appears that it is not the only polymorphism segregating at this locus (Kuhn et al. 2004). In this report, sires that are homozygous for the A allele of the DGAT1 gene still segregate a QTL in this region for fat percentage in their descendants, and the authors present evidence that polymorphisms in the promoter region of the DGAT1 gene may be responsible for the QTL in these sires. They implicate a variable nucleotide repeat (VNTR) region in the promoter, with three to seven repeats of the sequence AGGCCCCGCCCTCCCCGG, as potentially responsible for the additional QTL effects in this region. This sequence contains an SP1 transcription factor binding site and increases transcription of a reporter gene in mammary gland epithelial cells (Furbass et al. 2006), although transcription of the reporter gene did not vary with the number of repeats. However, another report did not confirm the effect of the VNTR region on fat percentage, although the authors do indicate the presence of additional QTL beyond that explained by the K232A polymorphism in DGAT1 (Gautier et al. 2007). This may be analogous to the different FecX gene alleles in sheep, all resulting in the impairment of the BMP15 function. Thus, other polymorphisms that alter the activity of DGAT1 in the mammary gland could have similar effects on fat percentage. These alterations could be increased transcription of the gene or increased translation of the protein. Alternatively, another nearby gene could be responsible for the QTL effects not explained by the K232A polymorphism in DGAT1.

Most of the original lactation QTL studies indicate interrelationships among various milk traits. Although the two previously described loci primarily affect protein (chromosome 6) and fat (chromosome 14), these loci have effects on the measured milk traits. In addition, Kaupe et al. (2007) reported a significant negative effect of the DGAT1 K allele on nonreturn rates, a measure of cow fertility. Allan et al. (2007) reported a significant association between polymorphisms in the osteopontin gene with calf birth weights, suggesting possible correlations between the chromosome 6 chromosomal region affecting milk protein with aspects of pregnancy. These are all examples of pleitropic effects of specific genetic loci and support the concept that marker-assisted selection for one trait may have consequences for other traits. It seems very unlikely, given their expression and likely function in other tissues, that polymorphisms in genes like ABCG2 and DGAT1 will have effects solely on milk production. It is well established that selection for increased milk production in dairy cattle has resulted in impaired fertility (Lucy 2001). This is at least in part due to antagonistic pleiotropy between genes affecting both traits. Fortunately, loci are likely to vary in the degree of multiple effects, and the most efficient use of whole genome association technology will be its ability to take these multiple effects of individual loci into account. This will require the collection of multiple phenotypes on the same animals.

2.5.3 QTL mapping for litter size in swine

Application of genomics to pig reproduction has been slower, due to lack of mutations with large effects as in sheep, and the lack of availability of suitable populations with broad phenotypic collections as in cattle. As previously indicated, genetic loci associated with reproduction in pigs (litter size) have been reported using a candidate gene approach and significant associations with litter size have been found for polymorphisms in the estrogen receptor (Rothschild et al. 1996; Short et al. 1997; Muñoz et al. 2007), retinol-binding protein (Rothschild et al. 2000), prolactin receptor (Drogemuller et al. 2001), and erythropoietin receptor (Vallet et al. 2005b). Genomic scans for various reproductive traits in pigs have been reported (Rohrer et al. 1999; Wilkie et al. 1999; Cassady et al. 2001; Holl et al. 2004; Tribout et al. 2008). The early genome scans were performed in defined populations that were crosses of lines with divergent phenotypes (e.g., Meishan and European crosses, crosses between lines selected for ovulation rate, embryo survival, and a randomly selected contemporary control line). Use of these populations maximized detection of reproduction QTL, but because they were done in lines that were not fully relevant to production lines used in the swine industry, validation of the regions in industry relevant pigs is an extra step to utilization of results of these genomics experiments in swine production.

2.6 Future research directions

As previously mentioned, two innovations will revolutionize genetic selection based on genotyping of individual animals, the routine availability of less expensive genotyping of sufficient density to successfully capture a large share of the genetic variation within each animal, and development of statistical approaches that optimize the use of this information for prediction of individual animal breeding values. Dense genotyping will be a natural by-product of livestock genome sequencing efforts, combined with genome-wide SNP discovery research making use of new high-throughput sequencing technologies. Because of the expense and complexity of this technology, it will initially be applied to research herds and elite livestock to improve the use of these animals in genetic selection schemes, and will therefore be the domain of scientists and livestock breeding companies. However, as more becomes known regarding regions associated with specific traits and as genotyping technology becomes cheaper, subsets of markers for specific traits will become available to producers to help fine-tune livestock for specific environments or specific markets.

Comprehensive genotyping and WGS studies on populations of animals for which a variety of traits have been measured will provide needed information on the genetic loci that explain various negative genetic correlations between traits and possibly between the mother and offspring for the same trait. Information on the genetic architecture of these correlated traits will provide genetic loci that can be selected to manipulate these negatively correlated traits independently of each other, or at least allow balanced selection procedures taking into account effects of the various loci on traits.

Selection for negatively correlated traits is now typically done using an index, to balance selection for the traits. Index selection puts selection pressure on all loci affecting both traits, both those with multiple antagonistic effects and those that independently affect each trait. The net effect of this selection is slower progress in changing the independent loci. Determination of the effects of loci influencing various traits will allow a more direct and controllable approach to selection for correlated traits.

Much of genomic analysis currently deals with independent additive effects of loci on associated traits. Dominance and imprinting effects at loci could also be incorporated into selection schemes as these effects can be readily detected in association studies. Imprinting is defined as differences in the expression of an allele depending on its parental origin. More difficult are epistatic interactions between loci and environment by genetic locus interactions. A special case that is similar to these effects but specific to pregnancy associated traits is the interactions between maternal, paternal, and fetal genetic influences on a trait. Epistatic gene interaction occurs when the influence of a genetic locus on a trait is dependent on the presence or absence of alleles at other loci. A useful reproductive example might be loci influencing ovulation rate, loci influencing uterine capacity, and their combined influence on litter size. Because of the sequential nature of expression of ovulation rate and uterine capacity, the influence of genes associated with differences in uterine capacity will only be observed if the alleles needed for high ovulation rate are present. Similarly, given the existence of environment by genotype interactions, some gene allele effects may only be observed under the appropriate environmental conditions. Finally, in the case of multiple genotype interactions on a trait, it seems likely that there may be fetal genes that only affect pregnancy outcome given an appropriate maternal environment, which would be controlled by genes of the dam. Analysis for epistatic, environmental, and multiple genome interactions will be a future direction of QTL analysis of reproductive traits.

Perhaps the brightest future for the genomics of female reproductive traits will be the information derived from the identification of the genes and the genetic variation within genes that are responsible for differences in reproductive traits. While not essential for the primary utility of the technology, elucidation of genes and polymorphisms responsible for differences in performance will almost certainly follow from the identification of QTL affecting these traits, once the regions are sufficiently narrowed to enable utility among livestock populations at large. However, using the milk production QTL on chromosome 6 as an example, proving that a specific polymorphism within a specific gene is actually responsible for the QTL can be difficult. Any polymorphism will have nearby DNA variation more or less associated with it. Proof that a specific polymorphism is responsible for the difference in the trait will require elimination of the contribution of other linked loci and/or corroborating physiological studies that support the effect of a specific polymorphism. This could take the form of transgenic incorporation of the polymorphism into unaffected individuals, but such experiments in livestock are currently very difficult. In addition, the lesson from sheep ovulation rate QTL and dairy cattle milk production QTL on chromosome 6 suggest that establishing the identity of the gene and the polymorphism does not necessarily immediately lead to an understanding of the role of that gene in the trait. Nevertheless, identification of genes responsible and elucidation of the physiological mechanisms could lead to other nongenetic means of improving reproductive and other traits. This represents the overlap between QTL genomics and so-called functional genomics, or the elucidation of how gene function translates into differences between animals. This will provide perhaps the greatest benefit that may arise from QTL studies, information about the genes, and gene mechanisms that affect reproductive traits, leading to a variety of strategies to improve those traits.

References

Abecasis, G.R., Noguchi, E., Heinzmann, A., Traherne, J.A., Bhattacharyya, S., Leaves, N.I., Anderson, G.G., Zhang, Y., Lench, N.J., Carey, A., Cardon, L.R., Moffatt, M.F., and Cookson, W.O.C. 2001. Extent and distribution of linkage disequilibrium in three genomic regions. *The American Journal of Human Genetics* **68**(1): 191–197.

Akashi, H. 2001. Gene expression and molecular evolution. *Current Opinion in Genetics & Development* **11**(6): 660–666.

Allan, M.F., Thallman, R.M., Cushman, R.A., Echternkamp, S.E., White, S.N., Kuehn, L.A., Casas, E., and Smith, T.P.L. 2007. Association of a single nucleotide polymorphism in SPP1 with growth traits and twinning in a cattle population selected for twinning rate. *Journal of Animal Science* **85**(2): 341–347.

Arango, J., Misztal, I., Tsuruta, S., Culbertson, M., Holl, J.W., and Herring, W. 2006. Genetic study of individual preweaning mortality and birth weight in Large White piglets using threshold-linear models. *Livestock Science* **101**(1): 208–218.

Ashwell, M.S., Heyen, D.W., Sonstegard, T.S., Van Tassell, C.P., Da, Y., VanRaden, P.M., Ron, M., Weller, J.I., and Lewin, H.A. 2004. Detection of quantitative trait loci affecting milk production, health, and reproductive traits in Holstein cattle. *Journal of Dairy Science* **87**(2): 468–475.

Ashwell, M.S., Van Tassell, C.P., and Sonstegard, T.S. 2001. A genome scan to identify quantitative trait loci affecting economically important traits in a US Holstein population. *Journal of Dairy Science* **84**(11): 2535–2542.

Beckmann, J.S., Estivill, X., and Antonarakis, S.E. 2007. Copy number variants and genetic traits: Closer to the resolution of phenotypic to genotypic variability. *Nature Reviews Genetics* **8**(8): 639–646.

Bennett, G.L. and Gregory, K.E. 2001. Genetic (co)variances for calving difficulty score in composite and parental populations of beef cattle: I. Calving difficulty score, birth weight, weaning weight, and postweaning gain. *Journal of Animal Science* **79**(1): 45–51.

Bogdan, M. and Doerge, R.W. 2005. Biased estimators of quantitative trait locus heritability and location in interval mapping. *Heredity* **95**(6): 476–484.

Boichard, D., Grohs, C., Bourgeois, F., Cerqueira, F., Faugeras, R., Neau, A., Rupp, R., Amigues, Y., Boscher, M.Y., and Levéziel, H. 2003. Detection of genes influencing economic traits in three French dairy cattle breeds. *Genetics Selection Evolution* **35**(1): 77–101.

Bormann, J.M., Totir, L.R., Kachman, S.D., Fernando, R.L., and Wilson, D.E. 2006. Pregnancy rate and first-service conception rate in Angus heifers. *Journal of Animal Science* **84**(8): 2022–2025.

Buske, B., Sternstein, I., and Brockmann, G. 2006. QTL and candidate genes for fecundity in sows. *Animal Reproduction Science* **95**(3–4): 167–183.

Byrne, M., Jones, G., and Warner, C. 2007. Preimplantation embryo development (Ped) gene copy number varies from 0 to 85 in a population of wild mice identified as *Mus musculus domesticus*. *Mammalian Genome* **18**(11): 767–778.

Calus, M.P.L. and Veerkamp, R.F. 2007. Accuracy of breeding values when using and ignoring the polygenic effect in genomic breeding value estimation with a marker density of one SNP per cM. *Journal of Animal Breeding and Genetics* **124**(6): 362–368.

Canario, L., Roy, N., Gruand, J., and Bidanel, J.P. 2006. Genetic variation of farrowing kinetics traits and their relationships with litter size and perinatal mortality in French Large White sows. *Journal of Animal Science* **84**(5): 1053–1058.

Cassady, J.P., Johnson, R.K., Pomp, D., Rohrer, G.A., Van Vleck, L.D., Spiegel, E.K., and Gilson, K.M. 2001. Identification of quantitative trait loci affecting reproduction in pigs. *Journal of Animal Science* **79**(3): 623–633.

Cohen-Zinder, M., Seroussi, E., Larkin, D.M., Loor, J.J., Everts-van der Wind, A., Lee, J.H., Drackley, J.K., Band, M.R., Hernandez, A.G., Shani, M., Lewin, H.A., Weller, J.I., and Ron, M. 2005. Identification of a missense mutation in the bovine ABCG2 gene with a major effect on the QTL on chromosome 6 affecting milk yield and composition in Holstein cattle. *Genome Research* **15**(7): 936–944.

Conley, A.J., Jung, Y.C., Schwartz, N.K., Warner, C.M., Rothschild, M.F., and Ford, S.P. 1988. Influence of SLA haplotype on ovulation rate and litter size in miniature pigs. *Journal of Reproduction and Fertility* **82**(2): 595–601.

Coppieters, W., Riquet, J., Arranz, J.-J., Berzi, P., Cambisano, N., Grisart, B., Karim, L., Marcq, F., Moreau, L., Nezer, C., Simon, P., Vanmanshoven, P., Wagenaar, D., and Georges, M. 1998. A QTL with major effect on milk yield and composition maps to bovine chromosome 14. *Mammalian Genome* **9**(7): 540–544.

Crawford, A.M. and Cuthbertson, R.P. 1996. Mutations in sheep microsatellites. *Genome Research* **6**(9): 876–879.

Davis, G.H., McEwan, J.C., Fennessy, P.F., Dodds, K.G., and Farquhar, P.A. 1991. Evidence for the presence of a major gene influencing ovulation rate on the X chromosome of sheep. *Biology of Reproduction* **44**(4): 620–624.

de Koning, D.-J. 2006. Conflicting candidates for cattle QTLs. *Trends in Genetics* **22**(6): 301–305.

Drogemuller, C., Hamann, H., and Distl, O. 2001. Candidate gene markers for litter size in different German pig lines. *Journal of Animal Science* **79**(10): 2565–2570.

Du, F.-X., Clutter, A.C., and Lohuis, M.M. 2007. Characterizing linkage disequilibrium in pig populations. *International Journal of Biological Sciences* **3**(3): 166–178.

Dunning, A.M., Durocher, F., Healey, C.S., Teare, M.D., McBride, S.E., Carlomagno, F., Xu, C.-F., Dawson, E., Rhodes, S., Ueda, S., Lai, E., Luben, R.N., Van Rensburg, E.J., Mannermaa, A., Kataja, V., Rennart, G., Dunham, I., Purvis, I., Easton, D., and Ponder, B.A.J. 2000. The extent of linkage disequilibrium in four populations with distinct demographic histories. *The American Journal of Human Genetics* **67**(6): 1544–1554.

Ellis, S.A., Holmes, E.C., Staines, K.A., Smith, K.B., Stear, M.J., McKeever, D.J., MacHugh, N.D., and Morrison, W.I. 1999. Variation in the number of expressed

MHC genes in different cattle class I haplotypes. *Immunogenetics* **50**(5): 319–328.

Farnir, F., Grisart, B., Coppieters, W., Riquet, J., Berzi, P., Cambisano, N., Karim, L., Mni, M., Moisio, S., Simon, P., Wagenaar, D., Vilkki, J., and Georges, M. 2002. Simultaneous mining of linkage and linkage disequilibrium to fine map quantitative trait loci in outbred half-sib pedigrees: Revisiting the location of a quantitative trait locus with major effect on milk production on bovine chromosome 14. *Genetics* **161**(1): 275–287.

Fedorova, L. and Fedorov, A. 2003. Introns in gene evolution. *Genetica* **118**(2): 123–131.

Furbass, R., Winter, A., Fries, R., and Kuhn, C. 2006. Alleles of the bovine DGAT1 variable number of tandem repeat associated with a milk fat QTL at chromosome 14 can stimulate gene expression. *Physiological Genomics* **25**(1): 116–120.

Galloway, S.M., McNatty, K.P., Cambridge, L.M., Laitinen, M.P.E., Juengel, J.L.T., Jokiranta, S., McLaren, R.J., Luiro, K., Dodds, K.G., Montgomery, G.W., Beattie, A.E., Davis, G.H., and Ritvos, O. 2000. Mutations in an oocyte-derived growth factor gene (BMP15) cause increased ovulation rate and infertility in a dosage-sensitive manner. *Nature Genetics* **25**(3): 279–283.

Gautier, M., Capitan, A., Fritz, S., Eggen, A., Boichard, D., and Druet, T. 2007. Characterization of the DGAT1 K232A and variable number of tandem repeat polymorphisms in French dairy cattle. *Journal of Dairy Science* **90**(6): 2980–2988.

Georges, M., Nielsen, D., Mackinnon, M., Mishra, A., Okimoto, R., Pasquino, A.T., Sargeant, L.S., Sorensen, A., Steele, M.R., Zhao, X., Womack, J.E., and Hoeschele, I. 1995. Mapping quantitative trait loci controlling milk production in dairy cattle by exploiting progeny testing. *Genetics* **139**(2): 907–920.

Gregory, K.E., Bennett, G.L., Van Vleck, L.D., Echternkamp, S.E., and Cundiff, L.V. 1997. Genetic and environmental parameters for ovulation rate, twinning rate, and weight traits in a cattle population selected for twinning. *Journal of Animal Science* **75**(5): 1213–1222.

Grisart, B., Coppieters, W., Farnir, F., Karim, L., Ford, C., Berzi, P., Cambisano, N., Mni, M., Reid, S., Simon, P., Spelman, R., Georges, M., and Snell, R. 2002. Positional candidate cloning of a QTL in dairy cattle: Identification of a missense mutation in the bovine DGAT1 gene with major effect on milk yield and composition. *Genome Research* **12**(2): 222–231.

Grisart, B., Farnir, F., Karim, L., Cambisano, N., Kim, J.J., Kvasz, A., Mni, M., Simon, P., Frere J.M., Coppieters, W., and Georges, M. 2004. Genetic and functional confirmation of the causality of the DGAT1 K232A quantitative trait nucleotide in affecting milk yield and composition. *Proceedings of the National Academy of Sciences of the United States of America* **101**(8): 2398–2403.

Guillaume, F., Gautier, M., Ben Jemaa, S., Fritz, S., Eggen, A., Boichard, D., and Druet, T. 2007. Refinement of two female fertility QTL using alternative phenotypes in French Holstein dairy cattle. *Animal Genetics* **38**(1): 72–74.

Gutierrez, J.P., Goyache, F., Fernandez, I., Alvarez, I., and Royo, L.J. 2007. Genetic relationships among calving ease, calving interval, birth weight, and weaning weight in the Asturiana de los Valles beef cattle breed. *Journal of Animal Science* **85**(1): 69–75.

Hanford, K.J., Van Vleck, L.D., and Snowder, G.D. 2002. Estimates of genetic parameters and genetic change for reproduction, weight, and wool characteristics of Columbia sheep. *Journal of Animal Science* **80**(12): 3086–3098.

Hanrahan, J.P., Gregan, S.M., Mulsant, P., Mullen, M., Davis, G.H., Powell, R., and Galloway, S.M. 2004. Mutations in the genes for oocyte-derived growth factors GDF9 and BMP15 are associated with both increased ovulation rate and sterility in Cambridge and Belclare sheep (*Ovis aries*). *Biology of Reproduction* **70**(4): 900–909.

Harmegnies, N., Farnir, F., Davin, F., Buys, N., Georges, M., and Coppieters, W. 2006. Measuring the extent of linkage disequilibrium in commercial pig populations. *Animal Genetics* **37**(3): 225–231.

Holl, J.W., Cassady, J.P., Pomp, D., and Johnson, R.K. 2004. A genome scan for quantitative trait loci and imprinted regions affecting reproduction in pigs. *Journal of Animal Science* **82**(12): 3421–3429.

Holm, B., Bakken, M., Klemetsdal, G., and Vangen, O. 2004. Genetic correlations between reproduction and production traits in swine. *Journal of Animal Science* **82**(12): 3458–3464.

Holm, B., Bakken, M., Vangen, O., and Rekaya, R. 2005. Genetic analysis of age at first service, return rate, litter size, and weaning-to-first service interval of gilts and sows. *Journal of Animal Science* **83**(1): 41–48.

Hu, Z.-L., Dracheva, S., Jang, W., Maglott, D., Bastiaansen, J., Rothschild, M.F., and Reecy, J.M.. 2005. A QTL resource and comparison tool for pigs: PigQTLDB. *Mammalian Genome* **16**(10): 792–800.

Hu, Z.-L., Fritz, E.R., and Reecy, J.M. 2007. AnimalQTLdb: A livestock QTL database tool set for positional QTL information mining and beyond. *Nucleic Acids Research* **35**(Database issue): D604–D609.

Hu, Z.-L. and Reecy, J. 2007. Animal QTLdb: Beyond a repository. *Mammalian Genome* **18**(1): 1–4.

Iida, Y. and Masuda, T. 1996. Strength of translation initiation signal sequence of mRNA as studied by quantification method: Effect of nucleotide substitutions upon translation efficiency in rat preproinsulin mRNA. *Nucleic Acids Research* **24**(17): 3313–3316.

Janssens, S., Vandepitte, W., and Bodin, L. 2004. Genetic parameters for litter size in sheep: Natural *versus* hormone-induced oestrus. *Genetics Selection Evolution* **36**(5): 543–562.

Kappes, S.M., Bennett, G.L., Keele, J.W., Echternkamp, S.E., Gregory, K.E., and Thallman, R.M. 2000. Initial results of genomic scans for ovulation rate in a cattle population selected for increased twinning rate. *Journal of Animal Science* **78**(12): 3053–3059.

Kaupe, B., Brandt, H., Prinzenberg, E.-M., and Erhardt, G. 2007. Joint analysis of the influence of CYP11B1 and DGAT1 genetic variation on milk production, somatic cell score, conformation, reproduction, and productive lifespan in German Holstein cattle. *Journal of Animal Science* **85**(1): 11–21.

Kiefer, J.C. 2006. MicroRNAs under the microscope. *Developmental Dynamics* **235**(3): 846–853.

Komisarek, J. and Dorynek, Z. 2002. Genetic aspects of twinning in cattle. *Journal of Applied Genetics* **43**(1): 55–68.

Kuhn, C., Thaller G., Winter, A., Bininda-Emonds, O.R.P., Kaupe, B., Erhardt, G., Bennewitz, J., Schwerin, M., and Fries, R. 2004. Evidence for multiple alleles at the DGAT1 locus better explains a quantitative trait locus with major effect on milk fat content in cattle. *Genetics* **167**(4): 1873–1881.

Kurland, C.G. 1991. Codon bias and gene expression. *FEBS Letters* **285**(2): 165–169.

Leonard, S., Khatib, H., Schutzkus, V., Chang, Y.M., and Maltecca, C. 2005. Effects of the osteopontin gene variants on milk production traits in dairy cattle. *Journal of Dairy Science* **88**(11): 4083–4086.

Lucy, M.C. 2001. ADSA Foundation Scholar Award. Reproductive loss in high-producing dairy cattle: Where will it end? *Journal of Dairy Science* **84**(6): 1277–1293.

MacNeil, M.D., Geary, T.W., Perry, G.A., Roberts, A.J., and Alexander, L.J. 2006. Genetic partitioning of variation in ovulatory follicle size and probability of pregnancy in beef cattle. *Journal of Animal Science* **84**(7): 1646–1650.

Maniatis, T. and Reed, R. 2002. An extensive network of coupling among gene expression machines. *Nature* **416**(6880): 499–506.

Martinez, G.E., Koch, R.M., Cundiff, L.V., Gregory, K.E., Kachman, S.D., and Van Vleck, L.D. 2005. Genetic parameters for stayability, stayability at calving, and stayability at weaning to specified ages for Hereford cows. *Journal of Animal Science* **83**(9): 2033–2042.

McCarroll, S.A. and Altshuler, D.M. 2007. Copy-number variation and association studies of human disease. *Nature Genetics* **39**(Supplement 7): S37–S42.

McKay, S.D., Schnabel, R.D., Murdoch, B.M., Matukumalli, L.K., Aerts, J., Coppieters, W., Crews, D., Dias Neto, E., Gill, C.A., Gao, C., Mannen, H., Stothard, P., Wang, Z., Van Tassell, C.P., Williams, J.L., Taylor, J.F., and Moore, S.S. 2007. Whole genome linkage disequilibrium maps in cattle. *BMC Genetics* **8**: 74.

McNatty, K.P., Moore, L.G., Hudson, N.L., Quirke, L.D., Lawrence, S.B., Reader, K., Hanrahan, J.P., Smith, P., Groome, N.P., Laitinen, M., Ritvos, O., and Juengel, J.L. 2004. The oocyte and its role in regulating ovulation rate: A new paradigm in reproductive biology. *Reproduction* **128**(4): 379–386.

McRae, A.F., McEwan, J.C., Dodds, K.G., Wilson, T., Crawford, A.M., and Slate, J. 2002. Linkage disequilibrium in domestic sheep. *Genetics* **160**(3): 1113–1122.

Mesa, H., Safranski, T.J., Cammack, K.M., Weaber, R.L., and Lamberson, W.R. 2006. Genetic and phenotypic relationships of farrowing and weaning survival to birth and placental weights in pigs. *Journal of Animal Science* **84**(1): 32–40.

Mesa, H., Safranski, T.J., Fischer, K.A., Cammack, K.M., and Lamberson, W.R. 2005. Selection for placental efficiency in swine: Genetic parameters and trends. *Journal of Animal Science* **83**(5): 983–991.

Meuwissen, T.H.E., Hayes, B.J., and Goddard, M.E. 2001. Prediction of total genetic value using genome-wide dense marker maps. *Genetics* **157**(4): 1819–1829.

Montgomery, G.W., Crawford, A.M., Penty, J.M., Dodds, K.G., Ede, A.J., Henry, H.M., Pierson, C.A., Lord, E.A., Galloway, S.M., Schmack, A.E., Sise, J.A., Swarbrick, P.A., Hanrahan, V., Buchanan, F.C., and Hill, D.F. 1993. The ovine Booroola fecundity gene (FecB) is linked to markers from a region of human chromosome 4q. *Nature Genetics* **4**(4): 410–414.

Montgomery, G.W., Lord, E.A., Penty, J.M., Dodds, K.G., Broad, T.E., Cambridge, L., Sunden, S.L.F., Stone, R.T., and Crawford, A.M. 1994. The Booroola fecundity (FecB) gene maps to sheep chromosome 6. *Genomics* **22**(1): 148–153.

Mote, B.E., Serenius, T., Stalder, K.J., and Rothschild, M.F. 2006. The holy grail for pigs: Candidate genes affecting sow productive live. *Proceedings of the 8th World Congress on Genetics Applied to Livestock Production*, August Belo Horizonte, MG Brasil, August 13–18.

Muncie, S.A., Cassady, J.P., and Ashwell, M.S. 2006. Refinement of quantitative trait loci on bovine chromosome 18 affecting health and reproduction in US Holsteins. *Animal Genetics* **37**(3): 273–275.

Muñoz, G., Ovilo, C., Estellé, J., Silió, L., Fernández, A., and Rodriguez, C. 2007. Association with litter size of new polymorphisms on ESR1 and ESR2 genes in a Chinese-European pig line. *Genetics Selection Evolution* **39**(2): 195–206.

Nachman, M.W. and Crowell, S.L. 2000. Estimate of the mutation rate per nucleotide in humans. *Genetics* **156**(1): 297–304.

Nsengimana, J., Baret, P., Haley, C.S., and Visscher, P.M. 2004. Linkage disequilibrium in the domesticated pig. *Genetics* **166**(3): 1395–1404.

Okut, H., Bromley, C.M., Van Vleck, L.D., and Snowder, G.D. 1999. Genotypic expression at different ages: I. Prolificacy traits of sheep. *Journal of Animal Science* 77(9): 2357–2365.

Olsen, H.G., Lien, S., Gautier, M., Nilsen, H., Roseth, A., Berg, P.R., Sundsaasen, K.K., Svendsen, M., and Meuwissen, T.H.E. 2005. Mapping of a milk production quantitative trait locus to a 420-kb region on bovine chromosome 6. *Genetics* **169**(1): 275–283.

Olsen, H.G., Nilsen, H., Hayes, B., Berg, P.R., Svendsen, M., Lien, S., and Meuwissen, T. 2007. Genetic support for a quantitative trait nucleotide in the *ABCG2* gene affecting milk composition of dairy cattle. *BMC Genetics* **8**: 32.

Polineni, P., Aragonda, P., Xavier, S.R., Furuta, R., and Adelson, D.L. 2006. The bovine QTL viewer: A web accessible database of bovine quantitative trait loci. *BMC Bioinformatics* 7: 283.

Rathje, T.A., Rohrer, G.A., and Johnson, R.K. 1997. Evidence for quantitative trait loci affecting ovulation rate in pigs. *Journal of Animal Science* **75**(6): 1486–1494.

Redon, R., Ishikawa, S., Fitch, K.R., Feuk, L., Perry, G.H., Andrews, T.D., Fiegler, H., Shapero, M.H., Carson, A.R., Chen, W., Cho, E.K., Dallaire, S., Freeman, J.L., González, J.R., Gratacòs, M., Huang, J., Kalaitzopoulos, D., Komura, D., MacDonald, J.R., Marshall, C.R., Mei, R., Montgomery, L., Nishimura, K., Okamura, K., Shen, F., Somerville, M.J., Tchinda, J., Valsesia, A., Woodwark, C., Yang, F., Zhang, J., Zerjal, T., Zhang, J., Armengol, L., Conrad, D.F., Estivill, X., Tyler-Smith, C., Carter, N.P., Aburatani, H., Lee, C., Jones, K.W., Scherer, S.W., and Hurles, M.E. 2006. Global variation in copy number in the human genome. *Nature* **444**(7118): 444–454.

Renard, C., Hart, E., Sehra, H., Beasley, H., Coggill, P., Howe, K., Harrow, J., Gilbert, J., Sims, S., Rogers, J., Ando, A., Shigenari, A., Shiina, T., Inoko, H., Chardon, P., and Beck, S. 2006. The genomic sequence and analysis of the swine major histocompatibility complex. *Genomics* **88**(1): 96–110.

Roeder, R.G. 1991. The complexities of eukaryotic transcription initiation: Regulation of preinitiation complex assembly. *Trends in Biochemical Sciences* **16**(11): 402–408.

Roehe, R. 1999. Genetic determination of individual birth weight and its association with sow productivity traits using Bayesian analyses. *Journal of Animal Science* 77(2): 330–343.

Rohrer, G.A., Ford, J.J., Wise, T.H., Vallet, J.L., and Christenson, R.K. 1999. Identification of quantitative trait loci affecting female reproductive traits in a multigeneration Meishan-White composite swine population. *Journal of Animal Science* **77**(6): 1385–1391.

Ron, M. and Weller, J.I. 2007. From QTL to QTN identification in livestock—Winning by points rather than knock-out: A review. *Animal Genetics* **38**(5): 429–439.

Rosendo, A., Druet, T., Gogue, J., and Bidanel, J.P. 2007. Direct responses to six generations of selection for ovulation rate or prenatal survival in Large White pigs. *Journal of Animal Science* **85**(2): 356–364.

Rothschild, M.F., Hu, Z.-L., and Jiang, Z. 2007. Advances in QTL mapping in pigs. *International Journal of Biological Sciences* **3**(3): 192–197.

Rothschild, M., Jacobson, C., Vaske, D., Tuggle, C., Wang, L., Short, T., Eckardt, G., Sasaki, S., Vincent, A., McLaren, D., Southwood, O., van der Steen, H., Mileham, A., and Plastow, G. 1996. The estrogen receptor locus is associated with a major gene influencing litter size in pigs. *Proceedings of the National Academy of Sciences of the United States of America* **93**(1): 201–205.

Rothschild, M.F., Messer, L., Day, A., Wales, R., Short, T., Southwood, O., and Plastow, G. 2000. Investigation of the retinol-binding protein 4 (RBP4) gene as a candidate gene for increased litter size in pigs. *Mammalian Genome* **11**(1): 75–77.

Ruiz-Echevarria, M.J., Munshi, R., Tomback, J., Kinzy, T.G., and Peltz, S.W. 2001. Characterization of a general stabilizer element that blocks deadenylation-dependent mRNA decay. *Journal of Biological Chemistry* **276**(33): 30995–31003.

Sato, S., Atsuji, K., Saito, N., Okitsu, M., Sato, S., Komatsuda, A., Mitsuhashi, T., Nirasawa, K., Hayashi, T., Sugimoto, Y., and Kobayashi, E. 2006. Identification of quantitative trait loci affecting corpora lutea and number of teats in a Meishan x Duroc F2 resource population. *Journal of Animal Science* **84**(11): 2895–2901.

Schier, A.F. 2007. The maternal-zygotic transition: Death and birth of RNAs. *Science* **316**(5823): 406–407.

Schnabel, R.D., Kim, J.J., Ashwell, M.S., Sonstegard, T.S., Van Tassell, C.P., Connor, E.E., and Taylor, J.F. 2005b. Fine-mapping milk production quantitative trait loci on BTA6: Analysis of the bovine osteopontin gene. *Proceedings of the National Academy of Sciences of the United States of America* **102**(19): 6896–6901.

Schnabel, R.D., Sonstegard, T.S., Taylor, J.F., and Ashwell, M.S. 2005a. Whole-genome scan to detect QTL for milk production, conformation, fertility and functional traits in two US Holstein families. *Animal Genetics* **36**(5): 408–416.

Schrooten, C., Bink, M.C.A.M., and Bovenhuis, H. 2004. Whole genome scan to detect chromosomal regions affecting multiple traits in dairy cattle. *Journal of Dairy Science* **87**(10): 3550–3560.

Serenius, T. and Stalder, K.J. 2004. Genetics of length of productive life and lifetime prolificacy in the Finnish Landrace and Large White pig populations. *Journal of Animal Science* **82**(11): 3111–3117.

Serenius, T. and Stalder, K.J. 2006. Selection for sow longevity. *Journal of Animal Science* **84**(13 Electronic Supplement 1): E166–E171.

Shifman, S., Kuypers, J., Kokoris, M., Yakir, B., and Darvasi, A. 2003. Linkage disequilibrium patterns of the human genome across populations. *Human Molecular Genetics* **12**(7): 771–776.

Short, T.H., Rothschild, M.F., Southwood, O.I., McLaren, D.G., de Vries, A., van der Steen, H., Eckardt, G.R., Tuggle, C.K., Helm, J., Vaske, D.A., Mileham, A.J., and Plastow, G.S. 1997. Effect of the estrogen receptor locus on reproduction and production traits in four commercial pig lines. *Journal of Animal Science* **75**(12): 3138–3142.

Souza, C.J.H., González-Bulnes, A., Campbell, B.K., McNeilly, A.S., and Baird, D.T. 2004. Mechanisms of action of the principal prolific genes and their application to sheep production. *Reproduction, Fertility, and Development* **16**(4): 395–401.

Souza, C.J.H., MacDougall, C., Campbell, B.K., McNeilly, A.S., and Baird, D.T. 2001. The Booroola (FecB) phenotype is associated with a mutation in the bone morphogenetic receptor type 1 B (BMPR1B) gene. *Journal of Endocrinology* **169**(2): R1–R6.

Sterning, M., Rydhmer, L., and Eliasson-Selling, L. 1998. Relationships between age at puberty and interval from weaning to estrus and between estrus signs at puberty and after the first weaning in pigs. *Journal of Animal Science* **76**(2): 353–359.

Stitzel, M.L. and Seydoux, G. 2007. Regulation of the oocyte-to-zygote transition. *Science* **316**(5823): 407–408.

Thaller, G., Kramer, W., Winter, A., Kaupe, B., Erhardt, G., and Fries, R. 2003. Effects of DGAT1 variants on milk production traits in German cattle breeds. *Journal of Animal Science* **81**(8): 1911–1918.

Tribout, T., Iannuccelli, N., Druet, T., Gilbert, H., Riquet, J., Gueblez, R., Mercat, M.-J., Bidanel, J.-P., Milan, D., and Le Roy, P. 2008. Detection of quantitative trait loci for reproduction and production traits in Large White and French Landrace pig populations. *Genetics Selection Evolution* **40**(1): 61–78.

Tsunoda, T., Lathrop, G.M., Sekine, A., Yamada, R., Takahashi, A., Ohnishi, Y., Tanaka, T., and Nakamura, Y. 2004. Variation of gene-based SNPs and linkage disequilibrium patterns in the human genome. *Human Molecular Genetics* **13**(15): 1623–1632.

Vallet, J.L., Freking, B.A., Leymaster, K.A., and Christenson, R.K. 2005a. Allelic variation in the secreted folate binding protein gene is associated with uterine capacity in swine. *Journal of Animal Science* **83**(8): 1860–1867.

Vallet, J.L., Freking, B.A., Leymaster, K.A., and Christenson, R.K. 2005b. Allelic variation in the *erythropoietin receptor* gene is associated with uterine capacity and litter size in swine. *Animal Genetics* **36**(2): 97–103.

van der Steen, H.A.M. 1985. The implication of maternal effects for genetic improvement of litter size in pigs. *Livestock Production Science* **13**:159–168.

van Herwaarden, A.E. and Schinkel, A.H. 2006. The function of breast cancer resistance protein in epithelial barriers, stem cells and milk secretion of drugs and xenotoxins. *Trends in Pharmacological Sciences* **27**(1): 10–16.

Walling, G.A., Dodds, K.G., Galloway, S.M., Beattie, A.E., Lord, E.A., Lumsden, J.M., Montgomery, G.W., and McEwan, J.C. 2000. The consequences of carrying the Booroola fecundity (FecB) gene on sheep liveweight. *Proceedings of the British Society of Animal Science*, March, p. 43.

White, I.M.S., Roehe, R., Knap, P.W., and Brotherstone, S. 2006. Variance components for survival of piglets at farrowing using a reduced animal model. *Genetics Selection Evolution* **38**(4): 359–370.

Wienholds, E. and Plasterk, R.H.A. 2005. MicroRNA function in animal development. *FEBS Letters* **579**(26): 5911–5922.

Wilkie, P.J., Paszek, A.A., Beattie, C.W., Alexander, L.J., Wheeler, M.B., and Schook, L.B. 1999. A genomic scan of porcine reproductive traits reveals possible quantitative trait loci (QTLs) for number of corpora lutea. *Mammalian Genome* **10**(6): 573–578.

Wilson, T., Wu, X.-Y., Juengel, J.L., Ross, I.K., Lumsden, J.M., Lord, E.A., Dodds, K.G., Walling, G.A., McEwan, J.C., O'Connell, A.R., McNatty, K.P., and Montgomery, G.W. 2001. Highly prolific Booroola sheep have a mutation in the intracellular kinase domain of bone morphogenetic protein IB receptor (ALK-6) that is expressed in both oocytes and granulosa cells. *Biology of Reproduction* **64**(4): 1225–1235.

Winter, A., Krämer, W., Werner, F.A.O., Kollers, S., Kata, S., Durstewitz, G., Buitkamp, J., Womack, J.E., Thaller, G, and Fries, R. 2002. Association of a lysine-232/alanine polymorphism in a bovine gene encoding acyl-CoA:diacylglycerol acyltransferase (*DGAT1*) with variation at a quantitative trait locus for milk fat content. *Proceedings of the National Academy of Sciences of the United States of America* **99**(14): 9300–9305.

Yan, C., Wang, P., DeMayo, J., DeMayo, F.J., Elvin, J.A., Carino, C., Prasad, S.V., Skinner, S.S., Dunbar, B.S., Dube, J.L., Celeste, A.J., and Matzuk, M.M. 2001. Synergistic roles of bone morphogenetic protein 15 and growth differentiation factor 9 in ovarian function. *Molecular Endocrinology* **15**(6): 854–866.

Yi, S.E., Daluiski, A., Pederson, R., Rosen, V., and Lyons, K.M. 2000. The type I BMP receptor BMPRIB is required for chondrogenesis in the mouse limb. *Development* **127**(3): 621–630.

Zhao, Y. and Srivastava, D. 2007. A developmental view of microRNA function. *Trends in Biochemical Sciences* **32**(4): 189–197.

3

Quantitative Genomics of Male Reproduction

Eduardo Casas, J. Joe Ford, and Gary A. Rohrer

3.1 Introduction

An individual male that is selected as a sire impacts the output of a livestock enterprise proportionately to a much greater extent than a single female. Adoption of artificial insemination combined with improved methods to estimate genetic merit narrowed the number of sires that provide semen for the dairy industry (Funk 2006). Similarly, incorporation of marker-assisted selection by beef, sheep, and swine producers will lead to greater use of fewer sires with the most desirable genetic worth, thereby magnifying the impact of the fertility of these selected individuals upon productivity. In spite of these obvious changes in animal production, investment of research into male reproduction has diminished steadily during the past decade.

Genome is understood as all the genetic material contained within the chromosomes. The term was used by H. Winkler for the first time in 1920, when "gene" and "chromosome" were fused in a single word. Genomics is the discipline that focuses on the study of the genome. This term, proposed by Thomas R. Roderick in 1986 (Editorial Perspective 1997), is used to identify the location of underlying genetic variation. Limited studies have been focused on male reproduction in livestock species. Thus, the objective of this chapter will be to establish the current status of quantitative genomics for male reproduction.

3.2 Male reproduction phenotypes

3.2.1 *Testes*

The testes are the primary organ of male reproduction, responsible for producing male gametes (spermatozoa) and hormones (steroids and proteins). Within the testes, Sertoli cells provide support to germ cells as they mature into spermatozoa, while Leydig cells are responsible for a wide range of steroidal hormones (Reeves 1987). Production of spermatozoa is the primary function of males in most stud farms; it is desirable for males to produce spermatozoa as young and as much as possible. Semen collection in livestock production systems (dairy and beef cattle, and swine industries) is an expensive process. Therefore, it is necessary to identify males with larger testis to increase the number of doses per ejaculate and to reduce costs (Ford et al. 2006).

3.2.2 *Puberty*

The age at which males reach puberty is less well defined than in females. Puberty is the age at which fertile spermatozoa are present in the ejaculate. However, spermatozoa are formed in seminiferous tubules prior to being observed in the ejaculate. In bulls, spermatozoa are present in seminiferous tubules at least 10 weeks prior to being observed in an ejaculate (Bearden et al. 2004). Lunstra et al. (1978) defined puberty in bulls as the age at which the male produces an ejaculate containing at least 50×10^6 spermatozoa with more than 10% progressive motility observed in an ejaculate. Genetic variation exists for age at puberty, given that diverse cattle breeds reach puberty at different ages. Lunstra and Cundiff (2003) and Casas et al. (2007) estimated age at puberty for crossbred animals derived from different cattle breeds. The Angus breed reaches puberty at an earlier age (238 day of age), while Brahman reach puberty later in life (320 day of age). Thus, age at puberty is an important component of a livestock production system that can be modified through use of appropriate breeds.

3.2.3 *Testicular volume*

Testicular volume is a desirable characteristic in males. Large testes are associated with increased sperm production in bovine and swine (Lunstra et al. 1988; Ford et al. 2006). Sperm production is directly associated with the number of Sertoli cells in the testis. Similarly, the number of Sertoli cells in the testis is directly proportional with testicular size (Huang and Johnson 1996; Lunstra et al. 2003; Ford et al. 2006). Lunstra et al. (1988) proposed a simple method to estimate testis volume, scrotal circumference (SC) that represents the connected circumference of two apposed circles of equal radius (r), using the formula

$$SC = 4r + 2\pi r.$$

3.2.4 *Average testicular length (ATL)*

ATL is the mean length of both testicles. Then, assuming each testicle is a prolate spheroid, paired testicular volume (PTV) may be calculated as follows:

$$PTV = 0.0396(ATL)(SC)^2.$$

3.2.5 *Semen evaluation*

Physical and physiological characteristics of semen that are considered as important traits in semen evaluation include total sperm production, sperm motility, sperm viability, and percentage of abnormal spermatozoa (Hafez 1987). A male is considered fertile at puberty when important traits in semen evaluation achieve thresholds for these characteristics. However, in intensive production systems (such as dairy), where there is a constant need for artificial insemination, males are considered fertile when sperm characteristics reach the age at freezable semen. This is the age when the bull produces an ejaculate containing at least 500 million sperm with more than 50% progressive motility. The age at freezable semen represents a threshold after which freezing of semen becomes economically feasible (Hafez 1987; Lunstra et al. 1993).

3.3 Genetics, genomics, and quantitative trait loci (QTL)

3.3.1 Genetic variation of male reproduction

Selection programs directed toward improving male reproduction are lacking in all livestock species of economic importance. Limited attempts have been made to establish the genetic variation underlying male reproduction. Toelle et al. (1984) reported moderate heritability estimates for testes traits in Duroc and Yorkshire boars (Table 3.1). Repeatability is the upper limit of the heritability, given that it includes the total genetic variance and the proportion of the environmental variance unique to the individual. Huang and Johnson (1996) and Smital et al. (2005) estimated that repeatability for semen traits ranged between 0.16 and 0.74 in pigs. If the environmental variance was to be negligible, the heritability of semen traits would range between the values indicated by Huang and Johnson (1996). Huang and Johnson (1996) indicated that repeatability for total sperm production ranges between 0.37 and 0.40, and Oh et al. (2006) estimated the heritability for this trait between 0.27 and 0.48 in swine. Lunstra et al. (1988) and Kealey et al. (2006) reported heritabilities of moderate magnitude for % motility in cattle (Table 3.1), indicating that semen-related traits have an underlying genetic component.

Genetic components also contribute to the expression of testicular physical attributes. Young et al. (1986) indicated that testicular volume has a heritability ranging between 0.12 and 0.55 in swine. In cattle, the heritability for the same trait has been estimated to be 0.37 (Lunstra et al. 1988). The magnitude of the heritability estimates indicates that inclusion of testicular physical attributes in selection programs would be effective.

3.3.2 Genomics approaches

An important component of genomics is the development of genomic maps. The first

Table 3.1 Genetic parameters for male reproductive traits in swine and cattle.

Species	Trait	$H^2 \pm SE$	Reference
Swine	Testicular volume (140 d)	0.21 ± 0.11	Toelle et al. (1984)
	Testicular volume (168 d)	0.30 ± 0.12	Toelle et al. (1984)
	Testicular volume (98 d)	0.12 ± 0.14	Young et al. (1986)
	Testicular volume (154 d)	0.55 ± 0.12	Young et al. (1986)
	% spermatogenesis	0.22 ± 0.22	Young et al. (1986)
	Tubular diameter	0.50 ± 0.25	Young et al. (1986)
	Total sperm cells/ejaculate	0.27–0.48	Oh et al. (2006)
Cattle	Testis length (cm)	0.34 ± 0.06	Lunstra et al. (1988)
	Paired testis volume (cm^3)	0.37 ± 0.06	Lunstra et al. (1988)
	Motility (%)	0.41 ± 0.06	Lunstra et al. (1988)
	Motility (%)	0.22 ± 0.09	Kealey et al. (2006)
	Scrotal circumference (cm)	0.57 ± 0.09	Kealey et al. (2006)
	Scrotal circumference (cm)	0.31 ± 0.10	Quirino et al. (2004)
	Concentration	0.16 ± 0.08	Kealey et al. (2006)
	Ejaculate volume (mL)	0.09 ± 0.08	Kealey et al. (2006)
	Ejaculate concentration	0.23–0.36	Carabano et al. (2007)

developed maps, known as linkage maps, consisted mostly of microsatellites. Linkage maps have been developed for most economically relevant species: porcine (Rohrer et al. 1994, 1996; Archibald et al. 1995), bovine (Barendse et al. 1994, 1997; Bishop et al. 1994; Kappes et al. 1997), ovine (Crawford et al. 1995; De Gortari et al. 1998), equine (Penedo et al. 2005), and caprine (Vaiman et al. 1996). The current effort is to produce the complete sequence of the genome for most economically relevant species and to develop single nucleotide polymorphism (SNP) maps (Snelling et al. 2007; Li et al. 2008). These maps have been used to identify markers associated with, and to assess the existence of genes involved in the expression of economically important traits in livestock species.

3.3.3 QTL basics

Economically important traits in livestock are considered quantitative traits because they are controlled by several genes. Although quantitative traits are regulated by several genes (Geldermann 1975), it has been postulated that not all genes have similar influence in their expression, and that few genes contribute to a greater extent to the expression of genetic variation (Lande 1981). We now have the technology to identify the regions where genes influencing economically important traits reside in the genome; however, this is not a new concept (Smith 1967). These approaches require phenotypic information on large populations of animals with known parentage.

The genomic regions where genes influencing the expression of economically important traits reside are known as QTL. Given that multiple genes influence quantitative traits, several chromosomal regions will influence a specific trait. Identification of QTL for production traits has been done in most livestock species producing a wealth of information regarding searches for QTL for growth traits like birth weight, weaning weight, final weight, and growth rate. There have also been searches for QTL for milk production and its components (Georges et al. 1995; Zhang et al. 1998; Rodriguez-Zas et al. 2002; Ashwell et al. 2004) and for carcass traits that are economically important and expensive to measure (Rohrer and Keele 1998a,b; Casas et al. 2001, 2003, 2004a; Rohrer et al. 2001; Li et al. 2004; Mizoshita et al. 2004; Walling et al. 2004). Detection of informative QTL for these traits allows producers to identify animals with the most genetic potential at an earlier age than in traditional selection schemes.

3.4 QTL identified for male reproduction traits

Male reproductive traits in livestock have been infrequently studied. Several studies have established the existence of genetic variation for male reproductive traits in livestock (Table 3.1). However, a limited number of studies identified chromosomal regions where genes associated with male reproductive traits reside. Table 3.2 lists the chromosomes in which QTL for male reproductive traits have been detected.

3.4.1 QTL mapping for boar reproduction traits

In boars, several chromosomal regions likely contain genes associated with male reproductive traits. Evidence suggests the presence of QTL for these traits on swine chromosomes X, 3, and 8 that harbor genes associated with plasma follicle-stimulating hormone (FSH), and testicular weight in

Table 3.2 Quantitative trait loci for male reproductive traits in livestock.

Species	Chromosome	Trait	Reference
Swine	SSC3	Plasma FSH	Rohrer et al. (2001)
	SSC3	Testicular weight	Sato et al. (2003)
	SSC8	Plasma FSH	Rohrer et al. (2001)
	SSC10	Plasma FSH	Rohrer et al. (2001)
	SSCX	Plasma FSH	Rohrer et al. (2001)
	SSCX	Testicular weight	Sato et al. (2003)
	SSCX	Testicular weight	Ford et al. (2001)
Cattle	BTA5	Plasma FSH	Casas et al. (2004b)
	BTA29	Paired testis weight	Casas et al. (2004b)
	BTA29	Paired testis volume	Casas et al. (2004b)
	BTA29	Age at puberty	Casas et al. (2004b)
	BTA29	Body weight at castration	Casas et al. (2004b)

FSH, follicle-stimulating hormone.

males. In sows, similar chromosomal regions have been associated with ovulation rate (Rohrer et al. 1999), but it remains to be determined if the same genes are involved in both traits in both sexes.

There is evidence that a gene or cluster of genes, residing on swine chromosome X, is involved in the expression of male reproductive traits (Figure 3.1; Nonneman et al. 2005). Lunstra et al. (1997) indicated that Meishan sires exhibited smaller testes when compared with conventional swine breeds. Ford et al. (2001) and Rohrer et al. (2001), using a population derived from Meishan and White Composite, determined that animals inheriting this specific region of the X chromosome from the Meishan breed had smaller testicles and greater plasma FSH concentrations identified when compared with the White Composite. Sato et al. (2003), using a crossbred population from Meishan and Duroc, confirmed these findings. A QTL for testicular weight has also been detected in the X chromosome in mice (Le Roy et al. 2001). In cattle, Casas et al. (2004a) evaluated a paternal half-sib family obtained from an F1 sire (Brahman × Hereford) but were unable to analyze the X chromosome due to the family structure of the population studied. No additional studies have been conducted in other livestock species showing evidence of a QTL on chromosome X for male reproduction traits.

Swine chromosome 3 harbors genes associated with male reproduction (Table 3.2). Sato et al. (2003) identified a chromosomal region associated with testes weight spanning the interval between marker SWR1637 and S0094. These markers are located in centimorgans 28 and 58 of the swine linkage map, respectively (Rohrer et al. 1996). The maximum evidence for the presence of the QTL for testes weight was at centimorgan 47. In the same chromosomal region, Rohrer et al. (2001) identified a QTL associated with plasma FSH in males. The location of the QTL from Rohrer et al. (2001) resided between markers SW2527 and SW2618. These markers are located at centimorgan 42 and 51 of the linkage map, respectively (Rohrer et al. 1996). For this region of chromosome 3, Sato et al. (2003) indicated that animals inheriting the Meishan allele had heavier testes weight, while Rohrer et al. (2001) found that in this region, animals with the Meishan allele had less plasma FSH

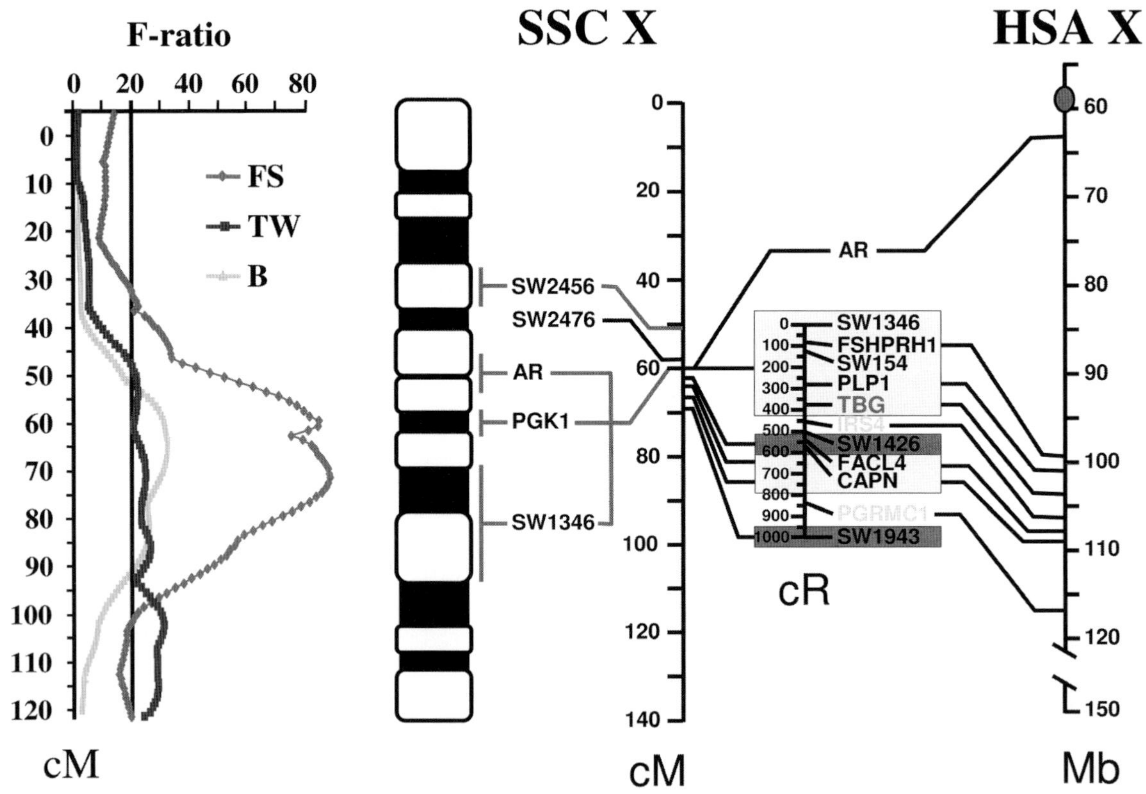

Figure 3.1 F-ratio profiles on swine chromosome X indicating evidence of QTL for plasma follicle-stimulating hormone (FS), testicular weight (TW), and backfat (B). Genetic markers are aligned in their relative position on the porcine cytogenetic, genetic, and physical maps and compared with the human physical sequence map. Units are in centimorgans (cM), centirays (cR), and megabases (Mb). Figure was reproduced from Nonneman et al. (2005).

concentration. It is possible that a gene in this chromosomal region has an antagonistic effect. That is, for this chromosomal region, animals with the Meishan allele exhibit lower plasma FSH and lighter testes as opposed to what has been observed on swine chromosome X (Ford et al. 2001; Rohrer et al. 2001; Sato et al. 2003) and what has been observed in the Meishan breed (Lunstra et al. 1997).

Regions of swine chromosome 3, 8, and X, where QTL for plasma FSH have been identified, reside in similar locations where QTL for ovulation rate have been detected. Rohrer et al. (2001) indicated that it remains to be established whether similar genes may be influencing the same trait in males and females.

3.4.2 QTL mapping for bull reproduction traits

In cattle, a QTL for plasma FSH was identified on chromosome 5 (Casas et al. 2004b). This QTL resides between centimorgans 47 and 82 of the bovine chromosome 5 linkage map (Kappes et al. 1997). Several studies have detected QTL for ovulation rate or twinning rate in cattle in this chromosome. Lien et al. (2000) and Cruickshank et al.

(2004) detected a QTL for twinning rate in a similar region of bovine chromosome 5. Similarly, Kappes et al. (2000) identified a QTL for ovulation rate on a similar region of the chromosome. The position of the QTL in the three studies is similar to the position where the QTL for plasma FSH was identified by Casas et al. (2004b). If the QTL for ovulation rate in females and the QTL for plasma FSH in males are caused by a single gene, then the mechanism behind the QTL for ovulation rate is possibly related to regulation of FSH in the female with a similar effect on FSH expression in males.

3.4.3 Candidate genes associated with male reproduction

Several studies have attempted to establish association between molecular markers and male reproductive traits. Table 3.3 shows a summary of these associations. Markers have been developed and evaluated in diverse populations for their association with the expression of male reproductive traits based on the gene in which they reside. These genes have been selected based on their location on the genome (candidate genes under a QTL), based on their putative role in the expression of male reproductive traits, or at random.

Wimmers et al. (2005) provided some evidence for the effect of a marker at the gamma-actin 2 (ACTG2) gene for sperm volume in boars (Table 3.3). This gene resides on swine chromosome 3, making it a potential candidate associated with male fertility traits. Lin et al. (2006b) evaluated markers at the gonadotropin-releasing hormone receptor gene and reported an SNP in this gene that associates with % motility and abnormal sperm rate (Table 3.3). This gene could be considered a putative candidate gene for the QTL detected in this chromosome. Rohrer et al. (2001) postulated a candidate gene for the QTL detected on swine chromosome X. They indicated that androgen receptor (AR) is one potential candidate for this QTL. Lin et al. (2006c) evaluated a marker developed in this gene and found it to be unassociated with any male reproductive trait. However, Nonneman et al. (2005) postulated that a marker in the thyroxine-binding globulin (TBG) could be used in marker-assisted selection. Differences in plasma FSH concentration and

Table 3.3 Candidate genes associated with male reproductive traits in swine.

Chromosome	Gene	Trait	Reference
1	*ESR1*	Sperm volume Sperm concentration % sperm alive	Terman et al. (2006) Terman et al. (2006) Terman et al. (2006)
3	*ACTG2*	Sperm volume	Wimmers et al. (2005)
5	*ACR*	Sperm concentration	Lin et al. (2006c)
7	*PGK2*	Semen volume	Chen et al. (2004)
8	*GNRHR*	% motility Abnormal sperm rate	Lin et al. (2006b) Lin et al. (2006b)
X	*TBG*	Testis weight FSH concentration	Nonneman et al. (2005) Nonneman et al. (2005)
Unknown	*ACTB*	% motility Abnormal sperm rate	Lin et al. (2006a) Lin et al. (2006a)

ESR1, estrogen receptor; *ACTG2*, gamma-actin 2; *ACR*, acrosin; *PGK2*, phosphoglycerate kinase 2; *GNRHR*, gonadotropin-releasing hormone receptor; *TBG*, thyroxine-binding globulin; *ACTB*, beta-actin.

testis weight were observed when comparing alternative alleles in this gene. The TBG gene resides in the same region where the QTL for plasma FSH was detected (Figure 3.1), making TBG a likely candidate gene for this QTL.

Terman et al. (2006) evaluated a marker for estrogen receptor gene (ESR1) and found an association with male fertility traits (Table 3.3). Interest in ESR1 was stimulated by the report of Rothschild et al. (1996) implicating an association of allelic variants for this gene with litter size. Acrosin is a trypsin-like serine proteinase extrinsically associated with membranes of the mammalian sperm acrosome (Straus and Polakoski 1982). Acrosin is required during the acrosome reaction, to facilitate sperm penetration to the oocyte (Westbrook-Case et al. 1994) and acrosin activity has been associated with infertility in humans (Nakagawa et al. 1997). Lin et al. (2006c) evaluated a significant marker near this gene on swine chromosome 5 and observed a significant allele substitution effect for sperm concentration (Table 3.3). Phosphoglycerate kinase 2 (PGK2) is an enzyme that modulates sperm metabolism during epididymal transport (Salisbury et al. 1977). Chen et al. (2004) evaluated an SNP in this gene on swine chromosome 8 (Table 3.3) and observed that homozygous animals for one allele had the tendency to produce a smaller sperm volume than homozygous animals with the alternative genotype. Additionally, Lin et al. (2006a) evaluated haplotypes developed from SNPs in the beta-actin (ACTB) gene (Table 3.3) and observed that different haplotypes affected the variation of % motility and abnormal sperm rate in populations of Pietrain and Pietrain × Hampshire boars.

Markers throughout the genome have been used to detect chromosomal regions associated with boar sperm viability. Thurston et al. (2002), using amplified fragment length polymorphisms, found 16 candidate genetic markers associated with differences in semen freezability within a small population of boars previously classified as good and poor freezers. Thurston et al. (2002) proposed that these findings demonstrate the existence of a genetic basis to this variation. No indication is given to the location in the genome where these markers reside. Therefore, the information presented by Thurston et al. (2002) is of limited use in genomic studies.

Chanock et al. (2007) indicated that it is unlikely that a single study may establish a genotype–phenotype association without the need for replication. This is the case for studies where candidate genes are selected based on their location in the genome (under QTL) and followed by the selection of candidate genes. The QTL scan is presented as initial evidence of the presence of a gene in a specific chromosomal region, and evaluation of markers in genes under this QTL in additional populations is a replicate of the study. Conclusions drawn from these studies are useful and productive in the implementation of programs where this information is to be used. Results from candidate genes based on their putative role in the expression of traits studied, imply the existence of genetic variation for male reproductive traits; however, most studies showed weak association between markers and traits, and no replication was pursued. This may lead to incorrect conclusions about the role of selected genes based exclusively on their role in the expression of a trait.

3.5 Future research directions

The objective of improvement programs is to identify those individuals with the best genetics, to become the founders of the following generation. Artificial insemination is the most used tool to disseminate the improvement in a species or population. Male reproduction is one of the most important components in this process. A single male will have a greater impact in the improvement of a trait than any female. Despite this fact, insufficient emphasis is placed on the study of male reproduction. Few studies have focused on the study of genetic variation behind the expression of male reproductive traits. This variation can be exploited to improve reproduction in males through breeding programs. However, the cost and time required to obtain appropriate phenotypic data from large populations of males hamper this research.

Genomics is a thriving area of research that will assist in understanding the genetic basis of physiological mechanisms. Of the limited number of studies where genomics of male reproductive traits are analyzed, most have shown weak associations between genomics and the expression of male reproductive traits. Replication is needed to validate these findings. Causative changes in the genome have not been identified yet, but results indicate that differences in male reproduction are linked to differences in fertility rate in females. Genomic regions in chromosomes 3, 8, and X in pigs and chromosome 5 in cattle influence ovulation rate in females and differences in male reproduction traits. Further studies are needed to ascertain the existence of genes in these regions, as well as in additional genomic regions in livestock.

References

Archibald, A.L., Haley, C.S., Brown, J.F., Couperwhite, S., McQueen, H.A., Nicholson, D., Coppieters, W., Van de Weghe, A., Stratil, A., Wintero, A.K., Fredholm, M., Larsen, N.J., Nielsen, V.H., Milan, D., Woloszyn, N., Robic, A., Dalens, M., Riquet, J., Gellin, J., Caritez, J.-C., Buraud, G., Ollivier, L., Bidanel, J.-P., Vaiman, M., Renard, C., Geldermann, H., Davoli, R., Ruyter, D., Verstege, E.J.M., Groenen, M.A.M., Davies, W., Hoyheim, B., Kieserud, A., Andersson, L., Ellegren, H., Johansson, M., Marklund, L., Miller, J.R., Anderson Dear, D.V., Signer, E., Jeffreys, A.J., Moran, C., Le Tissier Muldano, P., Rothschild, M.F., Tuggle, C.K., Vaske, D., Helm, J., Liu, H.-C., Rahman, A., Yu, T.-P., Larson, R.G., and Schmitz, C.B. 1995. The PiGMaP consortium linkage map of the pig (*Sus scrofa*). *Mammalian Genome* **6**: 157–175.

Ashwell, M.S., Heyen, D.W., Sonstegard, T.S., Van Tassell, C.P., Da, Y., VanRaden, P.M., Ron, M., Weller, J.I., and Lewin, H.A. 2004. Detection of quantitative trait loci affecting milk production, health, and reproductive traits in Holstein cattle. *Journal of Dairy Science* **87**: 468–475.

Barendse, W., Armitage, S.M., Kossarek, L.M., Shalom, A., Kirkpatrick, B.W., Ryan, A.M., Clayton, D., Li, L., Neibergs, H.L., Zhang, N., Grosse, W.M., Weiss, J., Creighton, P., McCarthy, F., Ron, M., Teale, A.J., Fries, R., McGraw, R.A., Moore, S.S., Georges, M., Soller, M., Womack, J.E., and Hetzel, D.J.S. 1994. A genetic linkage map of the bovine genome. *Nature Genetics* **6**: 227–235.

Barendse, W., Vaiman, D., Kemp, S.J., Sugimoto, Y., Armitage, S.M., Williams, J.L., Sun, H.S., Eggen, A., Agaba, M., Aleysin, S.A., Band, M., Bishop, M.D., Buitkamp, J., Byrne, K., Collins, F., Cooper, L., Coppettiers, W., Denys, B., Drinkwater, R.D., Easterday, K., Elduque, C., Ennis, S., Erhardt, G., Ferreti, L.,

Flavin, N., Gao, Q., Georges, M., Gurung, R., Harlizius, B., Hawkins, G., Hetzel, D.J.S., Hirano, T., Hulme, D., Jorgensen, C., Kessler, M., Kirkpatrick, B.W., Konfortov, B., Kostia, S., Kuhn, C., Lenstra, J.A., Leveziel, H., Lewin, H.A., Leyhe, B., Lil, L., Martin Burriel, I., McGraw, R.A., Miller, J.R., Moody, D.E., Moore, S.S., Nakane, S., Nijman, I.J., Olsaker, I., Pomp, D., Rando, A., Ron, M., Shalom, A., Teale, A.J., Thieven, U., Urquhart, B.G.D., Vage, D.-I., Van der Weghe, A., Varvio, S., Velmala, R., Vilkki, J., Weikard, R., Woodside, C., Womack, J.E., Zannoti, M., and Zaragoza, P. 1997. A medium-density genetic linkage map of the bovine genome. *Mammalian Genome* **8**: 21–28.

Bearden, H.J., Fuquay, J.H., and Willard, S.T. 2004. Spermatogenesis and maturation of spermatozoa. In: Bearden, H.J., Fuquay, J.H., and Willard, S.T. (eds.), *Applied Animal Reproduction*, 6th Edition. Upper Saddle River, NJ: Pearson Prentice Hall, pp. 75–86.

Bishop, M.D., Kappes, S.M., Keele, J.W., Stone, R.T., Sunden, S.L., Hawkins, G.A., Toldo, S.S., Fries, R., Grosz, M.D., Yoo, J., and Beattie, C.W. 1994. A genetic linkage map for cattle. *Genetics* **136**: 619–639.

Carabano, M.J., Diaz, C., Ugarte, C., and Serrano, M. 2007. Exploring the use of random regression models with Legendre polynomials to analyze measures of volume of ejaculate in Holstein bulls. *Journal of Dairy Science* **90**: 1044–1057.

Casas, E., Keele, J.W., Shackelford, S.D., Koohmaraie, M., and Stone, R.T. 2004a. Identification of quantitative trait loci for growth and carcass composition in cattle. *Animal Genetics* **35**: 2–6.

Casas, E., Lunstra, D.D., Cundiff, L.V., and Ford, J.J. 2007. Growth and pubertal development of F1 bulls from Hereford, Angus, Norwegian Red, Swedish Red and White, Friesian, and Wagyu sires. *Journal of Animal Science* **85**: 2904–2909.

Casas, E., Lunstra, D.D., and Stone, R.T. 2004b. Quantitative trait loci for male reproductive traits in beef cattle. *Animal Genetics* **35**: 451–453.

Casas, E., Shackelford, S.D., Keele, J.W., Koohmaraie, M., Smith, T.P.L., and Stone, R.T. 2003. Detection of quantitative trait loci for growth and carcass composition in cattle. *Journal of Animal Science* **81**: 2976–2983.

Casas, E., Stone, R.T., Keele, J.W., Shackelford, S.D., Kappes, S.M., and Koohmaraie, M. 2001. A comprehensive search for quantitative trait loci affecting growth and carcass composition of cattle segregating alternative forms of myostatin. *Journal of Animal Science* **79**: 854–860.

Chanock, S.J., Manolio, T., Boehnke, M., Boerwinkle, E., Hunter, D.J., Thomas, G., Hirschhorn, J.N., Abecasis, G., Altshuler, D., Bailey-Wilson, J.E., Brooks, L.D., Cardon, L.R., Daly, M., Donnelly, P., Fraumeni, J.F. Jr., Freimer, N.B., Gerhard, D.S., Gunter, C., Guttmacher, A.E., Guyer, M.S., Harris, E.L., Hoh, J., Hoover, R., Kong, C.A., Merkingas, K.R., Morton, C.C., Palmer, L.J., Phimister, E.G., Rice, J.P., Roberts, J., Rotimi, C., Tucker, M.A., Vogan, K.J., Wacholder, S., Wijsman, E.M., Winn, D.M., and Collins, F.S. 2007. Replicating genotype-phenotype associations. *Nature* **447**: 655–660.

Chen, K., Knorr, C., Moser, G., Gatphayak, K., and Brenig, B. 2004. Molecular characterization of the porcine testis-specific phosphoglycerate kinase 2 (PGK2) gene and its association with male fertility. *Mammalian Genome* **15**: 996–1006.

Crawford, A.M., Dodds, K.G., Ede, A.J., Pierson, C.A., Montgomery, G.W., Garmonsway, H.G., Beattie, A.E., Davies,

K., Maddox, J.F., Kappes, S.M., Stone, R.T., Nguyen, T.C., Penty, J.M., Lord, E.A., Broom, J.E., Buitkamp, J., Schwaiger, W., Epplen, J.T., Matthew, P., Mattews, M.E., Hulme, D.J., Beh, K.J., McGraw, R.A., and Beattie, C.W. 1995. An autosomal genetic linkage map of the sheep genome. *Genetics* **140**: 703–724.

Cruickshank, J., Dentine, M.R., Berger, P.J., and Kirkpatrick, B.W. 2004. Evidence for quantitative trait loci affecting twinning rate in North American Holstein cattle. *Animal Genetics* **35**: 206–212.

De Gortari, M.J., Freking, B.A., Cuthbertson, R.P., Kappes, S.M., Keele, J.W., Stone, R.T., Leymaster, K.A., Dodds, K.G., Crawford, A.M., and Beattie, C.W. 1998. A second generation linkage map of the sheep genome. *Mammalian Genome* **9**: 204–209.

Editorial Perspective. 1997. Genomics: Structural and functional studies of genomes. *Genomics* **45**: 244–249.

Ford, J.J., McCoard, S.A., Wise, T.H., Lunstra, D.D., and Rohrer, G.A. 2006. Genetic variation in sperm production. *Society of Reproduction and Fertility Supplement* **62**: 99–112

Ford, J.J., Wise, T.H., Lunstra, D.D., and Rohrer, G.A. 2001. Interrelationships of porcine X and Y chromosomes with pituitary gonadotropins and testicular size. *Biology of Reproduction* **65**: 906–912.

Funk, D.A. 2006. Major advances in globalization of the artificial insemination industry. *Journal of Dairy Science* **89**: 1362–1368.

Geldermann, H. 1975. Investigations on inheritance of quantitative characters in animals by gene markers. I. Methods. *Theoretical and Applied Genetics* **46**: 319–330.

Georges, M., Nielsen, D., Mackinnon, M., Mishra, A., Okimoto, R., Pasquino, A.T., Sargeant, L.S., Sorensen, A., Steele, M.R., Zhao, X., Womack, J.E., and Hoeschele, I. 1995. Mapping quantitative trait loci controlling milk production in dairy cattle by exploiting progeny testing. *Genetics* **139**: 907–920.

Hafez, E.S.E. 1987. Semen evaluation. In: Hafez, E.S.E. (ed.), *Reproduction in Farm Animals*, 5th Edition. Philadelphia: Lea and Febiger, pp. 455–480.

Huang, Y.T. and Johnson, R.K. 1996. Effect of selection for size of testis traits. *Journal of Animal Science* **74**: 750–760.

Kappes, S.M., Bennett, G.L., Keele, J.W., Echternkamp, S.E., Gregory, K.E., and Thallman, R.M. 2000. Initial results of genomic scans for ovulation rate in a cattle population selected for increased twinning rate. *Journal of Animal Science* **78**: 3053–3059.

Kappes, S.M., Keele, J.W., Stone, R.T., McGraw, R.A., Sonstegard, T.S., Smith, T.P.L., Lopez-Corrales, N., and Beattie, C.W. 1997. A second-generation linkage map of the bovine genome. *Genome Research* **7**: 235–249.

Kealey, C.G., MacNeil, M.D., Tess, M.W., Geary, T.W., and Bellows, R.A. 2006. Genetic parameter estimates for scrotal circumference and semen characteristics of Line 1 Hereford bulls. *Journal of Animal Science* **84**: 283–290.

Lande, R. 1981. The minimum number of genes contributing to quantitative variation between and within populations. *Genetics* **99**: 541–553.

Le Roy, I., Tordjman, S., Migliore-Samour, D., Degrelle, H., and Roubertoux, P.L. 2001. Genetic architecture of testis and seminal vesicle weights in mice. *Genetics* **158**: 333–340.

Li, C., Basarab, J., Snelling, W.M., Benkel, B., Kneeland, J., Murdoch, B., Hansen, C., and Moore, S.S. 2004. Identification and fine mapping of quantitative trait loci for backfat on bovine chromosomes 2, 5, 6, 19, 21, and 23 in a commercial line of *Bos taurus*. *Journal of Animal Science* **82**: 967–972.

Li, X.P., Hu, Z.L., Moon, S.J., Do, K.T., Ha, Y.K., Kim, H., Byun, M.J., Choi, B.H., Rothschild, M.F., Reecy, J.M., and Kim, K.S. 2008. Development of an in silico coding gene SNP map in pigs. *Animal Genetics* **39**(4): 446–450.

Lien S., Karlsen, A., Klemetsdal, G., Vage, D.I., Olsaker, I., Klungland, H., Aasland, M., Heringstad, B., Ruane, J., and Gomez-Raya, L. 2000. A primary screen of the bovine genome for quantitative trait loci affecting twinning rate. *Mammalian Genome* **11**: 877–882.

Lin, C.L., Jennen, D.G.J., Ponsuksili, S., Tholen, E., Tesfaye, D., Schellander, K., and Wimmers, K. 2006a. Haplotype analysis of beta-actin gene for its association with sperm quality and boar fertility. *Journal of Animal Breeding and Genetics* **123**: 384–388.

Lin, C.L., Ponsuksili, S., Tholen, E., Jennen, D.G.J., Schellander, K., and Wimmers, K. 2006b. Candidate gene markers for sperm quality and fertility of boar. *Animal Reproductive Science* **92**: 349–363.

Lin, C.L., Tholen, E., Jennen, D., Ponsuksili, S., Schellander, K., and Wimmers, K. 2006c. Evidence for effects of testis and epididymis expressed genes on sperm quality and boar fertility traits. *Reproduction in Domestic Animals* **41**: 538–543.

Lunstra, D.D., Crouse, J.D., and Cundiff, L.V. 1993. Pages 90–92 in puberty occurs at the same testis size in both *Bos taurus* and *bos indicus* crossbred beef bulls. *USDA ARS-71. RLHUSMARC Beef Cattle Research Program, Progress Report No. 4*, ARS/USDA, Clay Center, NE.

Lunstra, D.D. and Cundiff, L.V. 2003. Growth and pubertal development in Brahman-, Boran-, Tuli-, Belgian Blue-, Hereford- and Angus-sired F1 bulls. *Journal of Animal Science* **81**: 1414–1426.

Lunstra, D.D., Ford, J.J., and Echternkamp, S.E. 1978. Puberty in beef bulls: Hormone concentrations, growth, testicular development, sperm production and sexual aggressiveness in bulls of different breeds. *Journal of Animal Science* **46**: 1054–1062.

Lunstra, D.D., Ford, J.J., Klindt, J., and Wise, T.H. 1997. Physiology of the Meishan boar. *Journal of Reproduction and Fertility* **52**: 181–193.

Lunstra, D.D., Gregory, K.E., and Cundiff, L.V. 1988. Heritability estimates and adjustment factors for the effects of bull age and age of dam on yearling testicular size in breeds of beef bulls. *Theriogenology* **30**: 127–136.

Lunstra, D.D., Wise, T.H., and Ford, J.J. 2003. Sertoli cells in the boar testis: Changes during development and compensatory hypertrophy after hemicastration at different ages. *Biology of Reproduction* **68**: 140–150.

Mizoshita, K., Watanabe, T., Hayashi, H., Kubota, C., Yamakuchi, H., Todoroki, J., and Sugimoto, Y. 2004. Quantitative trait loci analysis for growth and carcass traits in a half-sib family of purebred Japanese Black (Wagyu) cattle. *Journal of Animal Science* **82**: 3415–3420.

Nakagawa, H., Okada, H., Fujisawa, M., Matsumoto, O., and Kamidono, S. 1997. Relationship of acrosin activity to sperm function tests. *Andrology* **29**: 103–108.

Nonneman, D., Rohrer, G.A., Wise, T.H., Lunstra, D.D., and Ford, J.J. 2005. A variant of porcine thyroxine-binding globulin has reduced affinity for thyroxine and is associated with testis size. *Biology of Reproduction* **72**: 214–220.

Oh, S.H., See, M.T., Long, T.E., and Galvin, J.M. 2006. Genetic parameters of various random regression models to describe total sperm cells per ejaculate over the reproductive lifetime of boars. *Journal of Animal Science* **84**: 538–545.

Penedo, M.C., Millon, L.V., Bernoco, D., Binns, M., Cholewinski, G., Ellis, N., Flynn, J., Gralak, B., Guthrie, A., Hasegawa, T., Lindgren, G., Lyons, L.A., Roed, K.H., Swinburne, J.E., and Tozaki, T. 2005. International equine gene mapping workshop report: A comprehensive linkage map constructed with data from new markers and by merging four mapping resources. *Cytogenetic Genome Research* **111**: 5–15.

Quirino, C.R., Bergmann, J.A.G., Vale Filho, V.R., Andrade, V.J., Reis, S.R., Mendoca, R.M., and Fonseca, C.G. 2004. Genetic parameters of libido in Brazilian Nellore bulls. *Theriogenology* **62**: 1–7.

Reeves, J.J. 1987. Endocrinology of reproduction. In: Hafez, E.S.E. (ed.), *Reproduction in Farm Animals,* 5th Edition. Philadelphia: Lea and Febiger, pp. 85–106.

Rodriguez-Zas, S.L., Southey, B.R., Heyen, D.W., and Lewin, H.A. 2002. Interval and composite interval mapping of somatic cell score, yield, and components of milk in dairy cattle. *Journal of Dairy Science* **85**: 3081–3091.

Rohrer, G.A., Alexander, L.J., Hu, Z., Smith, T.P.L., Keele, J.W., and Beattie, C.W. 1996. A comprehensive map of the porcine genome. *Genome Research* **6**: 371–391.

Rohrer, G.A., Alexander, L.J., Keele, J.W., Smith, T.P.L., and Beattie, C.W. 1994. A microsatellite linkage map of the porcine genome. *Genetics* **136**: 231–245.

Rohrer, G.A., Ford, J.J., Wise, T.H., Vallet, J.L., and Christensen, R.K. 1999. Identification of quantitative trait loci affecting female reproductive traits in a multigeneration Meishan-White Composite swine population. *Journal of Animal Science* 77: 1385–1391.

Rohrer, G.A. and Keele, J.W. 1998a. Identification of quantitative trait loci affecting carcass composition in swine: I. Fat deposition traits. *Journal of Animal Science* **76**: 2247–2254.

Rohrer, G.A. and Keele, J.W. 1998b. Identification of quantitative trait loci affecting carcass composition in swine: II. Muscling and wholesale product yield traits. *Journal of Animal Science* **76**: 2255–2262.

Rohrer, G.A., Wise, T.H., Lunstra, D.D., and Ford, J.J. 2001. Identification of genomic regions controlling plasma FSH concentrations in Meishan-White Composite boars. *Physiology Genomics* **6**: 145–151.

Rothschild, M., Jacobson, C., Vaske, D., Tuggle, C., Wang, L., Short, T., Eckhardt, G., Sasaki, S., Vincent, A., McLaren, M., Southwood, O., Van Der Steen, H., Mileham, A., and Plastow, G. 1996. The estrogen receptor locus is associated with major gene influencing litter size in pigs. *Proceedings of the National Academy of Science of the United States of America* **93**: 201–205.

Salisbury, G.W., Hart, R.G., and Lodge, J.R. 1977. The spermatozoon. *Perspective Biology and Medicine* **20**: 372–393.

Sato, S., Oyamada, Y., Atsuji, K., Nade, T., Sato, S.-I., Kobayashi, E., Mitsuhashi, T., Nirasawa, K., Komatsuda, A., Sato, Y., Terai, S., Hayashi, T., and Sugimoto, Y. 2003. Quantitative trait loci analysis for growth and carcass traits in a Meishan x Duroc F2 resource population. *Journal of Animal Science* **81**: 2938–2949.

Smital, J., Wolf, J., and De Sousa, L.L. 2005. Estimation of genetic parameters of semen characteristics and reproductive traits of AI boars. *Animal Reproductive Science* **86**: 119–130.

Smith, C. 1967. Improvement of metric traits through specific genetic loci. *Animal Production* **9**: 349–358.

Snelling, W.M., Chiu, R., Schein, J.E., Hobbs, M., Abbey, C.A., Adelson, D.L., Aerts, J., Bennett, G.L., Bosdet, I.E., Boussaha, M., Brauning, R., Caetano, A.R., Costa, M.M., Crawford, A.M., Dalrymple, B.P., Eggen, A., Everts-van der Wind, A., Floriot, S., Gautier, M., Gill, C.A., Green, R.D., Holt, R., Jann, O., Jones, S.J.M., Kappes, S.M., Keele, J.W., de Jong, P.J., Larkin, D.M., Lewin, H.A., McEwan, J.C., McKay, S., Marra, M.A., Mathewson, C.A., Matukumalli, L.K., Moore, S.S., Murdoch, B., Nicholas, F.W., Osoegawa, K., Roy, A., Salih, H., Schibler, L., Schnabel, R.D., Silveri, L., Skow, L.C., Smith, T.P.L., Sonstegard, T.S., Taylor, J.F., Tellam, R., Van Tassell, C.P., Williams, J.L., Womack, J.E., Wye, N.H., Yang, G., Zhao, S., and the International Bovine BAC Mapping Consortium. 2007. A physical map of the bovine genome. *Genome Biology* **8**: R165.

Straus, J.W. and Polakoski, K.L. 1982. Acrosin inhibition. Comparison of membrane-associated and -solubilized enzyme. *Journal of Biological Chemistry* **257**: 7962–7964.

Terman, A., Kmiec, M., and Polasik, D. 2006. Estrogen receptor gene (ESR) and semen characteristics of boars. *Archiv für Tierzucht* **1**: 71–76.

Thurston, L.M., Siggins, K., Mileham, A.J., Warson, P.F., and Holt, W.V. 2002. Identification of amplified restriction fragment length polymorphism markers linked to genes controlling boar sperm viability following cryopreservation. *Biology of Reproduction* **66**: 545–554.

Toelle, V.D., Johnson, B.H., and Robison, O.W. 1984. Genetic parameters for testes traits in swine. *Journal of Animal Science* **59**: 967–973.

Vaiman, D., Schibler, L., Bourgeois, F., Oustry, A., Amigues, Y., and Cribiu, E.P. 1996. A genetic linkage map of the male goat genome. *Genetics* **144**: 279–305.

Walling, G.A., Visscher, P.M., Wilson, A.D., McTeir, B.L., Simm, G., and Bishop, S.C. 2004. Mapping of quantitative trait loci for growth and carcass traits in commercial sheep populations. *Journal of Animal Science* **82**: 2234–2245.

Westbrook-Case, V.A., Winfrey, V.P., and Olson, G.E. 1994. A domain-specific 50-kilodalton structural protein of the acrosomal matrix is processed and released during the acrosome reaction in the guinea pig. *Biology of Reproduction* **51**: 1–13.

Wimmers, K., Lin, C.L., Tholen, E., Jennen, D.G., Schellander, K., and Ponsuksili, S. 2005. Polymorphisms in candidate genes as markers for sperm quality and boar fertility. *Animal Genetics* **36**: 152–155.

Young, L.D., Leymaster, K.A., and Lunstra, D.D. 1986. Genetic variation in testicular development and its relationship to female reproductive traits in swine. *Journal of Animal Science* **63**: 17–26.

Zhang, Q., Boichard, D., Hoeschele, I., Ernst, C., Eggen, A., Murkve, B., Pfister-Genskow, M., Witte, L.A., Grignola, F.E., Uimari, P., Thaller, G., and Bishop, M.D. 1998. Mapping quantitative trait loci for milk production and health of dairy cattle in a large outbred pedigree. *Genetics* **149**: 1959–1973.

4

Genetics and Genomics of Reproductive Disorders

Peter Dovc, Tanja Kunej, and Galen A. Williams

4.1 Introduction

Genetics of reproductive traits in farm animals is becoming an increasingly important field of research of late and certainly has been significantly propelled by the advances in genomic research. The literature describing reproductive disorders in farm animals is densest during two periods of time. The first period (late 1970s, early 1980s) described mostly anatomical and pathophysiological characteristics of disorders without referencing possible causal mutations in the candidate genes. Publications from the second period (late 1990s through the present) attempted to either confirm the role of candidate genes by identifying causal mutations or apply genomic approaches in order to identify quantitative trait loci (QTL) regions, associated markers, or homologous regions across the species barrier, which could be involved in the genesis of reproductive disorders. However, the traditional assumption of the polygenic nature of reproductive traits, in addition to likely environmental effects, makes the identification of causal mutations for reproductive disorders a difficult task. In addition, the frequency of the majority of reproductive disorders is rather low, thus making acquisition of appropriate material quite often problematic. Some reproductive disorders are caused by complex developmental mechanisms, which are difficult to explain by simple genetic means. An example for such a disorder is cross-contamination of fetal bloodstream with cell populations and sex hormones through placental anastomoses causing formation of XX/XY chimeras in the case of dizygotic pregnancies with fetuses of different sex. XX/XY chimeras appear in many species with rather different frequencies with equally varied effects. The clinical consequences are by far the most severe in cattle, less so in sheep and pigs, and virtually no negative effects in horses. From this example, it is obvious that the same phenomenon (bloodstream communication among fetuses of different sex) can result in completely different clinical consequences. Apparently, the frequency and extent of hormone and blood cell exchange between feti is a result of species specific architecture of placental circulation in cattle; however, the real cause for the formation of anastomoses remains unclear.

In this chapter, we will review relevant literature and present the current status of relevant information related to reproductive disorders in farm animals. In some cases, we will refer to similar disorders in man or in model organisms (mainly mouse) in order to give some hints for necessary further research in this important but difficult field.

4.2 Reproductive disorders associated with the ovary

4.2.1 Ovarian subfunction

Ovarian subfunction can result in several undesired phenotypic traits that affect reproductive capacity in virtually all species. The most frequently studied traits relating to this disorder are ovulation rate, silent heat, and litter size. Therefore, selection strategies and different methods of genetic screening have been applied to identify the major genetic causes underlying ovarian subfunction. Due to the economic importance of ovulation rate, this is the most systematically studied trait related to ovarian function in farm animals. Certainly, for selection purposes, genes enhancing ovulation rate are of central interest; however, allelic counterparts of positive alleles could be considered as alleles that have negative effect on the ovulation rate.

Ovulation is the final event in the process of ovarian follicle maturation and describes the discharge of the ovum from the graafian follicle. Conversely, ovulation failure is a situation when the ripe follicle does not rupture and discharge its ovum.

In mouse, Simon et al. (1997) described the crucial role of connexin 37 in oogenesis and ovulation. Connexin 37 is a member of the family of gap junction proteins, which are structurally related transmembrane proteins that assemble to form vertebrate gap junctions. Connexin 37 is present in gap junctions between oocyte and granulosa cells and is involved in cell–cell signaling, which critically regulates complex cellular interactions that are required for oogenesis and ovulation. Mice lacking connexin 37 lack graafian follicles, resulting in arrested oocyte development before achieving meiotic competence. These mice then fail to ovulate and develop numerous corpora lutea.

In the Mouse Genome Informatics (MGI) database, 11 genes associated with the gene ontology term "ovulation from ovarian follicle" were found, including *Adamts1*, a disintegrin-like and metallopeptidase (reprolysin type) with thrombospondin type 1 motif, 1; *Afp*, alpha fetoprotein; *Agt*, angiotensinogen (serpin peptidase inhibitor, clade A, member 8); *Bfo*, bell flash ovulation; *Foxo3*, forkhead box O3; *Nos2*, nitric oxide synthase 2, inducible, macrophage; *Nos3*, nitric oxide synthase 3, endothelial cell; *Nrip1*, nuclear receptor interacting protein 1; *Oas1d*, 2′-5′ oligoadenylate synthetase 1D; *Pgr*, progesterone receptor; and *Sirt1*, sirtuin 1 (silent mating type information regulation 2, homolog) 1 (*Saccharomyces cerevisiae*). These candidate genes may be associated with ovarian subfunction due to their pleiotropic mode of action in different tissues.

A comparative mapping approach based on 11 genes from human chromosome 4p16-p15 and exploitation of porcine large-insert genomic libraries revealed 11 potential candidate genes for the ovulation rate in the porcine homologous region at SSC8 (Campbell et al. 2003). Seven genes (*GNRHR*, *IDUA*, *MAN2B2*, *MSX1*, *PDE6B*, *PPP2R2C*, and *RGS12*) were mapped using informative microsatellite markers, three genes (*LRPAP1*, *GPRK2L*, and *FLJ20425*) were mapped using single nucleotide polymorphisms (SNPs), and two genes were identified using marker information in selected genomic clones assigned since they were present in clones that contained mapped markers (*HGFAC* and *HMX1*). The resulting linkage map contains markers associated with 14 genes in the first 27 cM of the porcine chromosome 8. In the region with the highest F-ratio were markers closest to the *MAN2B2* gene. In a later study, Campbell et al. (2008) identified positional candidate gene, coding mannosidase 2B2 (MAN2B2) for ovulation rate within the targeted QTL region on the p-terminal end of pig chromosome 8 (Figure 4.1) in a Meishan-cross resource population. Eleven nonsynonymous mutations in the coding region for mannosidase 2B2 were identified and tested for statistical associations with ovulation rate in a resource population over three generations. The most significant effect was associated with a polymorphism located at position 1574 of the mRNA (1574A>G) where the additive effect of the 1574A allele was estimated to be –0.89 ova. Due to the fact that this polymorphism was not associated with

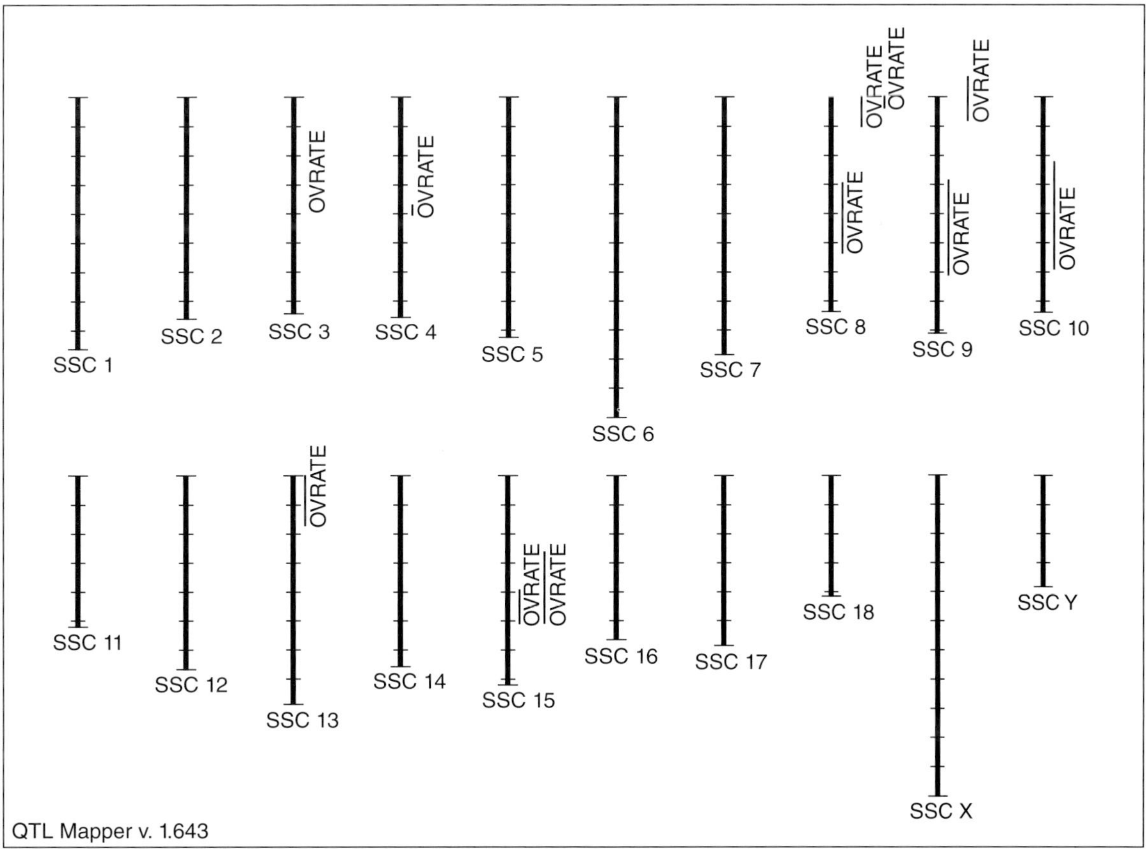

Figure 4.1 Genomic distribution of QTL for the ovulation rate in pigs (AnimalQTL database; www.animalgenome.org/QTLdb/). Courtesy of the NAGRP Bioinformatics Project Team.

ovulation rate in the occidental population, the authors concluded that MAN2B2 has either a unique epistatic interaction within the Meishan-cross population or the 1574A>G SNP is in linkage disequilibrium with the causative genetic variant in the Meishan-cross population.

Further studies revealed a number of candidate genes possibly associated with ovulation rate in pigs. Among them were the most promising gene for the aldo-keto reductase 1C (*AKR1C*) gene, which was also associated with age at puberty and nipple number (Nonneman et al. 2006). Four candidate genes (*PIP5K2A*, *ITIH2*, *GAD2*, and *AKR1C2*) were identified in the vicinity of the QTL region at SSC10 (Nonneman and Rohrer 2003). A physiological candidate gene, *SPP1*, was found in the QTL region on the short arm of the SSC8 (King et al. 2003).

Galloway et al. (2002) found an interesting effect of the bone morphogenic protein (BMP15) locus in sheep. BMP15, also known as growth and differentiation factor 9B (GDF9B), is a member of the transforming growth factor beta superfamily (TGFbeta), which is, in humans, rodents, and sheep, expressed only in the oocyte. Inactivation of the *BMP15* gene in mice has only minor effects on fertility, whereas in sheep heterozygous for a *BMP15* mutation, an increase in ovulation rate has been observed, but homozygote animals are infertile. However, the discovery that a point mutation in the BMP1B receptor in Booroola sheep is responsible for increased ovulation rate confirms the importance of the TGFbeta signaling molecules in early folliculogenesis.

McNatty et al. (2005b) confirmed the involvement of BMP15, GDF9, and activin receptor-like kinase 6 (ALK6), otherwise known as the BMP receptor type IB (BMPRIB), on the ovulation rate in sheep. As previously shown, animals homozygous for the *BMP15* or *GDF9* mutations were anovulatory, whereas animals heterozygous for BMP15 or GDF9 or heterozygous or homozygous for ALK6 had elevated ovulation rates (Notter 2008). The authors were able to show by immunizing ewes against BMP15 or GDF9 that both proteins are essential for follicular development and control of ovulation rate. Several point mutations in two growth factor genes (*BMP15* and *GDF9*) and a related receptor (*ALK6*) have been found to be associated with ovulation rate in different sheep breeds (McNatty et al. 2005a). As well, heterozygotes for mutations in BMP15 or GDF9 or homozygotes for the ALK6 mutation had higher ovulation rates (i.e., +0.6–10) than their wild-type contemporaries. The expression of BMP15 and GDF9, which is restricted to the oocyte, supports a new paradigm in reproductive biology, presenting the oocyte as a major player in the regulation of ovulation rate.

4.2.2 Ovarian cyst

Cystic ovarian disease (COD) is a common disease in cattle, particularly in dairy breeds and less common in sows, mares, dogs, and cats. The disease is characterized by gross estrus abnormalities, either anestrus or more frequent and prolonged estrus. COD includes the formation of the cystic follicle (CF), luteal cyst (LC), and cystic corpus luteum (CCL). Mature follicle ovulation failure is a result of exogenous or endogenous disruption of the hypothalamo-hypophyseal-ovarian axis. The anovulatory follicular structure can regress or persist as a follicular cyst or LC. CFs do not rupture, are significantly enlarged, and may appear as multiple CFs on both ovaries. They are usually caused by insufficient level of luteinizing hormone. The two pathological forms of bovine COD, follicular cysts, and LCs are etiologically and pathogenetically related but differ clinically.

It is a common belief that COD is caused by high milk production. However, this observation is biased since higher-producing cows are more likely to be examined, more likely to be treated if found to have COD, and more likely to remain in the herd despite some decrease in reproductive performance. Multiple evidences suggest that COD increases milk production, rather than high production causing cows to develop COD. The incidence of COD increases with age with the reported herd incidence of 5–25% per lactation.

The pathogenesis of ovarian cyst development is still poorly understood, but the general hypothesis is that COD results from an imbalance of neuroendocrine hormones involving the hypothalamic-pituitary-gonadal axis, by endogenous or exogenous factors. The lack of the preovulatory surge of luteinizing hormone (LH) in cystic cows seems to be associated with a lowered gonadotropin-releasing hormone (GnRH) content in the hypothalamic area (Hooijcr et al. 2003). Secretion of GnRH/LH from the hypothalamus-pituitary is aberrant, which is caused by insensitivity of the hypothalamus-pituitary to the positive feedback effect of estrogens. Cysts occurring within the ovary include follicular cysts, LCs (luteinized follicular cyst), cystic corpora lutea, cystic rete ovarii, inclusion cysts derived from the surface epithelium, and cysts of the subsurface epithelial structures. Luteal and follicular cysts are derived from anovulatory graafian follicles and most likely represent different manifestations of the same condition. In cattle, follicular cysts develop most commonly in heavily producing animals during the winter period. In some cases of cystic ovarian degeneration, mucometra, a uterus distended with a fluid containing much mucin, can occur.

At the ovarian level, cellular and molecular changes in the growing follicle may contribute to anovulation and cyst formation. Differences in receptor expression between cystic ovarian follicles and dominant follicles may be an indication of the pathways involved in cyst formation (Vanholder et al. 2006).

Although the genetic background of COD etiology is unclear, there are several reports in the literature revealing the association of different loci with the appearance of COD. Sharif et al. (1998) reported the association of bovine leukocyte antigen (BoLA) alleles (*BoLA-DRB3.2*22*, **2*, and**16*) with a lower risk of COD in Holstein cattle. Increased expression of LH receptor and 3b-hydroxysteroid dehydrogenase mRNAs in granulosa cells and increased follicular estradiol-17b concentrations were associated with dominant cysts compared with normal follicles (Calder et al. 2001). In transgenic female mice overexpressing plasminogen activator inhibitor-1, increased incidence of polycystic ovarian changes was found (Devin et al. 2007).

In the MGI database, 22 genes associated with the GO term "ovary cysts" were found: *Amhr2*, anti-Müllerian hormone type 2 receptor; *Bmp15*, bone morphogenetic protein 15; *Brca1*, breast cancer 1; *Cyp19a1*, cytochrome P450, family 19, subfamily a, polypeptide 1; *Esr1*, estrogen receptor 1 (alpha); *Fanca*, Fanconi anemia, complementation group A; *Fancl*, Fanconi anemia, complementation group L; *Foxc1*, forkhead box C1; *Fshr*, follicle-stimulating hormone receptor; *Gdf9*, growth differentiation factor 9; *Gja1*, gap junction protein, alpha 1; *Kiss1*, KiSS-1 metastasis suppressor; *Kit*, kit oncogene; *Lhb*, luteinizing hormone beta; *Mos*, Moloney sarcoma oncogene; *Nobox*, NOBOX oogenesis homeobox; *Ots1*, ovarian teratoma susceptibility 1; *Pdcd4*, programmed cell death 4; *repro46*, reproductive mutant 46, JAX Reproductive Mutagenesis Program; *Rspo2*, R-spondin 2 homolog (*Xenopus laevis*); *Tom1l2*, target of myb1-like 2 (chicken); and *Ybx2*, Y box protein 2.

The increased permeability of microvessels, causing the accumulation of follicular fluid in CFs may be caused by expression of vascular endothelial growth factor (VEGF) receptors in the granulosa and theca interna layers (Isobe et al. 2008). Ortega et al. (2008) suggested an important role for the insulin-like growth factor (IGF)-I in the regulation of folliculogenesis and also its involvement in the pathogenesis of COD in cattle. However, in the rat model, considerable changes in the ovarian expression of IGF-I, fibroblast growth factor (FGF)-2, and VEGF were detected in induced COD (Ortega et al. 2007).

In high-yielding dairy cows with COD, a significantly lower insulin response to a standard glucose load was observed (Opsomer et al. 1999), and therefore, insulin was considered as a factor in the pathogenesis of COD. In animals with COD, the follicular cysts synthesized a significantly higher amount of estrogen receptor alpha in all follicular layers than secondary, tertiary, and atretic follicles in healthy animals (Salvetti et al. 2007). Ovaries of animals with COD exhibited altered estrogen receptor expression compared with that in normal animals. Dissen et al. (2000) reported that an abnormally elevated production of nerve growth factor (NGF) within the ovary suffices to initiate several structural and functional alterations associated with the development of follicular cysts in the rat ovary.

4.2.3 Silent heat

In the literature, silent heat is frequently referred to as silent estrus, silent ovulation, anaphrodisia, or anestrus. The prevailing characteristic is that female animals do not give any behavioral signal that an ovarian follicle is maturing and rupturing although the follicle maturation and ovulation occurred normally. In the case of silent heat, ovulation can be detected by palpation or by measuring estrogen levels in the blood. The frequency of silent heat decreases with the progress of lactation, so that incidence is relatively low by 4 months postpartum. Palpation and measuring of estrogen in milk or in plasma are the only methods allowing detection of cows with true silent heat. Analysis of the most common factors involved in silent heat syndrome revealed a number of possible causes including negative energy balance postpartum, high milk yield, age, breed, season, stress, and other diseases, and also the quality of estrus detection and housing (Hoedemaker 2008), placing the incidence of silent heat between 10% and 40% in different herds (Zdunczyk et al. 2005). The fact that there were observed differences in silent heat frequency among breeds and that disturbance of the hypothalamo-hypophysial ovarian system, which is under genetic as well as environmental control, supports the assumption that a genetic component is also involved in the silent heat syndrome.

4.2.4 Retained corpus luteum

Retained corpus luteum is characterized by the failure of corpus luteum resorption at the appropriate time in the reproductive cycle. As a consequence, the animal remains anestral. Abnormal persistence of the corpus luteum occurs in several species. In the bitch, the corpus luteum is normally retained for a prolonged period after ovulation, but in other species, retention of the corpus luteum in undesirable, because it frequently prevents normal cycling. Corpus luteum retention in cattle usually occurs postpartum, frequently in association with disorders such as fetal mummification, endometritis, pyometra, or hydrometra. These disorders often disrupt normal cyclic luteolysis, most likely because of impaired transfer of prostaglandin $F_2\alpha$ from the uterus to the ovary. In the mare, retention of the corpus luteum can occur spontaneously, in the absence of uterine disorders and affected mares cease cycling. In some species, the corpus luteum may persist in the absence of pregnancy, which causes "pseudopregnancy" with clinical signs of pregnancy.

4.3 Reproductive disorders associated with the vagina and uterus

4.3.1 Pyometra and puerperal metritis

Pyometra is a hormonally mediated diestrual disorder characterized by cystic endometrial hyperplasia with secondary bacterial infection (Figure 4.2). Typical is the accumulation of purulent or mucopurulent material within the uterine lumen, persistent corpus luteum is present, and the cervix is closed (Sheldon et al. 2006). Pyometra is frequent in older bitches, 4–6 weeks after estrus, in cows; it is invariably accompanied by the persistence of an active corpus luteum and interruption of the estrous cycle. In affected mares, the cervix is often fibrotic, inelastic, or otherwise impaired. Mares may continue to cycle normally, or the cycle may be interrupted. Discharge from the genital tract may be absent or intermittent. As a rule, affected animals do not exhibit any systemic signs of illness, but affected mares may be in poor condition. Metritis is an inflammation of the uterus, while puerperal metritis is an infection of the pregnant uterus. Cows failing to eliminate infection more than 21 days after calving develop endometritis,

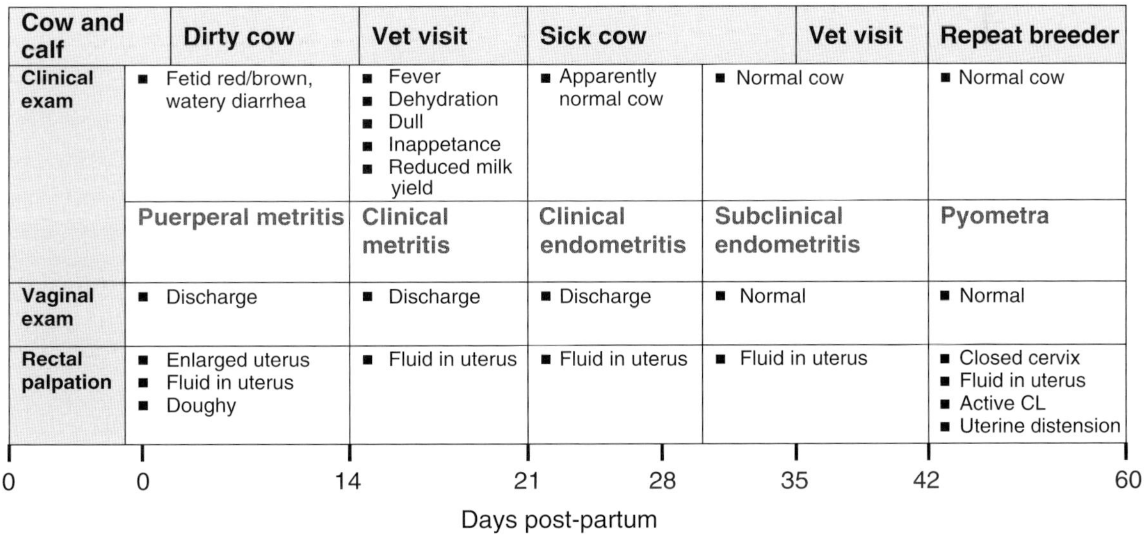

Figure 4.2 Classification and diagnosis of uterine disease in postpartum cows. Metritis develops between 1 and 21 days after calving, endometritis between 22 and 42 days. Pyometra occurs 43 days onward after calving in cows with an active corpus luteum (CL) and closed cervix (Chapwanya 2008). Reprinted with permission from IFP Media, Irish Veterinary Journal.

which is either a clinical or a subclinical inflammation of the uterine endometrium characterized by a purulent vulval discharge up to 42 days postpartum with no signs of systemic illness. Unresolved endometritis often progresses to pyometra.

Acute puerperal metritis occurs in all species within the first postpartum week. It results from the infection of the reproductive tract at parturition and often follows complicated parturition. The causative organisms in cattle are most frequently *Arcanobacterium pyogenes* in association with gram-negative anaerobic bacteria such as *Fusobacterium necrophorum*. The condition is acute in onset. Affected cows, mares, ewes, does, or sows are depressed and may be febrile without appetite. A fetid, watery uterine discharge is characteristic for cows but may not be present in other species. Puerperal metritis is often associated with retained placenta, dystocia, and stillbirth, and usually occurs toward the end of the first week postpartum.

Expression of lactoferrin in canine uterus has been investigated during the estrous cycle in normal bitches and bitches exhibiting pyometra (Kida et al. 2006). Lactoferrin is a nonspecific antimicrobial agent, synthesized in the canine uterus during the normal estrous cycle. Real-time reverse transcription polymerase chain reaction (RT-PCR) analysis revealed the presence of lactoferrin gene transcripts in the endometrium at all stages of the estrous cycle, reaching the highest levels in estrus. In normal bitches, endometrial lactoferrin mRNA increased from proestrus to estrus followed by dramatic reduction from estrus to day 10 of diestrus. Levels of lactoferrin mRNA were higher in bitches with pyometra than in healthy animals. In the canine uterus, lactoferrin expression is related to the blood concentration of estrogen and a dramatic reduction in lactoferrin observed in early diestrus may impair antimicrobial defense. Also, enhanced expression of lactoferrin mRNA in the endometrium with pyometra may be associated with neutrophil invasion into the uterus to combat the infection.

Ishiguro et al. (2007) studied the relationship between adherence of *Escherichia coli* and expression of mucin-1 mRNA in the endometrium of beagle bitches at different stages of the estrous cycle and in those with cystic endometrial hyperplasia/pyometra complex. Bitches with pyometra had a lower level of expression of the *MUC1* gene, and the number of *E. coli* adhering to the endometrial epithelial cells was inversely correlated with the level of *MUC1* transcription. Sugiura et al. (2004) demonstrated suppressed activity of cellular immunity in the first half of the diestrous stage, characterized by significantly decreased response of peripheral blood mononuclear cells (PBMNCs) to infection, but increased in proestrus/estrus. This is probably the consequence of increased progesterone concentration and minimal estrogen release. This marked decrease of immune resistance allows the expansion of *E. coli*, which enters the uterine cavity through the loosened cervical canal during estrus, leading to pyometra onset. In some rat strains, chronic administration of exogenous estrogens induces pyometritis, suggesting that there is genetic variation in susceptibility to estrogen-induced inflammation and pyometritis. Using two inbred rat strains, Pandey et al. (2005) demonstrated significant strain-specific differences in the incidence of pyometra after 10 weeks of treatment with synthetic estrogen diethylstilbestrol (DES). In addition, they could also show that a congenic rat strain carrying the RNO5 segment from a susceptible line in the genetic background of the pyometra resistant line consistently developed pyometra, supporting the assumption that a strong candidate gene for pyometra susceptibility is located within this region. The susceptibility to 17β-estradiol induced pyometritis appears to segregate as a recessive trait in crosses between rat strains, supporting evidence for a major genetic determinant of susceptibility to 17β-estradiol induced pyometritis on rat chromosome 5 (Gould et al. 2005). The presence of potent proteinase inhibitors has also been associated with the incidence of pyometra in mare. Pemberton et al. (1994) investigated the possibility that the severity of endometritis in thoroughbred mares correlates with the haplotypes of plasma alpha 1-proteinase inhibitor (alpha 1-PI). The frequency of the N haplotype was much higher in mares with pyometra compared with the rest of the population. This finding supports the hypothesis that other two haplotypes (S and T), in contrast to haplotype N, may have protective function.

4.3.2 *Hydrometra*

Hydrometra is a collection of watery or mucoid fluid in the uterus. Postmating noninfectious hydrometra and hydrovagina of unknown etiology, leading to a scrotum-like swelling of the perineum, were observed in mice (Kunstýr et al. 1982). The mice were otherwise clinically healthy, and the disease could not be transmitted to other females. Hydrometra was observed also in goats where the diagnosis can be easily made by ultrasound. The incidence in older goats is normally significantly higher than in yearlings (Hesselink 1993).

4.3.3 *Vaginitis, cervicitis, and endometritis*

Vaginitis (inflammation of the vagina, colpitis), cervicitis (inflammation of the cervix uteri), and endometritis (inflammation of the endometrium) represent the most common forms of inflammation of the female urogenital tract. In mare, Troedsson (1999) reported an interesting finding that spermatozoa trigger peripheral mononuclear cell chemotaxis into the uterine lumen, which would suggest that transient endometritis is a normal physiological response to breeding. In mares with impaired uterine defense mechanisms, the condition may develop into persistent endometritis and subsequently lead to reduced fertility.

4.3.4 *Uterine torsion*

Uterine torsion is torsion of the body and uterus in cows and mares and of a horn of the uterus in the sow. It causes dystocia characterized by the nonappearance of any part of the fetus in the vulva. Uterine torsion has been defined as a rotation of more than 45 degrees of the uterus around its long axis that occurs at the junction between the cervix and the corpus. The extent of the rotation is usually 180 degrees, although cases with torsion from 60 to 720 degrees have been reported. Heifers and cows bearing twins have lower risk of uterine torsion. In cattle, most uterine torsions are to the left (counterclockwise), and with the severe torsion, circulatory embarrassment occurs (Drost 2007). In heifers, the odds of a uterine torsion are higher in animals that receive calcium in order to prevent milk fever than in nontreated animals. However, there is no association between milk fever and uterine torsion in multiparous animals. Very little is known about the genetic background for uterine torsion, but it appears that large term fetuses predispose a cow to uterine torsion (Frazer et al. 1996).

4.3.5 Vaginal prolapse

Vaginal prolapse (estral eversion, vaginal hyperplasia) is an edematous enlargement of vaginal tissue during estrus. Usually, the prolapse contains only the mucosa of the ventral floor, but it may also contain the urinary bladder or the cervix. In mouse, Connell et al. (2008) found that Hoxa11-null mice had no detectable uterosacral ligaments. Nikolova et al. (2007) suggested linkage of familial pelvic organ prolapse in human to HSA1q31 and identified an SNP in the *LAMC1* promoter region for which the rare T variant segregated with the phenotype. This SNP affected the binding site for NFIL3, a transcription factor coexpressed with LAMC1 in the vaginal tissue.

4.4 Reproductive disorders associated with pregnancy and placenta

4.4.1 Abortion

Abortion is a premature expulsion from the uterus of the products of conception; termination of the pregnancy before fetus is viable (Blood et al. 2007). In cattle, cytogenetic abnormalities were found in aborted calves. Hanada and Geshi (1995) reported abnormal 60, XX, rob(7;12) karyotype in aborted Japanese black cattle fetuses. The deleterious effect of the Robertsonian translocation 7/21 was confirmed cytogenetically in unbalanced embryos (Hanada et al. 1995). Schmutz et al. (1996) performed cytogenetic analysis in aborted and stillborn calves and found association between spontaneous abortions and neonatal losses with chromosomal aneuploidy. The deleterious effect of the 14/20 Robertsonian translocation was confirmed in cattle, having stronger effect on embryo mortality than a lowered conception rate, compared with 1/29 translocation (Schmutz et al. 1997).

4.4.2 Prolonged gestation

Prolonged gestation is most frequently a result of defective function of fetal hypothalamic-pituitary-adrenal axis, which is no longer able to initiate parturition. The absence or developmental abnormality of the fetal adrenal or pituitary glands is characteristic for all forms of prolonged gestation (Graves et al. 1991). The prolonged gestation may be caused by genetic as well as other factors.

In cattle, the inherited form of prolonged gestation is characterized by pregnancy, prolonged for 3 weeks to 3 months. The phenotype of the calf can be normal except for the great size, which requires cesarean section (fetal giantism). It has been reported in Holstein, Ayrshire, and Swedish breeds of cattle. The calf weighs 48–80 kg at birth and shows signs of postmaturity. Breathing is difficult as a result of failure of surfactant release, and the calf may die from hypoglycemia. At necropsy, hypoplasia of the anterior pituitary and adrenal glands is seen. In another type, it is characteristic that the fetus does not develop beyond the 6-month stage, is much smaller than normal, and has severe developmental abnormalities, like Cyclops calves with only one eye. Such cases have been reported in Ayrshire, Guernsey, and Jersey breeds. The pedigree data suggest that this type of defect is caused by a recessive gene. Calves are usually dead when delivered; however, there is no spontaneous parturition in affected Guernsey animals due to nonfunctional pituitary gland in the fetus. The third type is characterized by multiple skeletal deformities and cleft palate and has most frequently been reported in Hereford cattle. Affected calves show evidence of pituitary aplasia or hypoplasia, arthrogryposis, torticollis, kyphosis, and scoliosis.

In pigs, Wilkie et al. (1999) suggested QTL for gestation length on SSC9, SSC15, and SSC1, which are associated with the number of corpora lutea. In the MGI database, there were seven genes associated with the term abnormal gestational length; six associated with long and one with short gestational period.

MGI genes associated to the term "abnormal gestational length" include *Akp5*, long gestation period; *Akr1c18*, long gestation period; *B4galt1*, long gestation period; *Cdkn1c*, short gestation period; *Cenpb*, abnormal gestational length; *Inhbb*, long gestation period; and *Lpar3*, long gestation period.

4.4.3 Dropsy of fetal membrane

Dropsy of fetal membrane is characterized by abnormal accumulation of serous fluid in the allantoic sac. It occurs in cows, rarely in mares, and is often associated with dystocia, uterine inertia, and death or abortion of the fetus (Blood et al. 2007). It has been frequently observed in cattle–bison hybrids. Genetic background is not clear.

4.4.4 Dystocia

Dystocia is defined as calving difficulty resulting from prolonged spontaneous calving or prolonged or severe assisted extraction (Mee 2008). Calving difficulties are often categorized in three categories: easy calving, slight problems, and difficult calving. The overall frequency of difficult calvings in Norwegian red cattle was estimated to be 2–3% in heifers and 1% in cows at second or later calvings (Heringstad et al. 2007). Posterior means of direct and maternal heritabilities for dystocia were 0.13 and 0.09, respectively. The genetic correlation between direct and maternal effects were close to zero, and during the last 20 years, only slight genetic improvement of calving difficulty was detected. Kühn et al. (2003) found significant maternal effect on calving difficulty in the central part of the BTA8 chromosome in German Holstein cows. In addition, a QTL for maternal effect on dystocia was found in the middle part of BTA18, whereas direct effects on dystocia were found on BTA6 at 44 cM. QTL on BTA18 were also detected in Swedish dairy cattle (Holmberg and Andersson-Eklund 2006) as were QTL on BTA6. In Danish Holstein cattle, four significant QTL were found for calving difficulty on chromosomes 8, 18, 25, and 28. Analysis of the family material in Holstein Frisian cattle identified three significant QTL influencing calving ease (Ashwell et al. 2005) on chromosomes 8, 17, and 27.

In general, all studies support the finding that there is little correlation between direct and maternal effect; however, there is a slight correlation between dystocia and stillbirth. Probably due to low heritability, there was little selection progress in calving difficulties over the last few decades.

4.4.5 Retained placenta

Retained placenta is failure to pass the placenta within 24 h postpartum (Kelton et al. 1998). Retained placenta affects 5–10% of calvings and greatly increases the risk of metritis and endometritis. Due to the physiological basis of placenta expulsion, the genes related to major histocompatibility complex (MHC) were suggested as possible candidate genes involved in retained placenta. Sharif et al. (1998) found the association of the bovine MHC DRB3 (BoLA-DRB3) allele *3 with a lower risk of retained placenta. In the study of Joosten et al. (1991), the MHC class I compatibility between dam and calf increased the risk of retained placenta. They suggested that compatibility of MHC products between dam and calf might negatively influence placental maturation and expulsion, and therefore increase the risk of retained placenta. Induction of tolerance against noninherited maternal antigens (NIMAs) might be implicated in the occurrence of the disorder, suggesting a tolerance-inducing effect of NIMA in cattle in relation to retained placenta. In addition to genetic factors, circulating PgF2alpha and nutritional parameters at parturition in dairy cows were associated with retained placenta. In horses, Sevinga et al. (2004) suggested, based on MHC data, a negative effect of inbreeding on the incidence of retained placenta in Friesian horses.

4.5 Reproductive disorders associated with male reproductive organs

4.5.1 Hernia inguinalis and scrotalis

Inguinal and scrotal hernias are caused by the weakness of the inguinal canal that embraces different organs and acts as a natural corset. Due to the pressure in the abdominal cavity, a rupture in the inguinal canal can develop, and internal organs, most commonly intestines, can be pushed through the rupture. Hernias are considered to be congenital defects, which are caused by connective tissue weakness. Depending on location, hernias can be classified as diaphragmatic, scrotal (inguinal), or umbilical (abdominal). In the context of reproduction, two types of hernias are of relevance: inguinal and scrotal hernia. By definition, in the case of inguinal hernia, the contents of the abdominal cavity, mainly fat tissue and intestines are present in the inguinal canal, whereas scrotal hernia refers to a situation where hernial contents are present in the scrotum. As a consequence of hernia, the surrounding connective tissue can exert strong pressure on the soft hernial material, reducing the blood flow due to the strong pressure on the veins, which may develop into gangrena and sepsis. Because of familial incidence, hernias are considered a hereditary disease of the connective tissue in man and animals (Smith and Sparkes 1968). Hernias also often occur in patients with Marfan or Ehlers–Danlos syndrome. The most frequently affected species in farm animals is pig, where *hernia inguinalis* and *hernia scrotalis* occur, depending on population, at frequencies from 1.7% to 6.7% (Thaller et al. 1996). Scrotal hernia especially is believed to be a genetic disorder with recessive inheritance (Jubb et al. 2007).

Several anatomical characteristics such as abnormally wide inguinal canal and not obliterated processus vaginalis are considered as risk factors in the development of inguinal and scrotal hernia. Most frequently, the distal jejunum and ileum slide through the vaginal ring and enter the inguinal canal. Herniation of the small colon and omentum are less common. In addition, abnormalities during the process of testicular descent may also contribute to the predisposition for the development of inguinal and scrotal hernia in male pigs. Testicular migration is characterized by rapid development of the gubernaculum, which is the key anatomical structure controlling descent of testes from the abdomen, where it develops into the scrotum. Swelling of the gubernaculum, caused by deposition of hyaluronan, extends the inguinal canal. Testicular descent occurs after biodegradation of this structure by proteolytic enzymes.

The involvement of genetic factors in the development of inguinal and scrotal hernias has been demonstrated in several studies (Cook et al. 2000; Koskimies et al. 2003); however, the mode of inheritance has not been clarified. In humans, an autosomal dominant inheritance with incomplete penetrance and involvement of genomic imprinting has been proposed (Gong et al. 1994). Collagen matrix has also been studied as a target in recurrent inguinal hernia in man (Zheng et al. 2002). In this context, expression of procollagen type I/III, matrix metalloproteinases (MMPs) 1 and 13 were investigated. Zheng et al. (2002) found in patients with recurrent inguinal hernia decreased ratio of collagen types I to III and increased expression of MMP-1 and MMP-13, suggesting that recurrent inguinal hernias should be considered as a disease of the collagen matrix due to the involvement of connective tissue. The estimated h^2 values for inguinal and scrotal hernias range from 0.20 to 0.86 (Mikami and Fredeen 1979). Several candidate genes coding for proteins involved in the gubernacular growth such as insulin-like receptor 3, Müllerian inhibiting substance (MIS), and relaxin, as well as calcitonin gene-related peptide released from genitofemoral nerve, have been considered as possible candidates for hernia development.

Taking into account the physiology of testicular descent, it has been hypothesized that mutations affecting genes coding for the hyaluronan degrading enzymes (hyaluronidase, β-hexosaminidase, and β-glucuronidase) could prevent obliteration of the processus vaginalis and so indirectly increase the chance for hernia formation (Beck et al. 2006). Using a genome-wide linkage scan, the β-glucuronidase gene (*GUSB*) was mapped within the genomic region on porcine chromosome 3 (SSC3) associated with congenital inguinal and scrotal hernia (Beck et al., 2006). Therefore, *GUSB* became a positional candidate for inguinal and scrotal hernia. In order to test the association between *GUSB* polymorphisms and incidence of inguinal/scrotal hernia, extensive sequence analysis of the porcine *GUSB* gene and SNP-based association analysis were conducted. However, due to the polygenic character of the trait and only one candidate gene from the targeted genomic region in this analysis, no association could be confirmed (Beck et al. 2006).

Following the physiology of testicular descent in humans, extensive apoptosis has been proposed in the smooth muscles around the *processus vaginalis* after testicular descent. In pigs, a genome scan has revealed five chromosomal regions associated with hernia inguinalis/scrotalis (Knorr et al. 2006). In order to determine whether a disturbed apoptosis might be responsible for hernia development in pig, Germerodt et al. (2008) determined chromosomal positions of genes involved in apoptosis, isolated sequence tagged site (STS) markers specific for the disorder-associated chromosomal regions, and evaluated the role of apoptosis in the inguinal occlusion and testicular descent. Interestingly, all identified porcine apoptotic genes have been mapped to genomic regions associated with porcine inguinal/scrotal hernia. Physiological data based on significant decrease of Ca^{2+} concentration in pigs with hernia, which might be a consequence of perturbed apoptosis in affected pigs, as well as the assignment of apoptotic genes to chromosomal regions associated with the disorder, support the involvement of apoptotic genes in hernia development in pigs.

Based on a whole genome scan (Figure 4.3), Grindflek et al. (2006) identified suggestive QTL for inguinal and scrotal hernias. Several promising candidate genes are located within these regions: collagen type IXα (*COL9A1*), estrogen receptor 1 (*ESR1*), calcitonin gene-

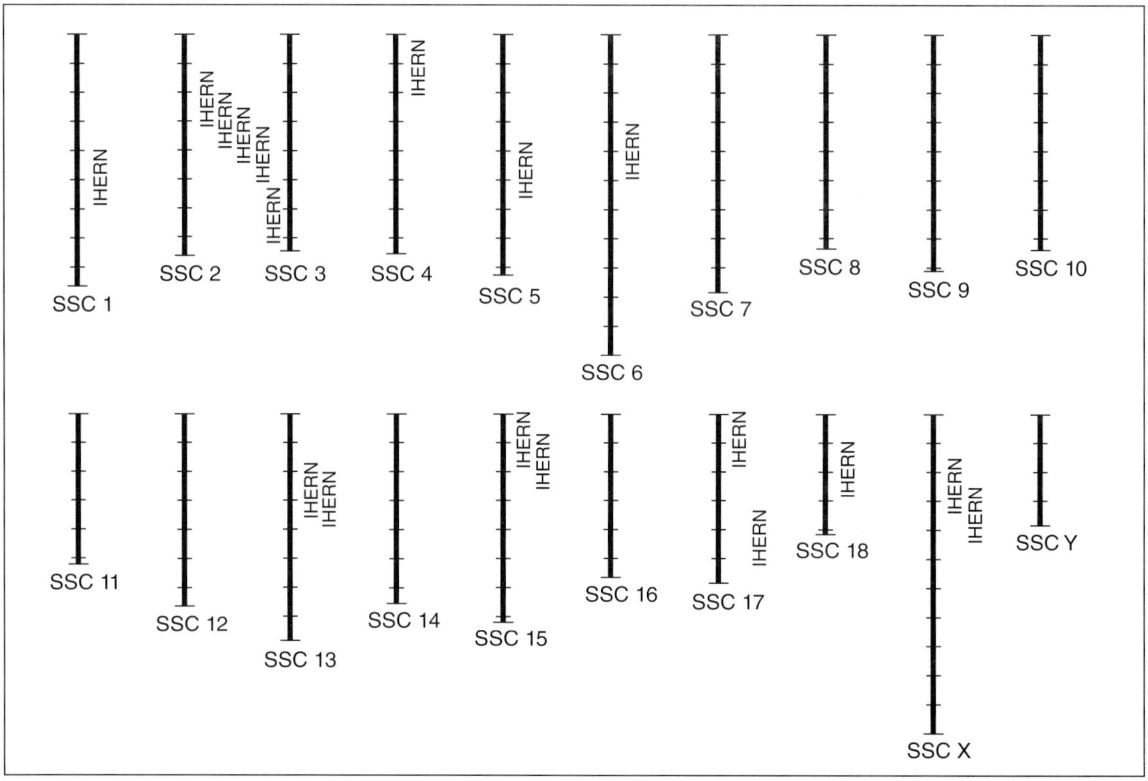

Figure 4.3 QTL for inguinal hernia in pigs (AnimalQTL database; www.animalgenome.org/QTLdb/). Courtesy of the NAGRP Bioinformatics Project Team.

related peptide (*CGRP*), insulin-like hormone 3 (*INSL3*), *MIS*, collagen type II (*COL2A1*), insulin-like hormone 5 (*INSL5*), and cytochrome P450 family19A1 (*CYP19A1*). Some of these candidate genes (*MIS*, *INSL3*, relaxin, and *CGRP*) were already identified in other studies (Clarnette and Hutson 1997; Kubota et al. 2002). Grindflek et al. (2006) revealed significant QTL for inguinal and scrotal hernia on 8 out of 19 porcine chromosomes, the most promising being located on SSC1, SSC2, SSC5, SSC6, SSC15, SSC17, and SSCX. One haplotype on SSC5 has been found to be transmitted to hernia pigs with four times higher frequency than to healthy pigs.

In a recent study, Germerodt et al. (2008) applied 15 porcine STS markers to fine-map chromosomal regions associated with hernia inguinalis/scrotalis. However, further studies are required in order to further narrow down the suggestive QTL regions, to investigate the candidate genes, and to confirm the suggestive QTL in other populations. Similarly, as with other complex traits, the identification of haplotypes associated with inguinal and scrotal hernias may be helpful in selection programs against the disorder.

Abnormal collagen metabolism is thought to play an important role in the development of primary inguinal hernia (Rosch et al. 2002). The altered ratio of the collagen subtypes can result either by a modified synthesis or by an imbalanced breakdown. The cleavage is regulated by the activity of the MMPs, proteins of a family of zinc-dependent endopeptidases. Among them, MMP-1 and MMP-13 are the principal matrix enzymes cleaving fibrillar types I, II, and III collagen. A defective collagen metabolism contributes to decreased tensile strength and mechanical stability of both the connective tissues and the induced scar tissue. Therefore, these alterations in collagen formation should be of central relevance in the pathophysiology of hernias (Rosch et al. 2002).

Knowledge of the transcriptional regulation of collagen in patients with primary inguinal hernia may help to elucidate the pathogenesis of primary inguinal hernia. In normal skin, types I and III collagen are known to exist in a ratio of up to 4:1. The results indicated that the ratio of type I to type III procollagen mRNA was decreased in patients with primary hernia. This decrease was mainly due to the increase of type III procollagen mRNA. They concluded that abnormal change of type I and type III collagen mRNAs contributes to the development of primary inguinal hernia (Rosch ct al. 2002).

It has been shown that recurrent inguinal hernias are a disease of the collagen matrix. An increase of MMPs MMP-1 and MMP-13 mRNAs and proteins was observed in the recurrent hernia group and showed significant differences compared with the control group (Zheng et al. 2002).

4.5.2 Cryptorchidism

Cryptorchidism is defined as the incomplete descent of the testis and associated structures from the abdomen through the inguinal canal into the scrotum. It is common in humans, pigs, horses, and companion animals (2–12%) but rare in cattle, sheep, and goats (≤1%) (Amann and Veeramachaneni 2007). Descent of the testis, epididymis, and spermatic cord, together with the testicular artery and vein, is a complex series of events, which require concerted action of hormones, constitutive mechanisms, and the nervous system. The complete descent of the testis occurs in most species prenatally with the exception of the dog, where it occurs postnatally. Defects in the testis descent cause several problems ranging from impaired spermatogenesis and reduced fertility to increased rate of testicular neoplasia and testicular torsion. It is widely accepted that cryptorchidism can be caused by genetic or environmental factors, but the genetic component seems to prevail, and therefore, breeding from affected individuals is not recommended. However, when outbreaks of cryptorchidism occur, the genetic component seems to be less likely; therefore, in such cases, hormonal and environmental factors (endocrine disruptors) should be considered as possible causes.

Testicular descent is divided in three phases: relative transabdominal migration phase, the intra-inguinal migration and extra-inguinal migration. From studies of hernia cases, there is strong evidence that MIS is involved in the regulation of the first phase of migration, the second phase requires increased intra-abdominal pressure, and finally, the extra-inguinal migration is controlled by androgenic hormones, calcitonin gene-related protein, genitofemoral nerve, and other factors.

In the literature, cryptorchidism has been associated with at least 393 different syndromes, and the list of clinical syndromes with known genetic mutations that feature cryptorchidism, published by Barthold (2008), was expanded for the purpose of this review with additional clinical syndromes extracted from the Online Mendelian Inheritance in Man (OMIM) and Disease databases (Table 4.1). The location of loci in cattle was

Table 4.1 Selected clinical syndromes with known genetic mutations that may cause or feature cryptorchidism.

Syndrome	OMIM (syndrome)	Gene	Location (human)	Location (cattle)	Gene name
Aarskog	100050	*FGD1*	Xp11.21	X (59,725K–59,813K)	FYVE, RhoGEF, and PH domain containing 1
Amelogenesis imperfecta—polycistic renal disease—cl/p	/	*MSX2*	5q34-q35	20 (6,569K–6,645K)	msh homeobox 2
Apert—acrocephalosyndactyly type 1	101200	*FGFR2*	10q26	26 (42,064K–42,116K)	Fibroblast growth factor receptor 2
Cardiofaciocutaneous	115150	*BRAF*	7q34	4 (107,677K–107,870K)	v-raf murine sarcoma viral oncogene homolog B1
Cardiofaciocutaneous Noonan	/	*KRAS*	12p12.1	5 (89,959K–90,084K)	v-Ki-ras2 Kirsten rat sarcoma viral oncogene homolog
		MAP2K1	15q22.1-q22.33	10 (13,089K–13,277K)	Mitogen-activated protein kinase kinase 1
		MAP2K2	19p13.3	7 (18,493K–18,545K)	Mitogen-activated protein kinase kinase 2
Costello	218040	*HRAS*	11p15.5	/	v-Ha-ras Harvey rat sarcoma viral oncogene homolog
Distal arthrogryposis type 2A	193700	*MYH3*	17p13.1	19 (30,149K–30,201K)	Myosin, heavy chain 3, skeletal muscle, embryonic
Distal arthrogryposis type 2B	601680	*TNNI2*	11p15.5	29 (51,474K–51,526K)	Troponin I type 2 (skeletal, fast)
		TNNT3	11p15.5	29 (51,421K–51,459K)	Troponin T type 3 (skeletal, fast)
Fanconi anemia	227650	*FANCE*	6p22-p21	23 (9,662K–9,715K)	Fanconi anemia, complementation group E
		FANCED2	3p26	/	Fanconi anemia, complementation group D
Gorlin—fronto-metaphyseal dysplasia	305620	*FLNA*	Xq28	X (23,656K–23,718K)	Filamin A, alpha (actin-binding protein 280)
Kallman syndrome	308700	*KAL1*	Xp22.3	/	Kallmann syndrome 1 sequence
Noonan	163950	*SOS1*	2p22-p21	11 (22,475K–22,585K)	Son of sevenless homolog 1
Prader–Willi	176270	*SNRPN*	15q11.2	/	Small nuclear ribonucleoprotein polypeptide N
Sotos	117550	*NSD1*	5q35.2-q35.3	7 (37,932K–38,044K)	Nuclear receptor binding SET domain protein 1
Weaver	277590				
Beckwith–Wiedemann	130650				

Table 4.2 Transgenic and knockout murine models related to cryptorchidism.

Gene	Chromosome	Location	Location (human)	Gene name
EPHA4	1	43 cM	2q36.1	Eph receptor A4
GLI2	1	63 cM	2q14	GLI-Kruppel family member GLI2
LBR	1	97.3 cM	1q42.1	Lamin B receptor
DNAJC5I	2	106 cM	/	DnaJ (Hsp40) homolog, subfamily C, member 5
LRP2	2	40 cM	2q24-q31	Low-density lipoprotein receptor-related protein 2
SCG5	2	64 cM	15q13-q14	Secretogranin V
NHLH2	3	01814078-101818429 bp	1p12-p11	Nescient helix loop helix 2
RXFP1	3	79448638-79541716 bp	4q32.1	Relaxin/insulin-like family peptide receptor 1
CRSP	5	84 cM	/	Cryptorchidism with white spotting, deletion region
GNRHR	5	44 cM	4q21.2	Gonadotropin-releasing hormone receptor
RXFP2	5	84 cM	13q13.1	Relaxin/insulin-like family peptide receptor 2
FKBP4	6	128379753-128388695 bp	12p13.33	FK506 binding protein 4
HOXA10	6	26.33 cM	7p15-p14	Homeo box A10
HOXA11	6	26.33 cM	7p15-p14	Homeo box A11
RET	6	53.2 cM	10q11.2	Ret proto-oncogene
HMGB2	8	31 cM	4q31	High mobility group box 2
INSL3	8	33 cM	19p13.2-p12	Insulin-like 3
ARID5B	10	10 cM	10q21.2	AT rich interactive domain 5B (Mrf1 like)
GLI1	10	69 cM	12q13.2-q13.3	GLI-Kruppel family member GLI1
GNRH1	14	30.5 cM	8p21-p11.2	Gonadotropin-releasing hormone 1
DHH	15	57.4 cM	12q12-q13.1	Desert hedgehog
PARL	16	14 cM	3q27.1	Presenilin associated, rhomboid-like
LHCGR	17	46.5 cM	2p21	Luteinizing hormone/choriogonadotropin receptor
SOX8	17	8 cM	16p13.3	SRY-box containing gene 8
AMH	33	10 cM	19p13.3	Anti-Müllerian hormone
BMP5	42	9 cM	6p12.1	Bone morphogenetic protein 5
AR	X	36 cM	Xq11.2-q12	Androgen receptor
FOXP3	X	2.1 cM	Xp11.23	Forkhead box P3
MECP2	X	29.6 cM	Xq28	Methyl CpG binding protein 2
WW1	UN	UN	/	Small papilla 1

determined using the bovine–human synteny map (Razpet 2007).

The search in the MGI database revealed 30 mouse gene knockout models that result in phenotypes associated with cryptorchidism (Table 4.2).

Five genes showed positive association between sequence variation/mutation screening and cryptorchidism in humans: *ESR1*, *NR5A1*, *RXFP2*, *INSL3*, and *AR* (Gorlov et al. 2002; Yoshida et al. 2005; Ferlin et al. 2006; Wada et al. 2006; and Wang et al. 2008), as well as *INSL3* in sheep (Williams et al. 2007) and dogs (Cassata et al. 2008) (Table 4.3). Ferlin et al. (2005) found no difference between the numbers of CAG and GGC repeats, resulting in variable lengths of PolyGln/PolyGly in the androgen receptor (*AR*) gene and cryptorchidism; however, it has been proposed that a particular combination of the PolyGln/PolyGly polymorphisms may be linked to cryptorchidism. Studies of

Table 4.3 Genes tested for association with cryptorchidism.

Gene	Species	Chromosome	Gene name	Reference
LHCGR	Human	2p21	Luteinizing hormone/ choriogonadotropin receptor	(–) Simoni et al. (2008)
ESR1	Human	6q25.1	Estrogen receptor 1	(+) Wang et al. (2008); (+) Yoshida et al. (2005); (–) Galan et al. (2007)
NR5A1 (SF-1)	Human	9q33	Nuclear receptor subfamily 5, group A, member 1	(+) Wada et al. (2006)
RXFP2 (LGR8/GREAT)	Human	13q13.1	Relaxin/insulin-like family peptide receptor 2	(+) Gorlov et al. (2002); (+) Bogatcheva et al. (2007); (–) Nuti et al. (2008)
INSL3	Human	19p13.2-p12	Insulin-like 3 (Leydig cell)	(+) Canto et al. (2003); (+) Ferlin et al. (2006); (+) El Houate et al. (2007); (+) Yamazawa et al. (2007); (–) Krausz et al. (2000); (–) Baker et al. (2002); (–) Takahashi et al. (2001)
AR	Human	Xq11.2-q12	Androgen receptor	(+) Ferlin et al. (2005); (+) Silva-Ramos et al. (2006)
INSL3	Sheep	ND	Insulin-like 3 (Leydig cell)	(+) Williams et al. (2007)
INSL3	Dog	20	Insulin-like 3 (Leydig cell)	(+) Cassata et al. (2008)[1]

[1] Association was found but not statistically evaluated; mutation screening (case report).
ND, not defined; +, statistically significant association; –, no association.

insulin-like 3 (*INSL3*), relaxin/insulin-like family peptide receptor 2 (*RXFP2*) and estrogen receptor 1 (*ESR1*) showed opposing results; for instance. Galan et al. (2007) found no association, while Yoshida et al. (2005) and Wang et al. (2008) reported association between *ESR1* sequence polymorphisms and cryptorchidism. In addition, no association between luteinizing hormone/choriogonadotropin receptor (*Lhcgr*) and cryptorchidism could be shown (Simoni et al. 2008).

Due to the complex character of cryptorchidism, the phenotype has been observed together with different types of chromosomal abnormalities. In rams, cryptorchidism has been associated with an autosomal recessive or possibly autosomal dominant gene with incomplete penetrance, and an increased prevalence of cryptorchidism has been found in polled animals (Claxton and Yeates 1972). Similarly, in goats, the gene causing polledness has been shown to be associated with a number of abnormalities during the development of the reproductive tract, including also cryptorchidism. The maldescent of the right testis is a common feature of the goat polled/ intersex syndrome (Soller et al. 1969).

Recently, transcriptomic analysis has been applied for the detection of candidate genes associated with cryptorchidism in humans (Nguyen et al. 2009) and in rats (Barthold et al. 2008). Comparison of cryptorchid and normal samples revealed a number of genes differentially expressed in both groups, where the majority of identified genes were underexpressed in cryptorchid samples, which is most likely the reason for impaired germ cell maturation and sperm tail formation in cryptorchid testis. On the other hand, an antiapoptotic gene (*TNFAIP3*) was highly overexpressed in cryptorchid samples. The transcriptomic approach revealed a number of differentially expressed genes, which can at least in part explain the cryptorchid phenotype, but the reason for differences in expression profile might be very complex including genetic and nongenetic factors.

Our literature search revealed 140 gene loci associated with cryptorchidism (associations based on gene mutations and polymorphisms or specific expression profiles), showing that the most common strategies for identification of candidate genes for cryptorchidism are expression studies and knockout experiments identifying 53 and 30 loci, respectively.

However, the most reliable candidate genes seem to be those that were identified using different approaches. Among 140 candidate loci, 11 loci were identified by two different approaches and therefore seem to be strong candidates for association with cryptorchidism: *LHCGR*, *RXFP2*, *INSL3*, *MSX1*, *CYP19A1*, *ESR1*, *AR*, *WT1*, *HRAS*, *TNNI2*, and *TNNT3*.

4.6 Reproductive disorders associated with embryos and fetuses

4.6.1 Freemartin syndrome

The fact that most female calves born cotwin to a male calf are sterile belongs for centuries to the traditional knowledge of cattle breeders. Such females have, in the vast majority of cases, an underdeveloped female genital system and show signs of masculinization. Due to this characteristic phenotype, they are called freemartins. The same condition has also been recognized in other species, although in much lower frequencies. The term freemartin is now used to describe sterile females born cotwin to a male in any species. As already mentioned, the cases of freemartin are much more frequent in cattle than in sheep, goats, or pigs.

The reason for masculinization of the female calf *in utero* is the formation of the chorionic placental blood vessels that enable common circulation between the feti prior to sexual differentiation, allowing anti-Müllerian duct hormone and testosterone secreted by the male fetus to inhibit the development of the female reproductive tract. In about 92% of cases, the females from mixed-sex twin pregnancies are sterile. The females (in similar proportion, this is true for male calves too) from mixed-sex twin pregnancies are also erytrocytic chimeras and can be diagnosed as freemartin based on blood group typing. The XX/XY mosaicism can also be diagnosed by cytogenetic techniques due to the fact that, probably as a consequence of exchange of blood stem cells, individuals from mixed-sex pregnancies produce white cells with XY as well as cells with XX chromosomes. In freemartin females, the structural changes of the female reproductive system are diagnostic but inconsistent. The tubular genital organs in affected animals range from cord-like bands to near-normal uterine horns. Freemartins have a short blind end vagina without communication with the uterus. The cervix is absent, and the ovaries usually fail to develop and remain hypoplastic. Normal and freemartin cattle can be differentiated on the basis of length of the vagina and on the presence or absence of a cervix. In 1- to 4-week-old normal heifers, the vagina is 13–15 cm long, while in freemartin heifers, the length of the vagina does not exceed 5–8 cm in length. However, some freemartins are quite normal clinically, but they are nearly all sterile.

Due to the exposure of the female fetus to male hormones as well as hypoplastic ovaries in freemartins, the hormonal profile in freemartin heifers differs significantly from normal females. The estradiol production in freemartin ovarian tissue is lower than in normal heifers (Shore and Shemesh 1981), and the response to intravenous injection of hCG has not been detected (Cavalieri and Farin 1999), whereas plasma concentration of testosterone was not different from normal heifers (Saba et al. 1975). Freemartins were also detected using H-Y antigen detection technique (Wachtel et al. 1980). MIS plays a special role in the development of freemartin heifers and is produced by the fetal Sertoli cells during mammalian male sexual differentiation, also known as anti-Müllerian hormone. The newborn males and freemartins have similar levels of MIS in plasma (>700 ng/mL), whereas normal females have much lower levels (<120 ng/mL) (Rota et al. 2002).

Karyotyping, Y chromosome-specific PCR, and fluorescent *in situ* hybridization (FISH) are commonly used methods for the detection of freemartinism. In cases where XX/XY chimerism has occurred with low frequency, a relatively high number of analyzed mitoses is required (26 and 168 for 95% and 99% confidence, respectively) for cytogenetic analysis (Dunn et al. 1981). The same is true for the number of mitoses examined by FISH. There are some other chromosomal abnormalities that occasionally coincide with XX/XY chimerism: 1/29 Robertsonian translocation (Zhang et al. 1994), 4/21 tandem fusion (Pinheiro et al. 1995), and 61, XXY trisomy (Zhang et al. 1994). Different types of tissue (spleen bone marrow, lung, connective tissue, gonad interstitial tissue, lymphoid tissue, and liver) have been used in addition to lymphocytes for the detection of chimerism in freemartins; however, the percentage of XY cells in these tissues in freemartins was normally lower and more variable than in leukocytes (Marcum 1974). Due to its higher sensitivity, PCR is the most commonly used technique for the identification of Y chromosome-specific DNA in freemartins.

In addition to mixed-sex twins, there are also freemartins born as singletons. They can result from mixed-sex pregnancies where *in utero* death of the male cotwin occurs. The evidence for this comes from experiments where transfer of more than one embryo in a recipient cow resulted in a single birth of a female calf, which was characterized as an XX/XY chimera and phenotypically almost a normal freemartin. The incidence of single-born freemartins in cattle is not known, but is likely to be relatively rare due to the fact that the percentage of twins in cattle that survive to full term when one twin dies is very low. According to some studies, about 5% of all singleton births in cattle are calves that survived *in utero* death of their cotwins. However, freemartin syndrome will remain a limiting factor for the introduction of strategies aiming to increase the frequency of twins either through genetic selection or through multiple ovulation and embryo transfer.

The freemartin syndrome is also present in other species such as sheep, goats, camels, pigs, and horses. However, in no other species than cattle and sheep, abnormalities of the female sexual tract are so common and deleterious for normal reproduction. In sheep, freemartins normally have a higher degree of masculinization of the reproductive tract than in cattle, which might be the result of the formation of anastomoses during the early developmental stages compared with the situation in cattle (Parkinson et al. 2001). In sheep, the discovery and introduction of alleles at the high fecundity locus (booroola) led to significantly increased rate of twins, triplets, and even quadruplets. This consequently also increased the risk for multiple pregnancies with mixed-sex fetuses. In the case of intersexuality in a Dorset horn ewe, the animal had female external genitalia but had a male internal reproductive tract with inguinal testes, epididymides, vasa deferentia, and seminal vesicles; no cervix; no uterus; and only the caudal part of the vagina. In peripheral leukocytes, chimerism of the type 54, XX/XY was found, while other tissues revealed the normal female karyotype 54, XX. The ewe was born in a set of triplets with one dead male fetus and one living male, supporting the evidence that the animal was a freemartin (Wilkes et al. 1978)

In other species like goats and pigs, the freemartins seem to be relatively rare, whereas in horses, twins of different sex in approximately 50% of cases show vascular anastomoses and consequently XX/XY lymphocyte mosaicism. In horses, clinical signs of abnormal female sexual tract are very rare, due to the late establishment of anastomoses during embryonal development. In the literature, there are only a few reports of masculinization of the female genital tract as in an example of a hermaphrodite Welsh pony with an XX/XY karyotype and bilateral ovotestes (Bouters and Vandeplassche 1972). Similar to horses, in marmoset monkeys, the exchange of cells between embryos is common, but the resultant chimerism does not lead to infertility. Interestingly, in birds, anastomoses within double-yolked eggs are common, but they lead to the feminization of the male reproductive tract, which is the situation opposite that of mammals but in agreement with the basic mechanisms of sex differentiation in birds.

Based on findings in humans, where the fetus releases cell-free DNA through the fetoplacental unit into maternal circulation in sufficient number of copies, the concentration remains high enough that the presence of Y chromosome-specific DNA in maternal circulation can be detected by PCR. This is in spite of the rush to turn over fetal DNA with an average half-life time of 16 min in maternal circulation. In farm animals, the detectable amounts of fetal Y chromosome-specific DNA remains to be confirmed in maternal circulation. On the other hand, the presence of MIS in circulation offers the production of MIS as a therapeutic tool for tumors with expressed MIS receptors in humans such as ovarian cancers (Teixeira et al. 2001).

Although the genetic background of freemartnism seems to be complex and at least in cattle to a large extent caused by the architecture and developmental pace of the placenta, there is some evidence that cytogenetic abnormalities, referred to as fragile Xq chromosome, might be associated with the incidence of freemartin syndrome in cattle (Llambi and Postiglioni 1994).

4.6.2 Embryonic and fetal death

Embryonic death occurs most frequently very early during embryonic development, quite often in the preimplantation phase. In cattle, the majority of embryo losses occur within the first 2 weeks of development. This is also one of the most important reasons for low success rates of modern reproductive techniques (*in vitro* fertilization, embryo transfer). It is often overlooked that less than half of inseminated bovine and human oocytes reach the blastocyst stage (Betts and King 2001). The term fetal death is a common name for different defects of prenatal development, which then lead to resorption, fetal maceration, mummification, or abortion. In the case of abortion and fetal maceration, the hormonal support of pregnancy is lost, and the animal will show signs of terminated pregnancy. Sometimes, the aborted fetus can be found, the dam may show an abnormal vaginal discharge, and she may return to estrus. In the case of fetal mummification, the fetal death is not apparent immediately, corpus luteum persists in the ovary, and there is no vaginal discharge. Such abnormal pregnancy can persist for an indefinite period of time.

In vitro as well as *in vivo* studies convincingly showed that chromosomal aberrations are frequently a cause for embryonic death. In some species, cell death in the early developing embryo is first observed during blastocyst formation, predominantly in the inner cell mass (Betts and King 2001).

Embryonic and fetal death is an important cause of economic losses in animal production; therefore, a number of studies were performed to identify QTL regions or candidate genes causing embryonic and fetal death. Holl et al. (2004) identified one Mendelian QTL for the number of mummified piglets and four additional QTL in a multi-QTL model. The QTL for the number of mummified piglets was identified at SSC6 at position 81 with a significant logarithm of the odds (LOD) score of 4.03. These QTL could also be confirmed in the two and three QTL models and are characterized by the presence of genetic imprinting.

The mouse model has tremendous potential for the detection of developmental disorders due to the large number of mutations and knockout models that identify candidate genes regulating different important developmental stages. It is surprising that relatively few developmental disturbances lead to death of the embryo and early fetus. It seems that crucial disturbances are failure to establish and to maintain vascular circularization and failure to make the transition from yolk sac-based to liver-based hematopoiesis (Copp 1995). Conversely, it seems that other embryonic and organ systems have little effect on survival *in utero*. Cross (2001) reported the selection of genes that are critical for major developmental events in mice. Candidate genes were grouped into nine functional groups governing different developmental stages: implantation (*Lif, Ets2*), trophoblast differentiation (*Hand1, Mash2, Gcm1*), allantoic differentiation (*Dnmt1*), chorioallantoic fusion (*Mrj, Vcam1, Integrin α4, Fgfr2*), villous morphogenesis (*Gcm1, Hsp90β*), placental vascularization (*Esx1, JunB, Arnt*), regulation of maternal blood flow (*Gata2, Gata3*), vascular development (*Vegf, Flk1, Flt1*), and cardiac development (*Nkx2-5, Hand1*). In addition, a search of the MGI database revealed two additional loci involved in fetal death: prostaglandin F receptor (*Ptgfr*) and steroid 5 alpha reductase 1 (*Srd5a1*).

4.6.3 Stillbirth

Stillbirth is often observed in cattle and pigs and may be a consequence of late fetal death or fetus injury during birth. Kühn et al. (2003) analyzed maternal and direct effects on stillbirth in the German Holstein cattle. They identified suggestive QTL for maternal effect on BTA8 and in the telomeric region of the BTA18. In addition, the indication of QTL for direct effects on stillbirth was found on BTA6. In Swedish dairy cattle, two QTL for maternal effect on stillbirth were found on BTA7 and 11 (Holmberg and Andersson-Eklund 2006). When they included cofactors in the analysis, two additional QTL were detected on BTA4 and 19. Five QTL for stillbirth were identified on chromosomes BTA3, 7, 12, 18, and 26 in Danish Holstein cattle (Thomasen et al. 2008). In the Norwegian red cattle, Heringstad et al. (2007) reported 3% of stillbirth at first calving and about 1.5% at second and subsequent calvings. Posterior means for direct and maternal heritabilities were 0.07 and 0.08, respectively. They also found that genetic correlations between direct and maternal effects within trait were close to zero.

Using a three generation resource population developed by crossing low and high indexing pigs for ovulation rate and embryonic survival, Cassady et al. (2001) identified two QTL for the number of stillborn pigs on SSC5 and SSC13. Holl et al. (2004) confirmed both QTL for the number of stillborn piglets on SSC5 ($P < 0.10$) and SSC13 ($P < 0.05$) and identified an additional QTL on SSC12 ($P < 0.10$). For the QTL on SSC5, the overdominance expression was observed, whereas dominant and additive expression modes were proposed for the QTL on SSC13 and SSC12, respectively. However, Wilkie et al. (1999) found suggestive linkage at a 5% genome-wide level for the number of stillborn piglets on SSC4. In a recent study, performed by Tribout et al. (2008), additional QTL for the number of stillborn pigs at a chromosome-wide significance level of 5% were found on SSC6, SSC11, and SSC14. The involvement of SSC14 was already suggested based on cytogenetical abnormalities in the offspring of a boar carrying a translocation: rcp(14;15)(q29;q24), which developed from zygotes partly monosomic for chromosome 14 (14q29-qter) and partly trisomic for chromosome 15 (15q24→qter) (Gustavsson and Jönsson 1992). In stillborn piglets with these cytogenetic abnormalities, cleft palate and cardiac septal defect were frequently found.

4.7 Future research directions

The classical way to identify candidate genes for disease phenotypes is based on two strategies, that is, genome-wide scanning and candidate gene approach. Each of both approaches has specific advantages and disadvantages; however, the successful examples of finding causal genes for complex phenotypes in the past do not provide a reliable strategy for a systematic exploration of genetic network causing phenotypic variation in complex diseases. Genome-wide scanning typically does not require any assumption about physiological, biochemical, or developmental causes for the disease phenotype, but requires the establishment of expensive biological resources (reference populations). The typical output of genome-wide scanning experiments is the identification of QTL regions at cM level, which can harbor dozens or even hundreds of genes but allow further search for candidate genes in the targeted regions. In contrast to genome-wide scanning strategy, the candidate gene approach requires profound physiological, biochemical, and developmental knowledge for the identification of promising candidate genes. However, the lack of profound knowledge about the factors involved in pathogenesis represent the major obstacle for further use of candidate gene approach. Especially in cases of very complex, rare diseases, the poor knowledge about factors involved in pathogenesis seriously hampers the dissection of molecular anatomy of the disease. It will be necessary to solve this information bottleneck in order to make a faster progress in the identification of causal mutations for inherited reproductive disorders in the future.

Recently, several strategies have been developed in order to overcome information bottleneck in the classical candidate gene approach. One of the most commonly used strategies combines genome scans with candidate gene approach leading to position-dependent strategy, where search for candidate genes is focused on genomic regions, already identified by the QTL approach. This combined strategy already showed good results in some quantitative traits; however, in a number of situations, no candidate gene could be found in the highly significant QTL interval. This is typical for the cases where multiple genes with low penetrance contribute to the phenotype. Another strategy that can help to resolve the information bottleneck is comparative genomics approach. Especially in the situation when fully sequenced genomes from more or less related species appear in public databases, the use of functional genomics data across species could be very helpful. However, we have to be aware that very similar phenotypes might have quite different genetic background in different species, and this represents the major obstacle for wider use of comparative genomics strategy. Further development of function-dependent strategy, including study of signaling pathways, regulatory networks, transcriptional profiles, and knockout and transgenic animal models, is an additional strategy to integrate different pieces of information in order to elucidate molecular architecture of hereditary reproductive disorders. These approaches, combined with high-throughput genomic technologies, could significantly contribute to efficient widening of the information bottlenecks. One of the most promising strategies proposed just recently is the so-called digital candidate gene approach (DigiCGA) (Zhu and Zhao 2007), a combination of the above-mentioned approaches with bioinformatics tools, which allow efficient data mining and comprehensive integration of different types of evidence. Using this approach and its further development will shed new light in the complex genetic architecture of hereditary reproductive disorders in domestic animals.

References

Amann, R.P. and Veeramachaneni, D.N. 2007. Cryptorchidism in common eutherian mammals. *Reproduction* **133**(3): 541–561.

Ashwell, M.S., Heyen, D.W., Weller, J.I., Ron, M., Sonstegard, T.S., Van Tassell, C.P., and Lewin, H.A. 2005. Detection of quantitative trait loci influencing conformation traits and calving ease in Holstein-Friesian cattle. *Journal of Dairy Science* **88**(11): 4111–4119.

Baker, L.A., Nef, S., Nguyen, M.T., Stapleton, R., Pohl, H., and Parada, L.F. 2002. The insulin-3 gene: Lack of a genetic basis for human cryptorchidism. *Journal of Urology* **167**: 2534–2537.

Barthold, J.S. 2008. Undescended testis: Current theories of etiology. *Current Opinion in Urology* **18**(4): 395–400.

Barthold, J.S., McCahan, S.M., Singh, A.V., Knudsen, T.B., Si, X., Campion, L., Akins, R.E. 2008. Altered expression of muscle- and cytoskeleton-related genes in a rat strain with inherited cryptorchidism. *Journal of Andrology* **29**: 352–366.

Beck, J., Bornemann-Kolatzki, K., Knorr, C., Taeubert, H., and Brenig, B. 2006. Molecular characterization and exclusion of porcine *GUSB* as a candidate gene for congenital hernia inguinalis/scrotalis. *BMC Veterinary Research* **2**: 14.

Betts, D.H. and King, W.A. 2001. Genetic regulation of embryo death and senescence. *Theriogenology* **55**(1): 171–191.

Blood, D.C., Studdert, V.P., and Gay, C.C. 2007. *Saunders Comprehensive Veterinary Dictionary*, 3rd ed. St. Louis, MO: Saunders Elsevier.

Bogatcheva, N.V., Ferlin, A., Feng, S., Truong, A., Gianesello, L., Foresta, C., and Agoulnik, A.I. 2007. T222P mutation of the insulin-like 3 hormone receptor LGR8 is associated with testicular maldescent and hinders receptor expression on the cell surface membrane. *American Journal of Physiology-Endocrinology and Metabolism* **292**: E138–E144.

Bouters, R. and Vandeplassche, M. 1972. Twin gestation in the mare: The incidence of placental vascular anastomoses and their influence on the reproductive performance of heterosexual equine twins. *Journal of Reproduction and Fertility* **29**(1): 149.

Calder, M.D., Manikkam, M., Salfen, B.E., Youngquist, R.S., Lubahn, D.B., Lamberson, W.R., and Garverick, H.A. 2001. Dominant bovine ovarian follicular cysts express increased levels of messenger RNAs for luteinizing hormone receptor and 3b-hydroxysteroid dehydrogenase D4,D5 isomerase compared to normal dominant follicles. *Biology of Reproduction* **65**(2): 471–476.

Campbell, E.M., Nonneman, D.J., Kuehn, L.A., and Rohrer, G.A. 2008. Genetic variation in the mannosidase 2B2 gene and its association with ovulation rate in pigs. *Animal Genetics* **39**(5): 515–519.

Campbell, E.M., Nonneman, D., and Rohrer, G.A. 2003. Fine mapping a quantitative trait locus affecting ovulation rate in swine on chromosome 8. *Journal of Animal Science* **81**(7): 1706–1714.

Canto, P., Escudero, I., Söderlund, D., Nishimura, E., Carranza-Lira, S., Gutierrez, J., Nava, A., and Mendez, J.P. 2003. A novel mutation of the insulin-like 3 gene in patients with cryptorchidism. *J Human Genetics* **48**(2): 86–90.

Cassady, J.P., Johnson, R.K., Pomp, D., Rohrer, G.A., Van Vleck, L.D., Spiegel, E.K., and Gilson, K.M. 2001. Identification of quantitative trait loci affecting reproduction in pigs. *Journal of Animal Science* **79**(3): 623–633.

Cassata, R., Iannuzzi, A., Parma, P., De Lorenzi, L., Peretti, V., Perucatti, A., Iannuzzi, L., and Di Meo, G.P. 2008. Clinical, cytogenetic and molecular evaluation in a dog with bilateral cryptorchidism and hypospadias. *Cytogenetic and Genome Research* **120**(1–2): 140–143.

Cavalieri, J. and Farin, P.W. 1999. Birth of a holstein freemartin calf co-twinned to a schistosomus reflexus fetus. *Theriogenology* **52**(5): 815–826.

Chapwanya, A. 2008. Uterine disease in dairy cows: Classification, diagnosis and key roles for veterinarians. *Irish Veterinary Journal* **61**(3): 183–186.

Clarnette, T.D. and Hutson, J.M. 1997. Exogenous calcitonin gene-related peptide can induce the testis to cross the scrotal septum. *British Journal of Urology* **79**(4): 623–627.

Claxton, J.H. and Yeates, N.T. 1972. The inheritance of cryptorchism in a small crossbred flock of sheep. *The Journal of Heredity* **63**(3): 141–144.

Connell, K.A., Guess, M.K., Chen, H., Andikyan, V., Bercik, R., and Taylor, H.S. 2008. HOXA11 is critical for development and maintenance of uterosacral ligaments and deficient in pelvic prolapse. *The Journal of Clinical Investigation* **118**(3): 1050–1055.

Copp, A.J. 1995. Death before birth: Clues from gene knockouts and mutations. *Trends in Genetics* **11**(3): 87–93.

Cook, B.J., Hasthorpe, S., and Hutson, J.M. 2000. Fusion of childhood inguinal hernia induced by HGF and CGRP via an epithelial transition. *Journal of Pediatric Surgery* **35**(1): 77–81.

Cross, J.C. 2001. Genes regulating embryonic and fetal survival. *Theriogenology* **55**(1): 193–207.

Devin, J.K., Johnson, J.E., Eren, M., Gleaves, L.A., Bradham, W.S., Bloodworth, J.R. Jr., and Vaughan, D.E. 2007. Transgenic overexpression of plasminogen activator inhibitor-1 promotes the development of polycystic ovarian changes in female mice. *Journal of Molecular Endocrinology* **39**(1): 9–16.

Dissen, G.A., Lara, H.E., Leyton, V., Paredes, A., Hill, D.F., Costa, M.E., Martinez-Serrano, A., and Ojeda, S.R. 2000. Intraovarian excess of nerve growth factor increases androgen secretion and disrupts estrous cyclicity in the rat. *Endocrinology* **141**(3): 1073–1082.

Drost, M. 2007. Complications during gestation in the cow. *Theriogenology* **68**(3): 487–491.

Dunn, H.O., Johnson, R.H. Jr., and Quaas, R.L. 1981. Sample size for detection of Y-chromosomes in lymphocytes of possible freemartins. *Cornell Veterinarian* **71**(3): 297–304.

El Houate, B., Rouba, H., Sibai, H., Barakat, A., Chafik, A., Chadli, E.B., Imken, L., Bogatcheva, N.V., Feng, S., Agoulnik, A.I., and McElreavey, K. 2007. Novel mutations involving the INSL3 gene associated with cryptorchidism. *Journal of Urology* **177**: 1947–1951.

Ferlin, A., Bogatcheva, N.V., Gianesello, L., Pepe, A., Vinanzi, C., Agoulnik, A.I., and Foresta, C. 2006. Insulin-like factor 3 gene mutations in testicular dysgenesis syndrome: Clinical and functional characterization. *Molecular Human Reproduction* **12**(6): 401–406.

Ferlin, A., Garolla, A., Bettella, A., Bartoloni, L., Vinanzi, C., Roverato, A., and Foresta, C. 2005. Androgen receptor gene CAG and GGC repeat lengths in cryptorchidism. *European Journal of Endocrinology* **152**(3): 419–425.

Frazer, G.S., Perkins, N.R., and Constable, P.D. 1996. Bovine uterine torsion: 164 hospital referral cases. *Theriogenology* **46**(5): 739–758.

Galan, J.J., Guarducci, E., Nuti, F., Gonzalez, A., Ruiz, M., Ruiz, A., and Krausz, C. 2007. Molecular analysis of estrogen receptor alpha gene AGATA haplotype and SNP12 in European populations: Potential protective effect for cryptorchidism and lack of association with male infertility. *Human Reproduction* **22**(2): 444–449.

Galloway, S.M., Gregan, S.M., Wilson, T., McNatty, K.P., Juengel, J.L., Ritvos, O., and Davis, G.H. 2002. Bmp15 mutations and ovarian function. *Molecular and Cellular Endocrinology* **191**(1): 15–18.

Germerodt, M., Beuermann, C., Rohrer, G.A., Snelling, W.M., Brenig, B., and Knorr, C. 2008. Characterization and linkage mapping of 15 porcine STS markers to fine-map chromosomal regions associated with hernia inguinalis/scrotalis. *Animal Genetics* **39**(6): 671–672.

Gong, Y., Shao, C., Sun, Q., Chen, B., Jiang, Y., Guo, C., Wei, J., and Guo, Y. 1994. Genetic study of indirect inguinal hernia. *Journal of Medical Genetics* **31**(3): 187–192.

Gorlov, I.P., Kamat, A., Bogatcheva, N.V., Jones, E., Lamb, D.J., Truong, A., Bishop, C.E., McElreavey, K., and Agoulnik, A.I. 2002. Mutations of the GREAT gene cause cryptorchidism. *Human Molecular Genetics* **11**(19): 2309–2318.

Gould, K.A., Pandey, J., Lachel, C.M., Murrin, C.R., Flood, L.A., Pennington, K.L., Schaffer, B.S., Tochacek, M., McComb, R.D., Meza, J.L., Wendell, D.L., and Shull, J.D. 2005. Genetic mapping of *Eutr1*, a locus controlling E2-induced pyometritis in the Brown Norway rat, to RNO5. *Mammalian Genome* **16**(11): 854–864.

Graves, T.K., Hansel, W., and Krook, L. 1991. Prolonged gestation in a Holstein cow: Adenohypophyseal aplasia and skeletal pathology in the offspring. *Cornell Veterinarian* **81**(3): 277–294.

Grindflek, E., Moe, M., Taubert, H., Simianer, H., Lien, S., and Moen, T. 2006. Genome-wide linkage analysis of inguinal hernia in pigs using affected sib pairs. *BMC Genetics* **7**: 25.

Gustavsson, I. and Jönsson, L. 1992. Stillborns, partially monosomic and partially trisomic, in the offspring of a boar carrying a translocation: rcp(14;15)(q29;q24). *Hereditas* **117**(1): 31–37.

Hanada H. and Geshi, M. 1995. An aborted fetus with a presumptive 60,XX,rob(7;21) karyotype in Japanese black cattle. *Hereditas* **123**(1): 91–93.

Hanada, H., Geshi, M., and Suzuki, O. 1995. Additional evidence of the formation of unbalanced embryos in cattle with the 7/21 Robertsonian translocation. *Theriogenology* **44**(4): 499–505.

Heringstad, B., Chang, Y.M., Svendsen, M., and Gianola, D. 2007. Genetic analysis of calving difficulty and stillbirth in Norwegian Red cows. *Journal of Dairy Science* **90**(7): 3500–3507.

Hesselink, J.W. 1993. Incidence of hydrometra in dairy goats. *Veterinary Record* **132**(5): 110–112.

Hoedemaker, M. 2008. Anoestrus in dairy cows: Causes and solutions. *Der Praktische Tierarzt* **89**(5): 402.

Holl, J.W., Cassady, J.P., Pomp, D., and Johnson, R.K. 2004. A genome scan for quantitative trait loci and imprinted regions affecting reproduction in pigs. *Journal of Animal Science* **82**(12): 3421–3429.

Holmberg, M. and Andersson-Eklund, L. 2006. Quantitative trait loci affecting fertility and calving traits in Swedish dairy cattle. *Journal of Dairy Science* **89**(9): 3664–3671.

Hooijer, G.A., van Oijen, M.A.A.J., Frankena, K., and Noordhuizen, J.P.T.M. 2003. Milk production parameters in early lactation: Potential risk factors of cystic ovarian disease in Dutch dairy cows. *Livestock Production Science* **81**(1): 25–33.

Ishiguro, K., Baba, E., Torii, R., Tamada, H., Kawate, N., Hatoya, S., Wijewardana, V., Kumagai, D., Sugiura, K., Sawada, T., and Inaba, T. 2007. Reduction of mucin-1 gene expression associated with increased *Escherichia coli* adherence in the canine uterus in the early stage of dioestrus. *Veterinary Journal* **173**(2): 325–332.

Isobe, N., Kitabayashi, M., and Yoshimura, Y. 2008. Expression of vascular endothelial growth factor receptors in bovine cystic follicles. *Reproduction in Domestic Animals* **43**(3): 267–271.

Joosten, I., Sanders, M.F., and Hensen, E.J. 1991. Involvement of major histocompatibility complex class I compatibility between dam and calf in the aetiology of bovine retained placenta. *Animal Genetics* **22**(6): 455–463.

Jubb, K.V.F, Kennedy, P.C., and Palmer, N. 2007. The female genital system. In: Maxie, M.G. (ed.), *Pathology of Domestic Animals*, 5th Edition, Vol. **3**. Edinburgh: Saunders Elsevier, pp. 429–565.

Kelton, D.F., Lissemore, K.D., and Martin, R.E. 1998. Recommendations for recording and calculating the incidence of selected clinical diseases of dairy cattle. *Journal of Dairy Science* **81**(9): 2502–2509.

Kida, K., Baba, E., Torii, R., Kawate, N., Hatoya, S., Wijewardana, V., Sugiura, K., Sawada, T., Tamada, H., and Inaba, T. 2006. Lactoferrin expression in the canine uterus during the estrous cycle and with pyometra. *Theriogenology* **66**(5): 1325–1333.

King, A.H., Jiang, Z.H., Gibson, J.P., Haley, C.S., and Archibald, A.L. 2003. Mapping quantitative trait loci affecting female reproductive traits on porcine chromosome 8. *Biology of Reproduction* **68**(6): 2172–2179.

Knorr, C., Beuermann, C., Laenoi, W., Beck, J., and Brenig, B. 2006. Molecular decipherment of porcine hernia inguinalis/scrotalis. *Proceedings of the 30th ISAG Conference, Porto Seguro, Brazil*, p. 79.

Koskimies, P., Suvanto, M., Nokkala, E., Huhtaniemi, I.T., McLuskey, A.,

Themmen, A.P.N., and Poutanen, M. 2003. Female mice carrying a ubiquitin promoter-Insl3 transgene have descended ovaries and inguinal hernias but normal fertility. *Molecular and Cellular Endocrinology* **206**(1–2): 159–166.

Krausz, C., Quintana-Murci, L., Fellous, M., Siffroi, J.P., and McElreavey, K. 2000. Absence of mutations involving the INSL3 gene in human idiopathic cryptorchidism. *Molecular Human Reproduction* **6**: 298–302.

Kubota, Y., Temelcos, C., Bathgate, R.A., Smith, K.J., Scott, D., Zhao, C., and Hutson, J.M. 2002. The role of insulin 3, testosterone, Müllerian inhibiting substance and relaxin in rat gubernacular growth. *Molecular Human Reproduction* **8**(10): 900–905.

Kühn, Ch., Bennewitz, J., Reinsch, N., Xu, N., Thomsen, H., Looft, C., Brockmann, G.A., Schwerin, M., Weimann, C., Hiendleder, S., Erhardt, G., Medjugorac, I., Förster, M., Brenig, B., Reinhardt, F., Reents, R., Russ, I., Averdunk, G., Blümel, J., and Kalm, E. 2003. Quantitative trait loci mapping of functional traits in the German Holstein cattle population. *Journal of Dairy Science* **86**(1): 360–368.

Kunstýr, I., Matthiesen, T., Gärtner, K., Maess, J., and Heimann, W. 1982. Post-mating non-infectious hydrometra in BALB/c: Bom mice. *Laboratory Animals* **16**(1): 51–55.

Llambi, S. and Postiglioni, A. 1994. Localization of the fragile X chromosome break points in Holstein-Friesian cattle (*Bos taurus*). *Theriogenology* **42**(5): 789–794.

Marcum, J.B. 1974. The Freemartin syndrome. *Animal Breeding Abstracts* **42**: 227–242.

McNatty, K.P., Galloway, S.M., Wilson, T., Smith, P., Hudson, N.L., O'Connell, A., Bibby, A.H., Heath, D.A., Davis, G.H., Hanrahan, J.P., and Juengel, J.L. 2005a. Physiological effects of major genes affecting ovulation rate in sheep. *Genetics Selection Evolution* **37**(Supplement 1): S25–S38.

McNatty, K.P., Smith, P., Moore, L.G., Reader, K., Lun, S., Hanrahan, J.P., Groome, N.P., Laitinen, M., Ritvos, O., and Juengel, J.L. 2005b. Oocyte-expressed genes affecting ovulation rate. *Molecular and Cellular Endocrinology* **234**(1–2): 57–66.

Mee, J.F. 2008. Prevalence and risk factors for dystocia in dairy cattle: a review. *The Veterinary Journal* **176**(1): 93–101.

Mikami, H. and Fredeen, H.T. 1979. A genetic study of cryptorchidism and scrotal hernia in pigs. *Canadian Journal of Genetics and Cytology* **21**(1): 9–19.

Nguyen, M.T., Delaney, D.P., and Kolon, T.F. 2009. Gene expression alterations in cryptorchid males using spermatozoal microarray analysis. *Fertility Sterility* **92**(1): 182–187.

Nikolova, G., Lee, H., Berkovitz, S., Nelson, S., Sinsheimer, J., Vilain, E., and Rodríguez, L.V. 2007. Sequence variant in the laminin gamma1 (*LAMC1*) gene associated with familial pelvic organ prolapse. *Human Genetics* **120**(6): 847–856.

Nonneman, D.J. and Rohrer, G.A. 2003. Comparative mapping of a region on chromosome 10 containing QTL for reproduction in swine. *Animal Genetics* **34**(1): 42–46.

Nonneman, D.J., Wise, T.H., Ford, J.J., Kuehn, L.A., and Rohrer, G.A. 2006. Characterization of the aldo-keto reductase 1C gene cluster on pig chromosome 10: Possible associations with reproductive traits. *BMC Veterinary Research* **2**: 28.

Notter, D.R. 2008. Genetic aspects of reproduction in sheep. *Reproduction in*

Domestic Animals **43**(Supplement 2): 122–128.

Nuti, F., Marinari, E., Erdei, E., El-Hamshari, M., Echavarria, M.G., Ars, E., Balercia, G., Merksz, M., Giachini C., Shaeer K.Z., Forti, G., Ruiz-Castane, E., and Krausz C. 2008. The leucine-rich repeat-containing G protein-coupled receptor 8 gene T222P mutation does not cause cryptorchidism. *Journal of Clinical Endocrinology and Metabolism* **93**: 1072–1076.

Opsomer, G., Wensing, T., Laevens, H., Coryn, M., and de Kruif, A. 1999. Insulin resistance: The link between metabolic disorders and cystic ovarian disease in high yielding dairy cows? *Animal Reproduction Science* **56**(3–4): 211–222.

Ortega, H.H., Palomar, M.M., Acosta, J.C., Salvetti, N.R., Dallard, B.E., Lorente, J.A., Barbeito, C.G., and Gimeno, E.J. 2008. Insulin-like growth factor I in sera, ovarian follicles and follicular fluid of cows with spontaneous or induced cystic ovarian disease. *Research in Veterinary Science* **84**(3): 419–427.

Ortega, H.H., Salvetti, N.R., Amable, P., Dallard, B.E., Baravalle, C., Barbeito, C.G., and Gimeno, E.J. 2007. Intraovarian localization of growth factors in induced cystic ovaries in rats. *Anatomia Histologia Embryologia* **36**(2): 94–102.

Pandey, J., Gould, K.A., McComb, R.D., Shull, J.D., and Wendell, D.L. 2005. Localization of Eutr2, a locus controlling susceptibility to DES-induced uterine inflammation and pyometritis, to RNO5 using a congenic rat strain. *Mammalian Genome* **16**(11): 865–872.

Parkinson, T.J., Smith, K.C., Long, S.E., Douthwaite, J.A., Mann, G.E., and Knight, P.G. 2001. Inter-relationships among gonadotrophins, reproductive steroids and inhibin in freemartin ewes. *Reproduction* **122**(3): 397–409.

Pemberton, A.D., John, H.A., Ricketts, S.W., Rossdale, P.D., and Scott, A.M. 1994. Investigation of association between alpha-1 proteinase inhibitor haplotype and endometritis in the thoroughbred mare. *Equine Veterinary Journal* **26**(2): 122–124.

Pinheiro, L.E., Carvalho, T.B., Oliveira, D.A., Popescu, C.P., and Basrur, P.K. 1995. A 4/21 tandem fusion in cattle. *Hereditas* **122**(2): 99–102.

Razpet, A. 2007. *Using sequenced mammalian genomes for synteny block identification in unfinished genomes*. PhD thesis, University of Ljubljana.

Rosch, R., Klinge, U., Si, Z., Junge, K., Klosterhalfen, B., and Schumpelick, V. 2002. A role for the collagen I/III and MMP-1/-13 genes in primary inguinal hernia? *BMC Medical Genetics* **3**: 2.

Rota, A., Ballarin, C., Vigier, B., Cozzi, B., and Rey, R. 2002. Age dependent changes in plasma anti-Müllerian hormone concentrations in the bovine male, female, and freemartin from birth to puberty: Relationship between testosterone production and influence on sex differentiation. *General and Comparative Endocrinology* **129**(1): 39–44.

Saba, N., Cunningham, N.F., and Millar, P.G. 1975. Plasma progesterone, androstenedione and testosterone concentrations in freemartin heifers. *Journal of Reproduction and Fertility* **45**(1): 37–45.

Salvetti, N.R., Acosta, J.C., Gimeno, E.J., Müller, L.A., Mazzini, R.A., Taboada, A.F., and Ortega, H.H. 2007. Estrogen receptors alpha and beta and progesterone receptors in normal bovine ovarian follicles and cystic ovarian disease. *Veterinary Pathology* **44**(3): 373–378.

Schmutz, S.M., Moker, J.S., Clark, E.G., and Orr, J.P. 1996. Chromosomal aneuploidy associated with spontaneous abortions and neonatal losses in cattle. *Journal of Veterinary Diagnostic Investigation* **8**(1): 91–95.

Schmutz, S.M., Moker, J.S., Pawlyshyn, V., Haugen, B., and Clark, E.G. 1997. Fertility effects of the 14;20 Robertsonian translocation in cattle. *Theriogenology* **47**(4): 815–823.

Sevinga, M., Vrijenhoek, T., Hesselinks, J.W., Barkema, H.W., and Groen, A.F. 2004. Effect of inbreeding on the incidence of retained placenta in Friesian horses. *Journal of Animal Science* **82**(4): 982–986.

Sharif, S., Mallard, B.A., Wilkie, B.N., Sargeant, J.M., Scott, H.M., Dekkers, J.C., and Leslie, K.E. 1998. Associations of the bovine major histocompatibility complex DRB3 (BoLA-DRB3) alleles with occurrence of disease and milk somatic cell score in Canadian dairy cattle. *Animal Genetics* **29**(3): 185–193.

Sheldon, I.M., Lewis, G.S., LeBlanc, S., and Gilbert, R.O. 2006. Defining postpartum uterine disease in cattle. *Theriogenology* **65**(8): 1516–1530.

Shore, L. and Shemesh, M. 1981. Altered steroidogenesis by the fetal bovine freemartin ovary. *Journal of Reproduction and Fertility* **63**(2): 309–314.

Silva-Ramos, M., Oliveira, J.M., Cabeda, J.M., Reis, A., Soares, J., and Pimenta, A. 2006. The CAG repeat within the androgen receptor gene and its relationship to cryptorchidism. *International Brazilian Journal of Urology* **32**: 330–334; discussion 335.

Simon, A.M., Goodenough, D.A., Li, E., and Paul, D.L. 1997. Female infertility in mice lacking connexin 37. *Nature* **385**(6616): 525–529.

Simoni, M., Tüttelmann, F., Michel, C., Böckenfeld, Y., Nieschlag, E., and Gromoll, J. 2008. Polymorphisms of the luteinizing hormone/chorionic gonadotropin receptor gene: Association with maldescended testes and male infertility. *Pharmacogenetics and Genomics* **18**(3): 193–200.

Smith, M.P. and Sparkes, R.S. 1968. Familial inguinal hernia. *Surgery* **57**: 809–812.

Soller, M., Padeh, B., Wysoki, M., and Ayalon, N. 1969. Cytogenetics of Saanen goats showing abnormal development of the reproductive tract associated with the dominant gene for polledness. *Cytogenetics* **8**: 51–67.

Sugiura, K., Nishikawa, M., Ishiguro, K., Tajima, T., Inaba, M., Torii, R., Hatoya, S., Wijewardana, V., Kumagai, D., Tamada, H., Sawada, T., Ikehara, S., and Inaba, T. 2004. Effect of ovarian hormones on periodical changes in immune resistance associated with estrous cycle in the beagle bitch. *Immunobiology* **209**(8): 619–627.

Takahashi, T., Takahashi, I., Komatsu, M., Matsuda, J., and Takada, G. 2001. Ala/Thr60 variant of the Leydig insulin-like hormone is not associated with cryptorchidism in the Japanese population. *Pediatrics International* **43**: 256–258.

Teixeira, J., Maheswaran, S., and Donahoe, P.K. 2001. Müllerian inhibiting substance: An instructive developmental hormone with diagnostic and possible therapeutic applications. *Endocrine Reviews* **22**(5): 657–674.

Thaller, G., Dempfle, L., and Hoeschele, I. 1996. Maximum likelihood analysis of rare binary traits under different modes of inheritance. *Genetics* **143**(4): 1819–1829.

Thomasen, J.R., Guldbrandtsen, B., Sørensen, P., Thomsen, B., and Lund, M.S. 2008. Quantitative trait loci affecting calving traits in Danish Holstein cattle. *Journal of Dairy Science* **91**(5): 2098–2105.

Tribout, T., Iannuccelli, N., Druet, T., Gilbert, H., Riquet, J., Gueblez, R.,

Mercat, M.J., Bidanel, J.P., Milan, D., and Le Roy, P. 2008. Detection of quantitative trait loci for reproduction and production traits in Large White and French Landrace pig populations. *Genetics Selection Evolution* **40**(1): 61–78.

Troedsson, M.H. 1999. Uterine clearance and resistance to persistent endometritis in the mare. *Theriogenology* **52**(3): 461–471.

Vanholder, T., Opsomer, G., and de Kruif, A. 2006. Aetiology and pathogenesis of cystic ovarian follicles in dairy cattle: A review. *Reproduction Nutrition Development* **46**(2): 105–119.

Wachtel, S.S., Hall, J.L., Müller, U., and Chaganti, R.S. 1980. Serum-borne H-Y antigen in the fetal bovine freemartin. *Cell* **21**(3): 917–926.

Wada, Y., Okada, M., Fukami, M., Sasagawa, I., and Ogata, T. 2006. Association of cryptorchidism with Gly146Ala polymorphism in the gene for steroidogenic factor-1. *Fertility and Sterility* **85**(3): 787–790.

Wang, Y., Barthold, J., Figueroa, E., González, R., Noh, P.H., Wang, M., and Manson, J. 2008. Analysis of five single nucleotide Polymorphisms in the ESR1 gene in cryptorchidism. *Birth Defects Research. Part A, Clinical and Molecular Teratology* **82**(6): 482–485.

Williams, G.A., Ott, T.L., Michal, J.J., Gaskins, C.T., Wright, R.W. Jr., Daniels, T.F., and Jiang, Z. 2007. Development of a model for mapping cryptorchidism in sheep and initial evidence for association of INSL3 with the defect. *Animal Genetics* **38**(2): 189–191.

Wilkie, P.J., Paszek, A.A., Beattie, C.W., Alexander, L.J., Wheeler, M.B., and Schook, L.B. 1999. A genomic scan of porcine reproductive traits reveals possible quantitative trait loci (QTLs) for number of corpora lutea. *Mammalian Genome* **10**(6): 573–578.

Wilkes, P.R., Munro, I.B., and Wijeratne, W.V. 1978. Studies on a sheep freemartin. *The Veterinary Record* **102**(7): 140–142.

Yamazawa, K., Wada, Y., Sasagawa, I., Aoki, K., Ueoka, K., and Ogata, T. 2007. Mutation and polymorphism analyses of INSL3 and LGR8/GREAT in 62 Japanese patients with cryptorchidism. *Hormone Research* **67**: 73–76.

Yoshida, R., Fukami, M., Sasagawa, I., Hasegawa, T., Kamatani, N., and Ogata, T. 2005. Association of cryptorchidism with a specific haplotype of the estrogen receptor alpha gene: Implication for the susceptibility to estrogenic environmental endocrine disruptors. *Journal of Clinical Endocrinology and Metabolism* **90**(8): 4716–4721.

Zdunczyk, S., Janowski, T., and Ras, M. 2005. Current views on the phenomenon of silent heat in cows. *Medycyna Weterynaryjna* **61**(7): 726–729.

Zhang, T., Buoen, L.C., Seguin, B.E., Ruth, G.R., and Weber, A.F. 1994. Diagnosis of freemartinism in cattle: The need for clinical and cytogenic evaluation. *Journal of the American Veterinary Medical Association* **204**(10): 1672–1675.

Zheng, H., Si, Z., Kasperk, R., Bhardwaj, R.S., Schumpelick, V., Klinge, U., and Klosterhalfen, B. 2002. Recurrent inguinal hernia: Disease of the collagen matrix? *World Journal of Surgery* **26**(4): 401–408.

Zhu, M. and Zhao, S. 2007. Candidate gene identification approach: Progress and challenges. *International Journal of Biological Science* **3**(7): 420–427.

5

Genomics of Reproductive Diseases in Cattle and Swine

Holly Neibergs and Ricardo Zanella

5.1 Introduction

Differences in host response to reproductive diseases have not been extensively studied, although the economic loss attributed to these diseases is large (Bishop et al. 2002; Morris 2007). Vaccination for many of these diseases is either not effective or interferes with the determination of disease status. Many of these diseases also do not respond well to available treatments, and so new approaches or alternatives to addressing reproductive diseases are needed.

It is common to observe a population of animals that have been exposed to an infectious disease and detect a spectrum of susceptibility and severity of disease. Cattle and pigs show considerable variability in their responses to disease challenges. Breed or line differences have been noted for bovine respiratory disease (BRD), bovine paratuberculosis, and Aujeszky's disease, providing evidence for a role for genetics in the infection of animals with these diseases (Lundeheim 1979, 1988; Muggli-Cockett et al. 1992; Halbur et al. 1998; Petry et al. 2005; Snowder et al. 2005, 2006; Gonda et al. 2006; Vincent et al. 2006). Many factors influence animal health and thus complicate the study of the role of the host's genetics in the infection process. In addition, defining the phenotype for a disease may be challenging, as most ill animals will present with an assortment of clinical signs that may actually represent one or more diseases. Not all ill animals will have clinical symptoms, and not all animals are exposed to the same level of pathogen. Therefore, thc identification of loci associated with disease traits is complex. The selection of animals for disease resistance or tolerance offers a new approach to understanding and reducing the prevalence of reproductive diseases in domestic animals. Although there are challenges in the identification of loci associated with reproductive diseases, there has been some initial progress in understanding the role of genetics in bovine brucellosis,

BRD, bovine paratuberculosis, porcine leptospirosis, porcine respiratory and reproductive syndrome (PRRS), and Aujeszky's disease in pigs.

5.2 Bovine paratuberculosis

5.2.1 *Causative agent*

Bovine paratuberculosis, also known as Johne's disease, was first described in the late 19th century. The name Johne's disease comes from the work of H.A. Johne and L. Frothingham, who demonstrated a connection between cattle enteritis and the presence of acid-fast microorganisms in sections of the intestinal mucosa (Cocito et al. 1994). In 1906, Bang distinguished between tuberculosis and non-tuberculosis enteritis and proposed the name pseudotuberculosis enteritis. The identification of the etiologic agent is attributed to F.W. Twort, who succeeded in cultivating and characterizing a mycobacterium, which, in 1914, was shown to produce experimental enteritis in cattle (Cocito et al. 1994). After the full characterization of *Mycobacterium avium* subspecies *paratuberculosis* (*MAP*) as a distinct species within the genus *Mycobacterium*, the disease was renamed paratuberculosis (Chiodini et al. 1984; Kreeger 1991; Hermon-Taylor et al. 2000).

Bovine paratuberculosis is a bacterial infectious disease that is estimated to cost U.S. Agriculture $1.5 billion annually (Sweeney 1996). These losses are primarily a result of reduced milk yield, reduced slaughter values, increased premature and involuntary culling, decreased fertility, increased mortality rate, and increased susceptibility to other diseases (Whittington and Sergeant 2001). *MAP* has been found in 25–75% of Crohn's disease patients but in less than 5% of individuals without Crohn's (Chiodini 1989; Autschbach et al. 2005). Both bovine paratuberculosis and Crohn's disease are increasing worldwide in industrialized countries (Food Standards Australia New Zealand 2004). It is not known if the association of Crohn's and *MAP* is the same as the association of *MAP* and paratuberculosis in cattle. However, there is a concern that virtually all known mycobacterial pathogens are transmissible to humans and have the ability to cause disease. If *MAP* could be transferred from cattle to humans, milk or meat might be the vehicle (Millar et al. 1996; Food Standards Agency, Advisory Committee on the Microbiological Safety of Food 2000). Conflicting studies have been reported as to the efficacy of pasteurization in killing the bacteria (Stabel et al. 1997).

5.2.2 *Prevalence*

Bovine paratuberculosis has been recognized as a major disease concern in many countries, including the United States, because there is no known cure and it is easily transmitted to other animals (Chiodini et al. 1984). In spite of efforts to decrease the prevalence of paratuberculosis, U.S. dairy herds with infected animals have increased from 22% in 1997 to 67% in 2007 (Animal and Plant Health Inspection Service, USDA 2008). In England, an epidemiological study demonstrated that approximately 17.4% of cattle presented with clinical paratuberculosis (Lilenbaum et al. 2007). Enzyme-linked immunosorbent assay (ELISA) testing identified that 6% of dairy cattle in Belgium were seropositive for paratuberculosis. In the Netherlands, 31–71% of dairy herds are infected (Lilenbaum et al. 2007). Of the infected herds, 2.7–6.9% of the animals are affected with paratuberculosis. A 2-year epidemiological study in Slovenia indicated a prevalence of paratuberculosis in 11.59%

among national cattle (Sockett et al. 1992). Seventy percent of herds in Denmark tested positive for paratuberculosis as determined by a serological study of bulk-tanked milk from 900 dairy herds (Juárez et al. 2001).

5.2.3 Transmission

Infection of calves can take place by oral ingestion of *MAP* from contaminated manure, colostrum or milk, pasture, water, or other feed (Chiodini et al. 1984; Sweeney 1996). Contamination of the environment occurs from the shedding of *MAP* primarily in the feces of infected animals. It is generally assumed that animals start shedding the bacteria at about 2 years of age, and therefore do not become infectious before that age (Chiodini et al. 1984). However, this is most likely a dose-dependent observation, as fecal shedding of *MAP* has been shown shortly after oral inoculation with highly infectious doses (Crossley et al. 2005). Cattle infected with *MAP* will commonly exhibit a delay in fecal shedding for a few months to years, and the concentration of *MAP* in the feces may extend from a few colony-forming units to millions of colony-forming units per gram of feces (R. Whitlock, pers. comm.). In addition to calves, adult cows may also become infected through oral exposure or through *MAP*-infected semen (Ayele et al. 2004). *MAP* infection in adult cattle is not well understood, but animals exposed for the first time as adults may develop clinical disease while others may become carriers of the organism without manifesting clinical signs (Larsen et al. 1975).

The eradication of this disease is exacerbated by the hardiness of *MAP*. It survives for 11 months in bovine feces and black soil and can exist in temperatures as low as 14°C for at least a year (Chiodini et al. 1984). American bison, tule elk, and white-tailed deer have been implicated as potential carriers that may influence incidence and spread among wild and domestic ruminants (Libke and Walton 1975; Chiodini and Van Kruiningen 1983). Monogastric animals and birds can become infected experimentally; however, clinical disease usually does not develop (Amand 1974; Libke and Walton 1975; Williams and Spraker 1979; Jessup et al. 1981; Chiodini and Van Kruiningen 1983).

5.2.4 Clinical presentation

Paratuberculosis is a chronic, progressive granulomatous enteric disease of primarily ruminants. Clinical disease is characterized by diarrhea, weight loss, debilitation, and eventual death. Paratuberculosis is a disease that typically exhibits a latent period after the animals have been infected and before there are clinical signs, fecal shedding, or antibody production. By the time a single infected animal is identified, 38–42% of the herd may be infected (Larsen et al. 1963; Delisle et al. 1980). Annual death losses within a herd may be as high as 3–10%. Clinical disease is usually associated with adult (>2-year-old) animals; however, animals as young as 4 months old may occasionally develop clinical signs (Smyth and Christie 1950). Generally, macroscopic and histological lesions are restricted to the intestines, associated lymph nodes, and, occasionally, the liver (Buergelt et al. 1978).

5.2.5 Genetics

Resistance or susceptibility to *MAP* has been shown to have a hereditary component in cattle and mice. Resistance to *MAP* has been shown in mice to be associated with the *Bcg* gene or *nramp1*, which encodes the natural resistance-associated macrophage protein (Skamene et al. 1982; Skamene 1989;

Frelier et al. 1990). C57/B6 and BALB/c mice have the susceptible allele of *Bcg* and are susceptible to *MAP* infections, while the C3H/HeJ strain is resistant to *MAP* (Chandler 1962; Chiodini and Buergelt 1993; Tanaka et al. 1994; Veazey et al. 1995a,b). In cattle, *MAP*-susceptible Holstein sire lines have been found to be infected twice as often as resistant lines (Gonda et al. 2006). Heritability studies have been conducted on the presence or absence of disease based on postmortem tissue, ELISA and combined ELISA–fecal culture tests. In a Dutch study, the heritability of paratuberculosis infection was evaluated among vaccinated and unvaccinated animals based on diagnoses of postmortem examinations (Koets et al. 2000). A heritability of 0.09, 0.01, and 0.06 was found for vaccinated, unvaccinated, and all cows, respectively. A second study estimated the heritability of antibody response using a bivariate model with daily milk yield and optical density values from milk ELISAs (Mortensen et al. 2004). Mortensen and coworkers (2004) estimated the heritability to be 0.102 with the bivariate model and 0.091 when a sire model was used. Gonda and colleagues (2006) estimated the heritability of paratuberculosis as 0.153 based on fecal culture diagnostic testing, 0.159 based on ELISA, and 0.102 from the combined antibody and fecal culture tests.

Consistent with the number of heritability studies conducted, limited investigations have been conducted to identify loci associated with bovine paratuberculosis. Two groups have reported searches for loci responsible for susceptibility of cattle to paratuberculosis. Gonda and coworkers (2006) undertook a genome-wide linkage study using ELISA, fecal culture, or both to diagnose infected animals. In this study, microsatellites were used to genotype three half-sib families. The number of informative markers ranged from 151 to 176 within the three families. Genotypes of "positive" and "negative" animals were pooled, and allele frequencies were estimated. Eight chromosomal regions were associated with the pooled samples (bovine chromosomes 7, 10, 12, 14, 15, 18, 20, and 25). The eight chromosomal regions associated with *MAP* infection in pooled genotypes were further tested. Individual genotypes of the daughters were determined for three to five microsatellites within 15 cM of the markers identified in the pooled samples. Subsequently, only BTA 20 was found to be linked ($P = 0.0319$) in a chromosome-wide analysis in one of the sire families. Taylor and coworkers (2006) evaluated the allele frequencies of *CARD15* in 30 unrelated unaffected animals and 11 affected animals without finding evidence for an association. Settles and coworkers (2009) demonstrated evidence of genetic association of loci to infection of paratuberculosis based on tissue culture for disease diagnosis and the use of the *Illumina BovineSNP50 BeadChip*.

5.3 BRD

5.3.1 Causative agent

BRD, also known as shipping fever, is a general term that describes infectious pneumonia resulting in pulmonary lesions. It is a complex of diseases with many types of infection, each having its own causes, clinical signs, and genetic factors. The viruses associated with BRD include bovine herpesvirus 1, bovine respiratory syncytial virus, bovine viral diarrhea virus (BVDV), bovine respiratory coronaviruses, and parainfluenza 3 virus. Bacteria also play a prominent role in this disease and include *Mannheimia haemolytica*, *Mycoplasma bovis*, *Pasteurella multocida*, and *Haemophilus somni*. Often, severe BRD is associated with concurrent infections of more than one of these pathogens. The more common pathogens of BRD, *M. haemolytica*, *P. multocida*, *M. bovis*, *H. somni*, bovine respiratory syncytial virus, bovine herpesvirus type 1, and BVDV, are briefly described below.

The major agent of BRD is *M. haemolytica* (Rice et al. 2008). It is the primary bacterium isolated from feedlot cattle with respiratory disease and also plays a prominent role in pneumonia in neonatal calves (Kiorpes et al. 1988; Van Donkersgoed et al. 1993; Ames 1997). This bacterium is most effective as a pathogen when host defenses are burdened by stress or infection with other pathogens. This is consistent with studies demonstrating that *M. haemolytica's* greatest effect is on recently weaned beef calves shortly after entry into feedlots (Mosier et al. 1989; Wilson 1989).

Another primary bacterial pathogen associated with BRD is *P. multocida*. It is a gram-negative bacterium that results in pneumonia of young dairy calves and recently weaned beef calves (Lillie 1974; Watts et al. 1994; Fulton et al. 2004; Welsh et al. 2004). Infection with *P. multocida* is associated with the combination of stress or other viral or bacterial infections (Dabo et al. 2008).

Histophilus somni is a gram-negative bacterium that causes BRD in cattle and respiratory disease in sheep, bison, and bighorn sheep (Corbeil 2008). It frequently exists in an asymptomatic state in the reproductive and respiratory mucosa (Humphrey et al. 1982; Humphrey and Stephens 1983). This pathogen is most problematic in the feedlot, although it sometimes manifests as BRD in young calves (Humphrey and Stephens 1983; Harris and Janzen 1989).

M. bovis, first associated with BRD in 1976, is increasingly linked with pneumonia in feedlot cattle and dairy calves (Gourlay et al. 1976; Caswell and Archambault 2007). *M. bovis* is a gram-positive facultative anaerobic bacterium that lacks the ability to form a cell wall resulting in its characteristic pleomorphic shapes.

Bovine respiratory syncytial virus is an important viral pathogen of the BRD complex. It has been estimated that more than 60% of epizootic respiratory diseases in dairy herds and 70% in beef herds may be due to bovine respiratory syncytial virus (Meyer et al. 2008). Bovine respiratory syncytial virus is closely related to human respiratory syncytial virus. This pathogen is a single negative-strand enveloped RNA virus. Bovine respiratory syncytial virus has been associated with concurrent infections with *M. haemolytica*, *P. multocida*, and *H. somni* (Gershwin 2008).

Bovine herpesvirus type 1 can be categorized into three subtypes (Metzler et al. 1985). Subtype 1 virus isolates are responsible for infectious bovine rhinotracheitis and are often found in the respiratory tract as well as in aborted fetuses. Subtype 2b is associated with BRD but not abortions, whereas subtype 2a is responsible for a wide range of clinical presentations, including abortions. Subtype 1 strains are found in Europe, North America, and South America and are more pathogenic than type 2 strains (Edwards et al. 1990; Jones and Chowdhury 2007). Subtype 2a is found in Brazil (Van Oirschot 1995). In feedlot cattle, subtype 1 strains are the most common with an incubation period of 2–6 days (Yates 1982; Jones and Chowdhury 2007).

BVDV is a common virus among cattle. BVDV is a small positive-sense RNA single-stranded, enveloped, pestivirus that is prone to high mutation rates (Ridpath 2003). BVDV occurs in domestic, wild ruminants and swine (Becher et al. 1997). Pestiviruses rarely survive in the environment for more than 2 weeks and are readily inactivated by disinfectants (Kelling 2007). Isolates of BVDV are divided into biogroups (cytopathic and non-cytopathic) based on their ability to cause vacuolization and lysis of host cells *in vitro*. Only non-cytopathic isolates result in persistently infected animals, although both biogroups cause acute disease. The genotype, determined by comparison of genetic sequences, divides BVDV into at least four groups: BVD1, BVDV2, BDV, and CSFV (Pellerin et al. 1994; Ridpath et al. 1994; Ridpath and Bolin 1995, 1997). It is estimated that up to 4% of herds in the United States have persistently infected calves, and that 0.1–0.3% of all cattle are persistently infected (Wittum et al. 2001; Loneragan et al. 2005; O'Connor et al. 2007).

5.3.2 *Prevalence*

BRD contributes significantly to beef cattle mortality in the United States. Of all beef calves that were born alive that did not survive to weaning, 29.6% were associated with BRD (National Agricultural Statistics Service, Agricultural Statistics Board, United States Department of Agriculture 2006). This represents an association of 28.7% of all cattle deaths with respiratory disease accounting for over 1.1 million animal deaths. In a study of over 10,000 beef calves in Nebraska, 24% experienced at least one episode of respiratory disease during the first year of life with frequencies of BRD varying from 14% to 38% over a 6-year study (Muggli-Cockett et al. 1992). In a larger study of 110,412 calves born from 1993 to 2001, the incidence of BRD ranged from 3.3% to almost 23.6% with an average annual incidence of 10.5% (Snowder et al. 2005).

The incidence rates of BRD have been shown to differ among beef cattle breeds. Snowder and colleagues (2005) found that the highest incidence rates of BRD were for Braunvieh (19%), and a quarter-Braunvieh composite (17%) among nine pure breeds (Angus, Braunvieh, Charolais, Gelbvieh, Hereford, Limousin, Pinzgauer, Red Poll, and Simmental) and three composite breeds. These results may be due to a higher incidence of calving difficulty, as calves resulting from births requiring assistance were shown to be more susceptible to BRD (Snowder et al. 2005). Although Braunvieh calves were more likely to suffer from BRD, they had a lower mortality rate from the disease (9%) than all other breeds with the exception of Limousin (7%). The overall average mortality rate of BRD was 13% across all breeds (Snowder et al. 2005).

5.3.3 *Transmission*

Most of the pathogens associated with BRD are commensal organisms present on mucosal surfaces, including the mammary gland, respiratory, intestinal, and genital tracts. When animals are exposed to environmental stressors, such as weaning, feed changes, comingling of animals from other sources, adverse weather, presence of other pathogenic organisms, or transport over long distances, a disease state may result (Farley 1932). Transmission may occur through direct contact of infected animals or through infected body fluids. Transplacental transmission of BVDV may occur when a pregnant cow is acutely infected during pregnancy or if the dam is chronically infected herself (Stokstad et al. 2003; Bielefeldt-Ohmann et al. 2008). Persistently infected cattle will shed large amounts of virus, which will serve to further infect the herd (Moerman et al. 1993). Transmission of BVDV for acute cases primarily occurs by inhalation or ingestion of material contaminated with infected body fluids from infected animals (Houe 1995; Grooms 2004).

5.3.4 Clinical presentation

Adams and coworkers (1959) defined BRD as an acute infection of cattle that was characterized by dyspnea, fever, and fibrinous pneumonia and was of unknown cause. It is now known that BRD may be caused by many pathogens. *M. haemolytica* infections present as respiratory infections with nasal discharge, loss of appetite, cough, respiratory distress, fibrinous pleuropneumonia, and weight loss (Friend et al. 1977). Clinical symptoms of BRD associated with *P. multocida* include depression, loss of appetite, cough, nasal discharge, and fever (Dabo et al. 2008). Lung lesions may result in an acute to subacute bronchopneumonia that may be associated with pleuritis. Cattle with bovine respiratory syncytial virus demonstrate pyrexia, anorexia, depression, cough, increased respiratory rate, and dyspnea with open-mouthed breathing and wheezing in severe cases. *H. somni* infection is characterized by septicemia, thrombotic meningoencephalitis, myocarditis, arthritis, abortion, and infertility (Corbeil 2008).

Clinical signs of infection with *M. bovis* are pneumonia, arthritis, tenosynovitis, mastitis, otitis media in calves, keratoconjunctivitis, decubital abscesses, metritis, abortion, infertility, seminal vesiculitis, and meningitis. These symptoms occur in 8–10 days in experimentally infected animals (Stipkovits et al. 2000).

Cattle infected with bovine herpesvirus type 1 can present with upper respiratory tract disorders, conjunctivitis, genital disorders, and immune suppression, which can lead to BRD (Jones and Chowdhury 2007). The upper respiratory disease may include high fever, anorexia, coughing, excessive salivation, nasal discharge, conjunctivitis with lacrimal discharge, inflamed nares, and, occasionally, dyspnea. Without the concurrence of bacterial pneumonia, recovery may occur in 4–5 days after the onset of symptoms. Abortion may occur during the respiratory phase of the disease or up to 100 days after infection (Jones and Chowdhury 2007). The clinical presentation of genital infections in cows includes frequent urination, swollen vulva, and ulcers on the mucosal surface. In bulls, ulcers may occur on the penis and prepuce. Without the concurrence of bacterial infections, animals usually recover within 2 weeks.

BVDV infections may result in acute illness (bovine viral diarrhea) or chronic disease (mucosal disease). Acute disease occurs postnatally in immunocompetent animals. The severity of the disease varies from mild forms having low mortality, minimal mucosal lesions, pyrexia, nasal discharge, and transient leucopenia, to more severe forms with thrombocytopenia, hemorrhages, and high mortality rates. Immunosuppression and enteritis are characteristic of this disease, which provides commensal pathogens an opportunity to develop into a disease state (Ellis et al. 1988; Welsh et al. 1995; Brodersen and Kelling 1998; Liu et al. 1999). Viremia lasts for 3–10 days on average. Chronic disease may result when susceptible pregnant cattle are exposed to the virus when the developing fetus is immunologically naive (Coria and McClunkin 1978). This typically occurs during the third or fourth month of gestation (Moennig and Liess 1995; Bielefeldt-Ohmann et al. 2008). Transplacental infections may lead to embryonic or fetal death, abortion, congenital malformations, or development of immunotolerance (Moennig and Liess 1995). Fetuses that develop immunotolerance will be chronically infected throughout their lifetimes and serve as reservoirs for transmission of the disease (McClurkin et al. 1984; Brock et al. 1991; Wittum et al. 2001). These animals are referred to as persistently infected or BVD-PI cattle. Mucosal disease is associated with high mortality rates in cattle between 6 months and 2 years of age (Kelling 2007). Clinical manifestations include anorexia; enteritis; thymus atrophy; enlarged lymph nodes; pyrexia; diarrhea; ulcerations of the muzzle, lips, buccal mucosa, esophagus, and tongue; and death.

The rate of morbidity due to BRD ranges widely (4–89%) with an estimated average of 14% (National Animal Health Monitoring System 1997; Storz et al. 2000a,b; Snowder et al. 2006). Mortality estimates range from 1% to 13% in cattle (National Animal Health Monitoring Service 1997; Storz et al. 2000a,b; Snowder et al. 2006). Deaths from BRD may be seen shortly after the initiation of the disease, with peak death loss at 16–35 days postinfection (Van Donkersgoed et al. 1990; Ribble et al. 1995).

5.3.5 *Genetics*

Differences have been identified in the apparent susceptibility of preweaned and feedlot cattle to BRD. The average age in days at which BRD was diagnosed differed among breeds of cattle (Snowder et al. 2005). Angus, Hereford, and Gelbvieh had the highest average number of days to disease with Simmental having the shortest average number of days before onset of the disease in preweaned calves among nine pure breeds (Angus, Braunvieh, Charolais, Gelbvieh, Hereford, Limousin, Pinzgauer, Red Poll, and Simmental) and three composite breeds. Herefords were reported to be more susceptible to BRD infection than composite breeds, but Red Poll calves had the highest rate of mortality (9%) compared with the other breeds (4%) studied in the feedlot (Snowder et al. 2006). Pinzgauers had a higher frequency of postweaning BRD than did the other eight purebred breeds or three composite breeds (Muggli-Cockett et al. 1992).

Heritability estimates for the incidence of BRD range from 0.00 to 0.26 (Muggli-Cockett et al. 1992; Snowder et al. 2005, 2006, 2007; Heringstad et al. 2008). Snowder and colleagues (2006) estimated heritability for resistance to BRD as 0.04–0.09 in 18,112 feedlot cattle representing nine breeds and three composite types. Using the same purebred and composite breeds in 110,412 preweaned calves from records obtained from 1983 to 2002, overall heritability estimates of BRD incidence were 0.07 and 0.19 (Snowder et al. 2005). Records obtained from 1983 to 1988 on 10,142 calves from the same herd resulted in heritability estimates of BRD incidence of 0.10 for preweaning and 0.06 for postweaning periods (Muggli-Cockett et al. 1992). Correlations of BRD with carcass traits were low or near zero suggesting that selection for animals resistant to BRD would not have appreciable negative effects on carcass traits (Muggli-Cockett et al. 1992; Snowder et al. 2007). Currently, loci have not been identified that are associated with resistance or susceptibility to BRD.

5.4 Brucellosis in cattle

5.4.1 *Causative agent*

Brucellosis is caused by gram-negative, facultative bacteria of the *Brucella* genus. There is a debate concerning *Brucella's* taxonomy. DNA hybridization analysis of *Brucella* has characterized the genus as containing *Brucella melitensis* with *Brucella abortus*, *Brucella ovis*, *Brucella suis*, *Brucella canis*, and *Brucella neotomae* as biovarieties of *B. melitensis* (Verger et al. 1985). It has been proposed that only one species, *B. melitensis*, is recognized in the genus *Brucella* and that the remaining classical species should be considered biovars (Corbel 1988). However, this has not been widely adopted. *B. abortus* is the primary brucellae found in cattle, with *B. suis* or *B. melitensis* occasionally causing brucellosis in cattle.

5.4.2 Prevalence

Brucellosis is present in all continents but is well controlled in most developed countries. The prevalence of brucellosis is highest in the Middle East, Asia, Africa, South and Central America, the Mediterranean Basin, and the Caribbean (Roth et al. 2003). Brucellosis is present in land and marine animal populations as well as humans. Brucellosis remains an important zoonotic disease worldwide.

5.4.3 Transmission

The herd prevalence of brucellosis is estimated to be 0.014% in a typical state in the United States down from 11.5% in 1934 (Ragan 2002; Ebel et al. 2008). Wild animals, such as bison and elk, may serve as reservoirs of infection for livestock. Most often, brucellosis is introduced into herds by infected animals through shedding of the bacteria in milk, aborted fetuses, semen, vaginal discharges, placental membranes, and birth fluids. Animals may become infected by ingestion of contaminated food or water or sexual contact. The incubation period varies by the stage of gestation in which a cow becomes infected. Abortions and stillbirths usually occur 2 weeks to 5 months after infection.

5.4.4 Clinical presentation

The cardinal clinical signs of brucellosis infection in cattle are abortion in the second half of the pregnancy and epididymitis in bulls (Hill 1983; Enright et al. 1984; American Veterinary Medical Association 2007). Other symptoms in the cow may include a retained placenta, reduced milk production, and loss of weight (Nicoletti 1980; Corner et al. 1987). After the first abortion, subsequent pregnancies are generally normal, but the cows may continue to shed the organism in milk and uterine or vaginal discharges. In bulls, seminal vesiculitis, ampullitis, decreased libido, orchitis, and testicular abscesses may be seen (Rankin 1965; Plant et al. 1976). Infertility may result in both sexes. Arthritis may develop in chronic infections.

5.4.5 Genetics

Several studies have investigated the role of the natural resistance-associated macrophage protein 1 (*NRAMP1*), also known as solute carrier family 11 member 1 gene (*Slc11A1*), with brucellosis. This gene plays a critical role in promoting intracellular pathogen killing by macrophages. It has been described that naturally resistant macrophages of cows have a greater ability to inhibit the *in vitro* intracellular replication of *B. abortus* after challenge exposure (Price et al. 1990). Others (Paixao et al. 2007) did not find differences in bacterial intracellular survival in macrophages from resistant or susceptible cattle. *Slc11A1* has been associated with resistance against *B. abortus* infection in cattle in some studies (Feng et al. 1996; Adams and

Templeton 1998; Horin et al. 1999; Barthel et al. 2001) but not in others (Kumar et al. 2005; Paixao et al. 2007). Polymorphisms within the 3′ untranslated region $(GT)_n$ microsatellite have been identified. Repeats of 13 to 16 (GT) have been reported with the $(GT)_{13}$ allele associated with natural resistance to brucellosis *in vivo* (Adams and Templeton 1998). Five variants within the coding regions of the *Slc11A1* have also been found, three of which are missense mutations, as well as one single nucleotide polymorphism in the promoter region and five in introns (Martinez et al. 2008). These variants have not yet been studied for their association with resistance or susceptibility to brucellosis.

5.5 Leptospirosis in swine

5.5.1 Causative agent

Leptospirosis is caused by small, motile aerobic spirochete bacteria of the genus *Leptospira*. Two groups are recognized within the genus: *interrogan* and *biflexa*. About 23 serogroups are recognized containing 212 serovars (Ellis 1995). *Leptospira* has been classified into genome species based on their genetic sequences. Currently, there are more than 15 genome species of *Leptospira*, many of which contain organisms pathogenic to pigs. There have also been over 200 different serovars of pathogenic *Leptospira* identified worldwide (Levett 2001).

5.5.2 Prevalence

Leptospirosis is arguably the most widespread zoonosis worldwide (World Health Organization 1999). The incidence of leptospirosis is highest in warm humid regions where the organism has longer survival rates (Everard and Everard 1993). In the United States, significantly higher prevalence of disease was found in the southeastern, south central, and Pacific coastal regions than in other regions (Miller et al. 1990). Typically, only a small number of serovars is endemic in a specific region (Geistfeld 1975; Mazzonelli et al. 1979; Hathaway et al. 1982; Ellis et al. 1986; Chappel et al. 1998). The *Leptospira* subspecies serovars most frequently isolated from swine are *pomona*, *tarassovi*, *bratislava*, *grippothyphosa*, and, with less frequency, *icterohaemorrhagiae* and *canicola* (Faine et al. 1999). Therefore, in any region, pigs will be infected by serovars maintained by pigs or by other animal species present in the area. The relative importance of these incidental infections is determined by the opportunity that prevailing social, management, and environmental factors provide for the contact and transmission of leptospires from other species to pigs (Ellis 1999). A study of 1264 animals from 55 herds in Iowa over a 3-year period demonstrated that 38% of the animals had antibodies to one or more of 12 *Leptospira* antigens (Miller et al. 1990). Of those animals with antibodies, 42% were seropositive to *bratislava*, 8% to *copenhageni*, 6% to *ballun*, 4% *autumnalis*, 3% to *hardjo*, and 2% to *pomona*. In the same study, leptospires were isolated from 1.6% of animals with reproductive failure.

5.5.3 Transmission

Leptospirosis infection is commonly acquired by skin or mucous membrane contact with the urine of an infected animal or by the intake of contaminated feed or water (Sawhney and Saxena 1967). Large outbreaks of leptospirosis have occurred following excess rainfall. Transmission of the disease can also occur through the ingestion of infected animals and sexual contact. Infections are readily established via the conjunctiva, vaginal mucosa, or skin lesions (Fennestad and Borg-Petersen 1966). However, the development of disease depends on multiple factors. Leptospires spread rapidly via the lymphatics to the bloodstream where they are transported to all tissues. In the immunologically naive animal, initial replication occurs in the lungs, followed by the liver and spleen, and then multiple organs. If the animal develops an immune response and survives, leptospires will be cleared from most organs as well as the bloodstream. However, infection persists in sites hidden from the immune system, such as the proximal renal tubules, brain, anterior chamber of the eyes, and genital tract (Hanson and Tripathy 1986). Persistence in the kidneys results in a carrier state where the animal may shed leptospires in the urine for over 1 year. If shedder animals are introduced into a herd previously free of the disease, leptospires are rapidly disseminated (Mitchell et al. 1966).

5.5.4 Clinical presentation

After infection, a 1- to 2-day acute or septicemic phase is followed by antibody production and excretion of the leptospires in the urine (Edwards and Domm 1960; Turner 1967; Kelley 1998). Animals will present with anorexia and pyrexia. Many animals will have a spontaneous recovery within a week (Morse et al. 1958; Hanson and Tripathy 1986). Chronic infection in swine with serovar *pomona* can result in fetal death and abortion, whereas the birth of weak piglets is associated with *icterohaemorrhagiae* (Burnstein and Baker 1954; Neto et al. 1997). Infertility is a feature of infection with the serovar *bratislava* (Hathaway and Little 1998). Abortions are often restricted to periods of declining immunity in the sow population (Ellis 1999). Hathaway (1985) demonstrated that the serovars *hardjo* and *canicola* are associated with reproductive disorders in swine. In endemically infected areas, it is expected that *Leptospira* infections will cause fewer obvious symptoms of reproductive failure due to immunity acquired earlier in life. It is not uncommon for pigs to become infected by several leptospiral serovars, due to exposure of reservoir–hosts, environment, and climate in the particular area (Faine et al. 1999). *Leptospira interrogans* serovar *bratislava* has also been associated with subfertility and a reduced number of piglets born per litter (Frantz et al. 1989; Van Til and Dohoo 1991; Mousing et al. 1995; Hathaway and Little 1998). Subfertility, as measured by nonproductive sow days per parity, has also been associated with serovar *pomona* (Van Til and Dohoo 1991).

5.5.5 Genetics

The major histocompatibility gene complex (MHC) plays both a role in immune responsiveness and disease resistance in animals. Przytulski and Porzeczkowska (1980) examined resistance to leptospirosis and estimated the heritability of resistance to be 0.20. Smith and coworkers (1962) estimated the heritability of lung lesions to be 0.14. Reports of disease resistance in the pig have generally consisted of breed differences or of heritability estimates of specific resistance. Przytulski and Porzeczkowska (1979) showed a relationship among various transferrin receptors in swine associated with resistance or susceptibility to leptospirosis mapping to pig chromosome 13q41 (Jørgensen et al. 2003; Python et al. 2005).

5.6 Aujeszky's disease (pseudorabies)

5.6.1 *Causative agent*

Aujeszky's disease, also known as pseudorabies virus, is considered an important cause of economic losses in the pig industry worldwide. It is caused by *Suid herpesvirus* 1 that belongs to the genus *Varicellovirus* family of *Herpesviridae*. It is a neuroinvasive alpha herpesvirus with a wide host range, but it does not include primates (Mettenleiter 2000; Zuckermann 2000). The pseudorabies virus is a double-stranded, linear DNA virus composed of 150 kilobase pairs that produces approximately 100 proteins (Cheung and Smith 1999; Mettenleiter 2000). Pseudorabies virus is closely related to *bovine herpesvirus 1*, *equine herpesvirus 1*, and *varicella zoster* virus (McGeoch and Cook 1994). Several strains of pseudorabies virus have been described. The strain of the virus influences the severity of the disease and the duration of viral shedding (Maes et al. 1983). For example, strains with deletions of the thymidine kinase gene are less virulent than those without the deletion (Kit 1999; Kluge and Truszcy'nski 2006).

5.6.2 *Prevalence*

Aujeszky's disease is found throughout the world, particularly in regions with high concentrations of swine (Schaefer et al. 2006). Norway, Finland, and Malta have never had a reported case of Aujeszky's disease, whereas in Germany, Sweden, and the United Kingdom, the disease has been eradicated (Kluge and Truszcy'nski 2006). The U.S. Department of Agriculture began a national pseudorabies eradication program in 1989, and as of 2001, domestic swine in 41 states and territories were considered free of pseudorabies virus (Marsh and Leafstedt 2001). However, Aujeszky's disease is well established in feral swine populations in the United States, and feral swine represent a potential reservoir of this virus for the infection of domestic swine and native wildlife (Corn et al. 2004).

5.6.3 *Transmission*

The pseudorabies virus is primarily transmitted between swine through nose to nose contact, but venereal, semen, and transplacental transmission during pregnancy has been known to occur (Romero et al. 2001). Infections in adult feral swine commonly occur by pseudorabies virus strains that appear to be attenuated, resulting in latent disease that does not cause morbidity or mortality (Romero et al. 2001). Once a population has become infected, it is possible that the virus can persist indefinitely (Pirtle et al. 1989; Van Der Leek et al. 1993).

5.6.4 *Clinical presentation*

The clinical presentation and severity of Aujeszky's disease depends on the age of the host and the virulence of the virus strain involved (Kluge and Truszcy'nski 2006; Ciacci-Zanella et al. 2008). Symptoms may range from respiratory distress, nervous and genital disorders, to death (Figure 5.1). The incubation of the virus ranges from 1 to 11 days, with young animals having a shorter incubation period in comparison with older pigs (Wittmann et al. 1980). The pseudorabies virus is highly neurotropic. It first replicates in the nasopharyngeal mucosa, tonsils, and the olfactory epithelia prior to the invasion of the central nervous system through the nerve ends of the tonsils and the upper respiratory tract (Wittman et al. 1980; Kit 1999). Highly virulent strains are able to

Figure 5.1 One-day-old piglet exhibiting clinical signs of ataxia, lateral recumbence, and paddling movements characteristic of Aujeszky's disease. Photograph courtesy of Professor David Barcellos, Federal University of Rio Grande do Sul, Porto Alegre, RS, Brazil.

extend to the rest of the central nervous system where they produce a non-suppurative meningoencephalitis that can be fatal (Card and Enquist 1995). Because pigs are the only species that are able to survive a pseudorabies virus infection, they may be considered to bc a reservoir for the disease (Enquist et al. 1998). Mortality in infected pigs is dependent on the age of the animals. Infection in the first 2 weeks of life results in 100% mortality but decreases to 50% when pigs are infected after the third and fourth week of life (Baskerville 1981; Kritas et al. 1999). Piglets born with the disease can become ataxic and convulsive within 24 h after birth (Lawhorn et al. 1994). Clinical manifestation of Aujeszky's disease in piglets results in fever, hypersalivation, ataxia, and nystagmus to opisthotonos (Kluge and Truszcy'nski 2006). Infected animals may assume a sitting position because of their respiratory distress. Sows and boars primarily develop respiratory signs. Transplacental transmission may occur in pregnant females resulting in death to the fetus. Sows infected in the first trimester will reabsorb the fetus, and the sow will return to estrus. Infection of the sow during the second trimester may result in abortion, or stillborn or weak fetuses (Kluge and Truszcy'nski 2006). Mortality in adult animals is less than <2% (Baskerville 1981).

5.6.5 Genetics

There are indications of genetic differences in serum neutralization titers of pigs after vaccination with pseudorabies vaccine (Reiner et al. 2002). Individual differences in cell-mediated and humoral immunity and in susceptibility to pseudorabies virus in pigs have also been observed (Rothschild et al. 1984; Meeker et al. 1987a,b; Hessing et al. 1994, 1995). In 2002, Reiner and coworkers reported loci associated with Aujeszky's disease on chromosomes 9, 5, 6, and 13. Loci associated with rectal temperature after pseudorabies virus infection were found on chromosomes 2, 4, 8, 10, 11, and 16. The *IL-12* gene, located on chromosome 2, is a possible candidate gene for Aujeszky's disease, as *IL-12* is located near the marker *Swr349* that was associated with elevated rectal temperature after infection. Grob and colleagues (1999) presented that specific antibodies against herpesviruses seems to be sustained by the IL-12/IFN-γ pathway.

5.7 PRRS

5.7.1 Causative agent

PRRS is caused by a virus referred to as ATCC VR2332-like strain of PRRS virus (PRRSV) in North America and as the Lelystad-like virus in Europe (Wensvoort et al. 1991; Collins et al. 1992). The etiologic agent of PRRS is a small, enveloped, single-strand RNA virus with morphological, physiochemical, and genetic properties similar to those of the floating genus *Arterivirus*, which includes lactate dehydrogenase-elevating virus, equine arteritis virus, and simian hemorrhagic fever virus (Benfield et al. 1992)

5.7.2 *Prevalence*

PRRS was first recognized in the United States in 1987. It peaked in prevalence in 1989–1990 and stabilized in 1991–1992 (Keffaber 1989; Bautista et al. 1993; Dee and Joo 1997). Bautista and coworkers (1993) presented data from 412 herds in the United States, indicating that 36% of herds from 17 states were identified as seropositive for PRRS. PRRS has become endemic and is an important cause of pneumonia in 3- to 24-week-old pigs in swine-producing countries around the world (Thanawongnuwech et al. 2000). Recent data from many laboratories indicate that subclinical infections with PRRSV may also be common. Some regions of Europe remain free from PRRS, such as Sweden and Finland (Elvander et al. 1997; Garner et al. 1997; Office International des Epizooties 1997; Rautiainen et al. 2001). Australia, South America, Argentina, Cuba, and Brazil are free of PRRS (Perfumo 2001; Alfonso and Frías-Lepoureau 2003; Ciacci-Zanella et al. 2004).

5.7.3 *Transmission*

PRRSV may be introduced to a herd by purchasing pigs infected with the virus or purchasing PRRS-vaccinated pigs, contaminated semen, or other body fluids, or the use of equipment contaminated with PRRSV (Mortensen et al. 2002). Transmission may also occur through contaminated visitors, vectors, or ingestion of contaminated meat originating from infected pigs (Van der Linden et al. 2003). Airborne transmission may also occur (Tomorremorell et al. 1997; Kristensen et al. 2004). PRRS may be transmitted from viremic dams through the placenta to the fetus, resulting in fetal death or birth of weak animals (Christianson et al. 1992). Some pigs in affected litters may escape from infection with PRRS. PRRS can replicate in the fetus after day 14 of gestation. However, infection during the first two-thirds of gestation is not common because most strains of PRRS are not transmitted efficiently across the placenta until the last trimester (Christianson et al. 1993).

5.7.4 *Clinical presentation*

The clinical presentation of PRRS varies greatly among herds, ranging from asymptomatic to devastating clinical disease. This is influenced by the virus strain, host immune status, host susceptibility, concurrent infections, and management practices (Hopper et al. 1992). The first stage of infection lasts for 2 or more weeks and is characterized by anorexia, fever, and lethargy in 5–75% of pigs (Collins et al. 1992). In sows exposed to the virus in the third trimester of pregnancy, PRRS causes abortion and poor litter quality. Adult animals may experience a 1–4% mortality rate (Hopper et al. 1992). In young pigs, PRRS may cause weakness, respiratory disease, and interstitial pneumonia and up to 60% mortality rate. Clinical signs are more severe when the animal is coinfected with other pathogens (Thacker et al. 2001). The PRRSV targets alveolar macrophages and induces apoptosis, resulting in ineffective elimination of the virus and persistence for several weeks (Murtaugh et al. 2002; Osorio 2002; Labarque et al. 2003;

Rowland et al. 2003). The morbidity and mortality associated with PRRS results in a cost of $560 million annually to U.S. pork producers (Neumann et al. 2005; Lewis et al. 2007).

5.7.5 Genetics

Despite substantial research efforts, the exact components of a protective anti-PRRS immune response are still unknown (Petry et al. 2005, 2007; Lewis et al. 2007). Breed and lines of pigs (within breeds) have shown differences in the prevalence of PRRS (Lundeheim 1979, 1988; Halbur et al. 1998; Petry et al. 2005; Vincent et al. 2006). Petry and colleagues (2005) used 100 animals from a Large White Landrace composite population (Nebraska Index Line [NEI]) and 100 Hampshire–Duroc (HD) cross animals to determine if there were differences in their responses to exposure to PRRSV. Uninfected HD animals had greater weight gain than NEI animals and also had higher basal rectal temperatures. After PRRSV exposure, NEI animals had greater weight gain, maintained lower rectal temperatures, and exhibited fewer interstitial lesions than the HD animals. These results may suggest that NEI animals may be more tolerant to PRRSV infection than HD animals. Genetic variation to immune response when challenged with PRRS has also been demonstrated (Edfors-Lilja et al. 1995; Mallard et al. 1998; Wilkie and Mallard 1999). However, because most seedstock breeding populations are maintained in a much different environment than commercial herds, natural selection for resistance to pathogens may not be occurring at the same rate in seedstock herds as in commercial herds. Loci associated or linked with PRRS resistance or susceptibility have not yet been identified in swine.

5.8 Future research directions

To date, the loci associated with these diseases have not been identified. Resources are now available for whole genome association analysis for cattle and swine, which will facilitate the identification of loci associated with complex traits such as disease resistance or susceptibility. The identification and characterization of loci associated with disease infection or transmission will allow for the selection of animals that are less susceptible to disease or is less likely to transmit disease. Once the loci are identified, functional studies will provide an opportunity to identify the causes and consequences of coevolution of the host and its pathogens, and an understanding of the pathogenesis of the disease. Identifying a means to interfere with the virulence of the pathogen may lead to new treatments and disease prevention. The outcome of this research for farm animals will be increased productivity, improved well-being, and healthier animals. For the livestock industry, selection of animals with disease resistance will result in increased profitability. For consumers, the impact is more affordable and safe food products.

References

Adams, L.G. and Templeton, J.W. 1998. Genetic resistance to bacterial diseases of animals. *Revue Scientifique et Technique (International Office of Epizootics)* **17**: 200–219.

Adams, O.R., Brown, W.W., Chow, T.L., Collier, J.R., Davis, R.W., Griner, L.A., Jensen, R., Pierson, R.E., and Wayt, L.K. 1959. Comparison of infectious bovine rhinotracheitis, shipping fever and calf diphtheria of cattle. *Journal of the*

American Veterinary Medical Association **134**: 85–89.

Alfonso, P. and Frías-Lepoureau, M.T. 2003. PRRS in Central America and the Caribbean region. In: Zimmerman, J. and Yoon, K.-J. (eds.), *PRRS Compendium*, 2nd Edition. Des Moines, IA: National Pork Board, pp. 217–220.

Amand, W.B. 1974. Paratuberculosis in a dromedary camel. *Annual Proceedings of the American Association of Zoo Veterinarians*, Atlanta, GA, pp. 150–153.

American Veterinary Medical Association. 2007. Backgrounder: Brucellosis. www.avma.org/reference/backgrounders/brucellosis_bgnd.asp.

Ames, T.R. 1997. Dairy calf pneumonia: The disease and its impact. *Veterinary Clinics of North America: Food Animal Practice* **13**: 379–391.

Animal and Plant Health Inspection Service, USDA. 2008. Johne's disease on U.S. dairies 1991–2007. nahms.aphis.usda.gov/dairy/dairy07/Dairy2007_Johnes.pdf.

Autschbach, F., Eisold, S., Hinz, U., Zinser, S., Linnebacher, M., Giese, T., Löffler, T., Büchler, M.W., and Schmidt, J. 2005. High prevalence of *Mycobacterium avium* subspecies *paratuberculosis* IS900 DNA in gut tissues from individuals with Crohn's disease. *Gut* **54**: 944–949.

Ayele, W.Y., Bartos, M., Svastova, P., and Pavlik, I. 2004. Distribution of *Mycobacterium avium* subsp. *paratuberculosis* in organs of naturally infected bull-calves and breeding bulls. *Veterinary Microbiology* **103**: 209–217.

Barthel, R., Feng, J., Piedrathia, J.A., McMurray, D.N., Templeton, J.W., and Adams, G.L. 2001. Stable transfection of the bovine *NRAMP1* gene into murine RAW264.7 cells: Effect on *Brucella abortus* survival. *Infection and Immunity* **69**: 3110–3119.

Baskerville, A. 1981. Aujeszky's disease: Recent advances and current problems. *New Zealand Veterinary Journal* **29**: 183–185.

Bautista, E.M., Morrison, R.B., Goyal, S.M., Collins, J.E., and Annelli, J.F. 1993. Seroprevalence of PRRS virus in United States. *Swine Health and Production* **1**: 4–8.

Becher, P., Orlich, M., Shannon, A.D., Horner, G., Konig, M., and Thiel, H.J. 1997. Phylogenetic analysis of pestiviruses from domestic and wild ruminants. *Journal of General Virology* **78**: 1357–1366.

Benfield, D.A., Nelson, E., Collins, J.E., Harris, L., Goyal, S.M., Robison, D., Christianson, W.T., Morrison, R.B., Gorcyca, D., and Chladek, D. 1992. Characterization of swine infertility and respiratory syndrome (SIRS) virus (isolate ATCC VR-2332). *Journal of Veterinary Diagnostic Investigation* **4**: 127–133.

Bielefeldt-Ohmann, H., Tolnay, A.E., Reisenhauer, C.E., Hansen, T.R., Smirnova, N., and Van Campen, H. 2008. Transplacental infection with non-cytopathic bovine viral diarrhoea virus types 1b and 2: Viral spread and molecular neuropathology. *Journal of Comparative Pathology* **138**: 72–85.

Bishop, S.C., Chesnais, J., and Stear, M.J. 2002. Breeding for disease resistance: Issues and opportunities. *Proceedings of the 7th World Congress on Genetics Applied to Livestock Production*, CD-ROM Communication No. 13-01.

Brock, K.V., Redman, D.R., Vickers, M.L., and Irvine, N.E. 1991. Quantitation of bovine viral diarrhea virus in embryo transfer flush fluids collected from a persistently infected heifer. *Journal of Veterinary Diagnostic Investigation* **3**: 99–100.

Brodersen, B.W. and Kelling, C.L. 1998. Effect of concurrent experimentally induced bovine respiratory syncytial virus and bovine viral diarrhea virus infection on respiratory tract and enteric diseases in calves. *American Journal of Veterinary Research* **59**: 1423–1430.

Buergelt, C.D., Hall, C.E., McEntee, K., and Duncan, J.R. 1978. Pathological evaluation of paratuberculosis in naturally infected cattle. *Veterinary Pathology* **15**: 196–207.

Burnstein, T. and Baker, J.A. 1954. Leptospirosis in swine caused by *Leptospira pomona*. *Journal of Infectious Disease* **94**: 53–64.

Card, J.P. and Enquist, L.W. 1995. Neurovirulence of pseudorabies virus. *Critical Reviews in Neurobiology* **9**: 137–162.

Caswell, J.L. and Archambault, M. 2007. *Mycoplasma bovis* pneumonia in cattle. *Animal Health Research Reviews* **8**: 161–186.

Chandler, R.L. 1962. Infection of laboratory animals with *Mycobacterium johnei*. V. Further studics on the comparative susceptibility of C57 black mice. *Journal of Comparative Pathology* **72**: 198–213.

Chappel, R.J., Prime, R.W., Millar, B.D., Jones, R.T., Cutler, R.S., and Adler, B. 1998. Prevalence and geographic origin of pigs with serological evidence of infection with *Leptospira interrogans* serovar *pomona* slaughtered in abattoirs in Victoria, Australia. *Veterinary Microbiology* **62**: 235–242

Cheung, A.K. and Smith, T.A. 1999. Analysis of the latency-associated transcript/UL1-3.5 gene cluster promoter complex of pseudorabies virus. *Archives of Virology* **144**: 381–391.

Chiodini, R.J. 1989. Crohn's disease and other mycobacteriosis: A review and comparison of two disease entities. *Clinical Microbiology Reviews* **2**: 90–117.

Chiodini, R.J. and Buergelt, C.D. 1993. Susceptibility of Balb/c C57/B6 and C57/B10 mice to infection with *Mycobacterium paratuberculosis*. *Journal of Comparative Pathology* **109**: 309–319.

Chiodini, R.J. and Van Kruiningen, H.J. 1983. Eastern white-tailed deer as a reservoir of ruminant paratuberculosis. *Journal of the American Veterinary Medical Association* **182**: 168–169.

Chiodini, R.J., Van Kruiningen, H.J., and Merkal, R.S. 1984. Ruminant paratuberculosis (Johne's disease): The current status and future prospects. *The Cornell Veterinarian* **74**: 218–262.

Christianson, W.T., Choi, C.S., Collins, J.E., Molitor, T.W., Morrison, R.B., and Joo, H.S. 1993. Pathogenesis of porcine reproductive and respiratory syndrome virus infection in mid-gestation sows and fetuses. *Canadian Journal of Veterinary Research* **57**: 262–268.

Christianson, W.T., Collins, J.E., Benfield, D.A., Harris, L., Gorcyca, D.E., Chladek, D.W., Morrison, R.B., and Joo, H.S. 1992. Experimental reproduction of swine infertility and respiratory syndrome in pregnant sows. *American Journal of Veterinary Research* **53**: 485–488.

Ciacci-Zanella, J.R., Amaral, A.L., Ventura, L.V., Morés, N., and Bortoluzzi, H. 2008. Aujeszky's disease eradication in Santa Catarina State: Relevance of sanitary status of replacement gilts. *Ciência Rural* **38**: 749–754.

Ciacci-Zanella, J.R., Trombetta, C., Vargas, I., and Costa, D.E.M. 2004. Lack of evidence of porcine reproductive and respiratory syndrome virus (PRRSV) infection in domestic swine in Brazil. *Ciência Rural* **34**: 449–455.

Cocito, C., Gilot, P., Coene, M., De Kesel, M., Poupart, P., and Vannuffel, P. 1994. Paratuberculosis. *Clinical Microbiology Reviews* **7**: 328–345.

Collins, J.E., Benfield, D.A., Christiason, W.T., Harris, L., Hennings, J.C., Shaw, D.P., Goyal, S.M. et al. 1992. Isolation of swine infertility and respiratory syndrome virus (isolate ATCC VR-2332) in North America and experimental reproduction of the disease in gnotobiotic pigs. *Journal of Veterinary Diagnostic Investigation* **4**: 117–126.

Corbeil, L.B. 2008. *Histophilus somni* host-parasite relationships. *Animal Health Research Review* **8**: 151–160.

Corbel, M.J. 1988. International Committee on Systematic Bacteriology Subcommittee on the taxonomy of *Brucella*. *International Journal of Systemic Bacteriology* **38**: 450–452.

Coria, M.F. and McClunkin, A.W. 1978. Specific immune tolerance in an apparently healthy bull persistently infected with BVD virus. *Journal of the American Veterinary Medical Association* **172**: 449–451.

Corn, J.L., Stallknecht, D.E., Mechlin, N.M., Luttrell, M.P., and Fischer, J.R. 2004. Persistence of pseudorabies virus in feral swine populations. *Journal of Wildlife Diseases* **40**: 307–310.

Corner, L.A., Alton, G.G., and Iyer, H. 1987. Distribution of *Brucella abortus* in infected cattle. *Australian Veterinary Journal* **64**: 241–244.

Crossley, B.M., Zagmutt-Vergara, F.J., Fyock, T.L., Whitlock, R.H., and Gardner, I.A. 2005. Fecal shedding of *Mycobacterium avium* subsp. *paratuberculosis* by dairy cows. *Veterinary Microbiology* **107**: 257–263.

Dabo, S.M., Taylor, J.D., and Confer, A.W. 2008. *Pasteurella multocida* and bovine respiratory disease. *Animal Health Research Review* **8**: 129–150.

Dee, S.A. and Joo, H. 1997. Strategies to control PRRS: A summary of field and research experiences. *Veterinary Microbiology* **55**: 347–353.

Delisle, G.W., Samagh, B.S., and Duncan, J.R. 1980. Bovine paratuberculosis. II. A comparison of fecal culture and antibody response. *Canadian Journal of Comparative Medicine* **44**: 183–191.

Food Standards Agency. Advisory Committee on the Microbiological Safety of Food. 2000. Preliminary results from the national study on the microbiological quality and heat processing of cows' milk: *Mycobacterium avium subsp. paratuberculosis*. ACM/485.

Ebel, E.D., Williams, M.S., and Tomlinson, S.M. 2008. Estimating herd prevalence of bovine brucellosis in 46 USA states using slaughter surveillance. *Preventive Veterinary Medicine.* March 21 [Epub ahead of print].

Edfors-Lilja, I., Gustafsson, U., Duval-Ilflah, Y., and Andersson, L.A. 1995. The porcine intestinal receptor for *Escherichia coli* K88ab, K88ac: Regional localization on chromosome 13 and influence of IgG response to the K88 antigen. *Animal Genetics* **26**: 237–242.

Edwards, G.A. and Domm, B.M. 1960. Human leptospirosis. *Medicine* **39**: 117–156.

Edwards, S., White, H., and Nixon, P. 1990. A study of the predominant genotypes of bovine herpesvirus 1 isolated in the UK. *Veterinary Microbiology* **22**: 213–223.

Ellis, J.A., Davis, W.C., Belden, E.L., and Pratt, D.L. 1988. Flow cytofluorimetric analysis of lymphocyte subset alterations in cattle infected with bovine viral diarrhea virus. *Veterinary Pathology* **25**: 231–236.

Ellis, W.A. 1995. International Committee on Systematic Bacteriology Subcommittee in the Taxonomy of *Leptospira*. *Inter-*

national Journal of Systematic Bacteriology **45**: 872–874.

Ellis, W.A. 1999. Leptospirosis. In: Straw, B.E., D'Allaire, S., Mengeling, W.L., and Taylor, D. (eds.), *Diseases of Swine*. London: Wolfe Publishing Ltd., pp. 483–493.

Ellis, W.A., McParland, P.J., Bryson, D.G., and Cassells, J.A. 1986. Prevalence of *Leptospira* infection in aborted pigs in Northern Ireland. *Veterinary Record* **118**: 63–65.

Elvander, M., Larsson, B., Engvall, A., Klingeborn, B., and Gunnarsson, A. 1997. Nationwide surveys of TGE/PRCV, CSF, PRRS, SVD, *L. pomona* and *B. suis* in pigs in Sweden. *Epidémiologie santé animale* pp. 31–32.

Enquist, L.W., Husak, P.J., Banfield, B.W., and Smith, G.A. 1998. Infection and spread of alphaherpesviruses in the nervous system. *Advances in Virus Research* **51**: 237–347.

Enright, F.M., Walker, J.V., Jeffers, G., and Doyoe, B.L. 1984. Cellular and humoral responses of *Brucella abortus*-infected bovine fetuses. *American Journal of Veterinary Research* **45**: 424–430.

Everard, J.D. and Everard, C.O.R. 1993. Leptospirosis in the Caribbean. *Reviews of Medical Microbiology* **4**: 114–122.

Faine, S. 1994. Taxonomy, classification, and nomenclature. In: *Leptospira and Leptospirosis*. Boca Raton, FL: CRC Press, pp. 119–137.

Farley, H. 1932. An epizoological study of shipping fever in Kansas. *Journal of the American Veterinary Medical Association* **52**: 165–172.

Feng, J., Li, Y., Hashad, M., Schurr, M., Gros, P., Adams, L.G., and Templeton, J.W. 1996. Bovine natural resistance associated macrophage protein 1 (*Nramp1*) gene. *Genome Research* **6**: 956–964.

Fennestad, K.L. and Borg-Petersen, C. 1966. Experimental leptospirosis in pregnant sows. *Journal of Infectious Disease* **116**: 57–66.

Food Standards Agency, Advisory Committee on the Microbiological Safety of Food. 2000. Preliminary results from the national study on the microbiological quality and heat processing of cows' milk: *Mycobacterium avium* subsp. *paratuberculosis*. ACM/485.

Food Standards Australia New Zealand. 2004. Association between Johne's disease and Crohn's disease. A microbiological review. *Technical Report Series No. 35*. www.foodstandards.gov.au/_srcfiles/edit_Report_JD%20and%20CD-%20Final%20Dec%202004.doc.

Frantz, J.C., Hanson, L.E., and Brown, A.L. 1989. Effect of vaccination with a bacteria containing *Leptospira interrogans* serovar *bratislava* on the breeding performance of swine herds. *American Journal of Veterinary Research* **50**: 1044–1047.

Frelier, P.F., Templetion, J.W., Estes, M., Whitford, H.W., and Kienle, R.D. 1990. Genetic regulation of *Mycobacterium paratuberculosis* infection in recombinant inbred mice. *Veterinary Pathology* **27**: 362–364.

Friend, S.C., Wilkie, B.N., Thomson, R.G., and Barnum, D.A. 1977. Bovine pneumonic pasteurellosis: Experimental induction in vaccinated and novaccinated calves. *Canadian Journal of Comparative Medicine* **41**: 77–83.

Fulton, R.W., Briggs, R.E., Payton, M.E., Confer, A.W., Saliki, J.T., Ridpath, J.F., Burge, L.J., and Duff, G.C. 2004. Maternally derived humoral immunity to bovine viral diarrhea virus (BVDV)1a, BVDV1b, BVDV2, bovine herpesvirus-1, parainfluenza-3 virus bovine respiratory syncytial virus, *Mannheimia haemolytica* and *Pasteurella multocida* in beef calves, antibody decline by half-life studies and effect on response to vaccination. *Vaccine* **22**: 643–649.

Garner, M.G., Gleeson, L.J., Holyoake, P.K., Cannon, R.M., and Doughy, W.J. 1997. A national serological survey to verify Australia's freedom from porcine reproductive and respiratory syndrome. *Australian Veterinary Journal* **75**: 596–600.

Geistfeld, J.G. 1975. Leptospirosis in the United States, 1971–1973. *Journal of Infectious Disease* **131**: 743–745.

Gershwin, L.J. 2008. Bovine respiratory syncytial virus infection: Immunopathogenic mechanisms. *Animal Health Research Reviews* **8**: 207–213.

Gonda, M.G., Chang, Y.M., Shook, G.E., Collins, M.T., and Kirkpatrick, B.W. 2006. Genetic variation of *Mycobacterium avium* ssp. *paratuberculosis* infection in US Holsteins. *Journal of Dairy Science* **89**: 1804–1812.

Gourlay, R.N., Thomas, L.H., and Howard, C.J. 1976. Pneumonia and arthritis in gnotobiotic calves following inoculation with *Mycoplasma agalactiae* subsp. *bovis*. *The Veterinary Record* **98**: 506–507.

Grob, P., Schijns, V.E.C.J., Van den Broek, M.F., Cox, S.P.J., Ackermann, M., and Suter, M. 1999. Role of the individual interferon systems and specific immunity in mice in controlling systemic dissemination of attenuated pseudorabies virus infection. *Journal of Virology* **73**: 4748–4754.

Grooms, D.L. 2004. Reproductive consequences of infection with bovine viral diarrhea virus. *Veterinary Clinics of North America: Food Animal Practice* **20**: 5–19.

Halbur, P., Rothschild, M., and Thacker, B. 1998. Differences in susceptibility of Duroc, Hampshire, and Meishan pigs to infection with a high-virulence strain (VR2385) of porcine reproductive and respiratory syndrome virus (PRRSV). *Journal of Animal Breeding and Genetics* **115**: 181–189.

Hanson, L.E. and Tripathy, D.N. 1986. Leptospirosis. In: Leman, A.D. (ed.), *Leptospirosis in Diseases of Swine*, 6th Edition. Ames, IA: Iowa State University Press, pp. 591–599.

Harris, F.W. and Janzen, E.D. 1989. The *Haemophilus sonmus* disease complex (Hemophilosis): A review. *Canadian Veterinary Journal* **30**: 816–822.

Hathaway, S.C. 1985. Porcine leptospirosis. *Pig News and Information* **6**: 31–34.

Hathaway, S.C. and Little, T.W. 1998. Prevalence and clinical significance of leptospiral antibodies in pigs in England. *The Veterinary Record* **108**: 224–228.

Hathaway, S.C., Little, T.W., and Stevens, A.E. 1982. Isolation of *Leptospira interrogans* serovar *muenchen* from a sow with a history of abortion. *The Veterinary Record* **111**: 100–102.

Heringstad, B., Chang, Y.M., Gianola, D., and Osteras, O. 2008. Genetic analysis of respiratory disease in Norwegian Red calves. *Journal of Dairy Science* **91**: 367–370.

Hermon-Taylor, J., Bull, T.J., Sheridan, J.M., Cheng, J., Stellakis, M.L., and Sumar, N. 2000. Causation of Crohn's disease by *Mycobacterium avium* subspecies *paratuberculosis*. *Canadian Journal of Gastroenterology* **14**: 521–539.

Hessing, M.J., Coenen, G.J., Vaiman, M., and Renard, C. 1995. Individual differences in cell-mediated and humoral immunity in pigs. *Veterinary Immunology and Immunopathology* **45**: 97–113.

Hessing, M.J., Scheepens, C.J., Schouten, W.G., Tielen, M.J., and Wiepkema, P.R. 1994. Social rank and disease susceptibility in pigs. *Veterinary Immunology and Immunopathology* **43**: 373–387.

Hill, D. 1983. The cultural and pathological examination of bulls serologically positive for brucellosis. *Australian Veterinary Journal* **60**: 7–9.

Hopper, S.A., White, M.E., and Twiddy, N. 1992. An outbreak of blue-eared pig disease (porcine reproductive and respiratory syndrome) in four pig herds in Great Britain. *The Veterinary Record* **131**: 140–144.

Horin, P., Rychlik, I., Templeton, J.W., and Adams, L.G. 1999. A complex pattern of microsatellite polymorphism within the bovine *NRAMP1* gene. *European Journal of Immunogenetics* **26**: 311–313.

Houe, H. 1995. Epidemiology of bovine viral diarrhea virus. *Veterinary Clinics of North America: Food Animal Practice* **11**: 521–547.

Humphrey, J. and Stephens, L. 1983. *Haemophilus somnus*: A review. *Veterinary Bulletin* **53**: 987–1004.

Humphrey, J.D., Little, P.B., Stephens, L.R., Barnum, D.A., Doig, P.A., and Thorsen, J. 1982. Prevalence and distribution of *Haemophilus somnus* in the male bovine reproductive tract. *American Journal of Veterinary Research* **43**: 791–795.

Jessup, D.A., Abbas, B., Behymer, D., and Gogan, P. 1981. Paratuberculosis in tule elk in California. *Journal of the American Veterinary Medical Association* **179**: 1252–1254.

Jones, C. and Chowdhury, S. 2007. A review of the biology of bovine herpesvirus type 1 (BHV-1), its role as a cofactor in the bovine respiratory disease complex and development of improved vaccines. *Animal Health Research Reviews* **8**: 187–205.

Jørgensen, C.B., Cirera, S., Anderson, S.I., Archibald, A.L., Raudsepp, T., Chowdhary, B., Edfors-Lilja, I., Andersson, L., and Fredholm M. 2003. Linkage and comparative mapping of the locus controlling susceptibility towards *E. coli* F4ab/ac diarrhoea in pigs. *Cytogenetic and Genome Research* **102**: 157–162.

Juárez, M.D., Torres, A., and Espitia, C. 2001. Characterization of *Mycobacterium tuberculosis* region containing the mpt83 and mpt70 genes. *FEMS Microbiology Letters* **203**: 95–102.

Keffaber, K.K. 1989. Reproductive failure of unknown etiology. *American Association of Swine Practitioners Newsletter* **1**: 1–9.

Kelley, P.W. 1998. Leptospirosis. In: Gorbach, S.L., Bartlett, J.G., and Blacklow, N.R. (eds.), *Infectious Diseases*, 2nd Edition. Philadelphia: W. B. Saunders, pp. 1580–1587.

Kelling, C. 2007. Viral diseases of the fetus. In: Youngquist, R.S. and Threlfall, W.R. (eds.), *Large Animal Theriogenology 2*. St. Louis, MO: Saunders Elsevier, pp. 399–408.

Kiorpes, I.L., Butler, D.G., Dubielzig, R.R., and Beck, K.A. 1988. Enzootic pneumonia in calves: Clinical and morphological features. *Compendium on Continuing Education for the Practicing Veterinarian* **10**: 248–260.

Kit, S. 1999. Pseudorabies virus (Herpesviridae). In: Granoff, A. and Webster, R.G. (eds.), *Encyclopedia of Virology*. New York: Academic Press, pp. 1421–1429.

Kluge, J.P. and Truszcy'nski M.J. 2006. Aujeszky's disease (pseudorabies). In: Straw, B.E., Dállaire, S., Mengeling, W.L., and Taylor, D.J. (eds.), *Diseases of Swine*. Oxford: Blackwell Science, pp. 419–433.

Koets, A.P., Adugna, G., Janss, L.L., Van Weering, H.J., Kalis, C.H., Wentink, G.H., Rutten, V.P., and Schukken, Y.H. 2000. Genetic variation of susceptibility to *Mycobacterium avium* subsp.

paratuberculosis infection in dairy cattle. *Journal of Dairy Science* **83**: 2702–2708.

Kreeger, J. 1991. Ruminant paratuberculosis—A century of progress and frustration. *Journal of Veterinary Diagnostic Investigation* **3**: 373–382.

Kristensen, C.S., Bøtner, A., Takai, H., Nielsen, J.P., and Jorsal, S.E. 2004. Experimental airborne transmission of PRRS virus. *Veterinary Microbiology* **99**: 197–202.

Kritas, S.K., Pensaert, M.B., Nauwynck, H.J., and Kyriakis, S.C. 1999. Neural invasion of two virulent suid herpesvirus 1 strains in neonatal pigs with or without maternal immunity. *Veterinary Microbiology* **69**: 143–156.

Kumar, N., Mitra, A., Ganguly, I., Singh, R., Deb, S.M., Srivastava, S.K., and Sharma, A. 2005. Lack of association of brucellosis resistance with (GT)(13) microsatellite allele at 3″UTR of *NRAMP1* gene in Indian zebu (*Bos indicus*) and crossbred (*Bos indicus* x *Bos taurus*) cattle. *Veterinary Microbiology* **111**: 139–143.

Labarque, G., Van Gucht, S., Van Reeth, K., Nauwynck, H., and Pensaert, M. 2003. Respiratory tract protection upon challenge of pigs vaccinated with attenuated porcine reproductive and respiratory syndrome virus vaccines. *Veterinary Microbiology* **95**: 187–197.

Larsen, A.B., Merkal, R.S., and Cutlip, R.C. 1975. Age of cattle as related to resistance to infection with *Mycobacterium paratuberculosis*. *American Journal of Veterinary Research* **36**: 255–257.

Larsen, A.B., Vardaman, T.H., and Merkal, R.S. 1963. An extended study of a herd naturally infected with Johne's disease. I. The significance of the intradermal Johnin test. *American Journal of Veterinary Research* **24**: 91–93.

Lawhorn, B., McConnell, S., Kit, M., and Kit, S. 1994. Vaccination of newborn pigs in the presence of low levels of pseudorabies colostral antibodies. *Vaccine* **12**: 601–606.

Levett, P.N. 2001. Leptospirosis. *Clinical Microbiology Reviews* **14**: 296–326.

Lewis, C.R.G., Ait-Ali, T., Clapperton, M., Archibald, A.L., and Bishop, S. 2007. Genetic perspectives on host responses to porcine reproductive and respiratory syndrome (PRRS). *Viral Immunology* **20**: 343–358.

Libke, K.G. and Walton, A.M. 1975. Presumptive paratuberculosis in a Virginia white-tailed deer. *Journal of Wildlife Diseases* **11**: 552–553.

Lilenbaum, W., Marassi, C.D., and Oelemann, W.M.R. 2007. Paratuberculosis: An update. *Brazilian Journal of Microbiology* **38**: 580–590.

Lillie, L.E. 1974. The bovine respiratory disease complex. *Canadian Veterinary Journal* **15**: 233–242.

Liu, L., Lehmkuhl, H.D., and Kaeberle, M.L. 1999. Synergistic effects of bovine respiratory syncytial virus and non-cytopathic bovine viral diarrhea virus infection on selected bovine alveolar macrophage functions. *Canadian Journal of Veterinary Research* **63**: 41–48.

Loneragan, G.H., Thomson, D.U., Montgomery, D.L., Mason, G.L., and Larson, R.L. 2005. Prevalence, outcome and health consequences associated with persistent infection with bovine viral diarrhea virus in feedlot cattle. *Journal of the American Veterinary Medical Association* **226**: 595–601.

Lundeheim, N. 1979. Genetic analysis of respiratory diseases in pigs. *Acta Agriculture Scandinavica* **29**: 209–215.

Lundeheim, N. 1988. Health disorders and growth performance at a Swedish pig progeny testing station. *Acta Agriculture Scandinavica* **38**: 77–88.

Maes, R.K., Kanitz, C.L., and Gustafson, D.P. 1983. Shedding patterns in swine of virulent and attenuated pseudorabies virus. *American Journal of Veterinary Research* **44**: 2083–2086.

Mallard, B.A., Wilkie, B.N., Kennedy, B.W., Gibson, J., and Quinton, M. 1998. Immune responsiveness in swine: Eight generations of selection for high and low immune response in Yorkshire pigs. *Proceedings of the 6th World Congress on Genetics Applied to Livestock Production* **27**: 295–302.

Marsh, B.D. and Leafstedt, J.W. 2001. Report of the committee on pseudorabies. *Proceedings of the United States Animal Health Association* **105**: 311–315.

Martinez, R., Dunner, S., Barrera, G., and Canon, J. 2008. Novel variants within the coding regions of the *Slc11A1* gene identified in *Bos taurus* and *Bos indicus* breeds. *Journal of Animal Breeding and Genetics* **125**: 57–62.

Mazzonelli, J., Jelambi, F., Alvarez, E., de la Canal, H., and Nava, B.O. 1979. Prospective studies of porcine leptospirosis in organized farms of Venezuela. *Boletín de la Oficina Sanitaria Panamericana* **87**: 60–71.

McClurkin, A.W., Littledike, E.T., Cutlip, R.C., Frank, G.H., Coria, M.F., and Bolin, S.R. 1984. Production of cattle immunotolerant to bovine viral diarrhea virus (BVDV). *Canadian Journal Comparative Medicine* **48**: 156–161.

McGeoch, D.J. and Cook, S. 1994. Molecular phylogeny of the Alphaherpesvirinae subfamily and a proposed evolutionary timescale. *Journal of Molecular Biology* **238**: 9–22.

Meeker, D.L., Rothschild, M.F., Christian, L.L., Warner, C.M., and Hill, H.T. 1987a. Genetic control of immune response to pseudorabies and atrophic rhinitis vaccines: I. Heterosis, general combining ability and relationship to growth and backfat. *Journal of Animal Science* **64**: 407–413.

Meeker, D.L., Rothschild, M.F., Christian, L.L., Warner, C.M., and Hill, H.T. 1987b. Genetic control of immune response to pseudorabies and atrophic rhinitis vaccines: II. Comparison of additive direct and maternal genetic effects. *Journal of Animal Science* **64**: 414–419.

Mettenleiter, T.C. 2000. Aujeszky's disease (pseudorabies) virus: The virus and molecular pathogenesis—State of the art, June 1999. *Veterinary Research* **31**: 99–115.

Metzler, A.E., Matile, H., Gasman, U., Engels, M., and Wyler, R. 1985. European isolates of bovine herpesvirus 1: A comparison of restriction endonuclease sites, polypeptides and reactivity with monoclonal antibodies. *Archives of Virology* **85**: 57–69.

Meyer, G., Deplanche, M., and Schelcher, F. 2008. Human and bovine respiratory syncytial virus vaccine research and development. *Comparative Immunology Microbiology and Infectious Diseases* **31**: 191–225.

Millar, D., Ford, J., Sanderson, J., Withey, S., Tizard, M., Doran, T., and Hermon-Taylor, J. 1996. IS900 PCR to detect *Mycobacterium paratuberculosis* in retail supplies of whole pasteurized cows' milk in England and Wales. *Applied and Environmental Microbiology* **62**: 3446–3452.

Miller, D.A., Wilson, M.A., Owen, W.J., and Beran, G.W. 1990. Porcine leptospirosis in Iowa. *Journal of Veterinary Diagnostic Investigation* **2**: 171–175.

Mitchell, D., Robertson, A., Corner, A.H., and Boulanger, P. 1966. Some observations on the diagnosis and epidemiology of leptospirosis in swine. *Canadian Journal of Comparative Medicine and Veterinary Science* **30**: 211–217.

Moennig, V. and Liess, B. 1995. Pathogenesis of intrauterine infections with bovine viral diarrhea virus. *Veterinary Clinics of North America: Food Animal Practice* **11**: 477–487.

Moerman, A., Straver, P.J., de Jong, M.C., Quak, J., Baanvinger, T., and van Oirschot, J.T. 1993. A long term epidemiological study of bovine viral diarrhoea infections in a large herd of dairy cattle. *The Veterinary Record* **132**: 622–626.

Morris, C.A. 2007. A review of genetic resistance to disease in *Bos taurus* cattle. *The Veterinary Journal* **174**: 480–491.

Morse, E.V., Bauer, D.C., Langham, R.F., Lang, R.W., and Ullrey, D.E. 1958. Experimental leptospirosis. IV. Pathogenesis of porcine *Leptospira pomona* infections. *American Journal of Veterinary Research* **19**: 388–394.

Mortensen, H., Nielsen, S.S., and Berg, P. 2004. Genetic variation and heritability of the antibody response to *Mycobacterium avium* subspecies *paratuberculosis* in Danish Holstein cows. *Journal of Dairy Science* **87**: 2108–2113.

Mortensen, S., Stryhn, H., Sogaard, R., Boklund, A., Stark, K.D.C., Christensen, J., and Willeberg, P. 2002. Risk factors for infection of sow herds with porcine reproductive and respiratory syndrome (PRRS) virus. *Preventive Veterinary Medicine* **53**: 83–101.

Mosier, D.A., Confer, A.W., and Panciera, R.J. 1989. The evolution of vaccines for bovine pneumonic pasteurellosis. *Research in Veterinary Science* **47**: 1–10.

Mousing, J., Christensen, J., Haugegaard, J., Schirmer, A.L., and Friis, N.F. 1995. A seroepidemiological survey of *Leptospira bratislava* infections in Danish sow herds. *Preventive Veterinary Medicine* **23**: 201–221.

Muggli-Cockett, N.E., Cundiff, L.V., and Gregory, K.E. 1992. Genetic analysis of bovine respiratory disease in beef calves during the first year of life. *Journal of Animal Science* **70**: 2013–2019.

Murtaugh, M.P., Xiao, Z., and Zuckermann, F. 2002. Immunological responses of swine to porcine reproductive and respiratory syndrome virus infection. *Viral Immunology* **15**: 533–547.

National Agricultural Statistics Service, Agricultural Statistics Board, United States Department of Agriculture. 2006. Cattle death loss. usda.mannlib.cornell.edu/usda/current/CattDeath/CattDeath-05-05-2006.pdf.

National Animal Health Monitoring Service. 1997. Johne's disease on U.S. dairy operations. *Report # N245.1097*. USDA: APHIS: VS, CEAH, National Animal Health Monitoring System, Fort Collins, CO. www.aphis.usda.gov/vs/ceah/ncahs/nahms/dairy/dairy96/DR96john.pdf.

Neto, J.S.F., Vasconcellos, S.A., Ito, F.H., Moretti, A.S., Camargo, C.A., Sakamoto, S.M., Marangon, S., Turilli, C., and Martini, M. 1997. *Leptospira interrogans* serovar *icterohaemorrhagiae* seropositivity and the reproductive performance of sows. *Preventive Veterinary Medicine* **31**: 87–93.

Neumann, E.J., Kliebenstein, J.B., Johnson, C.D., Mabry, J.W., Bush, E.J., Seitzinger, A.H., Green, A.L., and Zimmerman, J.J. 2005. Assessment of the economic impact of porcine reproductive and respiratory syndrome on swine production in the US. *Journal of American Veterinary Medical Association* **227**: 385–392.

Nicoletti, P. 1980. The epidemiology of bovine brucellosis. *Advances in Veterinary*

Science and Comparative Medicine **24**: 69–98.

O'Connor, A.M., Reed, M.C., Denagamage, T.N., Yoon, K.J., Sorden, S.D., and Cooper, V.L. 2007. Prevalence of calves persistently infected with bovine viral diarrhea virus in beef cow-calf herds enrolled in a voluntary screening project. *Journal of the American Veterinary Medical Association* **230**: 1691–1696.

Office International des Epizooties (OIE). 1997. *Part I: Reports on Animal Health Status and Disease Control Methods and List A Diseases Outbreaks—Statistics*, p. 249.

Osorio, F.A. 2002. Porcine reproductive and respiratory syndrome. *Proceedings of the 17th International Pig Veterinary Society Congress* **1**: 105–112.

Paixao, T.A., Poester, F.P., Neta, A.V.C., Borges, A.M., Lage, A.P., and Santos, R.L. 2007. NRAMP1 3′ untranslated region polymorphisms are not associated with natural resistance to *Brucella abortus* in cattle. *Infection and Immunity* **75**: 2493–2499.

Pellerin, C., Hurk, J.V.D., Lecomte, J., and Tijssen, P. 1994. Identification of a new group of bovine viral diarrhea virus strains associated with sever outbreaks and high mortalities. *Virology* **203**: 260–268.

Perfumo, C.J., and Sanguinetti H.R. 2003. Argentina: Serological studies on PRRS virus. In: Zimmerman, J., and Yoon, K.-J. (eds.), *The Porcine Reproductive and Respiratory Compendium* (2nd Ed.). Des Moines, IA: National Pork Board, pp. 209–211.

Petry, D.B., Holl, J.W., Weber, J.S., Doster, A.R., Osorio, F.A., and Johnson, R.K. 2005. Biological responses to porcine respiratory and reproductive syndrome virus in pigs of two genetic populations. *Journal of Animal Science* **83**: 1494–1502.

Petry, D.B., Lunney, J., Boyd, P., Kuhar, D., Blankenship, E., and Johnson, R.K. 2007. Differential immunity in pigs with high and low responses to porcine reproductive and respiratory syndrome virus infection. *Journal of Animal Science* **85**: 2075–2092.

Pirtle, E.C., Sacks, J.M., Nettles, V.F., and Rollor, E.A. 1989. Prevalence and transmission of pseudorabies virus in an isolated population of feral swine. *Journal of Wildlife Diseases* **25**: 605–607.

Plant, J.W., Claxton, P.D., Jakovljevic, D., and de Saram, W. 1976. *Brucella abortus* infection in the bull. *Australian Veterinary Journal* **52**: 17–20.

Price, R.E., Templeton, J.W., Smith, R., and Adams, L.G. 1990. Ability of mononuclear phagocytes from cattle naturally resistant or susceptible to brucellosis to control in vitro intracellular survival of *Brucella abortus*. *Infectious Immunity* **58**: 879–886.

Przytulski, T. and Porzeczkowska, D. 1979. Polymorphism of blood serum amylase and transferrin and leptospirosis in Large White Polish pigs. *British Veterinary Journal* **135**: 103–107.

Przytulski, T. and Pcrzeczkowska, D. 1980. Studies on genetic resistance to leptospirosis in pigs. *British Veterinary Journal* **136**: 25–32.

Python, P., Jorg, H., Neuenschwander, S., Asai-Coakwell, M., Hagger, C., Burgi, E., Bertschinger, H.U., Stranzinger, G., and Vogeli, P. 2005. Inheritance of the F4ab, F4ac and F4ad *E. coli* receptors in swine and examination of four candidate genes for F4acR. *Journal of Animal Breeding and Genetics* **122**: 5–14.

Ragan, V.E. 2002. The animal and plant health inspection service (APHIS) brucellosis eradication program in the United States. *Veterinary Microbiology* **90**: 11–18.

Rankin, J.E.F. 1965. *Brucella abortus* in bulls: A study of twelve naturally-infected cases. *The Veterinary Record* **77**: 132–135.

Plant, J.W., Claxton, P.D., Jakovljevic, D., and DeSaram, W. 1976. *Brucella abortus* infection in the bull. *Australian Veterinary Journal* **52**: 17–20.

Rautiainen, E., Konradsson, K., Lium, B., Mortensen, S., and Wallgren, P. 2001. Disease surveillance strategies in swine. *Acta Veterinaria Scandinavcia* **42**(Supplement 1): S31–S42.

Reiner, G., Melchinger, E., Kramarova, M., Pfaff, E., Büttner, M., Saalmüller, A., and Geldermann, H. 2002. Detection of quantitative trait loci for resistance/susceptibility to pseudorabies virus in swine. *Journal of General Virology* **83**: 167–172.

Ribble, C.S., Meek, A.H., Jim, G.K., and Guichon, P.T. 1995. The pattern of fatal fibrinous pneumonia (shipping fever) affecting calves in a large feedlot in Alberta (1985–1988). *Canadian Veterinary Journal* **36**: 753–757.

Rice, J.A., Carrasco-Medina, L., Hodgins, D.C., and Shewen, P.E. 2008. *Mannheimia haemolytica* and bovine respiratory disease. *Animal Health Research Reviews* **8**: 117–128.

Ridpath, J.F. 2003. BVDV genotypes and biotypes: Practical implications for diagnosis and control. *Biologicals* **31**: 127–131.

Ridpath, J.F. and Bolin, S.R. 1995. The genomic sequence of a virulent bovine viral diarrhea virus (BVDV) from the type 2 genotype: Detection of a large genomic insertion in a noncytopathic BVDV. *Virology* **212**: 39–46.

Ridpath, J.F. and Bolin, S.R. 1997. Comparison of the complete genomic sequence of the border disease virus, BD31, to other pestiviruses. *Virus Research* **50**: 237–243.

Ridpath, J.F., Bolin, S.R., and Dubovi, E.J. 1994. Segregation of bovine viral diarrhea virus into genotypes. *Virology* **205**: 66–74.

Romero, C.H., Meade, P.N., Shultz, J.E., Chung, H.Y., Gibbs, E.P., Hahn, E.C., and Lollis, G. 2001. Venereal transmission of pseudorabies viruses indigenous to feral swine. *Journal of Wildlife Diseases* **37**: 289–296.

Roth, F., Zinsstag, J., Orkhon, D., Ochir, G.C., Hutton, G., Cosivi, O., Carrin, G., and Otte, J. 2003. Human health benefits from livestock vaccination for brucellosis: Case study. *Bulletin of the World Health Organization* **81**: 867–876.

Rothschild, M.F., Hill, H.T., Christian, L.L., and Warner, C.M. 1984. Genetic differences in serum-neutralization titers of pigs after vaccination with pseudorabies modified live-virus vaccine. *American Journal of Veterinary Research* **45**: 1216–1218.

Rowland, R.R., Lawson, S., Rossow, K., and Benfield, D.A. 2003. Lymphoid tissue tropism of porcine reproductive and respiratory syndrome virus replication during persistent infection of pigs originally exposed to virus in utero. *Veterinary Microbiology* **96**: 219–235.

Sawhney, A.N. and Saxena, S.P. 1967. Leptospiral infections in domestic animals and man in the state of Madhya Pradesh. Incidence in sheep, dogs and pigs. (A serological study). *Indian Veterinary Journal* **44**: 1008–1101.

Schaefer, R., Ciacci-Zanella, J., Mores, N., Pan, K.A., Dambros, R.M.F., Schiochet, M.F., and Coldebella, M. 2006. Characterization of Aujeszky's disease virus isolated from South Brazil in the last twenty years by restriction enzyme analysis. *Brazilian Journal of Microbiology* **37**: 390–394.

Settles, M., Zanella, R., McKay, S., Taylor, J., Fyock, T., Whitlock, R., Schukken, Y. et al. 2009. A whole genome association analysis identifies loci associated with *Mycobacterium avium* subsp. *paratuberculosis* infection status in US Holstein cattle. *Animal Genetics* **40**: 655–662.

Skamene, E. 1989. Genetic control of susceptibility to mycobacterial infections. *Reviews of Infectious Diseases* **11**: S394–S399.

Skamene, E., Gros, P., Forget, A., Kongshavn, P.A., St. Charles, C., and Taylor, B.A. 1982. Genetic regulation of resistance to intracellular pathogens. *Nature* **297**: 506–509.

Smith, C., King W., and Gilbert, N. 1962. Genetic parameters of British Large White Bacon pigs. *Animal Production* **4**: 128.

Smyth, R.H. and Christie, G.J. 1950. Some observations on Johne's disease with a further note on the examination of fecal samples. *Veterinary Record* **62**: 429–450.

Snowder, G.D., Van Vleck, L.D., Cundiff, L.V., and Bennett, G.L. 2005. Influence of breed, heterozygosity and disease incidence on estimates of variance components of respiratory disease in preweaned beef calves. *Journal of Animal Science* **83**: 1247–1261.

Snowder, G.D., Van Vleck, L.D., Cundiff, L.V., and Bennett, G.L. 2006. Bovine respiratory disease in feedlot cattle: Environmental, genetic, and economic factors. *Journal of Animal Science* **84**: 1999–2008.

Snowder, G.D., Van Vleck, L.D., Cundiff, L.V., Bennett, G.L., Koohmaraie, M., and Dikeman, M.E. 2007. Bovine respiratory disease in feedlot cattle: Phenotypic, environmental and genetic correlations with growth, carcass, and longissimus muscle palatability traits. *Journal of Animal Science* **85**: 1885–1892.

Sockett, D.C., Conrad, T.A., Thomas, C.B., and Collins, T.M. 1992. Evaluation of four serological tests for bovine paratuberculosis. *Journal of Clinical Microbiology* **30**: 1134–1139.

Stabel, J.F., Steadham, E.M., and Bolin, C.A. 1997. Heat inactivation of *Mycobacterium paratuberculosis* in raw milk: Are current pasteurization conditions effective? *Applied Environmental Microbiology* **63**: 4975–4977.

Stipkovits, L., Glavits, R., Ripley, P., Molnar, T., Tenk, M., and Szeredi, L. 2000. Pathological and immunohistochemical studies of pneumonia in calves experimentally induced by *Mycoplasma bovis*. In: Bergonier, D., Berthelot, X., and Frey, J. (eds.), *Mycoplasmas of Ruminants: Pathogenicity, Diagnostics, Epidemiology and Molecular Genetics*. Brussels: European Commission, pp. 27–30.

Stokstad, M., Niskanen, R., Lindberg, A., Thren, P., Belak, S., Alenius, S., and Loken, T. 2003. Experimental infection of cows with bovine viral diarrhoea virus in early pregnancy—Findings in serum and foetal fluids. *Journal of Veterinary Medicine. B, Infectious Diseases and Veterinary Public Health* **50**: 424–429.

Storz, J., Lin, X., Purdy, C.W., Chouljenko, V.N., Kousoulas, K.G., Enright, F.M., Gilmore, W.C., Briggs, R.E., and Loan, R.W. 2000a. Coronavirus and *Pasteurella* infections in bovine shipping fever pneumonia and Evans' criteria for causation. *Journal of Clinical Microbiology* **38**: 3291–3298.

Storz, J., Purdy, C.W., Lin, X., Burrell, M., Truax, R.E., Briggs, R.E., Frank, G.H., and Loan, R.W. 2000b. Isolation of respiratory bovine coronavirus, other cytocidal viruses, and *Pasteurella* spp from cattle involved in two natural outbreaks of shipping fever. *Journal of the American*

Veterinary Medical Association **216**: 1599–1604.

Sweeney, R.W. 1996. Transmission of paratuberculosis. In: Sweeney, R.W. (ed.), *Paratuberculosis (Johne's disease)*. Philadelphia: Saunders, pp. 305–312.

Tanaka, S.M., Sato, M., Taniguichi, T., and Yokomizo, Y. 1994. Histopathological and morphometrical comparison of granulomatous lesions in BALB/c and C3H/HeJ mice inoculated with *Mycobacterium paratuberculosis*. *Journal of Comparative Pathology* **110**: 381–388.

Taylor, K.H., Taylor, J.F., White, S.N., and Womack, J.E. 2006. Identification of genetic variation and putative regulatory regions in bovine CARD15. *Mammalian Genome* **17**: 892–901.

Thacker, E.L., Thacker, B.J., and Janke, B.H. 2001. Interaction between *Mycoplasma hyopneumoniae* and swine influenza virus. *Journal of Clinical Microbiology* **39**: 2525–2530.

Thanawongnuwech, R., Brown, G.B., Halbur, P.G., Roth, J.A., Royer, R.L., and Thacker, B.J. 2000. Pathogenesis of porcine reproductive and respiratory syndrome virus-induced increase in susceptibility to *Streptococcus suis* infection. *Veterinary Pathology* **37**: 143–152.

Tomorremorell, M., Pijoan, C., Janni, K., Walker, R., and Joo, H.S. 1997. Airborne transmission of *Actinobacillus pleuropneumoniae* and porcine reproductive and respiratory syndrome virus in nursery pigs. *American Journal of Veterinary Research* **58**: 828–832.

Turner, L.H. 1967. Leptospirosis I. *Transactions of the Royal Society of Tropical Medicine and Hygiene* **61**: 842–855.

Van Der Leek, M.L., Becker, H.N., Pirtle, E.C., Humphrey, P., Adams, C.L., All, B.P., Erickson, G.A., Belden, R.C., Frandenberger, W.B., and Gibbs, E.P.J. 1993. Prevalence of pseudorabies (Aujeszky's disease) virus antibodies in feral swine in Florida. *Journal of Wildlife Diseases* **29**: 403–409.

Van der Linden, I.F.A., Van der Linde-Bril, E.M., Voermans, J.J.M., Van Rijn, P.A., Pol, J.M.A., Martin, R., and Steverink, P.J.G.M. 2003. Oral transmission of porcine reproductive and respiratory syndrome virus by muscle of experimentally infected pigs. *Veterinary* **97**: 45–54.

Van Donkersgoed, J., Janzen, E.D., and Harland, R.J. 1990. Epidemiological features of calf mortality due to hemophilosis in a large feedlot. *Canadian Veterinary Journal* **31**: 821–825.

Van Donkersgoed, J., Ribble, C.S., Boyer, L.G., and Townsend, H.G. 1993. Epidemiological study of enzootic pneumonia in dairy calves in Saskatchewan. *Canadian Journal of Veterinary Research* **57**: 247–254.

Van Oirschot, J.T. 1995. Bovine herpesvirus in semen of bulls and the risk of transmission: A brief overview. *Veterinary Quarterly* **17**: 29–33.

Van Til, L.D. and Dohoo, I.R. 1991. A serological survey of leptospirosis in Prince Edward island swine herds and its association with infertility. *Canadian Journal of Veterinary Research* **55**: 352–35.

Veazey, R.S., Horohov, D.W., Krahenbuhl, J.L., Taylor, H.W., Oliver, J.L. III, and Snider, T.G. III. 1995a. Comparison of the resistance of C5BL/6 and C3H/He mice to infection with *Mycobacterium paratuberculosis*. *Veterinary Microbiology* **47**: 79–87.

Veazey, R.S., Taylor, H.W., Horohov, D.W., Krahenbuhl, J.L., Oliver, J.L. III, and Snider, T.G. III. 1995b. Histopathology of C57BL/6 mice inoculated orally with *Mycobacterium paratuberculosis*.

Journal of Comparative Pathology **113**: 75–80.

Verger, J.M., Grimont, F., Grimont, P.A.D., and Grayon, M. 1985. *Brucella*, a monospecific genus as shown by deoxyribonucleic acid hybridization. *International Journal of Systemic Bacteriology* **35**: 292–295.

Vincent, A.L., Thacker, B.J., Halbur, P.G., Rothschild, M.F., and Thacker, E.L. 2006. An investigation of susceptibility to porcine reproductive and respiratory syndrome virus between two genetically diverse commercial lines of pigs. *Journal of Animal Science* **84**: 49–57.

Watts, J.L., Yancey, R.J. Jr., Salmon, S.A., and Case, C.A. 1994. A 4-year survey of antimicrobial susceptibility trends for isolates from cattle with bovine respiratory disease. *North American Journal of Clinical Microbiology* **32**: 725–731.

Welsh, M.D., Adair, B.M., and Foster, J.C. 1995. Effect of BVD virus infection on alveolar macrophage functions. *Veterinary Immunology and Immunopathology* **46**: 195–210.

Welsh, R.D., Dye, L.B., Payton, M.E., and Confer, A.W. 2004. Isolation and antimicrobial susceptibilities of bacterial pathogens from bovine pneumonia: 1994–2002. *Journal of Veterinary Diagnostic Investigation* **16**: 426–431.

Wensvoort, G., Terpstra, C., Pol, J.M.A., ter Laak, E.A., Bloemraad, M., de Kluyver, E.P., Kragten, C. et al. 1991. Mystery swine disease in the Netherlands: The isolation of Lelystad virus. *Veterinary Quarterly* **13**: 121–130.

Whittington, R.J. and Sergeant, E.S.G. 2001. Progress towards understanding the spread, detection and control of *Mycobacterium avium* subsp *paratuberculosis* in animal populations. *Australian Veterinary Journal* **79**: 267–278.

Wilkie, B. and Mallard, B. 1999. Selection for high immune response: An alternative approach to animal health maintenance. *Veterinary Immunology and Immunopathology* **72**: 231–235.

Williams, E.S. and Spraker, T.R. 1979. Paratuberculosis in free-ranging bighorn sheep and a Rocky Mountain goat with a brief review of the disease in wild species. *Annual Proceedings of the American Association of Zoo Veterinarians*, Denver, pp. 122–124a.

Wilson, S.H. 1989. Why are meaningful field trials difficult to achieve for bovine respiratory syncytial virus. *Canadian Veterinary Journal* **30**: 299–302.

Wittman, G., Jakubik, J., and Ahl, R. 1980. Multiplication and distribution of Aujeszky's disease (pseudorabies) virus in vaccinated and non-vaccinated pigs after intranasal infection. *Archives of Virology* **75**: 29–41.

Wittum, T.E., Grotelueschen, D.M., Brock, K.V., Kvasnicka, W.G., Floyd, J.G., Kelling, C.L., and Odde, K.G. 2001. Persistent bovine viral diarrhea virus infection in beef herds. *Preventive Veterinary Medicine* **49**: 83–94.

World Health Organization. 1999. Leptospirosis worldwide, 1999. *Weekly Epidemiology Record* **74**: 237–242.

Yates, W.D.G. 1982. A review of infectious bovine rhinotracheitis, shipping fever pneumonia, and viral-bacterial synergism in respiratory disease of cattle. *Canadian Journal of Comparative Medicine* **46**: 225–263.

Zuckermann, F.A. 2000. Aujeszky's disease virus: Opportunities and challenges. *Veterinary Research* **31**: 121–131.

6

Comparative Genomics of the Y Chromosome and Male Fertility

Wansheng Liu

6.1 Introduction

In mammals, sex determination is accomplished by the XY sex chromosome mechanism. The X chromosome is large and rich in genes, whereas the Y chromosome is small and heterochromatic and carried by males only. Since the initial report of the Y chromosome in 1921 (Painter 1921), there were debates on its biologic role. Until the late 1950s when the first XO females and XXY males were reported (Ford et al. 1959; Jacobs and Strong 1959), many biologists still considered the Y chromosome as a "genetic wasteland" with its only function for male sex determination. The wasteland model has been revised during the past two decades (Lahn and Page 1997), especially when the human Y chromosome was completely sequenced (Skaletsky et al. 2003). We now know that, in addition to the genes residing on the pseudoautosomal regions (PARs), there are 78 protein-coding genes in 27 gene families in the male-specific region (MSY) of the human Y chromosome. These MSY genes (or families) are clustered together, are expressed predominantly or exclusively in the testis, and play crucial roles in sex determination, differentiation, and maintenance of male-specific organs, spermatogenesis, and male fertility. In this chapter, we will discuss the genomic structure and gene content of the Y chromosome. We will review recent studies on functions and polymorphisms of the Y chromosome genes with a focus on the candidate genes for spermatogenesis and male fertility.

6.2 Characteristics of the mammalian Y chromosome

The Y chromosome is usually the smallest chromosome of the genome, comprising 2–3% of the haploid genome (Krausz and Degl'Innocenti 2006). It is an acrocentric chromosome and contains a short arm (Yp)

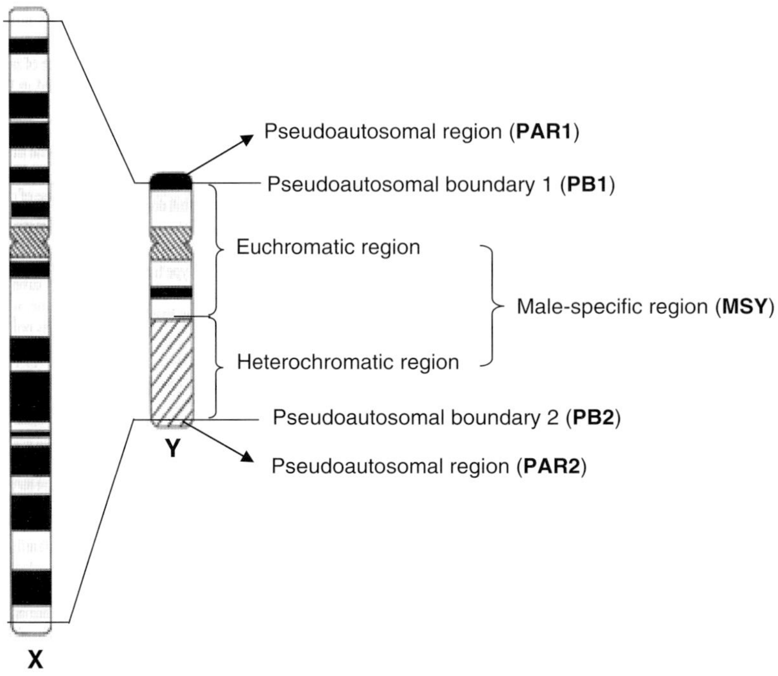

Figure 6.1 The human sex chromosomes. The G-banded ideogram of the X and Y chromosome is shown on the left. The different regions of the Y are listed on the right.

and long arm (Yq). A small region located in the distal part of either the Yp or the Yq (of both arms in human) is known as the PAR, and the rest of the Yp and Yq contain Y chromosome male-specific sequences (MSY), previously known as the "non-recombining region" (NRY) (Figure 6.1). These two regions have contrasting genetic properties. The X and Y chromosomes pair and recombine at the PAR during male meiosis that mediates X and Y segregation; the Y-specific and X-specific regions do not. The MSY region, comprising 95% of the DNA content of the Y chromosome, can be further divided into two regions: euchromatic and heterochromatic (Figure 6.1). According to the human Y chromosome sequence, the euchromatic region contains at least three different types of sequences (for details, see Section 6.3.1) and harbors all genes of the MSY, whereas the heterochromatic region contains Y-specific repetitive sequences that give it the special Giemsa-staining feature in C-banding.

There are several characteristics that make the Y chromosome unique among all other nuclear chromosomes. These include (1) the absence of recombination with the X chromosome in the MSY region at meiosis, (2) the abundance of Y-specific repetitive sequences, (3) the tendency of its genes to degenerate during evolution, (4) the presence of the massive palindromes (in humans and chimpanzees) within which high frequency of Y–Y gene conversion is evident, and (5) the accumulation and functional cluster of testis genes (gene families) in the MSY (Lahn and Page 1997; Tilford et al. 2001; Rozen et al. 2003). The absence of recombination makes genetic mapping of the MSY virtually impossible, and the depth, breadth, and complexity of the repetitive sequences make sequencing extremely difficult (Liu and Ponce de León 2007). Therefore, mapping and sequencing strategies applied successfully elsewhere in the genome have faltered in the MSY (Tilford et al. 2001), making the Y chromosome a difficult target for linkage mapping and, ultimately, sequencing. These difficulties led the mammalian genome sequencing projects, including humans (Lander et al. 2001), mice (Waterston et al. 2002), rats (Krzywinski et al. 2004), cows (www.ncbi.nlm.nih.gov/projects/genome/guide/cow/), and dogs (Lindblad-Toh et al. 2005), to choose to sequence DNA from female animals. To date, only the human Y chromosome has been sequenced (Skaletsky et al. 2003), and the chimpanzee Y chromosome has been partially sequenced (Kuroki et al. 2006).

6.3 Sequence and gene content of the Y chromosome

6.3.1 Sequence of the human Y chromosome

The human Y chromosome is roughly 58 megabases (Mb), within which PAR1 and PAR2 together is about 3 Mb. There are a total of 28 genes in the human PAR1 and 6 in PAR2, which have homologs on the X. The pseudoautosomal genes are typical of those found elsewhere in the genome, having diverse functions and not being significantly involved in reproduction. So, these genes are not discussed in the present review.

The MSY region is ~55 Mb (www.ncbi.nlm.nih.gov/projects/mapview/maps.cgi?taxid=9606&chr=Y). The euchromatic portion of the MSY (Figure 6.1) is roughly 24 Mb, whereas the heterochromatic region varies in length (polymorphic) among individuals, ranging from undetectable in some men to over half of the chromosome in some others (Krausz and Degl'Innocenti 2006). As the heterochromatic region is genetically inert and contains highly repetitive sequence families, this review will only focus on the euchromatic region of the MSY.

The euchromatic region contains three classes of sequences: X-transposed (3.4 Mb), X-degenerate (8.6 Mb), and ampliconic (10.2 Mb) (Skaletsky et al. 2003). The X-transposed sequences are 99% identical to the sequences in the Xq21 region as a result of a massive X to Y transposition, which happened ~3–4 million years ago after the divergence of the human and chimpanzee lineages (Page et al. 1984). These sequence blocks contain the highest density of interspersed repeat elements and harbor only two genes (TGIF2LY and PCDH11Y) homolog to the X-copies in Xq21 (Figure 6.2).

The X-degenerate sequences are 60–96% similar to the sequences on the X. It is believed that the X-degenerate sequences are relics of ancient autosomes from which the modern X and Y chromosomes evolved. These sequences contain 16 single-copy genes that have a homolog on the X. All of these genes, except for the SRY, are expressed ubiquitously. SRY is expressed predominantly in the testes (Figure 6.2). The discovery of X-degenerate sequences has been considered the most important fact in support of a theory proposed for the mammalian sex chromosome evolution. This theory states that as the X and Y chromosomes evolved from an autosomal pair, the X chromosome maintained most of its ancestor's genes, whereas the Y chromosome lost them (Ohno 1967; Graves and Schmidt 1992; Graves 1998, 2006; Lahn and Page 1999; Skaletsky et al. 2003). It explains why the Y chromosome genes tend to degenerate during evolution.

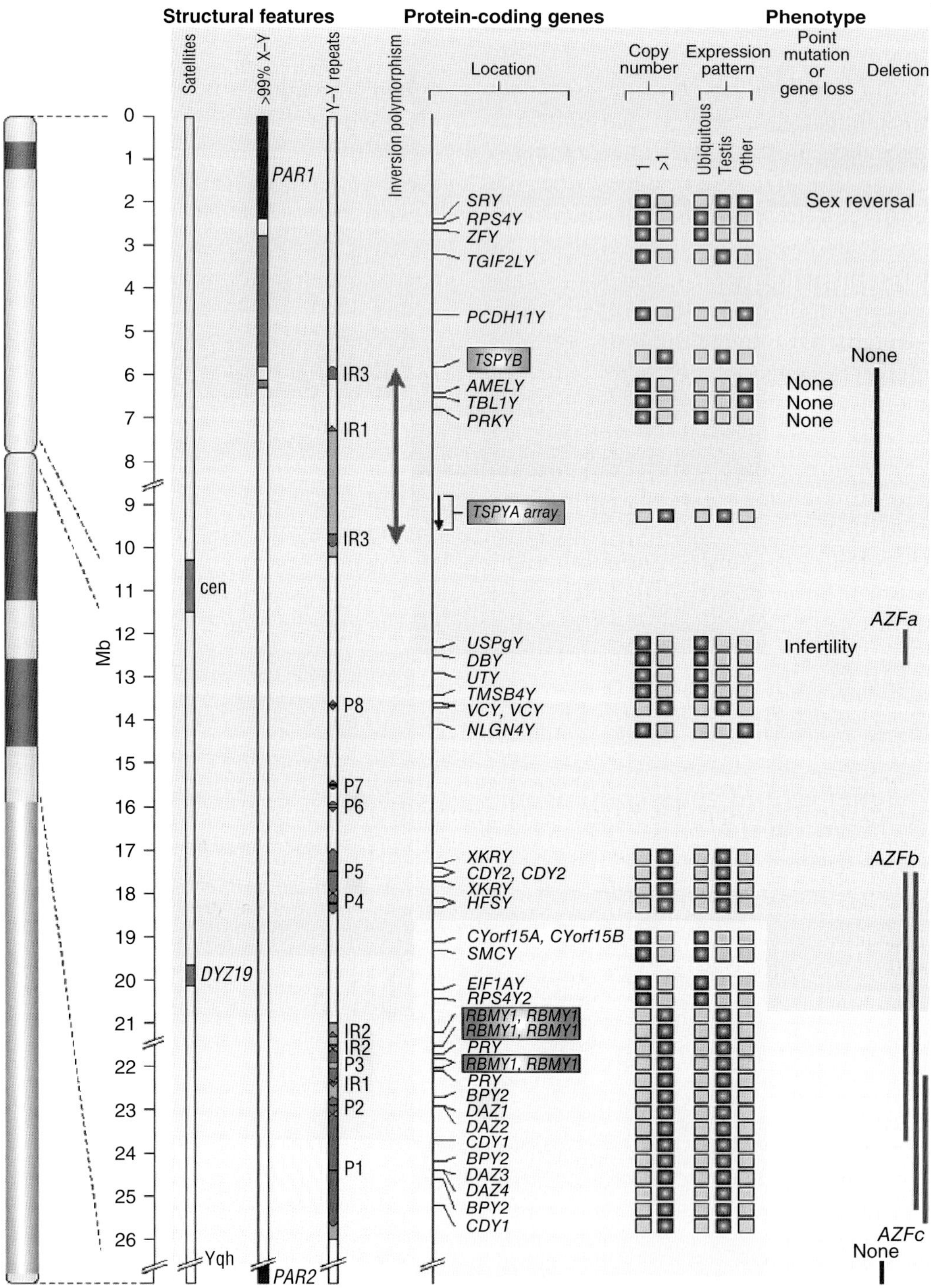
Structural features
Protein-coding genes
Phenotype
Satellites
>99% X–Y
Y–Y repeats
Inversion polymorphism
Location
Copy number
Expression pattern
Point mutation or gene loss
Deletion
1
>1
Ubiquitous
Testis
Other
Mb
PAR1
IR3
IR1
IR3
cen
P8
P7
P6
P5
P4
DYZ19
IR2
IR2
P3
IR1
P2
P1
Yqh
PAR2
SRY
RPS4Y
ZFY
TGIF2LY
PCDH11Y
TSPYB
AMELY
TBL1Y
PRKY
TSPYA array
USPgY
DBY
UTY
TMSB4Y
VCY, VCY
NLGN4Y
XKRY
CDY2, CDY2
XKRY
HFSY
CYorf15A, CYorf15B
SMCY
EIF1AY
RPS4Y2
RBMY1, RBMY1
RBMY1, RBMY1
PRY
RBMY1, RBMY1
PRY
BPY2
DAZ1
DAZ2
CDY1
BPY2
DAZ3
DAZ4
BPY2
CDY1
Sex reversal
None
None
None
None
Infertility
AZFa
AZFb
AZFc
None

The most significant finding from the human Y chromosome sequence project is the discovery of the ampliconic sequence blocks and the eight massive palindromes in the MSY. These so-called "amplicons" exhibit intrachromosomal identities of 99.9% or greater (Skaletsky et al. 2003). The ampliconic sequences contain nine distinct MSY-specific protein-coding gene families (~60 transcriptions), with copy numbers ranging from two (VCY, XKRY, HSFY, PRY) to three (BPY2) to four (CDY, DAZ) to six (RBMY) to ~35 (TSPY) (Figure 6.2). In addition, the ampliconic sequences also contain 75 putative non-coding transcription units. Together, the ampliconic sequences contain 135 of the 156 transcription units identified in the human MSY. All of the nine gene families and the majority of the noncoding transcription units are expressed predominantly or exclusively in the testes. It is thought that the genes in the ampliconic region were derived through three converging processes: (1) amplification of the X-degenerate genes (e.g., RBMY and VCY), (2) transposition and amplification of autosomal genes (DAZ), and (3) retroposition and amplification of autosomal genes (CDY) (Skaletsky et al. 2003). These processes were considered evidence for the Y chromosome's ability to accumulate and to maintain maleness and testes genes as these gene families are all involved in male reproduction (see below). Given the facts that the MSY does not recombine during meiosis, the molecular mechanism for conserving Y gene functions across evolutionary time is the Y–Y gene conversion, which is very common in the ampliconic sequences that exhibit intrachromosomal identities of 99.9% (Rozen et al. 2003).

All together, the human MSY contains a total of 156 transcription units, out of which 78 are protein-coding genes that collectively encode 27 (18 single-copy genes and 9 gene families) distinct proteins (Figure 6.2). The remaining 78 transcription units are noncoding transcripts (Skaletsky et al. 2003).

6.3.2 Genes on the Y chromosome are functionally clustered

Autosomes in mammals appear to contain randomly mixed collections of genes with extremely heterogeneous patterns of developmentally regulated expression in different tissues. The mammalian sex chromosomes,

Figure 6.2 Human Y chromosome structure and gene content. From left to right: cytogenetic features of the chromosome and their approximate locations, which are numbered from the Yp telomere. Structural features include three satellite regions (cen, DYZ19, and Yqh), segments of X–Y identity (PAR1 and PAR2) and high similarity, and Y–Y repeated sequences in which the regions with the greatest sequence identity are designated "IR" for "inverted repeat" and "P" for "palindrome." An inversion polymorphism on Yp that distinguishes haplogroup P from most other lineages is indicated. The locations of the 27 distinct Y-specific protein-coding genes are shown; some are present in more than one copy and their expression patterns are summarized. Pseudoautosomal genes and Y-specific noncoding transcripts are not shown. On the right, the phenotypes that are associated with gene inactivation or loss are indicated; some deletions produce no detectable phenotype (black) and represent polymorphisms in the population, whereas others result in infertility (AZFa, AZFb, and AZFc); contributions of the individual deleted genes are discussed in the text. Reproduced with permission from the Nature Publishing Group, Jobling, M.A. and Tyler-Smith, C. 2003. *Nature Reviews Genetics* **4**: 598–612.

however, are enriched for sex-biased genes related to sex development and reproduction (Lahn and Page 1997; Saifi and Chandra 1999; Wang et al. 2001; Khil et al. 2004, 2005). Analyses of the function of genes on the human and mouse X chromosomes have revealed that the X has significantly higher number of sex- and reproduction-related (SRR) genes, which are not subject to selection by meiotic sex chromosome inactivation (Khil et al. 2004). In contrast to the chromosomal distribution of most tissue-enriched genes which is not significantly different from randomness, ovary- and placenta-enriched genes are significantly overrepresented on the X chromosome. As to the testis-enriched genes, data from a few studies are inconsistent. Khil et al. (2004) reported underrepresented testis-enriched genes on the X, whereas Wang et al. discovered a roughly 15-fold enrichment on the X chromosome for male germ cell-specific spermatogonially expressed genes (Wang et al. 2001). In addition, the human X is associated with a high incidence of mental disability caused by mutations in genes on the X that are required for brain development or function (Zechner et al. 2001; Delbridge et al. 2008). This is supported by the findings of fivefold enrichment of the "intelligence" genes on the X. Therefore, the X chromosome is "smart and sexy" (Graves 2006).

Compared with the X, the Y chromosome is even more biased in its gene content with the highest density of testis-enriched genes. These testis genes are functionally coherent in the MSY and clustered together in the ampliconic and X-degenerate segments (Lahn and Page 1997).

Based on gene functions, the 27 protein-coding genes (families) in the human MSY can be classified into four groups. Group I contains only one gene (SRY) that is involved in sex determination. Group II contains 15 X-degenerate genes (EIF1AY, CYorf15A, and 15B; DDX3Y; NLGN4Y; PCDH11Y; PRKY; USP9Y; RPS4Y1; RPS4Y2; JARID1D; TBL1Y; TGIF2LY; TMSB4Y; ZFY; and UTY) that are single copy and expressed ubiquitously. Four genes (USP9Y, DDX3Y, UTY, and TMSB4Y) from this group are clustered together and play a significant role in spermatogenesis. The rest of the genes in this group have "housekeeping" functions. Group III contains AMELY and a proposed growth control Y (GCY) gene, which are associated with the control of embryonic growth, stature, and development of teeth (Fincham et al. 1990; Kirsch et al. 2004). However, the proposed GCY has not been confirmed transcriptionally. If the GCY gene is confirmed, it may have a potential value for growth selection in animal breeding. Group IV contains the nine multicopy genes (RBMY, DAZ, TSPY, CDY, BPY2, XKRY, PRY, HSFY, and VCY), of which eight families are localized in the palindromes of the ampliconic sequences. They are testis specific and are functionally coherent in spermatogenesis and fertility (Figure 6.2) (Lahn and Page 1997; Skaletsky et al. 2003).

6.3.3 Genes on the Y chromosome are not conserved between species

In contrast to the X, which is highly conserved between species in its size, gene content and gene order (except for the order in the rodent X) (O'Brien et al. 1999; Bourque et al. 2004; Raudsepp et al. 2004a), the Y chromosome is not conserved between species. It varies in size and gene content, and in homology relationships to the X (Graves 2006). Comparative mapping among several species, including human, mouse, cow, sheep, cat, dog, lemur, and wallaby, has demonstrated that the Y chromosome PAR genes are not conserved at all (Graves et al.

1998; Graves 2006). The variation of the PAR gene content and the origin and evolution of the PAR in mammals were explained by an "addition–attrition" theory, as proposed by Graves et al. (Graves 1995, 1998; Graves et al. 1998).

Other than human, chimpanzee, and mouse, the gene content of MSY is poorly known. Based on the currently available data, a comparative map of the MSY genes was constructed (Figure 6.3). From this map, we can conclude that (1) the SRY gene, the most important gene on the Y chromosome that triggers the male sex development, is conserved on all mammalian species studied so far. (2) Genes such as RBMY, TSPY, RPS4Y, SMCY (JARID1D), EIF2S3Y, AMELY, ZFY, UTY, DDX3Y, USP9Y, HSFY, and UBE1Y, which are homologous on the X, are either conserved in all species or lost in some species. For instance, UBE1Y is absent in primates, whereas AMELY is absent in the rodent Y chromosomes. These genes are thought to be present on the ancestral (or so-called proto) Y chromosome and are lost as a result of Y degradation during the mammalian evolution in different lineage (Graves 2006). (3) Genes that are acquired from autosomes by different mechanisms and amplified thereafter on the Y are not conserved. For instance, the DAZ gene family emerged on the Y chromosome

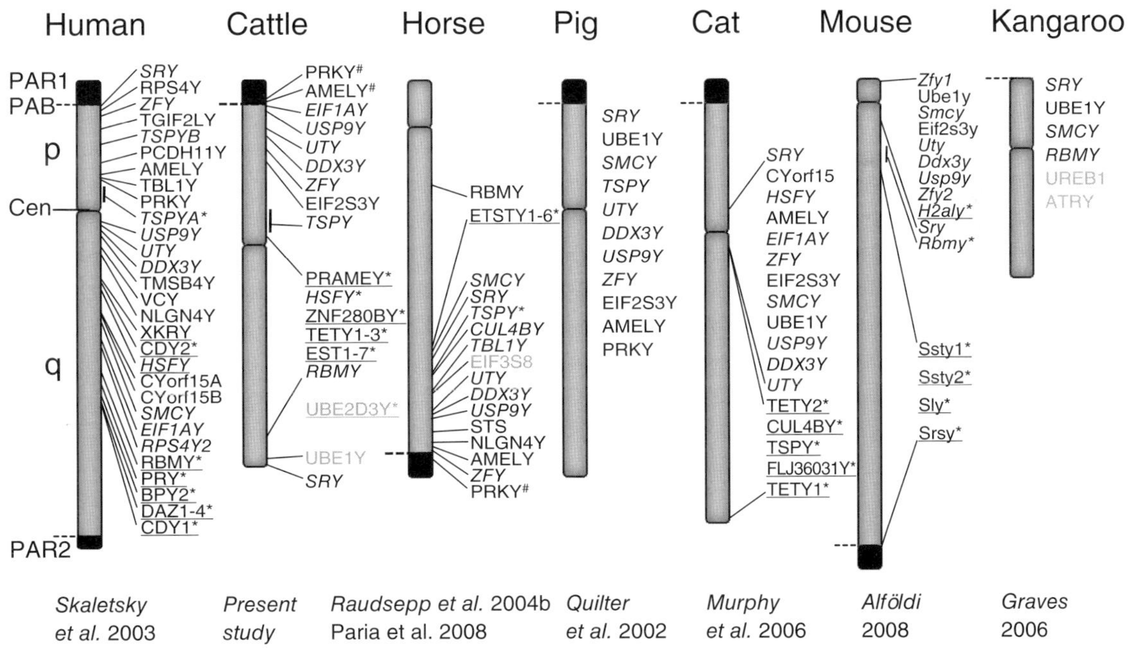

Figure 6.3 A comparison of MSY genes among several mammalian species. Active genes in MSY are marked for conserved (italic), species-specific (underlined), and not conserved on MSY (black). A few MSY genes are pseudoautosomal in cattle and horse (#). Pseudogenes are in gray. Multiple copy genes are labeled by an asterisk (*). Highly amplified human and bovine TSPY and the mouse Rbmy gene are indicated by a vertical bar. Where available, an order and/or map position of loci is provided. PAR, pseudoautosomal region (black box); PAB, pseudoautosomal boundary (dashed line); Cen, centromere; p, short arm; q, long arm.

by transposition of an autosomal DAZL gene (Saxena et al. 1996; Shan et al. 1996 Gromoll et al. 1999) and identified only in Old World monkeys and great apes, while DAZL is present in all vertebrates (Cooke et al. 1996; Saxena et al. 1996). Another example is the CDY gene that has been identified only in primates (Lahn and Page 1999; Kostova et al. 2002; Rottger et al. 2002; Wimmer et al. 2002; Dorus et al. 2003). Two members, CDYL and CDYL2 genes, map to autosomes and exist in most mammalian species (Lahn and Page 1999; Dorus et al. 2003; Wang et al. 2008). It is believed that the CDY gene arose by retroposition of a processed messenger RNA derived from an autosomal CDYL gene (Lahn and Page 1999; Dorus et al. 2003; Bhowmick et al. 2006). Recent efforts in searching for novel Y chromosome genes in cats, dogs (Murphy et al. 2006), horses (Paria et al. 2008), and cattle (our unpublished data) have resulted in a dozen species-specific Y-linked genes (termed TETY for testis expressed transcript on the Y) (Figure 6.3). As we expected, these novel genes appear to either have an autosomal origin (such as ZNF280BY in bovine and TETY1 in cat) or be relics of X-degeneration (cat TETY2 and CUL4BY) as described above (Figure 6.3). It is clear that more species-specific Y-linked genes will be identified once a complete MSY gene content (sequence) is available for most, if not all, mammalian species.

6.4 Function of Y chromosome genes in spermatogenesis and male fertility

6.4.1 Y chromosome deletion and infertility in men

Initial research before the 1950s suggested that the (human) Y chromosome was a wasteland of repetitive DNA carrying no genetic information apart from the sex-determining factor. The first association between Y chromosome function and spermatogenic failure was demonstrated by Tiepolo and Zuffardi in 1976 using a Y chromosome deletion mapping approach. Since patients with *de novo* microscopically detectable deletions of a region on Yq showed azoospermia and infertility, a spermatogenesis factor, designed AZoospermia Factor (AZF), was proposed to be located in the Yq deleted region (Tiepolo and Zuffardi 1976). However, the deleted region was not defined until the mid-1980s when Y chromosome-specific markers (especially ~200 sequence-tagged sites [STS]) and a fine deletion interval map were developed (Vergnaud et al. 1986; Vollrath et al. 1992). These markers have permitted simple deletion analysis in infertile men with azoospermia or severe oligozoospermia by polymerase chain reaction (PCR) to define AZF. By screening 76 Yq STS markers in a large group of 370 patients with azoospermia and severe oligozoospermia, Vogt et al. (1996) defined AZF to three non-overlapping regions, termed AZFa, AZFb, and AZFc (Figure 6.2). It was later found that the AZFb and AZFc regions overlapped on the basis of the Y sequence (Repping et al. 2002; Skaletsky et al. 2003).

The importance of Y chromosome microdeletions is underlined by the fact that they account for 10–18% of idiopathic primary testiculopathies (azoospermia and severe oligozoospermia) (Foresta et al. 2001; Kleiman et al. 2003; Krausz and Degl'Innocenti 2006). Y microdeletions have been found exclusively in patients with <1 million spermatozoa/mL, but are very rare in patients with >5 million spermatozoa/mL. The most frequent deletions occurred at AZFc (~60%), followed by AZFb, and AZFb+c, or AZFa+b+c (~35%), whereas deletions in AZFa are infrequent

(~5%) (Krausz and McElreavey 1999; Krausz and Degl'Innocenti 2006). However, reports on isolated gene-specific deletions within AZF regions are limited partially because most Y microdeletions involved more than one gene. To date, gene-specific deletions have been identified for DDX3Y (Foresta et al. 2000), HSFY (Vinci et al. 2005), and USP9Y (Sun et al. 1999), which are all associated with azoospermia and/or severe oligozoospermia and infertility.

6.4.2 Candidate genes for spermatogenesis and male fertility

As discussed above (in Section 6.2), genes on the Y chromosome are clustered together and functionally coherent in spermatogenesis and male fertility. These clusters correspond to the AZFa, AZFb, and AZFc regions (Figure 6.2). To date, the following seven candidate genes have been confirmed or proposed to play a role in spermatogenesis and male fertility: DDX3Y and USP9Y in the AZFa region; RBMY, PRY, HSFY, and CDY in AZFb; and DAZ and CDY in AZFc (reviewed in Krausz and Degl'Innocenti 2006).

DAZ gene family

The Deleted in Azoosermia (DAZ) gene has four copies on the human Y chromosome and two autosomal homologs, DAZL (DAZ-like) on chromosome 3 and BOLL (bol, boule-like [Drosophila]) on chromosome 2. It is believed that the ancestor member of the family is BOLL, which gave rise to DAZL via duplication prior to the divergence of vertebrates and invertebrates (Cauffman et al. 2005). Later, during primate evolution, the autosomal DAZL gene gave rise to DAZ on the Y chromosome by transposition, repeat amplification, and pruning (Saxena et al. 1996; Shan et al. 1996; Gromoll et al. 1999). As a result, DAZ has been identified only in Old World monkeys and great apes, while DAZL is present in all vertebrates (Cooke et al. 1996; Saxena et al. 1996).

The four copies of DAZ (DAZ1-4) form two pairs within palindromic duplications (Figure 6.2); one pair of genes (DAZ1-2) is part of the P2 palindrome and the second pair (DAZ3-4) is part of the P1 palindrome. Each gene contains a 2.4-kb repeat including a 72-bp exon, called the DAZ repeat. The number of DAZ repeats is variable, and there are several variations in the sequence of the DAZ repeat. Alternative splicing produces multiple transcript variants encoding different isoforms. The DAZ family is expressed exclusively in germ cells encoding proteins that contain a highly conserved RNA recognition motif (RRM) (Yen 2004). DAZ genes are expressed only in the testis (Reijo et al. 1995), while DAZL is expressed in both the testis and the ovary (Seligman and Page 1998; Dorfman et al. 1999; Cauffman et al. 2005). Both DAZ and DAZL are detected in the nuclei of primordial germ cells (PGCs) in fetal gonads (Xu et al. 2001) and are believed to function in the development of PGCs and in germ cell differentiation and maturation (reviewed in Yen 2004). Mutations in these genes have been linked to infertility in several species. In flies, loss of function of the boule gene, an ortholog of DAZ/DAZL, leads to male sterility (Eberhart et al. 1996). In frogs, inhibition of Xdazl (Xenopus daz-like) leads to defective migration and a reduction in PGCs (Houston and King 2000). In mice, a disruption of the Dazl leads to prenatal loss of all germ cells in both sexes during prenatal germ cell development, and hence infertility (Ruggiu et al. 1997). Further, the infertile phenotype of Dazl null mouse (Dazl–/–) can be partially rescued by a human DAZ transgene (Slee et al. 1999). As the gene name suggests, deletion or microdeletion in DAZ gene(s) in the AZFc region leads to severe spermatogenic failure and infertility in men (Reijo et al. 1995; Vogt et al. 1996; Ferlin et al. 1999), and DAZL transcripts in the testes are lower in men with spermatogenic failure compared with fertile men (Lin et al. 2001).

CDY gene family

Like the DAZ family, the human chromodomain protein, Y-linked (CDY) gene family also consists of three members. One member locates on the Y chromosome in the form of four highly related copies (CDY1 and CDY2), which are identified only in primates (Lahn and Page 1999; Kostova et al. 2002; Rottger et al. 2002; Wimmer et al. 2002; Dorus et al. 2003). Two members, CDYL and CDYL2 genes, map to chromosome 6 and 16, respectively, which exist in most mammalian species (Lahn and Page 1999; Dorus et al. 2003). It is believed that in the common ancestor of mammals, the progenitor of the gene family duplicated to result in the two autosomal genes, CDYL and CDYL2. The CDY gene instead arose by retroposition and subsequent amplification of a processed messenger RNA derived from an autosomal CDYL gene (Lahn and Page 1999; Dorus et al. 2003; Bhowmick et al. 2006). In the simian lineage, the CDY gene was retained and subsequently amplified into many copies. In other mammals, this gene has been lost (Lahn and Page 1999; Dorus et al. 2003).

The human chromosome Y has two identical copies of CDY1 in P1 palindrome and of CDY2 in P5 palindrome (Figure 6.2). Proteins encoded by genes of the CDY family contain two domains: the chromodomain and the enoyl-coenzyme A hydratase-isomerase catalytic domain (Lahn and Page 1999; Dorus et al. 2003). The chromodomain has been implicated in remodeling the chromatin structure (Briton-Jones and Haines 2000). Biochemical analysis revealed that human CDY/CDYL and mouse Cdyl proteins exhibit histone acetyltransferase activity *in vitro*. Both proteins can specifically acetylate H4 and H2A, with H4 being the strongly preferred substrate (Lahn et al. 2002). Chromodomain proteins are localized to the nucleus of late spermatids where histone hyperacetylation takes place. Histone hyperacetylation is thought to facilitate the transition in which protamines replace histones as the major DNA-packaging protein (Lahn and Page 1999; Kleiman et al. 2001, 2003; Kostova et al. 2002; Lahn et al. 2002; Dorus et al. 2003). In addition to the chromodomain, the catalytic domain in the carboxy-terminal portion of the CDY protein family can bind CoA and histone deacetylases, and acts as a corepressor of transcription in somatic cells, as well as during the early stages of spermatogenesis (Caron et al. 2003).

Members of the CDY gene family have different expression patterns. While the human Y-linked CDY genes are specifically expressed in the testis (Lahn and Page 1999; Kostova et al. 2002; Dorus et al. 2003), the autosomal homologs CDYL and CDYL2 are expressed ubiquitously. Two protein isoforms (540 and 554 amino acids [aa]) were identified for CDY and three isoforms (598, 544, and 309 aa) for CDYL as results of alternative spliced transcripts. In mice and rabbits, both CDYL and CDYL2 express a ubiquitous long transcript and a highly abundant testis-specific short transcript (Lahn and Page 1999; Kostova et al. 2002; Dorus et al. 2003). In cows, at least four transcript variants for the bovine CDYL gene have been identified (Wang et al. 2008). These transcripts are expressed predominantly or exclusively in the bovine testis (Wang et al. 2008).

RBMY gene family

The RNA-binding motif protein, Y-linked (RBMY) gene encodes a protein containing an RNA-binding motif in the N-terminus and four repetitions of a Ser-Arg-Gly-Tyr tetrapeptide motif (SRGY box) in the C-terminus (Ma et al. 1993). Multiple copies of this gene are found in the AZFb region of chromosome Y, and the encoded protein is thought to be involved in spermatogenesis. Most copies of this locus are pseudogenes, although six highly similar copies have full-length open reading frames (ORFs) and are considered functional (Ma et al. 1993; Skaletsky et al. 2003). Four functional copies of this gene are located within inverted repeat IR2, and the remaining two in P3 palindrome, along with two copies of PRY genes (Figure 6.2).

The mouse RBMY contains an RNA-binding motif with 74% similarity to the human RBMY, followed by only one SRGY box. RBMY-deficient mice do not show the same phenotype as in humans; they have abnormal sperm development but are not sterile (Mahadevaiah et al. 1998). The human RBMY gene family has an X-located member (RBMX), which encodes the widely expressed heterogeneous nuclear ribonucleoprotein G (hnRNPG) (Delbridge et al. 1999). There is also another member named hnRNPG-T, a functional retrogene on chromosome 11 (Elliott et al. 2000). The human RBMY is expressed specifically in the nuclei of adult male germ cells throughout all transcriptionally active stages of spermatogenesis, and deletion of the functional copies of RBMY is associated with an arrest of meiotic division I during spermatogenesis (Elliott et al. 1997; Elliott 2004).

PRY gene family

The PTPBL-related gene on Y (PRY, also known as PTPN13LY, PTPN13-like, Y-linked gene) has two nearly identical copies within inverted repeat IR2, along with RBMY genes on the human Y chromosome (Figure 6.2). PRY is expressed specifically in the testis. It encodes a protein, which has a low degree of similarity to the protein tyrosine phosphatase, non-receptor type 13. Since PRY is located in AZFb, and is often deleted in patients with severe infertility, it was proposed to play an important role in spermatogenesis (Stouffs et al. 2001). However, further study indicated that the role of the PRY gene in spermatogenesis could be questioned, but suggested its probable involvement in apoptosis of defective spermatozoa (Stouffs et al. 2004).

HSFY gene family

The heat shock transcription factor, Y-linked gene (HSFY) belongs to the HSF family. Two identical HSFY copies are situated in the P4 palindrome in proximal AZFb on Yq, whereas there are four pseudogenes mapping in two clusters in the P1 palindrome of AZFc and in P3. Sequences similar to few HSFY exons are also located in Yp, X, and 22 (Shinka et al. 2004; Tessari et al. 2004). The HSF family is a group of highly conserved regulators that play a role as transcriptional activators of heat shock protein (HSP) genes. This family consists of multiple genes in mammals, and it is thought to be involved in physiological pathways related to development and differentiation, other than in stress response (Wang et al. 2003; Tessari et al. 2004). HSFY is characterized by an HSF-type DNA-binding domain related to the HSF2 gene on chromosome 6 (Ferlin et al. 2003). HSF2 is expressed at high levels and is only active in embryogenesis and spermatogenesis (Pirkkala et al. 2001). HSF2 regulates the expression of many genes, and, in particular, it controls the hsp70 gene family promoter (Sistonen et al. 1992). There are three different transcripts and protein isoforms found in humans, each containing an HSF domain typical of HSF proteins. These HSFY transcripts are differentially expressed, transcript 1 being present in many tissues including the testis, and transcripts 2 and 3 being testis specific (Tessari et al. 2004). These observations suggest that HSFY could play an important role in spermatogenesis. HSFY gene-specific deletion has been reported in an azoospermic man, confirming its function in spermatogenesis (Vinci et al. 2005).

At least six different transcripts have been identified from an adult testis for the bovine HSFY gene, which can be classified into two groups. Group I contains three transcripts with a different size 3′UTR that encode a peptide of 207 aa. Group II contains the remaining three transcripts that encode a 417 aa. One of the group II transcripts shows a deletion of 9 bp, resulting in an isoform of 414 aa. All bovine HSFY isoforms contain the conserved HSF DNA-binding domain. Preliminary data indicated that multicopies of the bHSFY gene are present in the bovine Y chromosome (W.-S. Liu, T.-C. Chang, and Y. Yang, unpublished data).

DDX3Y

The DEAD box polypeptide 3, Y-linked gene (DDX3Y, also known as DBY, DEAD box gene on the Y) encodes a putative ATP-dependent RNA helicase (Abdelhaleem 2005). This gene belongs to the DEAD box protein family, which is characterized by the conserved motif Asp-Glu-Ala-Asp (DEAD) (Rosner and Rinkevich 2007). In humans, DDX3Y is located in the AZFa interval in MSY (Vogt et al. 1996; Lahn and Page 1997). This gene has a homolog on the X chromosome (DDX3X) that escapes X inactivation. DDX3Y may have both housekeeping and testis-specific functions as the gene produces two transcripts with different expression patterns. A long transcript (DDX3Y-L) is ubiquitously expressed, while a short transcript (DDX3Y-S) is testis specific (Foresta et al. 2000; Vong et al. 2006). DDX3Y protein was found only in male germ lines, while DDX3X protein was present in all testicular and nontesticular tissues, suggesting that DDX3Y is essential for human spermatogenesis (Ditton et al. 2004). The DDX3Y and its neighbor genes, USP9Y and UTY, in the AZFa region, as well as the gene order of USP9Y-DDX3Y-UTY, are similar in both human and mouse Y chromosomes, indicating that they represent a conserved synteny block (Mazeyrat et al. 1998; Vong et al. 2006). In mice, partial deletion in this conserved syntenic region on the short arm of the Y chromosome results in early failure of spermatogenesis and consequent sterility (Sutcliffe and Burgoyne 1989; Simpson and Page 1991; Wood et al. 1997; Mazeyrat et al. 1998). Although Ddx3y was proposed to be a Y chromosome gene essential for normal spermatogonial proliferation in the mouse (Mazeyrat et al. 2001), a recent study indicated that the Ddx3y gene may not be required for mouse spermatogenesis (Vong et al. 2006), signifying that there may be species-specific difference between the function of DDX3Y/Ddx3y in human and mice. A recent proteomics searching and functional study identified a number of spermatogenesis-enriched chromatin proteins with roles in fertility in *Caenorhabditis elegans* (Chu et al. 2006). One of these proteins, named glh-2, is an RNA helicase, which is the sole protein on the list that has an ortholog, the DDX3Y gene, on the human Y chromosome, suggesting a conserved function of DDX3Y in spermatogenesis and male fertility between *C. elegans* and humans (Chu et al. 2006).

Like the human DDX3Y and mouse Ddx3y gene, two transcripts were identified for the bovine DDX3Y gene, which are identical except for a three-base-pair insertion and an expanded 3′UTR in the bovine DDX3Y-L. The bovine DDX3Y is predominantly expressed in the testis. The bDDX3Y--S encodes a peptide of 660 aa, while the bDDX3Y-L encodes a 661 aa as a result of the insertion of a serine in the bDDX3Y-L peptide. Both bDDX3Y isoforms contain the conserved motifs of DEAD-box RNA helicases (Liu et al. 2008).

USP9Y

The ubiquitin-specific peptidase 9, Y-linked (USP9Y, previously known as DFFRY, Drosophila fat facets-related Y-linked) is a member of the peptidase C19 family. The human USP9Y is located in AZFa clustered with DDX3Y and UTY, which have a homolog on the X chromosome (USP9X). The gene contains 46 exons and has a transcript of 10048 bp, encoding a protein of 2555 aa (Brown et al. 1998). USP9Y protein does not contain known functional domains except for the Cys and His domains, which are present in ubiquitin-specific proteases. The latter cleave the ubiquitin moiety from ubiquitin-fused precursors and ubiquitinylated proteins (Lee et al. 2003). Both USP9Y and USP9X genes are expressed ubiquitously in different human tissues (Lahn and Page 1997).

USP9Y is one of the few Yq genes for which isolated gene-specific deletions/mutations have been reported: a massive deletion removing the entirety of USP9Y (Brown et al. 1998), a 4-bp splice-donor site deletion resulting in a severely truncated protein (Sun et al. 1999), and two cases of deletions removing part of the USP9Y gene, which were spontaneously transmitted from father to son (Krausz et al. 2006). These mutations are associated with Sertoli cell-only (SCO) syndrome, azoospermia, and male infertility.

Other genes that may be involved in spermatogenesis and male fertility

In addition to the major candidate genes/families for spermatogenesis and fertility described above, several other genes (families) including BPY2, JARID1D, XKRY, RPS4Y, eIF1AY, and CYorf15A and CYorf15B are also located within the AZFb region on the human Y chromosome (Figure 6.2). It is not clear if any of these genes play a role in spermatogenesis. The basic charge, Y-linked 2 (BPY2-, also known as VCY2) has three nearly identical copies; two of them are in the P1 palindrome. BPY2 is expressed specifically in the testis (Lahn and Page 1997; Tse et al. 2003). As the PBY2-encoded protein interacts with ubiquitin protein ligase E3A, it may be involved in male germ cell development and male infertility (Wong et al. 2002; Tse et al. 2003). The jumonji, AT-rich interactive domain 1D (JARID1D, previously known as SMCY or HY), encodes one human H-Y epitope. A short peptide derived from this protein is a minor histocompatibility (H-Y) antigen, which can lead to graft rejection of male donor cells in a female recipient (Wang et al. 1995). The XK, Kell blood group complex subunit-related, Y-linked (XKRY) gene has two identical copies located in the P5 palindrome in the proximal region of AZFb (Figure 6.2). XKRY encodes a putative membrane transport protein similar to the XK precursor and is expressed specifically in the testis (Lahn and Page 1997). The ribosomal protein S4, Y-linked (RPS4Y) gene has two copies on the human Y: one copy (RPS4Y1) maps to the Yp near SRY, the other (RPS4Y2) maps to Yq. The encoded ribosomal protein S4 is a component of the 40S subunit. The eukaryotic translation initiation factor 1A, Y-linked (EIF1AY) encodes a Y isoform of an eIF1A, an essential translation initiation factor. There is a homologous gene on the X

(EIF1AX). Both Y and X copies are expressed ubiquitously with a high level of expression in the testis (Lahn and Page 1997).

6.5 Polymorphisms of the Y chromosome and male fertility

6.5.1 Different types of polymorphisms on the Y chromosome

As the MSY region is inherited as a single haploid block in linkage from father to male offspring, the Y chromosome represents an important collection of all mutations that have occurred along male lineages during evolution. Therefore, Y chromosome DNA variations/polymorphisms are valuable for investigations on human evolution, forensic analysis, paternity test, and molecular medicine (Krausz et al. 2004)

There are at least four types of polymorphisms identified on the Y chromosome. The first type is the variation in tandemly repeated sequences including microsatellites, short interspersed nuclear elements (SINEs), and long interspersed nuclear elements (LINEs). This type of Y-specific markers, such as DYSl9 (a tetranucleotide microsatellite) and DYS287 (Y Alu insertional polymorphism or "YAP"), is applied for human evolution and migration studies (Hammer et al. 1997). The second type of variation on the Y chromosome is the changes in gene copy number, which is caused by gene duplication/amplification. A good example is the TSPY gene, which is estimated to be 30–60 copies on the human MSY, and 50–200 copies on the bovine MSY (Verkaar et al. 2004; Vodicka et al. 2007). Another type of polymorphism is single base DNA mutations including single nucleotide polymorphisms (SNPs) and single base insertions/deletions (indels). SNPs have been identified from the coding and noncoding sequences of the Y chromosome genes described in Section 6.3.2 of this chapter. Our recent work on the bovine Y chromosome found that single nucleotide indels are popular on BTAY (W.-S. Liu, unpublished data). Finally, the Y chromosome microdeletions are considered as polymorphisms (Machev et al. 2004). Although Y chromosome microdeletions are the most frequent genetic cause of severe oligozoospermia and azoospermia in infertile men (Krausz et al. 2003; Krausz 2005), some microdeletions, such as gr/gr deletions in AZFc region, were also observed in 3.5% of normal fertile men. It was suggested that most gr/gr deletions are neutral variants (Machev et al. 2004).

6.5.2 Polymorphisms and fertility—Genotype and phenotype correlation

The absence of recombination on the MSY means that polymorphisms within this region are in tight association with potential functional variations associated with Y-linked phenotypes. Thus, an indirect way to explore whether Y chromosome genes are involved in fertility/infertility is the characterization of Y chromosome haplogroups in infertile versus normal (fertile) men (Krausz et al. 2004). To date, approximately 600 binary markers have been characterized on the human Y chromosome, and the resultant haplogroups were standardized (Y-chromosome-consortium 2002; Vogt 2005; Karafet et al. 2008). The association between Y haplogroups and sperm counts and/or spermatogenic failure has been investigated in several populations (Table 6.1) (Kuroki et al. 1999; Paracchini et al. 2000, 2002; Quintana-Murci et al. 2001; Carvalho et al. 2003; Ferlin et al. 2005). A haplogroup (termed hg26) was found to be associated with reduced sperm count

Table 6.1 A summary of association studies dealing with Y chromosome polymorphisms and spermatogenic failure.

Phenotype	Population	No. of patients	No. of controls	No. of Y markers	Association found	References
AZF deletion	European	50	50	9	No	Quintana-Murci et al. (2001)
AZF deletion	European	73	299	11	No	Paracchini et al. (2000)
AZF deletion	Japanese	6	84	15	No	Carvalho et al. (2003)
Oligo/azoospermia	Japanese	51	57		No	Carvalho et al. (2003)
Azoospermia	Japanese	106	156	3	No	Kuroki et al. (1999)
Oligo/azoospermia Teratozoospermia	Italian	74	216	8	Yes	
Oligo/azoospermia	Danish	43	128	3	Yes	Krausz et al. (2001)
Sperm count	Italian	41	Not recorded	11	No	Paracchini et al. (2002)
Spermatogenic impairment	Italian	337	263	3	Yes	Ferlin et al. (2005)
Azoospermia	Brazilian	117	122	22	No	Carvalho et al. (2006)
Oligo/azoospermia	Chinese	285	515	12	Yes	Lu et al. (2007)
Oligo/azoospermia	Chinese	414	262	11	Yes	Yang et al. (2008)

($<20 \times 10^6$ spermatozoa/mL) in a Danish population (Krausz et al. 2001; McElreavey and Quintana-Murci 2003), while several "at-risk" Y haplogroups including hgK, hgC, and hgO3* were associated with spermatogenic failure in Chinese Han populations (Lu et al. 2007; Yang et al. 2008). However, similar findings have not been observed in other Japanese, Brazilian, and European populations (Table 6.1) (Paracchini et al. 2000, 2002; Quintana-Murci et al. 2001; Carvalho et al. 2003, 2006), indicating that an association between Y chromosome polymorphism and a phenotype in one population may not be relevant for another. This can be explained by the fact that some haplotypes are largely confined to particular populations.

It has been confirmed that decrease (deletion) in the number of DAZ and CDY copies in the AZFc region is associated with infertility (Moro et al. 2000). However, it was not clear until recently that increase in gene copy number has any effects on Y-linked phenotypes. By a quantitative fluorescence PCR approach, Vodicka et al. (2007) evaluated the relative number of TSPY copies compared with AMELY/X genes in 84 infertile men and 40 controls. Surprisingly, the number of TSPY copies was significantly higher in infertile men compared with the controls ($P = 0.002$). The potential of using the relative number of TSPY copies in clinical diagnostics was also evaluated, and the result indicated that the ratio TSPY/AMELY was a good marker to separate the infertile from the control samples (Vodicka et al. 2007). It remains to be seen whether evaluation of the TSPY copy number offers a new diagnostic approach in relation to the genetic cause of male infertility.

It is rare to find that a point mutation in the Y-linked genes causes spermatogenic failure and infertility. The only report was the identification of a 4-bp deletion in a splice-donor site of the human USP9Y. This deletion leads to an exon being skipped and protein truncation. The man with this mutation showed non-obstructive azoospermia, but his fertile brother has a normal Y without the deletion, suggesting that the USP9Y mutation caused spermatogenic failure (Sun et al. 1999).

An SNP (A/G transition) occurring in the RNA-binding domain of human DAZL gene, leading to Thr54→Ala change (T54A) of DAZL protein, conferred susceptibility to severe spermatogenic failure in men (Teng et al. 2002). Interestingly, there is no report, to date, on whether a single point mutation (SNP or indel) in any Y-linked genes could directly lead to spermatogenic failure.

6.5.3 *The potential to use Y-linked markers for male fertility selection in farm animals*

In contrast to sequence data on humans, chimpanzees, and mice, Y chromosome sequence information is very limited for other mammals. Although considerable efforts have been made in the past few years by animal geneticists to characterize the Y chromosome of farm animals, progress in Y-specific marker development and association study is scarce. To date, over 300 DNA markers have been reported on the bovine Y chromosome (reviewed by Liu and Ponce de León 2007), which includes ~80 microsatellites (MSs), ~30 SNPs, and ~200 STSs. The majority of these markers have multicopies or are part of the repetitive sequences on BTAY as demonstrated by radiation hybrid (RH) mapping (Liu et al. 2002; Liu and de Leon 2004; Liu and Ponce de León 2007). Approximately 1800 Y-linked BACs were isolated and fingerprinted. A high-resolution BTAY physical map is under construction (W.S. Liu, unpublished data). The horse Y chromosome project has generated ~300 Y-unique STSs including 32 gene-specific markers (Raudsepp et al. 2004b; Wallner et al. 2004; Chowdhary et al. 2008). A BAC-based physical map of the equine Y chromosome is available (Raudsepp et al. 2004b). Compared with the bovine and equine Y chromosomes, there are fewer markers available for the porcine (~20) (McGraw et al. 1988; Mileham et al. 1988; Quilter et al. 2002) and ovine (<10) (Meadows et al. 2004) Y chromosomes. We have tested 38 bovine Y-linked MSs on male and female genomic DNAs of sheep and goat by PCR. The results indicated that 11 MSs were amplified in male sheep DNA, and 10 MSs in male goat DNA only, while 21 and 18 MSs amplified products from both male and female DNA in sheep and goat, respectively. All but one MS amplified multiple bands, indicating the existence of multiple copies of these markers in ovine and caprine Y chromosomes (W.-S. Liu and A.F. Ponce de León, unpublished data).

As far as Y-linked marker variation is concerned, 14 of the BTAY MSs were found to be polymorphic in a small population of 17 bulls. These polymorphic markers were combined to produce unique haplotypes that could be used to identify an individual or a group of bulls (Liu et al. 2003). It is expected that more BTAY polymorphic MSs could be discovered in a large population or in animals with more diversified genetic backgrounds, as data from an earlier investigation of four BTAY-specific MSs showed that they were all highly polymorphic in different bovid species including domestic cattle, bison, mithan, buffalo, and yak (Edwards et al. 2000). Furthermore, Y halotypes on the basis of an intronic SNP (A/C) in bUTY and an intronic 2-bp indel in bZFY have been successfully applied to the study of the origin of domestic cattle in Europe (Gotherstrom et al. 2005). In contrast to BTAY, the horse Y chromosome is very low in variability. One investigation found a single haplotype in the domestic horse after analyzing three Y-linked MSs in 49 male horses of 32 different breeds, whereas notable variation has been observed in the other members of the genus *Equus* (Wallner et al. 2004). A separate research obtained a similar result (Lindgren et al. 2004). Lindgren et al. investigated 34 equine Y-linked STSs in 52 male horses from 15 breeds and found that all stallions carried the same Y chromosome haplotype. Based on their sequence data, the authors further predicted that the Y chromosome polymorphism level of the domestic horse was at least 10–30 times lower than that of humans (Lindgren et al. 2004).

At present, there is no report on the association of Y-linked polymorphic marker(s) with male fertility in farm animals except for one report on the bovine DAZL gene (Liu et al. 2008). There are two main reasons for this. One is that the development of farm animal Y chromosome markers, especially in searching for testis gene polymorphic markers, is insufficient to carry out the association study with male fertility. The other reason is that, unlike humans, within which male patients of infertility or subfertility are recorded and analyzed in details, male animals of infertility or subfertility are usually eliminated from the breeding program without genetic evaluation. Therefore, the lack of adequate DNA samples further slows down the molecular study of male infertility (or subfertility) in farm animals.

Infertility or subfertility is a common problem in farm animals (Steffen 1997). For example, it was estimated in the beef industry that as high as 18–30% of beef bulls used in natural service were reproductive deficient (Coulter 1980; Coulter and Kozub 1980). Given the facts that the Y chromosome genes in humans and mice play an essential role in spermatogenesis and fertility, and the recent findings that species-specific genes are present on the Y chromosome of cattle, horses, cats, and dogs (Murphy et al. 2006; Paria et al. 2008), I believe that there is a huge potential to use Y-linked gene polymorphisms and Y chromosome haplotypes in farm animals for marker-assisted selection (MAS) of fertility-related traits to select sires at an early age (as early as newborn) in a breeding program. DNA-based MAS would allow us not only to eliminate those young male animals that may have genetic defects on the Y chromosome, but also to select those that have a potential to become a higher fertility male early on during the progeny test. This is particularly important for large animals such as cattle and horses as subfertile or infertile bulls/stallions are usually not identified until the age of 18–24 months when they are expected to breed. Most importantly, early decision on selecting/eliminating a young male in progeny test will significantly reduce breeding costs.

6.6 Future research directions

The mammalian Y chromosome is unique in many aspects. It is present in the males only and is full of repeated sequences. It is responsible for vital biologic roles in sex determination and male fertility. Since very limited numbers of autosomal genes associated with measures of male germ cell defects have been identified in mammals (Tung et al. 2006; Matzuk and Lamb 2008), the Y chromosome "testis genes" have become the most important source for molecular study of male fertility/infertility. The sequencing of the human Y chromosome has provided us with a big (but not complete) picture of the mammalian Y chromosome in terms of its structure and gene content. Functional studies of the human and mouse Y chromosome genes have identified several candidates for azoospermia/oligozoospermia and infertility. The association of Y chromosome polymorphisms (haplogroups) with azoospermia/oligozoospermia in humans has opened the door for a possibility of using Y chromosome haplotypes/haplogroups for male fertility selection in farm animals. But significant improvements in animal Y chromosome sequence information (such as gene content, polymorphisms, and haplotypes) and association analysis with fertility traits are desperately necessary before designing any Y chromosome gene-based MAS strategy for male fertility selection. The recent discovery of novel Y chromosome genes by "direct cDNA selection" in cattle, horses, and cats holds promise for the further identification and analysis of genetic factors that are involved in regulating spermatogenesis and fertility. This approach will eventually lead to the completion of a "mammalian Y gene catalog" even without sequencing the Y chromosome in all species. In conclusion, the Y chromosome is a "gold mine" in the field of molecular evolution and male fertility; the more we study the Y chromosome, the more knowledge we gain to understand the evolution, migration, and reproduction in mammalian species.

References

Abdelhaleem, M. 2005. RNA helicases: Regulators of differentiation. *Clinical Biochemistry* **38**: 499–503.

Bhowmick, B.K., Takahata, N., Watanabe, M., and Satta, Y. 2006. Comparative analysis of human masculinity. *Genetics and Molecular Research* **5**: 696–712.

Bourque, G., Pevzner, P.A., and Tesler, G. 2004. Reconstructing the genomic architecture of ancestral mammals: Lessons from human, mouse, and rat genomes. *Genome Research* **14**: 507–516.

Briton-Jones, C. and Haines, C.J. 2000. Microdeletions on the long arm of the Y chromosome and their association with male-factor infertility. *Hong Kong Medical Journal* **6**: 184–189.

Brown, G.M., Furlong, R.A., Sargent, C.A., Erickson, R.P., Longepied, G., Mitchell, M., Jones, M.H., Hargreave, T.B., Cooke, H.J., and Affara, N.A. 1998. Characterisation of the coding sequence and fine mapping of the human DFFRY gene and comparative expression analysis and mapping to the Sxrb interval of the mouse Y chromosome of the Dffry gene. *Human Molecular Genetics* **7**: 97–107.

Caron, C., Pivot-Pajot, C., van Grunsven, L.A., Col, E., Lestrat, C., Rousseaux, S., and Khochbin, S. 2003. Cdyl: A new transcriptional co-repressor. *EMBO Reports* **4**: 877–882.

Carvalho, C.M., Fujisawa, M., Shirakawa, T., Gotoh, A., Kamidono, S., Freitas Paulo, T., Santos, S.E., Rocha, J., Pena, S.D., and Santos, F.R. 2003. Lack of association between Y chromosome haplogroups and male infertility in Japanese men. *American Journal of Medical Genetics* **116A**: 152–158.

Carvalho, C.M., Zuccherato, L.W., Bastos-Rodrigues, L., Santos, F.R., and Pena, S.D. 2006. No association found between gr/gr deletions and infertility in Brazilian males. *Molecular Human Reproduction* **12**: 269–273.

Cauffman, G., Van de Velde, H., Liebaers, I., and Van Steirteghem, A. 2005. DAZL expression in human oocytes, preimplantation embryos and embryonic stem cells. *Molecular Human Reproduction* **11**: 405–411.

Chowdhary, B.P., Paria, N., and Raudsepp, T. 2008. Potential applications of equine genomics in dissecting diseases and fertility. *Animal Reproduction Science* **107**: 208–218.

Chu, D.S., Liu, H., Nix, P., Wu, T.F., Ralston, E.J., Yates, J.R. 3rd, and Meyer, B.J. 2006. Sperm chromatin proteomics identifies evolutionarily conserved fertility factors. *Nature* **443**: 101–105.

Cooke, H.J., Lee, M., Kerr, S., and Ruggiu, M. 1996. A murine homologue of the human DAZ gene is autosomal and expressed only in male and female gonads. *Human Molecular Genetics* **5**: 513–516.

Coulter, G.H. 1980. Testicular development: Its management and significance in young beef bulls. In *Proceedings of the 8th Technical Conference on Artificial Insemination and Reproduction (May 2–3, 1980)*. Columbia, MO: National Association of Animal Breeders, pp. 106–111.

Coulter, G.H. and Kozub, G.C. 1980. Efficiency of methods used to test fertility of beef bulls used for multiple-sire breeding under range conditions. *Journal of Animal Science* **67**: 1757–1766.

Delbridge, M.L., Lingenfelter, P.A., Disteche, C.M., and Graves, J.A. 1999. The candidate spermatogenesis gene RBMY has a homologue on the human X chromosome. *Nature Genetics* **22**: 223–224.

Delbridge, M.L., McMillan, D.A., Doherty, R.J., Deakin, J.E., and Graves, J.A. 2008. Origin and evolution of candidate mental retardation genes on the human X chromosome (MRX). *BMC Genomics* **9**: 65.

Ditton, H.J., Zimmer, J., Kamp, C., Rajpert-De Meyts, E., and Vogt, P.H. 2004. The AZFa gene DBY (DDX3Y) is widely transcribed but the protein is limited to the male germ cells by translation control. *Human Molecular Genetics* **13**: 2333–2341.

Dorfman, D.M., Genest, D.R., and Reijo Pera, R.A. 1999. Human DAZL1 encodes a candidate fertility factor in women that localizes to the prenatal and postnatal germ cells. *Human Reproduction* **14**: 2531–2536.

Dorus, S., Gilbert, S.L., Forster, M.L., Barndt, R.J., and Lahn, B.T. 2003. The CDY-related gene family: Coordinated evolution in copy number, expression profile and protein sequence. *Human Molecular Genetics* **12**: 1643–1650.

Eberhart, C.G., Maines, J.Z., and Wasserman, S.A. 1996. Meiotic cell cycle requirement for a fly homologue of human Deleted in Azoospermia. *Nature* **381**: 783–785.

Edwards, C.J., Gaillard, C., Bradley, D.G., and MacHugh, D.E. 2000. Y-specific microsatellite polymorphisms in a range of bovid species. *Animal Genetics* **31**: 127–130.

Elliott, D.J. 2004. The role of potential splicing factors including RBMY, RBMX, hnRNPG-T and STAR proteins in spermatogenesis. *International Journal of Andrology* **27**: 328–334.

Elliott, D.J., Millar, M.R., Oghene, K., Ross, A., Kiesewetter, F., Pryor, J., McIntyre, M. et al. 1997. Expression of RBM in the nuclei of human germ cells is dependent on a critical region of the Y chromosome long arm. *Proceedings of the National Academy of Sciences of the United States of America* **94**: 3848–3853.

Elliott, D.J., Venables, J.P., Newton, C.S., Lawson, D., Boyle, S., Eperon, I.C., and Cooke, H.J. 2000. An evolutionarily conserved germ cell-specific hnRNP is encoded by a retrotransposed gene. *Human Molecular Genetics* **9**: 2117–2124.

Ferlin, A., Moro, E., Garolla, A., and Foresta, C. 1999. Human male infertility and Y chromosome deletions: Role of the

AZF-candidate genes DAZ, RBM and DFFRY. *Human Reproduction* **14**: 1710–1716.

Ferlin, A., Moro, E., Rossi, A., Dallapiccola, B., and Foresta, C. 2003. The human Y chromosome's azoospermia factor b (AZFb) region: Sequence, structure, and deletion analysis in infertile men. *Journal of Medical Genetics* **40**: 18–24.

Ferlin, A., Tessari, A., Ganz, F., Marchina, E., Barlati, S., Garolla, A., Engl, B., and Foresta, C. 2005. Association of partial AZFc region deletions with spermatogenic impairment and male infertility. *Journal of Medical Genetics* **42**: 209–213.

Fincham, A.G., Hu, Y.Y., Lau, E., Pavlova, Z., Slavkin, H.C., and Snead, M.L. 1990. Isolation and partial characterization of a human amelogenin from a single fetal dentition using HPLC techniques. *Calcified Tissue International* **47**: 105–111.

Ford, C.E., Jones, K.W., Polani, P.E., De Almeida, J.C., and Briggs, J.H. 1959. A sex-chromosome anomaly in a case of gonadal dysgenesis (Turner's syndrome). *Lancet* **1**: 711–713.

Foresta, C., Ferlin, A., and Moro, E. 2000. Deletion and expression analysis of AZFa genes on the human Y chromosome revealed a major role for DBY in male infertility. *Human Molecular Genetics* **9**: 1161–1169.

Foresta, C., Moro, E., and Ferlin, A. 2001. Y chromosome microdeletions and alterations of spermatogenesis. *Endocrine Reviews* **22**: 226–239.

Gotherstrom, A., Anderung, C., Hellborg, L., Elburg, R., Smith, C., Bradley, D.G., and Ellegren, H. 2005. Cattle domestication in the Near East was followed by hybridization with aurochs bulls in Europe. *Proceedings. Biological Sciences* **272**: 2345–2350.

Graves, J.A. 1995. The evolution of mammalian sex chromosomes and the origin of sex determining genes. *Philosophical Transactions of the Royal Society of London. Series B, Biological Sciences* **350**: 305–311; discussion 311–302.

Graves, J.A. 1998. Evolution of the mammalian Y chromosome and sex-determining genes. *The Journal of Experimental Zoology* **281**: 472–481.

Graves, J.A. 2006. Sex chromosome specialization and degeneration in mammals. *Cell* **124**: 901–914.

Graves, J.A. and Schmidt, M.M. 1992. Mammalian sex chromosomes: Design or accident? *Current Opinion in Genetics & Development* **2**: 890–901.

Graves, J.A., Wakefield, M.J., and Toder, R. 1998. The origin and evolution of the pseudoautosomal regions of human sex chromosomes. *Human Molecular Genetics* **7**: 1991–1996.

Gromoll, J., Weinbauer, G.F., Skaletsky, H., Schlatt, S., Rocchietti-March, M., Page, D.C., and Nieschlag, E. 1999. The Old World monkey DAZ (Deleted in AZoospermia) gene yields insights into the evolution of the DAZ gene cluster on the human Y chromosome. *Human Molecular Genetics* **8**: 2017–2024.

Hammer, M.F., Spurdle, A.B., Karafet, T., Bonner, M.R., Wood, E.T., Novelletto, A., Malaspina, P. et al. 1997. The geographic distribution of human Y chromosome variation. *Genetics* **145**: 787–805.

Houston, D.W. and King, M.L. 2000. A critical role for Xdazl, a germ plasm-localized RNA, in the differentiation of primordial germ cells in Xenopus. *Development (Cambridge, England)* **127**: 447–456.

Jacobs, P.A. and Strong, J.A. 1959. A case of human intersexuality having a possible XXY sex-determining mechanism. *Nature* **183**: 302–303.

Karafet, T.M., Mendez, F.L., Meilerman, M.B., Underhill, P.A., Zegura, S.L., and Hammer, M.F. 2008. New binary polymorphisms reshape and increase resolution of the human Y chromosomal haplogroup tree. *Genome Research* **18**: 830–838.

Khil, P.P., Oliver, B., and Camerini-Otero, R.D. 2005. X for intersection: Retrotransposition both on and off the X chromosome is more frequent. *Trends in Genetics* **21**: 3–7.

Khil, P.P., Smirnova, N.A., Romanienko, P.J., and Camerini-Otero, R.D. 2004. The mouse X chromosome is enriched for sex-biased genes not subject to selection by meiotic sex chromosome inactivation. *Nature Genetics* **36**: 642–646.

Kirsch, S., Weiss, B., Zumbach, K., and Rappold, G. 2004. Molecular and evolutionary analysis of the growth-controlling region on the human Y chromosome. *Human Genetics* **114**: 173–181.

Kleiman, S.E., Bar-Shira Maymon, B., Yogev, L., Paz, G., and Yavetz, H. 2001. The prognostic role of the extent of Y microdeletion on spermatogenesis and maturity of Sertoli cells. *Human Reproduction* **16**: 399–402.

Kleiman, S.E., Yogev, L., Gal-Yam, E.N., Hauser, R., Gamzu, R., Botchan, A., Paz, G. et al. 2003. Reduced human germ cell-less (HGCL) expression in azoospermic men with severe germinal cell impairment. *Journal of Andrology* **24**: 670–675.

Kostova, E., Rottger, S., Schempp, W., and Gromoll, J. 2002. Identification and characterization of the cynomolgus monkey chromodomain gene cynCDY, an orthologue of the human CDY gene family. *Molecular Human Reproduction* **8**: 702–709.

Krausz, C. 2005. Y chromosome and male infertility. *Andrologia* **37**: 219–223.

Krausz, C. and Degl'Innocenti, S. 2006. Y chromosome and male infertility: Update, 2006. *Frontiers in Bioscience* **11**: 3049–3061.

Krausz, C., Degl'Innocenti, S., Nuti, F., Morelli, A., Felici, F., Sansone, M., Varriale, G., and Forti, G. 2006. Natural transmission of USP9Y gene mutations: A new perspective on the role of AZFa genes in male fertility. *Human Molecular Genetics* **15**: 2673–2681.

Krausz, C., Forti, G., and McElreavey, K. 2003. The Y chromosome and male fertility and infertility. *International Journal of Andrology* **26**: 70–75.

Krausz, C. and McElreavey, K. 1999. Y chromosome and male infertility. *Frontiers in Bioscience* **4**: E1–E8.

Krausz, C., Quintana-Murci, L., and Forti, G. 2004. Y chromosome polymorphisms in medicine. *Annals of Medicine* **36**: 573–583.

Krausz, C., Quintana-Murci, L., Rajpert-De Meyts, E., Jorgensen, N., Jobling, M.A., Rosser, Z.H., Skakkebaek, N.E., and McElreavey, K. 2001. Identification of a Y chromosome haplogroup associated with reduced sperm counts. *Human Molecular Genetics* **10**: 1873–1877.

Krzywinski, M., Wallis, J., Gosele, C., Bosdet, I., Chiu, R., Graves, T., Hummel, O. et al. 2004. Integrated and sequence-ordered BAC- and YAC-based physical maps for the rat genome. *Genome Research* **14**: 766–779.

Kuroki, Y., Iwamoto, T., Lee, J., Yoshiike, M., Nozawa, S., Nishida, T., Ewis, A.A. et al. 1999. Spermatogenic ability is different among males in different Y chromosome lineage. *Journal of Human Genetics* **44**: 289–292.

Kuroki, Y., Toyoda, A., Noguchi, H., Taylor, T.D., Itoh, T., Kim, D.S., Kim, D.W. et al. 2006. Comparative analysis of chimpanzee and human Y chromosomes unveils complex evolutionary pathway. *Nature Genetics* **38**: 158–167.

Lahn, B.T. and Page, D.C. 1997. Functional coherence of the human Y chromosome. *Science* **278**: 675–680.

Lahn, B.T. and Page, D.C. 1999. Four evolutionary strata on the human X chromosome. *Science* **286**: 964–967.

Lahn, B.T., Tang, Z.L., Zhou, J., Barndt, R.J., Parvinen, M., Allis, C.D., and Page, D.C. 2002. Previously uncharacterized histone acetyltransferases implicated in mammalian spermatogenesis. *Proceedings of the National Academy of Sciences of the United States of America* **99**: 8707–8712.

Lander, E.S., Linton, L.M., Birren, B., Nusbaum, C., Zody, M.C., Baldwin, J., Devon, K. et al. 2001. Initial sequencing and analysis of the human genome. *Nature* **409**: 860–921.

Lee, K.H., Song, G.J., Kang, I.S., Kim, S.W., Paick, J.S., Chung, C.H., and Rhee, K. 2003. Ubiquitin-specific protease activity of USP9Y, a male infertility gene on the Y chromosome. *Reproduction, Fertility, and Development* **15**: 129–133.

Lin, Y.M., Chen, C.W., Sun, H.S., Tsai, S.J., Hsu, C.C., Teng, Y.N., Lin, J.S., and Kuo, P.L. 2001. Expression patterns and transcript concentrations of the autosomal DAZL gene in testes of azoospermic men. *Molecular Human Reproduction* **7**: 1015–1022.

Lindblad-Toh, K., Wade, C.M., Mikkelsen, T.S., Karlsson, E.K., Jaffe, D.B., Kamal, M., Clamp, M. et al. 2005. Genome sequence, comparative analysis and haplotype structure of the domestic dog. *Nature* **438**: 803–819.

Lindgren, G., Backstrom, N., Swinburne, J., Hellborg, L., Einarsson, A., Sandberg, K., Cothran, G., Vila, C., Binns, M., and Ellegren, H. 2004. Limited number of patrilines in horse domestication. *Nature Genetics* **36**: 335–336.

Liu, W.S., Beattie, C.W., and Ponce de León, F.A. 2003. Bovine Y chromosome microsatellite polymorphisms. *Cytogenetic and Genome Research* **102**: 53–58.

Liu, W.S. and de Leon, F.A. 2004. Assignment of SRY, ANT3, and CSF2RA to the bovine Y chromosome by FISH and RH mapping. *Animal Biotechnology* **15**: 103–109.

Liu, W.S., Mariani, P., Beattie, C.W., Alexander, L.J., and Ponce De León, F.A. 2002. A radiation hybrid map for the bovine Y Chromosome. *Mammalian Genome* **13**: 320–326.

Liu, W.S. and Ponce de León, F.A. 2007. Mapping of the bovine Y chromosome. *Electronic Journal of Biology* **3**: 5–12.

Liu, W.S., Wang, A., and Zhang, H.B. 2008. *Polymorphisms of the bovine DAZL gene are associated with male fertility.* Conference Abstract, PAG-XVI, P174.

Lu, C., Zhang, F., Xia, Y., Wu, B., Gu, A., Lu, N., Wang, S., Shen, H., Jin, L., and Wang, X. 2007. The association of Y chromosome haplogroups with spermatogenic failure in the Han Chinese. *Journal of Human Genetics* **52**: 659–663.

Ma, K., Inglis, J.D., Sharkey, A., Bickmore, W.A., Hill, R.E., Prosser, E.J., Speed, R.M. et al. 1993. A Y chromosome gene family with RNA-binding protein homology: Candidates for the azoospermia factor AZF controlling human spermatogenesis. *Cell* **75**: 1287–1295.

Machev, N., Saut, N., Longepied, G., Terriou, P., Navarro, A., Levy, N., Guichaoua, M. et al. 2004. Sequence family variant loss from the AZFc interval of the human Y chromosome, but not gene copy loss, is strongly associated with male infertility. *Journal of Medical Genetics* **41**: 814–825.

Mahadevaiah, S.K., Odorisio, T., Elliott, D.J., Rattigan, A., Szot, M., Laval, S.H., Washburn, L.L. et al. 1998. Mouse homologues of the human AZF candidate gene RBM are expressed in spermatogonia and spermatids, and map to a Y chromosome deletion interval associated with a high incidence of sperm abnormalities. *Human Molecular Genetics* 7: 715–727.

Matzuk, M.M. and Lamb, D.J. 2008. The biology of infertility: Research advances and clinical challenges. *Nature Medicine* **14**: 1197–1213.

Mazeyrat, S., Saut, N., Sargent, C.A., Grimmond, S., Longepied, G., Ehrmann, I.E., Ellis, P.S., Greenfield, A., Affara, N.A., and Mitchell, M.J. 1998. The mouse Y chromosome interval necessary for spermatogonial proliferation is gene dense with syntenic homology to the human AZFa region. *Human Molecular Genetics* 7: 1713–1724.

Mazeyrat, S., Saut, N., Grigoriev, V., Mahadevaiah, S.K., Ojarikre, O.A., Rattigan, A., Bishop, C., Eicher, E.M., Mitchell, M.J., and Burgoyne, P.S. 2001. A Y-encoded subunit of the translation initiation factor Eif2 is essential for mouse spermatogenesis. *Nature Genetics* **29**: 49–53.

McElreavey, K. and Quintana-Murci, L. 2003. Y chromosome haplogroups: A correlation with testicular dysgenesis syndrome? *APMIS* **111**: 106–113; discussion 114.

McGraw, R.A., Jacobson, R.J., and Akamatsu, M. 1988. A male-specific repeated DNA sequence in the domestic pig. *Nucleic Acids Research* **16**: 10389.

Meadows, J.R., Hawken, R.J., and Kijas, J.W. 2004. Nucleotide diversity on the ovine Y chromosome. *Animal Genetics* **35**: 379–385.

Mileham, A.J., Siggens, K.W., and Plastow, G.S. 1988. Isolation of a porcine male specific DNA sequence. *Nucleic Acids Research* **16**: 11842.

Moro, E., Ferlin, A., Yen, P.H., Franchi, P.G., Palka, G., and Foresta, C. 2000. Male infertility caused by a de novo partial deletion of the DAZ cluster on the Y chromosome. *The Journal of Clinical Endocrinology and Metabolism* **85**: 4069–4073.

Murphy, W.J., Pearks Wilkerson, A.J., Raudsepp, T., Agarwala, R., Schaffer, A.A., Stanyon, R., and Chowdhary, B.P. 2006. Novel gene acquisition on carnivore Y chromosomes. *PLoS Genetics* **2**: e43.

O'Brien, S.J., Menotti-Raymond, M., Murphy, W.J., Nash, W.G., Wienberg, J., Stanyon, R., Copeland, N.G., Jenkins, N.A., Womack, J.E., and Marshall Graves, J.A. 1999. The promise of comparative genomics in mammals. *Science* **286**: 458–462, 479–481.

Ohno, S. 1967. *Sex Chromosomes and Sex-Linked Genes*. Berlin: Springer-Verlag.

Page, D.C., Harper, M.E., Love, J., and Botstein, D. 1984. Occurrence of a transposition from the X-chromosome long arm to the Y-chromosome short arm during human evolution. *Nature* **311**: 119–123.

Painter, T.S. 1921. The Y-chromosome in mammals. *Science* **53**: 503–504.

Paracchini, S., Stuppia, L., Gatta, V., De Santo, M., Palka, G., and Tyler-Smith, C. 2002. Relationship between Y-chromosomal DNA haplotype and sperm count in Italy. *Journal of Endocrinological Investigation* **25**: 993–995.

Paracchini, S., Stuppia, L., Gatta, V., Palka, G., Moro, E., Foresta, C., Mengua, L. et al. 2000. Y-chromosomal DNA haplotypes in infertile European males carrying Y-microdeletions. *Journal of Endocrinological Investigation* **23**: 671–676.

Paria, N., Wilkerson, A.J.P., Murphy, W.J., Chowdhary, B.P., and Raudsepp, T. 2008. *Identification of Y-linked candidate genes for male fertility in horses. Plant & Animal Genomes XVI Conference*, San Diego, CA, January 12–16, p. 35.

Pirkkala, L., Nykanen, P., and Sistonen, L. 2001. Roles of the heat shock transcription factors in regulation of the heat shock response and beyond. *FASEB Journal* **15**: 1118–1131.

Quilter, C.R., Blott, S.C., Mileham, A.J., Affara, N.A., Sargent, C.A., and Griffin, D.K. 2002. A mapping and evolutionary study of porcine sex chromosome genes. *Mammalian Genome* **13**: 588–594.

Quintana-Murci, L., Krausz, C., Heyer, E., Gromoll, J., Seifer, I., Barton, D.E., Barrett, T. et al. 2001. The relationship between Y chromosome DNA haplotypes and Y chromosome deletions leading to male infertility. *Human Genetics* **108**: 55–58.

Raudsepp, T., Lee, E.J., Kata, S.R., Brinkmeyer, C., Mickelson, J.R., Skow, L.C., Womack, J.E., and Chowdhary, B.P. 2004a. Exceptional conservation of horse-human gene order on X chromosome revealed by high-resolution radiation hybrid mapping. *Proceedings of the National Academy of Sciences of the United States of America* **101**: 2386–2391.

Raudsepp, T., Santani, A., Wallner, B., Kata, S.R., Ren, C., Zhang, H.B., Womack, J.E., Skow, L.C., and Chowdhary, B.P. 2004b. A detailed physical map of the horse Y chromosome. *Proceedings of the National Academy of Sciences of the United States of America* **101**: 9321–9326.

Reijo, R., Lee, T.Y., Salo, P., Alagappan, R., Brown, L.G., Rosenberg, M., Rozen, S. et al. 1995. Diverse spermatogenic defects in humans caused by Y chromosome deletions encompassing a novel RNA-binding protein gene. *Nature Genetics* **10**: 383–393.

Repping, S., Skaletsky, H., Lange, J., Silber, S., Van Der Veen, F., Oates, R.D., Page, D.C., and Rozen, S. 2002. Recombination between palindromes P5 and P1 on the human Y chromosome causes massive deletions and spermatogenic failure. *American Journal of Human Genetics* **71**: 906–922.

Rosner, A. and Rinkevich, B. 2007. The DDX3 subfamily of the DEAD box helicases: Divergent roles as unveiled by studying different organisms and in vitro assays. *Current Medicinal Chemistry* **14**: 2517–2525.

Rottger, S., Yen, P.H., and Schempp, W. 2002. A fiber-FISH contig spanning the non-recombining region of the human Y chromosome. *Chromosome Research* **10**: 621–635.

Rozen, S., Skaletsky, H., Marszalek, J.D., Minx, P.J., Cordum, H.S., Waterston, R.H., Wilson, R.K., and Page, D.C. 2003. Abundant gene conversion between arms of palindromes in human and ape Y chromosomes. *Nature* **423**: 873–876.

Ruggiu, M., Speed, R., Taggart, M., McKay, S.J., Kilanowski, F., Saunders, P., Dorin, J., and Cooke, H.J. 1997. The mouse Dazla gene encodes a cytoplasmic protein essential for gametogenesis. *Nature* **389**: 73–77.

Saifi, G.M. and Chandra, H.S. 1999. An apparent excess of sex- and reproduction-related genes on the human X chromosome. *Proceedings. Biological Sciences* **266**: 203–209.

Saxena, R., Brown, L.G., Hawkins, T., Alagappan, R.K., Skaletsky, H., Reeve, M.P., Reijo, R. et al. 1996. The DAZ gene cluster on the human Y chromosome arose from an autosomal gene that was transposed, repeatedly amplified and pruned. *Nature Genetics* **14**: 292–299.

Seligman, J. and Page, D.C. 1998. The Dazh gene is expressed in male and female embryonic gonads before germ cell sex differentiation. *Biochemical and Biophysical Research Communications* **245**: 878–882.

Shan, Z., Hirschmann, P., Seebacher, T., Edelmann, A., Jauch, A., Morell, J., Urbitsch, P., and Vogt, P.H. 1996. A SPGY copy homologous to the mouse gene Dazla and the Drosophila gene boule is autosomal and expressed only in the human male gonad. *Human Molecular Genetics* **5**: 2005–2011.

Shinka, T., Sato, Y., Chen, G., Naroda, T., Kinoshita, K., Unemi, Y., Tsuji, K., Toida, K., Iwamoto, T., and Nakahori, Y. 2004. Molecular characterization of heat shock-like factor encoded on the human Y chromosome, and implications for male infertility. *Biology of Reproduction* **71**: 297–306.

Simpson, E.M. and Page, D.C. 1991. An interstitial deletion in mouse Y chromosomal DNA created a transcribed Zfy fusion gene. *Genomics* **11**: 601–608.

Sistonen, L., Sarge, K.D., Phillips, B., Abravaya, K., and Morimoto, R.I. 1992. Activation of heat shock factor 2 during hemin-induced differentiation of human erythroleukemia cells. *Molecular and Cell Biology* **12**: 4104–4111.

Skaletsky, H., Kuroda-Kawaguchi, T., Minx, P.J., Cordum, H.S., Hillier, L., Brown, L.G., Repping, S. et al. 2003. The male-specific region of the human Y chromosome is a mosaic of discrete sequence classes. *Nature* **423**: 825–837.

Slee, R., Grimes, B., Speed, R.M., Taggart, M., Maguire, S.M., Ross, A., McGill, N.I., Saunders, P.T., and Cooke, H.J. 1999. A human DAZ transgene confers partial rescue of the mouse Dazl null phenotype. *Proceedings of the National Academy of Sciences of the United States of America* **96**: 8040–8045.

Steffen, D. 1997. Genetic causes of bull infertility. *The Veterinary Clinics of North America* **13**: 243–253.

Stouffs, K., Lissens, W., Van Landuyt, L., Tournaye, H., Van Steirteghem, A., and Liebaers, I. 2001. Characterization of the genomic organization, localization and expression of four PRY genes (PRY1, PRY2, PRY3 and PRY4). *Molecular Human Reproduction* **7**: 603–610.

Stouffs, K., Lissens, W., Verheyen, G., Van Landuyt, L., Goossens, A., Tournaye, H., Van Steirteghem, A., and Liebaers, I. 2004. Expression pattern of the Y-linked PRY gene suggests a function in apoptosis but not in spermatogenesis. *Molecular Human Reproduction* **10**: 15–21.

Sun, C., Skaletsky, H., Birren, B., Devon, K., Tang, Z., Silber, S., Oates, R., and Page, D.C. 1999. An azoospermic man with a de novo point mutation in the Y-chromosomal gene USP9Y. *Nature Genetics* **23**: 429–432.

Sutcliffe, M.J. and P.S. Burgoyne, P.S. 1989. Analysis of the testes of H-Y negative XOSxrb mice suggests that the spermatogenesis gene (Spy) acts during the differentiation of the A spermatogonia. *Development* **107**: 373–380.

Teng, Y.N., Lin, Y.M., Lin, Y.H., Tsao, S.Y., Hsu, C.C., Lin, S.J., Tsai, W.C., and Kuo, P.L. 2002. Association of a single-nucleotide polymorphism of the deleted-in-azoospermia-like gene with susceptibility to spermatogenic failure. *The Journal of Clinical Endocrinology and Metabolism* **87**: 5258–5264.

Tessari, A., Salata, E., Ferlin, A., Bartoloni, L., Slongo, M.L., and Foresta, C. 2004. Characterization of HSFY, a novel AZFb gene on the Y chromosome with a possible role in human spermatogenesis. *Molecular Human Reproduction* **10**: 253–258.

Tiepolo, L. and Zuffardi, O. 1976. Localization of factors controlling spermatogenesis in the nonfluorescent portion of the human Y chromosome long arm. *Human Genetics* **34**: 119–124.

Tilford, C.A., Kuroda-Kawaguchi, T., Skaletsky, H., Rozen, S., Brown, L.G., Rosenberg, M., McPherson, J.D. et al. 2001. A physical map of the human Y chromosome. *Nature* **409**: 943–945.

Tse, J.Y., Wong, E.Y., Cheung, A.N., O, W.S., Tam, P.C., and Yeung, W.S. 2003. Specific expression of VCY2 in human male germ cells and its involvement in the pathogenesis of male infertility. *Biology of Reproduction* **69**: 746–751.

Tung, J.Y., Rosen, M.P., Nelson, L.M., Turek, P.J., Witte, J.S., Cramer, D.W., Cedars, M.I., and Pera, R.A. 2006. Variants in Deleted in AZoospermia-Like (DAZL) are correlated with reproductive parameters in men and women. *Human Genetics* **118**: 730–740.

Vergnaud, G., Page, D.C., Simmler, M.C., Brown, L., Rouyer, F., Noel, B., Botstein, D., de la Chapelle, A., and Weissenbach, J. 1986. A deletion map of the human Y chromosome based on DNA hybridization. *American Journal of Human Genetics* **38**: 109–124.

Verkaar, E.L., Zijlstra, C., van 't Veld, E.M., Boutaga, K., van Boxtel, D.C., and Lenstra, J.A. 2004. Organization and concerted evolution of the ampliconic Y-chromosomal TSPY genes from cattle. *Genomics* **84**: 468–474.

Vinci, G., Raicu, F., Popa, L., Popa, O., Cocos, R., and McElreavey, K. 2005. A deletion of a novel heat shock gene on the Y chromosome associated with azoospermia. *Molecular Human Reproduction* **11**: 295–298.

Vodicka, R., Vrtel, R., Dusek, L., Singh, A.R., Krizova, K., Svacinova, V., Horinova, V. et al. 2007. TSPY gene copy number as a potential new risk factor for male infertility. *Reproductive Biomedicine Online* **14**: 579–587.

Vogt, P.H. 2005. AZF deletions and Y chromosomal haplogroups: History and update based on sequence. *Human Reproduction Update* **11**: 319–336.

Vogt, P.H., Edelmann, A., Kirsch, S., Henegariu, O., Hirschmann, P., Kiesewetter, F., Kohn, F.M. et al. 1996. Human Y chromosome azoospermia factors (AZF) mapped to different subregions in Yq11. *Human Molecular Genetics* **5**: 933–943.

Vollrath, D., Foote, S., Hilton, A., Brown, L.G., Beer-Romero, P., Bogan, J.S., and Page, D.C. 1992. The human Y chromosome: A 43-interval map based on naturally occurring deletions. *Science* **258**: 52–59.

Vong, Q.P., Li, Y., Lau, Y.F., Dym, M., Rennert, O.M., and Chan, W.Y. 2006. Structural characterization and expression studies of Dby and its homologs in the mouse. *Journal of Andrology* **27**: 653–661.

Wallner, B., Piumi, F., Brem, G., Muller, M., and Achmann, R. 2004. Isolation of Y chromosome-specific microsatellites in the horse and cross-species amplification in the genus *Equus*. *The Journal of Heredity* **95**: 158–164.

Wang, A., Yasue, H., Li, L., Takashima, M., de Leon, F.A., and Liu, W.S. 2008. Molecular characterization of the bovine chromodomain Y-like genes. *Animal Genetics* **39**: 207–216.

Wang, G., Zhang, J., Moskophidis, D., and Mivechi, N.F. 2003. Targeted disruption of the heat shock transcription factor (hsf)-2 gene results in increased embryonic lethality, neuronal defects, and reduced spermatogenesis. *Genesis* **36**: 48–61.

Wang, P.J., McCarrey, J.R., Yang, F., and Page, D.C. 2001. An abundance of X-linked genes expressed in spermatogonia. *Nature Genetics* **27**: 422–426.

Wang, W., Meadows, L.R., den Haan, J.M., Sherman, N.E., Chen, Y., Blokland, E., Shabanowitz, J. et al. 1995. Human H-Y: A male-specific histocompatibility antigen derived from the SMCY protein. *Science* **269**: 1588–1590.

Waterston, R.H., Lindblad-Toh, K., Birney, E., Rogers, J., Abril, J.F., Agarwal, P., Agarwala, R. et al. 2002. Initial sequencing and comparative analysis of the mouse genome. *Nature* **420**: 520–562.

Wimmer, R., Kuhl, H., Rottger, S., and Schempp, W. 2002. Comparative mapping of CDY and DAZ in higher primates. *Cytogenetic and Genome Research* **96**: 287–289.

Wong, E.Y., Tse, J.Y., Yao, K.M., Tam, P.C., and Yeung, W.S. 2002. VCY2 protein interacts with the HECT domain of ubiquitin-protein ligase E3A. *Biochemical and Biophysical Research Communications* **296**: 1104–1111.

Wood, S.A., Pascoe, W.S., Ru, K., Yamada, T., Hirchenhain, J., Kemler, R., and Mattick, J.S. 1997. Cloning and expression analysis of a novel mouse gene with sequence similarity to the Drosophila fat facets gene. *Mechanisms of Development* **63**: 29–38.

Xu, E.Y., Moore, F.L., and Pera, R.A. 2001. A gene family required for human germ cell development evolved from an ancient meiotic gene conserved in metazoans. *Proceedings of the National Academy of Sciences of the United States of America* **98**: 7414–7419.

Y-chromosome-consortium. 2002. A nomenclature system for the tree of human Y-chromosomal binary haplogroups. *Genome Research* **12**: 339–348.

Yang, Y., Ma, M., Li, L., Zhang, W., Xiao, C., Li, S., Ma, Y. et al. 2008. Evidence for the association of Y-chromosome haplogroups with susceptibility to spermatogenic failure in a Chinese Han population. *Journal of Medical Genetics* **45**: 210–215.

Yen, P.H. 2004. Putative biological functions of the DAZ family. *International Journal of Andrology* **27**: 125–129.

Zechner, U., Wilda, M., Kehrer-Sawatzki, H., Vogel, W., Fundele, R., and Hameister, H. 2001. A high density of X-linked genes for general cognitive ability: A run-away process shaping human evolution? *Trends in Genetics* **17**: 697–701.

7

Mitochondriomics of Reproduction and Fertility

Zhihua Jiang, Galen A. Williams, Jie Chen, and Jennifer J. Michal

7.1 Introduction

Mitochondria are membrane-enclosed organelles found in most eukaryotic cells but located in the cytoplasm outside their nuclei. In animals, a typical cell contains 1000–2000 mitochondria. Primarily, mitochondria convert nutrients into chemical energy, thus supporting a cell's activities and functions. In addition, mitochondria also play an important role in many other metabolic tasks, such as apoptosis-programmed cell death, glutamate-mediated excitotoxic neuronal injury, cellular proliferation, regulation of the cellular redox state, heme synthesis, and steroid synthesis (McBride et al. 2006; Nisoli et al. 2007; Sas et al., 2007; Schwarz et al. 2007).

It has been widely believed that the mitochondrial genomes in eukaryotic cells are of endosymbiotic origin; that is, they are derived from a once-free living bacterium, almost certainly an α-proteobacterium (Gray 1999). Interestingly, the mitochondrial genome size has gradually receded over a long period of symbiosis, being ~100-fold smaller than those of free-living bacteria and thus containing significantly fewer genes (Selosse et al. 2001). For example, in humans, only 37 genes are encoded by mitochondrial DNA (mtDNA) and transcribed within the mitochondrion. Most of the original mitochondrial genes have been transferred to the nucleus, leading to a marked loss of coding capacity compared with that of their closest bacterial relatives (Smith et al. 2005). Approximately 1000–1500 mitochondrial genes were functionally transferred to the nucleus, leading to the birth of a set of so-called nucleus-encoded mitochondrial genes (Chinnery 2003).

Therefore, in a broad sense, the mitochondrial genome encompasses genes located in both the cytoplasm and the nucleus. Recently, the scientific and medical community has found that the functional status of mitochondria contributes to the quality of oocytes and spermatozoa, and plays a part in the process of fertilization and embryo development. This chapter reviews recent information about involvements of both cytoplasm and nuclear mitochondrial genomes in fertility and reproduction.

7.2 Cytoplasm mitochondrial genomes in fertility and reproduction

7.2.1 Mammalian mitochondrial genomes in the cytoplasm

The mitochondrial genome is present in mitochondria as a chromosome or a circular DNA molecule. Unlike nuclear chromosomes that are paired in mammals (except X and Y sex chromosomes), there are many copies of the mitochondrial chromosome in every cell. However, the copy number can be extremely variable in different cells. For example, an egg contains 100,000–1,000,000 mtDNA molecules, while a sperm contains only 100–1000. During fertilization, the sperm mtDNA is degraded. Therefore, no matter whether we are male or female, we inherit our mitochondrial chromosome from our mother. In other words, the mitochondrial chromosome is transmitted in a matrilinear manner.

As the mitochondrial genomes are relatively small in size, it is relatively easy to sequence them. Up to date, mtDNA has been completely sequenced in species related to farming and agriculture, such as *Bison bison* (American bison, NC_012346), *Bos grunniens* (domestic yak, NC_006380), *Bos indicus* (zebu, AF492350), *Bos taurus* (domestic cattle, NC_006853), *Bubalus bubalis* (water buffalo, NC_006295), *Capra hircus* (goat, NC_005044), *Equus asinus* (donkey, NC_001788), *Equus caballus* (horse, NC_001640), *Lama glama* (llama, NC_012102), *Oryctolagus cuniculus* (rabbit, NC_001913), *Ovis aries* (sheep, NC_001941), and *Sus scrofa* (swine, NC_000845). The size of the mitochondrial genome varies from 16,319 bp (NC_012346) to 17,245 bp (NC_001913) among these 12 agriculturally related species.

In mammals, each circular mtDNA molecule harbors genetic material that encodes the same 37 genes: 13 for proteins (polypeptides), 22 for transfer RNA (tRNA), and one each for the small and large subunits of ribosomal RNA (rRNA). Among these 12 agriculturally related species, all 37 genes plus D-loop and replication of origin regions are placed on the circular DNA in the same orientation (Table 7.1). Of 13 coding mitochondrial genes, four genes—*COX1*, *ND4L*, *ND4*, and *ND6* remain evolutionally consistent in terms of their sizes of coding sequence. The D-loop region seems very variable, ranging from 888 bp in *B. bison* to 1800 bp in *O. cuniculus*. Interestingly, one-third of these genes/regions (13/39) in the mitochondrial genome have the potential to overlap with one of the adjacent neighbors (Table 7.1). In addition, the genetic code in mitochondria is different from the genetic code in the nucleus. For example, UGA is read as tryptophan rather than "stop," AGA and AGG as "stop" rather than arginine, AUA as methionine rather than isoleucine, and AUA or AUU is sometimes used as an initiation codon instead of AUG (Anderson et al. 1982).

7.2.2 Special features of mitochondrial genetics

The cytoplasmic location of mtDNA and the high copy number contribute to certain unique features of mitochondrial genetics. First, as mentioned above, mtDNA is maternally inherited. Three theories have been developed to explain the maternal control of mitochondrial inheritance in mammals. The mtDNA dilution theory is based on the

Table 7.1 Characterization of mitochondrial genomes in 12 agriculturally related animals.[1]

Gene or region	Orientation	Gene size (in bp)	Gene distance (in bp)
D-loop	+	888–1800	1
tRNA-Phe	+	67–71	1
s-rRNA	+	955–975	0–3
tRNA-Val	+	66–68	1
l-rRNA	+	1569–1581	1
tRNA-Leu	+	74–75	3–4
ND1	+	955–957	(−1)–2
tRNA-Ile	+	69–70	(−2)
tRNA-Gln	−	72–73	2–3
tRNA-Met	+	69–70	1
ND2	+	1041–1044	(−1)–1
tRNA-Trp	+	67–70	2–7
tRNA-Ala	−	67–69	1–2
tRNA-Asn	−	73–75	1
Rep-origin	+	23–35	(−5)–10
tRNA-Cys	−	66–68	0–1
tRNA-Tyr	−	66–68	2–8
COX1	+	1545	(−2)–4
tRNA-Ser	−	69–71	4–7
tRNA-Asp	+	67–70	1–2
COX2	+	684–688	1–4
tRNA-Lys	+	67–69	1–2
ATP8	+	201–204	(−42)–(−39)
ATP6	+	679–681	0–2
COX3	+	781–784	1–4
tRNA-Gly	+	68–70	(−1)–1
ND3	+	340–350	1–2
tRNA-Arg	+	67–69	1–2
ND4L	+	297	(−6)
ND4	+	1378	1
tRNA-His	+	69–71	1
tRNA-Ser	+	59–60	1–2
tRNA-Leu	+	70–71	1
ND5	+	1812–1821	(−4)–(−16)
ND6	−	528	1
tRNA-Glu	−	68–69	4–7
CYTB	+	1139–1140	1–5
tRNA-Thr	+	66–74	0–2
tRNA-Pro	−	64–70	1

[1] Gene distance ≤0 means that there is (are) ≥1 nucleotide(s) overlapped, while gene distance ≥2 means that there is (are) ≥1 intergenic nucleotide(s) between two adjacent neighbor genes. When gene distance is equal to 1, it means that no overlap or intergenic sequences exist between two adjacent neighbor genes.

fact that the mammalian egg contains about 100,000 copies of mtDNA, whereas the sperm harbors 100–1500 molecules of mtDNA (Manfredi et al. 1997). This means that the theoretic ratio of paternal versus maternal mitochondria is ~1 : 1000. However, this does not mean that paternal mitochondria cannot enter into the oocyte cytoplasm at fertilization. Using the polymerase chain reaction (PCR), Gyllensten and colleagues (1991) detected paternally inherited mtDNA molecules in mice at a frequency of 10(−4), relative to the maternal contributions. Therefore, the dilution theory cannot explain why mtDNA is only maternally inherited. The oxidative damage theory (Allen 1996) suggests that sperm mitochondria are recognized by oocyte cytoplasm because of the oxidative damage they suffer during passage through the female genital tract. However, *in vitro* fertilization experiments do not support this hypothesis because oxidative damage is minimal during the process (Aitken 1995). Sutovsky and colleagues (2000) proposed a feasible explanation for maternal mtDNA inheritance in mammals. Using bull semen and cow eggs, the authors observed that sperm mitochondria are ubiquitinated inside the oocyte cytoplasm and later subjected to proteolysis during preimplantation development. Therefore, paternal mitochondria are degraded during early embryogenesis, leaving only maternal mitochondria in the cytoplasm.

Second, mtDNA genes have a much higher mutation rate than nuclear DNA genes. Various surveys of mutation accumulation in mitochondrial and nuclear genomes of animals and plants have consistently found that the former genomes accumulate deleterious mutations at a higher rate than the latter genomes. In a recent review, Neiman and Taylor (2009) concluded that asexuality might not be the primary determinant of the high mutation load in mtDNA. Instead, the authors proposed that a high rate of accumulation of mildly deleterious mutations in mtDNA may result from the small effective population size associated with effectively haploid inheritance, because this type of transmission is nearly ubiquitous among mitochondrial genomes. In addition, polymorphisms accumulate approximately 10–17 times faster in mtDNA compared with nuclear DNA. This may be explained by the lack of an efficient DNA repair system in mitochondria and histones associated with mtDNA (Wallace et al. 1997).

Third, mitochondria undergo replicative segregation at cell division. Each cell contains hundreds of mitochondria, each having 2–10 copies of mtDNA molecules. This polyploid nature of the mitochondrial genome—up to several thousand copies per cell—gives rise to an important feature of mitochondrial genetics, homoplasmy, and heteroplasmy. In simple terms, homoplasmy is when all copies of the mitochondrial genome are identical; heteroplasmy is when there is a mixture of two or more mitochondrial genotypes. At cell division, the mitochondria and their genomes are randomly distributed to the daughter cells, a process known as replicative segregation. Therefore, when mutant and wild-type mtDNA coexist in affected individuals (the condition of heteroplasmy), the proportion of mutant mtDNA transmitted from mother to offspring is variable because of a genetic bottleneck, and the "dose" of mutant mtDNA received influences the severity of the phenotype (Marchington et al. 1998).

Fourth, clinical expression of mitochondrial disease or defect depends on the threshold effect. In the presence of heteroplasmy, there is a threshold level of mutation that is important for the clinical expression of disease and for biochemical defects. This means that the clinical expression of a pathogenic mtDNA mutation is largely determined by the relative proportion of normal and mutant mtDNA genomes in different tissues. A minimum critical number of mutant mtDNAs is required to cause mitochondrial dysfunction in a particular organ or tissue, resulting in a mitochondrial disease. This is the so-called threshold effect. For example, the threshold varies for different mtDNA mutation types and is about 60% for deleted mtDNA (Hayashi et al. 1991). For the mutation 8344A>G that causes the syndrome of myoclonic epilepsy and ragged-red fibers, the threshold level is about 85% mutated DNA (Chomyn 1998).

Fifth, somatic mtDNA mutations accumulate in post-mitotic tissues with age, reducing the ATP-generating capacity. At cell division, the proportion of mutant mtDNAs in daughter cells may shift and the phenotype may change accordingly. This phenomenon, called mitotic segregation, explains how certain patients with mtDNA-related disorders may actually manifest different mitochondrial diseases at different stages of their lives.

7.2.3 *Mitochondria and male fertility*

Faulty mitochondrial function and reduced copy numbers are linked to male subfertility and sperm dysfunction (Folgero et al. 1993; Kao et al. 1995). In addition, new assays for evaluating the functional status of mitochondria in sperm have been developed. Using mitochondria-specific stains and fluorochromes, it has been shown that higher mitochondrial function is correlated to increased fertilization rates *in vitro*. These stains also enable the assessment of sperm viability when using different semen extenders (Huo and Scarpulla 2001). Interestingly, increased mtDNA amplification was found in abnormal sperm from infertile patients (May-Panloup et al. 2003). Further, significantly higher mtDNA deletions have been detected in diabetic men when compared with normal controls (Agbaje et al. 2007). An analysis of human mtDNA polymerase revealed an association between male infertility and the absence of a common CAG microsatellite allele (Rovio et al. 2001). While differences in mtDNA are associated with male infertility, these measures are currently considered "proxys" for male fertility status (St. John et al. 1997). However, evidence suggests that defects in mitochondrial respiration result in an overall reduction in fertility and reduced sperm motility (Nakada et al. 2006). This may be a clue to the molecular basis for the relationship between mtDNA content and fertility status in the reports above. Despite what is known about mtDNA and its role in male fertility, this field is still in its infancy.

7.2.4 *Mitochondria and female fertility*

In many mammalian species, mitochondria are the most abundant organelle in fully grown oocytes detected by electron microscopy (Van Blerkom 2004). However, determining if mtDNA mutations cause pre- or postimplantation embryo demise can be difficult because blastocysts used for in vitro fertilization (IVF) may be unavailable and even when they are available, they may be an unsuitable model to study mtDNA defects (Van Blerkom 2004). It is unclear whether mtDNA plays a role in apoptosis of oocytes in adult life (Jansen and de Boer 1998). Overall, there are little data available relating to mitochondriomics of oogenesis. Using 422 beef cattle of two different breeds (purebred Hereford and composite multibreed), Sutarno and colleagues (2002) found a significant association between calving rate and mitochondrial polymorphisms in both breeds. Calving rate is defined as the mean number of live calves born per year over 4 years. As pointed out by the authors, the association may have implications for genetically improving cow fertility.

7.2.5 *Mitochondria and reproductive aging*

It has long been known that aging in mammals results in reduced fertility and increased frequency of abnormal development. The mitochondrial theory of aging is by no means universally accepted, but it is likely that mitochondria do play a role in male and female reproductive aging. An analysis of mtDNA samples from females widely varying in age found a considerably large (~5 kb) deletion in tissues from abnormal tissues in the reproductive tract as well as nonreproductive muscle. This deletion was predominately found in menopausal and postmenopausal women and suggested that aging may be closely related to ovarian dysfunction (Kitagawa et al. 1993). In addition to this common 5-kb mtDNA deletion, additional deletions in mtDNA are more common in oocytes from older women when compared with those from younger women (Keefe et al. 1995). Others reported a correlation between increased oocyte volume and maternal age, but to date this has not been associated with mtDNA mutations or functional defects (Müller-Höcker et al. 1996). Interestingly, mtDNA deletions are more frequent in oocytes than embryos, partially supporting a "bottleneck" theory that filters out mutated mtDNA in conjunction with perhaps another nuclear mechanism (Brenner et al. 1998). Similarly, rearrangements in mtDNA are more common in oocytes than in embryos (Barritt et al. 1999).

7.3 Nuclear mitochondrial genomes in fertility and reproduction

7.3.1 Mammalian mitochondrial genomes in the nucleus

A central component of the endosymbiotic theory is that mitochondria originated as bacterial intracellular symbionts. However, many of the original genes from bacteria were transferred to the nucleus. In what form did the transferred genes physically make that intracellular journey—as RNA, as cDNA, as pieces of organelle DNA, or as whole organelle chromosomes? Two opinions exist: direct DNA transfer and cDNA-mediated transfer. But Henze and Martin (2001) proposed that direct DNA transfer, rather than cDNA-mediated transfer, probably prevailed during the early phases of organelle evolution. Mathematical model analysis revealed that the rate of gene transfer from mitochondria to the nucleus could be affected by three factors: the intensity of intracellular competition, the probability of paternal organelle transmission, and the effective population size (Yamauchi 2005). Gene transfer rate tends to increase with decreasing intracellular competition, increasing paternal organelle transmission, and decreasing effective population size. Intense intracellular competition tends to suppress gene transfer because it is likely to exclude mutant mitochondria that lose the essential gene due to the production of lethal individuals. In fact, most researchers believe that functional transfer of mitochondrial genes has ceased in animals (Boore 1999).

There are several databases publicly available that provide information on nucleus-encoded mitochondrial genes/proteins. The Human Mitochondrial Protein Database (HMPDb) (bioinfo.nist.gov/) provides comprehensive data on mitochondrial and human nuclear encoded proteins involved in mitochondrial biogenesis and function. This database consolidates information from SwissProt, LocusLink, Protein Data Bank (PDB), GenBank, Genome Database (GDB), Online Mendelian Inheritance in Man (OMIM), Human Mitochondrial Genome Database (mtDB), MITOMAP, Neuromuscular Disease Center, and Human 2-D PAGE Databases. The database also provides tools for database search, mtDNA sequence visualization, mtDNA polymorphism, mitochondrial protein-related diseases, and 3D structures of mitochondrial proteins. The MitoP2 database (http://www.mitop.de:8080/mitop2/) integrates information on mitochondrial proteins, their molecular functions, and associated diseases (Prokisch et al. 2006). As stated by the developers, the MitoP2 enables (1) the identification of putative orthologous proteins between these species to study evolutionarily conserved functions and pathways, (2) the integration of data from systematic genome-wide studies such as proteomics and deletion phenotype screening, (3) the prediction of novel mitochondrial proteins using data integration and the assignment of evidence scores, and (4) systematic searches that aim to find the genes that underlie common and rare mitochondrial diseases. Other databases include the Human Genome Resources at the National Center for Biotechnology Information (NCBI), the MitoProteome—an object-relational mitochondrial protein sequence database at the University of California, San Diego, CA (Cotter et al. 2004), and the MitoRes—a bio-sequences resource for mitochondria-related genes, transcripts, and proteins at the Institute of Biomedical Technologies, CNR, Italy (Catalano et al. 2006).

Approximately 1300 nucleus-encoded mitochondrial protein-coding genes have been well annotated in the human genome, including 113 on HSA1, 75 on HSA2, 64 on HSA3, 46 on HSA4, 54 on HSA5, 53 on HSA6, 59 on HSA7, 36 on HSA8, 55 on HSA9, 61 on HSA10, 79 on HSA11, 68 on HSA12, 17 on HSA13, 51 on HSA14, 37 on HSA15, 57 on HSA16, 76 on HSA17, 16 on HSA18, 70 on HSA19, 21 on HSA20, 12 on HSA21, 44 on HSA22, and 50 on human × chromosome, respectively. However, most of these nucleus-encoded mitochondrial genes are not yet annotated in domestic animals. Here, we propose to use a comparative annotation approach to retrieve both cDNA and genomic DNA sequences for each nucleus-encoded mitochondrial gene in livestock species in three steps. In step 1, we use the cDNA sequences of the human orthologs as references for BLAST (basic local alignment search tool) searches to retrieve the orthologous cDNA sequences against the GenBank database "nr" or the orthologous expressed sequence tags (ESTs) against the GenBank database "est_others" with a species option limited to livestock species, for example, *B. taurus*. The cDNA sequences in the "nr" database represents three categories: cDNA sequences derived from a full-length cDNA library, known gene cDNA sequences, or annotated cDNA sequences compiled by the GenBank staff. In step 2, we choose the longest cDNA sequence retrieved from the "nr" database or one assembled from several ESTs retrieved from the "est_others" database to form a primary cDNA sequence for the cattle gene. This sequence will then be used to perform a species-specific BLAST search against the "est_others" database in order to expand the primary sequence to a full-length cDNA sequence. For step 3, we use the full-length cDNA sequence to search for genomic DNA contigs. Such a process will retrieve both cDNA and genomic DNA sequences of each nucleus-encoded mitochondrial gene from the public database of different livestock species.

For example, the annotation of the bovine mitochondrial transcription factor B1 (*TFB1M*) is shown in Figure 7.1 and demonstrates how to utilize the EST database for annotation as outlined above. First, a cDNA sequence of the human ortholog (NM_016020) was used as a reference, and a BLAST search retrieved more than 50 bovine

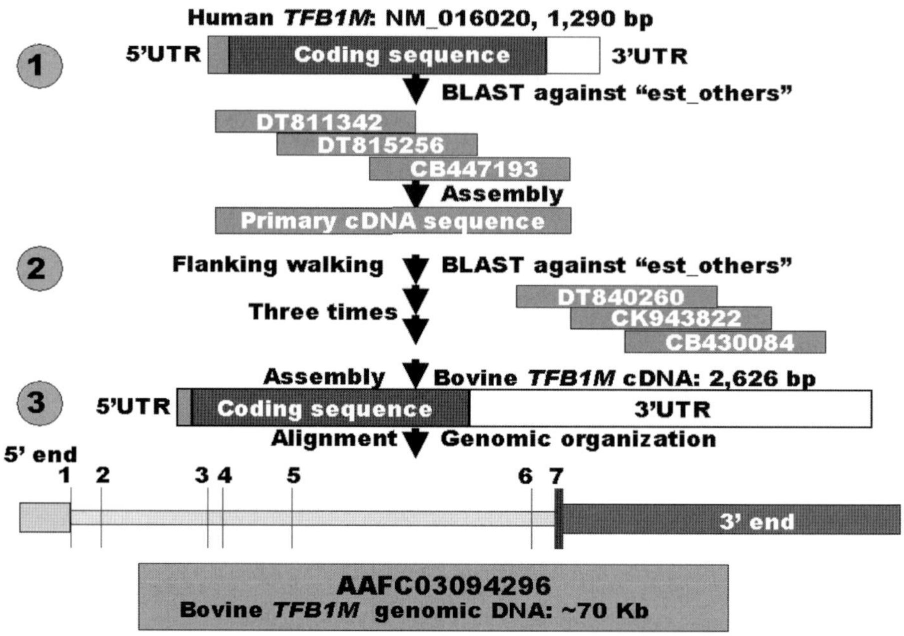

Figure 7.1 Compilation and annotation of both cDNA and genomic DNA sequences for the bovine *TFB1M* gene. Sequence length is not in the same scale.

ortholgous ESTs against the GenBank database "est_others." Second, three ESTs (DT811342, DT815256, and CB447193) were chosen and assembled to form a primary cDNA sequence for the bovine *TFB1M* gene, which was then used to perform a species-specific EST search for expansion. Three additional ESTs (DT840260, CK943822, and CB430084) were added to form a full-length cDNA sequence of 1626 bp. Finally, the full-length cDNA sequence retrieved a genomic DNA contig (AAFC03094296) from the 7.15X bovine genome sequence database and thus determined its genomic organization (Figure 7.1).

7.3.2 *Special features of mitochondrial genome in the nucleus*

The transfer of mitochondrial genes to the nucleus contributes to certain unique features. First, these mitochondrial genes are not evenly distributed in the nuclear genome. Based on the locations of 1298 human nucleus-encoded mitochondrial genes, the mitochondrial gene density is calculated as Mitochondrial Gene Density = $(N(mt)/N)/(T(mt)/T)$, where $N(mt)$ is the number of mitochondrial genes within a window of 5 Mbp on a chromosome, N is the number of genes within a window of 5 Mbp on a chromosome, $T(mt)$ is the number of total mitochondrial genes on all chromosome (1298), and T is the number of total genes on all chromosomes (28,644) (Build 33.0). The mitochondrial gene density plots are illustrated along each chromosome with an interval of 1 Mb (Figure 7.2). Overall, the mitochondrial genes tend to be clustered but are not evenly distributed along each human chromosome. Some chromosomes (e.g., human chromosome 1, HSA1) are very rich

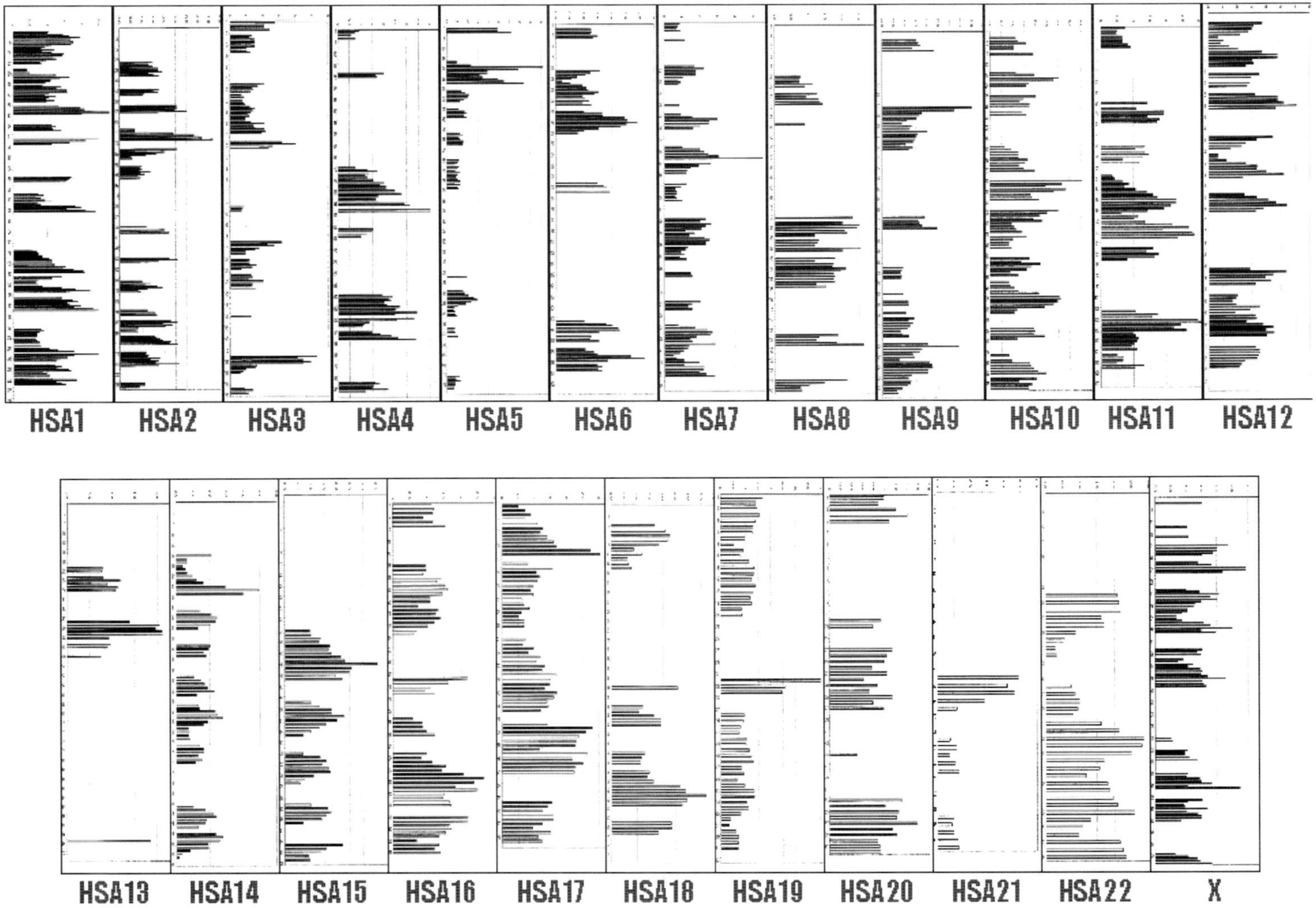

Figure 7.2 Distribution of nucleus-encoded mitochondrial genes in the human nuclear genome.

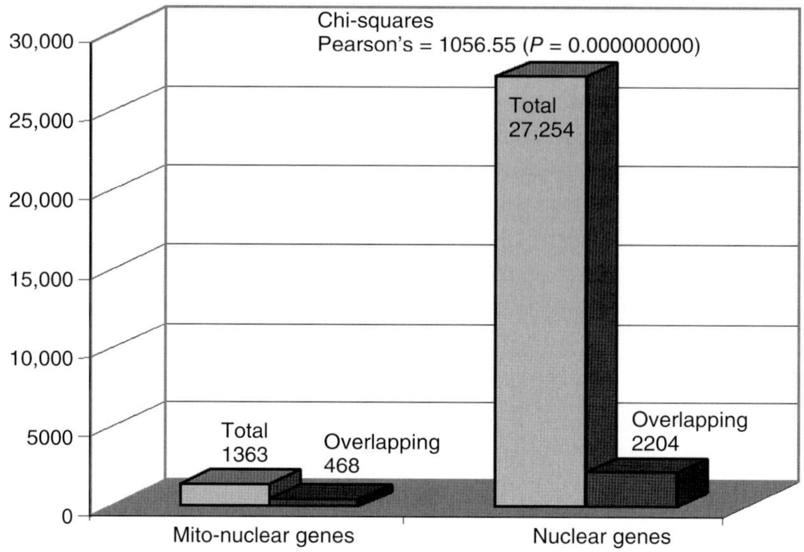

Figure 7.3 Overlapping genes associated with mitochondrial genes and nuclear genes.

in mitochondrial genes, while some (e.g., HSA13) contain few mitochondrial genes.

Second, nucleus-encoded mitochondrial genes tend to overlap adjacent neighbors or cause overlapping of adjacent genes with their neighbors. In mammalian genomes, hundreds of pairs of protein-coding overlapping genes have been reported so far. Overlapping genes might play an important role in various levels of gene expression control, such as transcription, mRNA processing, splicing, stability, transport, and translation. However, the evolutionary origin of such genes is not known, existing hypotheses can explain only selected cases of mammalian gene overlaps, which could originate as a result of rearrangements, overprinting and/or adoption of signals in the neighboring gene locus (Makalowska et al. 2005). Based on a total of 1363 mitochondrial coding genes and pseudogenes in the human nuclear genome, we found that 468 (34%) genes overlap adjacent neighbors or cause overlapping of their adjacent genes with neighbors (Figure 7.3). In contrast, among the rest of the nuclear genes in the human genome, only 8% (2204/27,254) contribute to the overlapping cases. Therefore, transfer of mitochondrial genes into the nucleus might play an important role in the evolutionary formation of overlapping genes in the nucleus.

Third, translation efficiency remains in nucleus-encoded mitochondrial genes. The Kozak consensus sequence is referred to as a sequence that occurs on eukaryotic mRNA and has the consensus (gcc)gccRccAUGG, where R is a purine (adenine or guanine) three bases upstream of the start codon (AUG), which is followed by another "G" (Kozak 1987). The Kozak consensus sequence plays a major role in the initiation of translation. Figure 7.4 shows frequencies of the Kozak consensus nucleotides between mitochondrial and nuclear genes in the human nuclear genome. The mitochondrial genes adapted the translation systems after transfer into the nucleus and even contain a

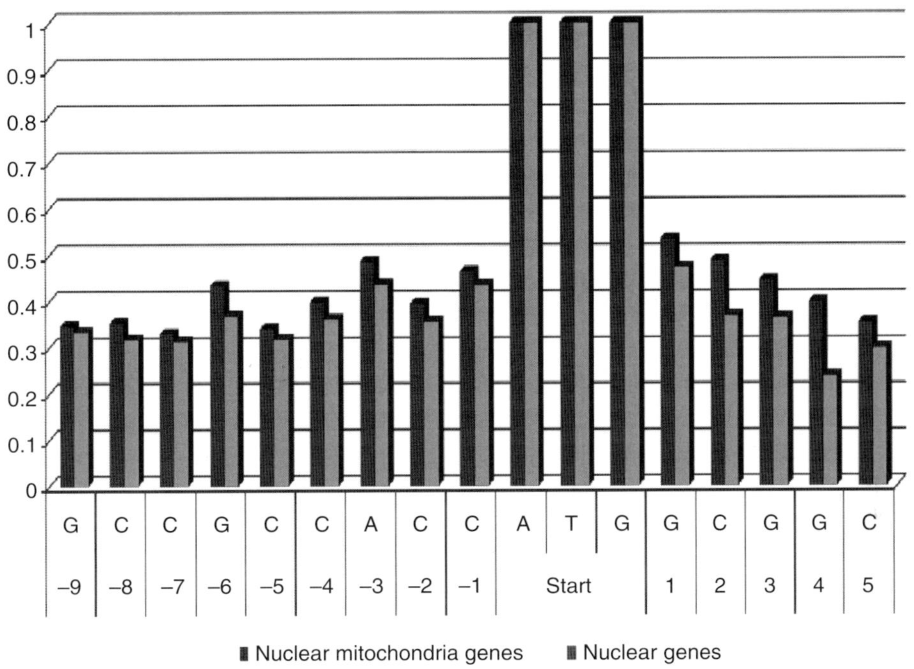

Figure 7.4 The Kozak sequence between mitochondrial genes in nucleus and other nuclear genes.

stronger consensus sequence within the translation binding site.

Fourth, transfer of mitochondrial genes into the nucleus might contribute to genome evolution. For example, the *TFAM* gene neighborhood is rearranged in both human and bovine genomes, causing a singleton (one gene) synteny between two species. For the *TFB1M* gene, intron 5 harbors a functional gene *CLDN20*, while its 3′UTR overlaps with the 3′UTR of *TIAM2* in human, while *CLDN20* might no longer be expressed in cattle. In humans, a gene (*LOC100132379*) similar to *TBC1* domain family member 3 is inserted into the adjacent region of *TFB2M*, which does not occur in the bovine genome.

7.3.3 Nucleus-encoded mitochondrial genes and male fertility

Spermatogenesis is a complex process that requires normal germ cell migration, testicular development, and Sertoli cell function, as well as appropriate maturation, motility, and penetration of the sperm. The following nucleus-encoded mitochondrial genes affect spermatogenesis in males based on knockout or deficiency mice models, including cannabinoid receptor 1 (*CNR1*), guanidinoacetate N-methyltransferase (*GAMT*), Huntingtin interacting protein 1 (*HIP1*), inner mitochondrial membrane peptidase 2-like (*IMMP2L*), pantothenate kinase 2 (*PANK2*), protein phosphatase 1, catalytic subunit, gamma isoform (*PPP1CC*), sirtuin (silent mating type information regulation 2 homolog) 1 (*Saccharomyces cerevisiae*) (*SIRT1*), sperm mitochondria-associated cysteine-rich protein (*SMCP*), TAF7-like RNA polymerase II, TATA box binding protein (TBP)-associated factor (*TAF7L*), and voltage-dependent anion channel 3 (*VDAC3*).

CNR1 encodes one of two cannabinoid receptors. The cannabinoids, basically delta-9-tetrahydrocannabinol and synthetic analogs, are psychoactive ingredients of marijuana. The cannabinoid receptors are members of the guanine-nucleotide-binding protein (G-protein) coupled receptor family, which inhibit adenylate cyclase activity in a dose-dependent, stereoselective and pertussis toxin-sensitive manner. It has been reported that *Cnr1* is expressed in mature sperm, and the sperm from *Cnr1*-knockout mice show a dramatic increase in motility in the caput epididymis (Rossato et al. 2008).

GAMT encodes a protein called methyltransferase, which catalyses the synthesis of creatine from guanidinoacetate and S-adensylmethionine. Schmidt et al. (2004) generated a knockout mouse model for *Gamt* deficiency by gene targeting in embryonic stem cells. Although some *Gamt*-deficient males are—at least temporarily—fertile, the authors observed severely attenuated spermatogenesis in knockout mice at the level of spermatid development; that is, elongated spermatids were hardly present and mature spermatozoa were almost absent in the lumen of seminiferous tubules. Furthermore, the authors found that the structure of the seminiferous tubules was highly unordered with a large number of resorption holes and a larger lumen. In addition, breeding between male *Gamt*$^{-/-}$ mice and wild-type or heterozygous mutant females produced no litters, indicating impaired fertility in *Gamt*-deficient males (Schmidt et al. 2004).

HIP1 is an endocytic adaptor protein with clathrin assembly activity that binds to cytoplasmic proteins, such as F-actin, tubulin, and huntingtin. Khatchadourian and colleagues (2007) performed a quantitative analysis of sperm counts from the caudal epididymis of *Hip1*$^{-/-}$mice and found a significant decrease compared with wild-type littermates. In addition, using computer-assisted sperm analyses, the authors also observed that velocities, amplitude of lateral head displacements, and numbers and percentages of sperm in the motile, rapid, and progressive categories were all significantly reduced in *Hip1*$^{-/-}$mice. The authors concluded that HIP1 plays an important role in stabilizing actin and microtubules, which are important cytoskeletal elements enabling normal spermatid and Sertoli cell morphology and function (Khatchadourian et al. 2007).

The mitochondrial inner membrane peptidase (IMP) complex generates mature, active proteins in the mitochondrial intermembrane space by proteolytically removing the mitochondrial targeting presequence of nuclear-encoded proteins. IMMP2L is one of the catalytic subunits of the IMP complex (Burri et al. 2005). Using a transgenic insertional mutagenesis strategy, Lu et al. (2008) generated a mouse mutant, *Immp2l*Tg(Tyr)979Ove, with a mutation in the *Immp2l* gene. The mutation affected the signal peptide sequence processing of mitochondrial proteins cytochrome c1 and glycerol phosphate dehydrogenase 2. The authors observed that the homozygous mutant males were severely subfertile due to erectile dysfunction.

PANK2 encodes a protein belonging to the pantothenate kinase family and is the only member of that family that is expressed in mitochondria. Pantothenate kinase catalyzes the first committed step in the universal biosynthetic pathway leading to CoA and is itself subject to regulation through feedback inhibition by acyl CoA species. Kuo and colleagues found that homozygous *Pank2* knockout male mice are infertile due to azoospermia. At 6 weeks, these *Pank2*$^{-/-}$males had smaller testes and lacked sperm in the epididymis. Histological analyses showed a complete absence of elongated spermatids in the seminiferous tubules of these males. In particular, the spermatocytes and spermatids are somewhat disordered in the *Pank2* testes tubule (Kuo et al. 2005).

Serine/threonine protein phosphatase 1 (PP1) consists of four ubiquitously expressed major isoforms, two of which, PP1gamma1 and PP1gamma2, are derived by alternative splicing of a single gene, protein phosphatase 1, catalytic subunit, gamma isoform (*PPP1CC*). Targeted disruption of the *Ppp1cc* gene causes male infertility in mice due to impaired spermiogenesis, as reported by Chakrabarti et al. (2007). Seminiferous tubules from *Ppp1cc*-null testes had multiple vacuoles, sloughing of germ cells into the lumen, and mislocated germ cells. In comparison with wild-type animals, the number of elongating spermatids in each seminiferous tubule was reduced in *Ppp1cc*-null testes. Furthermore, epididymides from *Ppp1cc*-null males contained immature germ cells.

In mammals, *SIRT1* is a member of the sirtuin family of proteins and plays an important physiological role in the regulation of glucose metabolism, cell survival, and mitochondrial respiration. Coussens et al. (2008) produced *Sirt1*$^{-/-}$mice and found that *Sirt1* deficiency markedly attenuates spermatogenesis. The authors observed that numbers of mature sperm and spermatogenic precursors were significantly reduced (~2- to 10-fold less; $P \leq 0.004$ in *Sirt1*$^{-/-}$mice. However, the proportion of mature sperm with elevated DNA damage (~7.5% of total epididymal sperm; $P = 0.02$) was significantly increased in adult *Sirt1*$^{-/-}$ males (Coussens et al. 2008).

SMCP is a cysteine- and proline-rich structural protein that is closely associated with the keratinous capsules of sperm mitochondria in the mitochondrial sheath surrounding the outer dense fibers and axoneme. Nayernia and coworkers (2002) reported that all *Smcp*$^{-/-}$ males with a 129/Sv background are infertile despite normal sexual behavior toward female mice and production of copulation plugs. *In vitro* fertilization revealed that only 24.5% of oocytes with intact zona pellucida were fertilized by *Smcp*-deficient sperm and 14% of all oocytes developed into normal blastocysts, while 82.6% of eggs were fertilized with sperm from the wild-type counterparts and 54% of them developed into blastocysts. These results indicate that the infertility of the male *Smcp*$^{-/-}$ mice with a 129/Sv background is due to reduced motility of the spermatozoa and decreased capability of the spermatozoa to penetrate oocytes (Nayernia et al. 2002).

TAF7L is an X-linked germ cell-specific paralogue of *TAF7*, which is a generally expressed component of TFIID. Cheng et al. (2007) generated *Taf7l* mutant mice by homologous recombination in embryonic stem cells. The authors observed that the weight of *Taf7l*$^{-/Y}$ testis was lower, and the amount of sperm in the epididymides was sharply reduced, although spermatogenesis is completed in mutant mice. Mutant epididymal sperm exhibited abnormal morphology, including folded tails. Overall, *Taf7l*$^{-/Y}$ males are fertile, but with reduced sperm motility and litter size (Cheng et al. 2007). Evidence has shown that there is a possible association of point mutation in exon 13 of the human *TAF7L* with infertility (Akinloye et al. 2007).

Voltage-dependent anion channels (VDACs), also known as mitochondrial porins, are small channel proteins involved in the translocation of metabolites across the mitochondrial outer membrane. Sampson et al. (2001) reported that mice lacking *Vdac3* are healthy, but males are infertile. When *Vdac3*-deficient male mice were mated with female mice, they demonstrated normal copulatory behavior, as evidenced by the presence of vaginal plugs in their mates, but no pregnancies were observed in over 100 matings. Although there are normal sperm numbers, the sperm exhibit markedly reduced motility. In particular, 68% of *Vdac3*$^{-/-}$ epididymal sperm axonemes (247/362) in cross section had some structural aberration, most commonly loss of one outer doublet from the normal 9 + 2 microtubule doublet arrangement. This compared with structural abnormalities in 9% of wild-type axonemes (37/423) (Sampson et al. 2001).

7.3.4 Nucleus-encoded mitochondrial genes and female fertility

Female fertility and reproduction is a complicated process. As described in Chapter 2, in order to produce offspring, females must efficiently reach puberty; display estrus; shed one or more competent ova; create the appropriate oviductal environment for fertilization to take place; undergo the necessary systemic, ovarian, and uterine modifications to support pregnancy; deliver the offspring; lactate; and successfully return to estrus after offspring are weaned. Gene knockout or deficiency mice revealed nucleus-encoded mitochondrial genes affecting female fertility as described below.

It has been widely recognized that CREB (the cyclic AMP responsive element-binding protein-1) regulated transcription coactivator 1 (*CRTC1*) is one the regulatory circuits that control some of the most fundamental aspects of metabolism, cell growth, proliferation, survival, and differentiation at both the cellular and organismal levels (Bhaskar and Hay 2007). Altarejos et al. (2008) reported that adult *Crtc1*$^{-/-}$mice are infertile; no offspring were obtained from either *Crtc1*$^{-/-}$ males or *Crtc1*$^{-/-}$ females mated with wild-type mice. *Crtc1*$^{-/-}$ female uteri appeared threadlike in appearance with noticeable thinning of the endometrium, and the ovaries contained no corpora lutea. In addition, circulating concentrations of pituitary luteinizing hormone were also reduced in *Crtc1* mutants compared with the controls.

DEAD (Asp-Glu-Ala-Asp) box polypeptide 20 (DDX20) is one of the DEAD box proteins that are implicated in a number of cellular processes involving alteration of RNA secondary structure such as translation initiation, nuclear and mitochondrial splicing, and ribosome and spliceosome assembly. Some members of this family are involved in embryogenesis, spermatogenesis, and cellular growth and division. Mouillet and colleagues (2008) disrupted *Ddx20* gene in mice and found that *Dp103*$^{+/-}$ females have heavier ovaries than wild-type mice. The weights of the testes or adrenal glands were similar between the two genotypes. Histological examination of ovarian sections showed that both wild-type and *Dp103* heterozygous mice have a similar number of follicles and corpora lutea, but the number of follicles that were devoid of oocytes was significantly higher in the latter mice compared with the former mice. In addition, the estrous cycle was altered and the basal secretion of adrenocorticotrophic hormone was reduced in the *Dp103*$^{+/-}$ females (Mouillet et al. 2008).

Nuclear receptor coactivator 3 (NCOA3) interacts with nuclear hormone receptors to enhance their transcriptional activator functions. Xu and colleagues (2000) reported that mouse *Ncoa3* is expressed in a tissue-specific fashion and distributed mainly in oocytes, mammary glands, hippocampus, olfactory bulb, smooth muscle, hepatocytes, and vaginal epithelium. The authors also found that genetic disruption of *Ncoa3* in mice results in a series of changes, such as dwarfism, delayed puberty, reduced female reproductive function, and blunted mammary gland development. Hormonal analysis further revealed that *Ncoa3* plays a role in both the growth hormone regulatory pathway and the production of estrogen (Xu et al. 2000).

Nitric oxide (NO) synthase 3 (NOS3) (endothelial cell) is one of the NO synthases that produce NO in almost all tissues and organs. NO is a reactive free radical that exerts a variety of biologic actions under both physiological and pathological conditions. Pallares and coworkers (2008) generated *Nos3*-knockout mice and found significant differences in mean number of corpora lutea (9.7 ± 1.2 in $Nos3^{-/-}$ vs. 14.2 ± 1.2 in $Nos3^{+/+}$; $P < 0.01$), rate of anovulation ($48.3 \pm 7.3\%$ in $Nos3^{-/-}$ vs. 29.7 ± 6.3 in $Nos3^{+/+}$; $P < 0.05$), total mean number of recovered oocytes/zygotes (4.0 ± 1.1 in $Nos3^{-/-}$ vs. 10.4 ± 1.6 in $Nos3^{+/+}$; $P < 0.01$), and non-fertilization rate (50.7 in $Nos3^{-/-}$ vs. 3.3% in $Nos3^{+/+}$; $P < 0.001$) between the knockout mice (groups $Nos3^{-/-}$ and the wild-type mice (groups $Nos3^{+/+}$).

In addition to affecting male fertility, both *IMMP2L* and *PANK2* affect female fertility. Lu et al. (2008) found that homozygous *Immp2l*Tg(Tyr)979Ove females are infertile due to defects in folliculogenesis and ovulation. Kuo et al. (2005) observed that female *Pank2* knockout mice produced viable offspring when they mated with wild-type males. However, the $Pank2^{-/-}$ females gave a much fewer number of offspring when compared with wild-type or heterozygote females probably due to an *in utero* effect of maternal *Pank2* deficiency.

7.3.5 Nucleus-encoded mitochondrial genes and embryonic development

The mitochondrion is involved in energy generation, apoptosis regulation, and calcium homeostasis. Deficiency in genes involved in mitochondrial processes and biogenesis often results in a severe embryonic lethality. The nucleus-encoded mitochondrial genes that are essential for embryonic development, include coenzyme Q7 homolog, ubiquinone (yeast) (COQ7), cullin 7 (*CUL7*), endonuclease G (*ENDOG*), ligase III, DNA, ATP dependent (*LIG3*), nuclear respiratory factor 1 (*NRF1*), optic atrophy 1 (autosomal dominant) (*OPA1*), v-raf-leukemia viral oncogene 1 (*RAF1*), succinate dehydrogenase complex, subunit D, integral membrane protein (*SDHD*), solute carrier family 25 (mitochondrial thiamine pyrophosphate carrier), member 19 (*SLC25A19*), and surfeit 1 (*SURF1*). In addition to infertility problems in females, both *DDX20* and *NOS3* are also important to embryonic development. Homozygous *Ddx20* null mice die early in embryonic development prior to the four-cell stage during early embryonic development (Mouillet et al. 2008). As for the *Nos3* gene, embryo losses were detected between days 8.5 and 13.5, in 62.5% of *Nos3*-knockout dams and, at days 10.5 and 11.5, in 16.7% of the control females ($P < 0.005$) (Pallares et al. 2008).

COQ7 encodes a protein similar to a mitochondrial di-iron-containing hydroxylase in *S. cerevisiae* that is involved with ubiquinone biosynthesis. Nakai et al. (2001) generated *Coq7*-deficient mice to investigate the biologic role of COQ7 in mammals. The authors observed the death of *Coq7*-deficient mouse embryos at days 10.5 with the neuroepithelial cells failing to show the radial arrangement in the developing cerebral wall, aborting neurogenesis. Electron microscopic analysis further indicated the enlarged mitochondria with vesicular cristae and enlarged lysosomes filled with disrupted membranes. Biochemical analysis demonstrated that *Coq7*-deficient embryos do not synthesize CoQ_9 but produce demethoxyubiquinone 9 (DMQ_9) instead. In addition, the *Coq7*-deficient embryos were smaller than normal (Nakai et al. 2001).

CUL7 is a member of the cullin family. Evidence has shown that *CUL7* is a new oncogene, which cooperates with Myc in transformation by blocking Myc-induced apoptosis in a p53-dependent manner, thus playing an essential role in numerous cellular and biologic activities (Kim et al. 2007). To further understand the physiological role of *CUL7*, Arai and colleagues (2003) generated mice lacking the gene. The authors observed that $Cul7^{-/-}$embryos are runted and die immediately after birth because of respiratory distress. Dermal and hypodermal hemorrhage was also detected in mutant embryos in late gestation. In addition, $Cul7^{-/-}$ placentas had defects in the differentiation of the trophoblast lineage with an abnormal vascular structure (Arai et al. 2003).

ENDOG encodes a nuclear-encoded endonuclease that is localized in the mitochondrion. This gene plays an important role in both nuclear DNA fragmentation during apoptosis by cleaving DNA at GC tracts and mtDNA replication by generating the RNA primers required by DNA polymerase gamma to initiate the replication of mtDNA. Zhang et al. (2003) found that *Endog* homozygous mutant embryos die between embryonic days 2.5 and 3.5. DNA fragmentation is also reduced in $Endog^{+/-}$ thymocytes and splenocytes compared with wild-type cells, as well as in $Endog^{+/-}$ thymus *in vivo* compared with that of the wild-type mice, on activation of apoptosis. These findings further confirmed that EndoG is essential during early embryogenesis and plays a critical role in normal apoptosis and nuclear DNA fragmentation.

DNA ligases catalyze the joining of strand breaks in the phosphodiester backbone of duplex DNA and play essential roles in DNA replication, recombination, repair, and maintenance of genomic integrity. *LIG3* gene is a member of the DNA ligase family. Puebla-Osorio et al. (2006) made a targeted interruption of *Lig3* gene in mice. Mice heterozygous for the *Lig3* mutation were fertile and did not display phenotypic abnormalities. The authors then bred heterozygous mice to generate homozygous *Lig3* mutant mice, which yielded no viable $Lig3^{-/-}$ pups in nearly 800 young mice genotyped by Southern blotting or PCR. These results clearly indicate that *Lig3* is essential for mouse embryonic development and survival. To further determine the effects of *Lig3* gene inactivation on embryonic development, the group collected embryos or yolk sacs from heterozygous mice crossbred at different days of gestation and genotyped by PCR. The authors found that the mutant embryonic developmental process stops at 8.5 days postcoitum (dpc), and excessive cell death occurs at 9.5 dpc.

NRF1 encodes a protein that homodimerizes and functions as a transcription factor, which activates the expression of some key metabolic genes regulating cellular growth and nuclear genes required for respiration, heme biosynthesis, and mtDNA transcription and replication. Huo and Scarpulla (2001) found that embryos homozygous for *Nrf1* disruption die between embryonic days 3.5 and 6.5. The authors detected the beta-galactosidase staining in growing oocytes as well as in 2.5- and 3.5-day-old embryos, indicating that the *Nrf1* gene is expressed during oogenesis and during early stages of embryogenesis.

OPA1 encodes a nuclear-encoded mitochondrial protein, which is similar to dynamin-related GTPases. Mutations in this gene have been associated with optic atrophy type 1, which is a dominantly inherited optic neuropathy resulting in progressive loss of visual acuity, leading in many cases to legal blindness (Gränse et al. 2003). Alavi and coworkers (2007) generated the first mouse model carrying a splice site mutation (c.1065 + 5G→A) in the *Opa1* gene. The mutation induces a skipping of exon 10 during transcript processing and leads to an in-frame deletion of 27 amino acid residues in the GTPase domain. Using magnetic resonance imaging, the authors reported that the homozygous mutant mice die *in utero* during embryogenesis with the first notable developmental delay at day 8.5 (Alavi et al. 2007).

RAF1 protein kinase has been identified as an integral component of the Ras/Raf/MEK/ERK signaling pathway in mammals. In the pathway, RAF1 is activated by its binding to the Ras family of membrane-associated GTPases and then the protein can phosphorylate to activate the dual specificity protein kinases MEK1 and MEK2. In turn, MEK1 and MEK2 phosphorylate to activate the serine/threonine-specific protein kinases, ERK1 and ERK2. Activated ERKs have pleiotropic effects on cell physiology and play an important role in the control of gene expression related to the cell division cycle, apoptosis, cell differentiation, and cell migration. Hüser et al. (2001) found that $Raf1^{-/-}$ mice die in embryogenesis and show vascular defects in the yolk sac and placenta as well as increased apoptosis of embryonic tissues.

Complex II of the respiratory chain includes four nuclear-encoded subunits and is localized in the mitochondrial inner membrane, which is specifically involved in the oxidation of succinate. The *SDHD* gene encodes one of the two membrane-anchoring proteins of succinate dehydrogenase (complex II) of the mitochondrial electron transport chain. Piruat and coworkers (2004) constructed a null allele of the *Sdhd* gene by target replacement of the wild-type allele with a nonfunctional allele lacking exons 2, 3, and 4. Among 152 animals born from mating between heterozygous parents, only 50 (33%) *Sdhd*$^{+/+}$ and 102 (66%) *Sdhd*$^{+/-}$ mice were found, indicating lethality for all homozygous mutant embryos. The authors then examined the embryos from heterozygous pregnant females mated with heterozygous males and found that at 9.5 dpc, approximately one-fourth of embryos are stalled. The use of specific primers for both the wild-type and mutant alleles confirmed that the stalled embryos are homozygous *Sdhd*$^{-/-}$ mice. These observations indicate that *Sdhd*$^{-/-}$ mutants die at early embryonic stages (Piruat et al. 2004).

SURF1 encodes a protein localized to the inner mitochondrial membrane and is thought to be involved in the biogenesis of the cytochrome c oxidase complex. Agostino et al. (2003) created a constitutive knockout mouse for *Surf1*, which is characterized by the high postimplantation embryonic lethality, affecting approximately 90% of the *Surf1*$^{-/-}$ individuals. The authors obtained homozygous *Surf1*$^{-/-}$ mice from *Surf1*$^{+/-}$ intercrosses and observed 10-fold lower (2.7%) *Surf1*$^{-/-}$ pups than expected by Mendelian transmission of a recessive trait (25%). PCR-genotyping revealed that the −/− genotype in blastocysts occurred 26% of the time, but dropped to 14% at E6.5–E7.5 dpc, to 10% at E8.5–E12, and to ~2% at E13–E18. These results show that the loss of *Surf1*$^{-/-}$ embryos started at a stage as early as gastrulation (E4–E7 dpc) and continued during organogenesis (E8.5–E12 dpc), body-mass growth, and organ maturation (E13–E18 dpc). Interestingly, *Surf1*$^{-/-}$ embryos at different developmental stages did not show gross morphological abnormalities, but were consistently smaller in size in comparison with their +/+ or −/+ littermates (Agostino et al. 2003).

SLC25A19 encodes a mitochondrial protein that belongs to the solute carrier family. Recent studies have shown that this protein functions as the mitochondrial thiamine pyrophosphate carrier, which transports thiamine pyrophosphates into mitochondria. Lindhurst et al. (2006) generated a knockout mouse model of *Slc25a19*. The authors observed that these mutant animals have 100% prenatal lethality by embryonic day 12. At embryonic day 10.5, these *Slc25a19*$^{-/-}$ embryos have a neural-tube closure defect with ruffling of the neural fold ridges, a yolk sac erythropoietic failure, and elevated α-ketoglutarate in the amniotic fluid (Lindhurst et al. 2006).

7.4 Future research directions

It has been well known that nucleus-encoded mitochondrial genes are required for the proper assembly of respiratory chain complexes, maintenance of mtDNA integrity and replication, transportation of nuclear-encoded proteins from the cytoplasm into mitochondria, synthesis of inner mitochondrial membrane phospholipids, and control of the abundance and quality of mtDNA. One example of future research is to determine how cross talk between the nuclear and mitochondrial genomes affects genetic complexity of fertility and reproduction in livestock species. In fact, mitochondrial biogenesis depends heavily on the coordinated expression of two genomes, nuclear and mitochondrial. As a consequence, the control of mitochondrial biogenesis and function depends on extremely complex processes that require a variety of well-orchestrated regulatory mechanisms (Garesse and Vallejo 2001). For example, significant cross talk exists between the nucleus and mitochondria, ultimately resulting in neuronal cell death (Dawson and Dawson 2004). On the other hand, regulation of mitochondrial biogenesis and proliferation is influenced by external factors, such as nutrients, hormones, temperature, and aging. Communications are also required for eliciting mitochondrial responses to specific stress pathways (Ryan and Hoogenraad 2007). Therefore, there will be a need in the future to study the mechanisms of mitochondrial biogenesis and the way cells respond to external signals to maintain mitochondrial function and thus maintain high reproductive efficiency in livestock species.

References

Agbaje, I.M., Rogers, D.A., McVicar, C.M., McClure, N., Atkinson, A.B., Mallidis, C., and Lewis, S.E. 2007. Insulin dependant diabetes mellitus: Implications for male reproductive function. *Human Reproduction* **22**: 1871–1877.

Agostino, A., Invernizzi, F., Tiveron, C., Fagiolari, G., Prelle, A., Lamantea, E., Giavazzi, A., Battaglia, G., Tatangelo, L., Tiranti, V., and Zeviani, M. 2003. Constitutive knockout of Surf1 is associated with high embryonic lethality, mitochondrial disease and cytochrome c oxidase deficiency in mice. *Human Molecular Genetics* **12**: 399–413.

Aitken, R.J. 1995. Free radicals, lipid peroxidation and sperm function. *Reproduction, Fertility, and Development* **7**: 659–668.

Akinloye, O., Gromoll, J., Callies, C., Nieschlag, E., and Simoni, M. 2007. Mutation analysis of the X-chromosome linked, testis-specific TAF7L gene in spermatogenic failure. *Andrologia* **39**: 190–195.

Alavi, M.V., Bette, S., Schimpf, S., Schuettauf, F., Schraermeyer, U., Wehrl, H.F., Ruttiger, L., Beck, S.C., Tonagel, F., Pichler, B.J., Knipper, M., Peters, T., Laufs, J., and Wissinger, B. 2007. A splice site mutation in the murine Opa1 gene features pathology of autosomal dominant optic atrophy. *Brain* **130**: 1029–1042.

Allen, J.F. 1996. Separate sexes and the mitochondrial theory of ageing. *Journal of Theoretical Biology* **180**: 135–140.

Altarejos, J.Y., Goebel, N., Conkright, M.D., Inoue, H., Xie, J., Arias, C.M., Sawchenko, P.E., and Montminy, M. 2008. The Creb1 coactivator Crtc1 is required for energy balance and fertility. *Nature Medicine* **14**: 1112–1117.

Anderson, S., de Bruijn, M.H., Coulson, A.R., Eperon, I.C., Sanger, F., and Young, I.G. 1982. Complete sequence of bovine mitochondrial DNA. Conserved features of the mammalian mitochondrial genome. *Journal of Molecular Biology* **156**: 683–717.

Arai, T., Kasper, J.S., Skaar, J.R., Ali, S.H., Takahashi, C., and DeCaprio, J.A. 2003. Targeted disruption of p185/Cul7 gene results in abnormal vascular morphogenesis. *Proceedings of the National Academy of Sciences of the United States of America* **100**: 9855–9860.

Barritt, J.A., Brenner, C.A., Cohen, J., and Matt, D.W. 1999. Mitochondrial DNA rearrangements in human oocytes and embryos. *Molecular Human Reproduction* **5**: 927–933.

Bhaskar, P.T. and Hay, N. 2007. The two TORCs and Akt. *Developmental Cell* **12**: 487–502.

Boore, J.L. 1999. Animal mitochondrial genomes. *Nucleic Acids Research* **27**: 1767–1780.

Brenner, C., Marzo, I., and Kroemer, G. 1998. A revolution in apoptosis: From a nucleocentric to a mitochondriocentric perspective. *Experimental Gerontology* **33**: 543–553.

Burri, L., Strahm, Y., Hawkins, C.J., Gentle, I.E., Puryer, M.A., Verhagen, A., Callus, B., Vaux, D., and Lithgow, T. 2005. Mature DIABLO/Smac is produced by the IMP protease complex on the mitochondrial inner membrane. *Molecular Biology of the Cell* **16**: 2926–2933.

Catalano, D., Licciulli, F., Turi, A., Grillo, G., Saccone, C., and D'Elia, D. 2006. MitoRes: A resource of nuclear-encoded mitochondrial genes and their products in Metazoa. *BMC Bioinformatics* **7**: 36.

Chakrabarti, R., Kline, D., Lu, J., Orth, J., Pilder, S., and Vijayaraghavan, S. 2007. Analysis of Ppp1cc-null mice suggests a role for PP1gamma2 in sperm morphogenesis. *Biology of Reproduction* **76**: 992–1001.

Cheng, Y., Buffone, M.G., Kouadio, M., Goodheart, M., Page, D.C., Gerton, G.L., Davidson, I., and Wang, P.J. 2007. Abnormal sperm in mice lacking the Taf7l gene. *Molecular and Cellular Biology* **27**: 2582–2589.

Chinnery, P.F. 2003. Searching for nuclear-mitochondrial genes. *Trends in Genetics* **19**: 60–62.

Chomyn, A. 1998. The myoclonic epilepsy and ragged-red fiber mutation provides new insights into human mitochondrial function and genetics. *American Journal of Human Genetics* **62**: 745–751.

Cotter, D., Guda, P., Fahy, E., and Subramaniam, S. 2004. MitoProteome: Mitochondrial protein sequence database and annotation system. *Nucleic Acids Research* **32**: D463–D467.

Coussens, M., Maresh, J.G., Yanagimachi, R., Maeda, G., and Allsopp, R. 2008. Sirt1 deficiency attenuates spermatogenesis and germ cell function. *PLoS One* **3**: e1571.

Dawson, V.L. and Dawson, T.M. 2004. Deadly conversations: Nuclear-mitochondrial cross-talk. *Journal of Bioenergetics and Biomembranes* **36**: 287–294.

Folgero, T., Bertheussen, K., Lindal, S., Torbergsen, T., and Oian, P. 1993. Mitochondrial disease and reduced sperm motility. *Human Reproduction* **8**: 1863–1868.

Garesse, R. and Vallejo, C.G. 2001. Animal mitochondrial biogenesis and function: A regulatory cross-talk between two genomes. *Gene* **263**: 1–16.

Gränse, L., Bergstrand, I., Thiselton, D., Ponjavic, V., Heijl, A., Votruba, M., and Andréasson, S. 2003. Electrophysiology and ocular blood flow in a family with dominant optic nerve atrophy and a mutation in the OPA1 gene. *Ophthalmic Genetics* **24**: 233–245.

Gray, M.W. 1999. Evolution of organellar genomes. *Current Opinion in Genetics & Development* **9**: 678–687.

Gyllensten, U., Wharton, D., Josefsson, A., and Wilson, A.C. 1991. Paternal inheritance of mitochondrial DNA in mice. *Nature* **352**: 255–257.

Hayashi, J., Ohta, S., Kikuchi, A., Takemitsu, M., Goto, Y., and Nonaka, I. 1991. Introduction of disease-related mitochondrial DNA deletions into HeLa cells lacking mitochondrial DNA results in mitochondrial dysfunction. *Proceedings of the National Academy of Sciences of the United States of America* **88**: 10614–10618.

Henze, K. and Martin, W. 2001. How do mitochondrial genes get into the nucleus? *Trends in Genetics* **17**: 383–387.

Huo, L. and Scarpulla, R.C. 2001. Mitochondrial DNA instability and peri-implantation lethality associated with targeted disruption of nuclear respiratory factor 1 in mice. *Molecular and Cellular Biology* **21**: 644–654.

Hüser, M., Luckett, J., Chiloeches, A., Mercer, K., Iwobi, M., Giblett, S., Sun, X.M., Brown, J., Marais, R., and Pritchard, C. 2001. MEK kinase activity is not necessary for Raf-1 function. *EMBO Journal* **20**: 1940–1951.

Jansen, R.P. and de Boer, K. 1998. The bottleneck: Mitochondrial imperatives in oogenesis and ovarian follicular fate. *Molecular and Cellular Endocrinology* **145**: 81–88.

Kao, S.H., Chao, H.T., and Wei, Y.H. 1995. Mitochondrial deoxyribonucleic acid 4977 bp deletion is associated with diminished fertility and motility of human sperm. *Biology of Reproduction* **52**: 729–736.

Keefe, D.L., Niven-Fairchild, T., Powell, S., and Buradagunta, S. 1995. Mitochondrial deoxyribonucleic acid deletions in oocytes and reproductive aging in women. *Fertility and Sterility* **64**: 577–583.

Khatchadourian, K., Smith, C.E., Metzler, M., Gregory, M., Hayden, M.R., Cyr, D.G., and Hermo, L. 2007. Structural abnormalities in spermatids together with reduced sperm counts and motility underlie the reproductive defect in HIP1-/- mice. *Molecular Reproduction and Development* **74**: 341–359.

Kim, S.S., Shago, M., Kaustov, L., Boutros, P.C., Clendening, J.W., Sheng, Y., Trentin, G.A., Barsyte-Lovejoy, D., Mao, D.Y., Kay, R., Jurisica, I., Arrowsmith, C.H., and Penn, L.Z. 2007. CUL7 is a novel antiapoptotic oncogene. *Cancer Research* **67**: 9616–9622.

Kitagawa, T., Suganuma, N., Nawa, A., Kikkawa, F., Tanaka, M., Ozawa, T., and Tomoda, Y. 1993. Rapid accumulation of deleted mitochondrial deoxyribonucleic acid in postmenopausal ovaries. *Biology of Reproduction* **49**: 730–736.

Kozak, M. 1987. An analysis of 5′-noncoding sequences from 699 vertebrate messenger RNAs. *Nucleic Acids Research* **15**: 8125–8148.

Kuo, Y.M., Duncan, J.L., Westaway, S.K., Yang, H., Nune, G., Xu, E.Y., Hayflick, S.J., and Gitschier, J. 2005. Deficiency of pantothenate kinase 2 (Pank2) in mice leads to retinal degeneration and azoospermia. *Human Molecular Genetics* **14**: 49–57.

Lindhurst, M.J., Fiermonte, G., Song, S., Struys, E., De Leonardis, F., Schwartzberg, P.L., Chen, A., Castegna, A., Verhoeven, N., Mathews, C.K., Palmieri, F., and Biesecker, L.G. 2006. Knockout of Slc25a19 causes mitochondrial thiamine pyrophosphate depletion, embryonic lethality, CNS malformations, and anemia. *Proceedings of the National Academy of Sciences of the United States of America* **103**: 15927–15932.

Lu, B., Poirier, C., Gaspar, T., Gratzke, C., Harrison, W., Busija, D., Matzuk, M.M., Andersson, K.E., Overbeek, P.A., and Bishop, C.E. 2008. A mutation in the inner mitochondrial membrane peptidase 2-like gene (Immp2l) affects mitochondrial function and impairs fertility in mice. *Biology of Reproduction* **78**: 601–610.

Makalowska, I., Lin, C.F., and Makalowski, W. 2005. Overlapping genes in vertebrate genomes. *Computational Biology and Chemistry* **29**: 1–12.

Manfredi, G., Thyagarajan, D., Papadopoulou, L.C., Pallotti, F., and Schon, E.A. 1997. The fate of human sperm-derived mtDNA in somatic cells. *American Journal of Human Genetics* **61**: 953–960.

Marchington, D.R., Macaulay, V., Hartshorne, G.M., Barlow, D., and Poulton, J. 1998. Evidence from human oocytes for a genetic bottleneck in an mtDNA disease. *American Journal of Human Genetics* **63**: 769–775.

May-Panloup, P., Chrétien, M.F., Savagner, F., Vasseur, C., Jean, M., Malthièry, Y., and Reynier, P. 2003. Increased sperm mitochondrial DNA content in male infertility. *Human Reproduction* **18**: 550–556.

McBride, H.M., Neuspiel, M., and Wasiak, S. 2006. Mitochondria: More than just a powerhouse. *Current Biology* **16**: R551.

Mouillet, J.F., Yan, X., Ou, Q., Jin, L., Muglia, L.J., Crawford, P.A., and Sadovsky, Y. 2008. DEAD-box protein-103 (DP103, Ddx20) is essential for early embryonic development and modulates ovarian morphology and function. *Endocrinology* **149**: 2168–2175.

Müller-Höcker, J., Schäfer, S., Weis, S., Münscher, C., and Strowitzki, T. 1996. Morphological-cytochemical and molecular genetic analyses of mitochondria in isolated human oocytes in the reproductive age. *Molecular Human Reproduction* **2**: 951–958.

Nakada, K., Sato, A., Yoshida, K., Morita, T., Tanaka, H., Inoue, S., Yonekawa, H., and Hayashi, J. 2006. Mitochondria-related male infertility. *Proceedings of the National Academy of Sciences of the United States of America* **103**: 15148–15153.

Nakai, D., Yuasa, S., Takahashi, M., Shimizu, T., Asaumi, S., Isono, K., Takao, T., Suzuki, Y., Kuroyanagi, H., Hirokawa, K., Koseki, H., and Shirsawa, T. 2001. Mouse homologue of coq7/clk-1, longevity gene in *Caenorhabditis elegans*, is essential for coenzyme Q synthesis, maintenance of mitochondrial integrity, and neurogenesis. *Biochemical and Biophysical Research Communications* **289**: 463–471.

Nayernia, K., Adham, I.M., Burkhardt-Göttges, E., Neesen, J., Rieche, M., Wolf,

S., Sancken, U., Kleene, K., and Engel, W. 2002. Asthenozoospermia in mice with targeted deletion of the sperm mitochondrion-associated cysteine-rich protein (Smcp) gene. *Molecular and Cellular Biology* **22**: 3046–3052.

Neiman, M. and Taylor, D.R. 2009. The causes of mutation accumulation in mitochondrial genomes. *Proceedings. Biological Sciences* **276**: 1201–1209.

Nisoli, E., Clementi, E., Carruba, M.O., and Moncada, S. 2007. Defective mitochondrial biogenesis: A hallmark of the high cardiovascular risk in the metabolic syndrome? *Circulation Research* **100**: 795–806.

Pallares, P., Garcia-Fernandez, R.A., Criado, L.M., Letelier, C.A., Esteban, D., Fernandez-Toro, J.M., Flores, J.M., and Gonzalez-Bulnes, A. 2008. Disruption of the endothelial nitric oxide synthase gene affects ovulation, fertilization and early embryo survival in a knockout mouse model. *Reproduction* **136**: 573–579.

Piruat, J.I., Pintado, C.O., Ortega-Sáenz, P., Roche, M., and López-Barneo, J. 2004. The mitochondrial SDHD gene is required for early embryogenesis, and its partial deficiency results in persistent carotid body glomus cell activation with full responsiveness to hypoxia. *Molecular and Cellular Biology* **24**: 10933–10940.

Prokisch, H., Andreoli, C., Ahting, U., Heiss, K., Ruepp, A., Scharfe, C., and Meitinger, T. 2006. MitoP2: The mitochondrial proteome database–now including mouse data. *Nucleic Acids Research* **34**(Database issue): D705–D711.

Puebla-Osorio, N., Lacey, D.B., Alt, F.W., and Zhu, C. 2006. Early embryonic lethality due to targeted inactivation of DNA ligase III. *Molecular and Cellular Biology* **26**: 3935–3941.

Rossato, M., Pagano, C., and Vettor, R. 2008. The cannabinoid system and male reproductive functions. *Journal of Neuroendocrinology* **20**(Supplement 1): 90–93.

Rovio, A.T., Marchington, D.R., Donat, S., Schuppe, H.C., Abel, J., Fritsche, E., Elliott, D.J., Laippala, P., Ahola, A.L., McNay, D., Harrison, R.F., Hughes, B., Barrett, T., Bailey, D.M., Mehmet, D., Jequier, A.M., Hargreave, T.B., Kao, S.H., Cummins, J.M., Barton, D.E., Cooke, H.J., Wei, Y.H., Wichmann, L., Poulton, J., and Jacobs, H.T. 2001. Mutations at the mitochondrial DNA polymerase (POLG) locus associated with male infertility. *Nature Genetics* **29**: 261–262.

Ryan, M.T. and Hoogenraad, N.J. 2007. Mitochondrial-nuclear communications. *Annual Review of Biochemistry* **76**: 701–722.

Sampson, M.J., Decker, W.K., Beaudet, A.L., Ruitenbeek, W., Armstrong, D., Hicks, M.J., and Craigen, W.J. 2001. Immotile sperm and infertility in mice lacking mitochondrial voltage-dependent anion channel type 3. *The Journal of Biological Chemistry* **276**: 39206–39212.

Sas, K., Robotka, H., Toldi, J., and Vecsei, L. 2007. Mitochondria, metabolic disturbances, oxidative stress and the kynurenine system, with focus on neurodegenerative disorders. *Journal of Neurological Sciences* **257**: 221–239.

Schmidt, A., Marescau, B., Boehm, E.A., Renema, W.K., Peco, R., Das, A., Steinfeld, R., Chan, S., Wallis, J., Davidoff, M., Ullrich, K., Waldschütz, R., Heerschap, A., De Deyn, P.P., Neubauer, S., and Isbrandt, D. 2004. Severely altered guanidino compound levels, disturbed body weight homeostasis and impaired fertility in a mouse model of guanidinoacetate N-methyltransferase (GAMT) deficiency. *Human Molecular Genetics* **13**: 905–921.

Schwarz, M., Andrade-Navarro, M.A., and Gross, A. 2007. Mitochondrial carriers and pores: Key regulators of the mitochondrial apoptotic program? *Apoptosis* **12**: 869–876.

Selosse, M., Albert, B., and Godelle, B. 2001. Reducing the genome size of organelles favours gene transfer to the nucleus. *Trends in Ecology & Evolution* **16**: 135–141.

Smith, L.C., Thundathil, J., and Filion, F. 2005. Role of the mitochondrial genome in preimplantation development and assisted reproductive technologies. *Reproduction, Fertility, and Development* **17**: 15–22.

St. John, J.C., Cooke, I.D., and Barratt, C.L.R. 1997. Mitochondrial mutations and male infertility [Letter]. *Nature Medicine* **3**: 124–125.

Sutarno, C.G., Cummins, J.M., Greeff, J., and Lymbery, A.J. 2002. Mitochondrial DNA polymorphisms and fertility in beef cattle. *Theriogenology* **57**: 1603–1610.

Sutovsky, P., Moreno, R.D., Ramalho-Santos, J., Dominko, T., Simerly, C., and Schatten, G. 2000. Ubiquitinated sperm mitochondria, selective proteolysis, and the regulation of mitochondrial inheritance in mammalian embryos. *Biology of Reproduction* **63**: 582–590.

Van Blerkom, J. 2004. Mitochondria in human oogenesis and preimplantation embryogenesis: Engines of metabolism, ionic regulation and developmental competence. *Reproduction* **128**(3): 269–280.

Wallace, K.B., Eells, J.T., Madeira, V.M., Cortopassi, G., and Jones, D.P. 1997. Mitochondria-mediated cell injury. Symposium overview. *Fundamental and Applied Toxicology* **38**: 23–37.

Xu, J., Liao, L., Ning, G., Yoshida-Komiya, H., Deng, C., and O'Malley, B.W. 2000. The steroid receptor coactivator SRC-3 (p/CIP/RAC3/AIB1/ACTR/TRAM-1) is required for normal growth, puberty, female reproductive function, and mammary gland development. *Proceedings of the National Academy of Sciences of the United States of America* **97**: 6379–6384.

Yamauchi, A. 2005. Rate of gene transfer from mitochondria to nucleus: Effects of cytoplasmic inheritance system and intensity of intracellular competition. *Genetics* **171**: 1387–1396.

Zhang, J., Dong, M., Li, L., Fan, Y., Pathre, P., Dong, J., Lou, D., Wells, J.M., Olivares-Villagómez, D., Van Kaer, L., Wang, X., and Xu, M. 2003. Endonuclease G is required for early embryogenesis and normal apoptosis in mice. *Proceedings of the National Academy of Sciences of the United States of America* **100**: 15782–15787.

Part II

Physiological Genomics of Reproduction

8

Functional Genomics Studies of Ovarian Function in Livestock: Physiological Insight Gained and Perspective for the Future

Beau Schilling and George W. Smith

8.1 Introduction

Reproductive efficiency is a major limiting factor in the success of livestock operations. The ovarian cycle is central to the reproductive process because only mature ovarian follicles release eggs competent to be fertilized. Once a follicle develops to the preovulatory stage, it undergoes one of two fates: ovulation or atresia. The corpus luteum (CL), formed from remnants of the ovulated follicle, secretes the steroid hormone progesterone, which sustains embryo growth and survival critical to pregnancy success. In the absence of appropriate embryonic signals, the CL will regress and a new reproductive cycle is initiated. An increased understanding of the intrinsic and extrinsic factors regulating ovarian function (follicular development and atresia, oocyte maturation and function, and CL growth and regression) is critical to improvements in reproductive efficiency of livestock species.

The dawning of the genomics and genome sequencing era in livestock has provided tremendous potential for advancement in understanding of the molecular mechanisms involved in the regulation of the above aspects of ovarian function. The recent increase in availability and economic feasibility of platforms for expressed sequence tag (EST) sequencing, microarrays, and proteomics appropriate for large-scale studies of RNA and protein expression in livestock species has provided reproductive biologists opportunities to characterize changes in RNA transcript (transcriptome) and protein (proteome) profiles of ovarian tissues/cell types at key developmental time points, in response to hormonal treatments and in animals of different reproductive potential. Such approaches hold considerable potential for advancing understanding of regulation of ovarian function and factors contributing to reproductive efficiency.

Expression profiling strategies have been successfully applied to the study of follicular growth and development, CL function, and oocyte biology in cattle and swine. While amounts of raw information obtained from the limited number of studies applying such technologies in livestock have been extensive, new insight of verified biologic/physiological relevance obtained has been more limited, and overall rate of utilization of profiling data in subsequent physiological or functional studies has been low. Some would argue that the limited physiological insight obtained from the application of such emerging technologies is a consequence of a move from hypothesis-driven research to descriptive discovery research. While technological and logistical impediments to elucidation of physiologically relevant information from expression profiling studies in livestock are still evident, rigid criteria for data analysis, utilization of the wealth of available bioinformatics tools for mining of expression profiling data, and a commitment to subsequent downstream functional studies necessary to prove biologic significance can greatly enhance the utility of data obtained.

The objective of this review is to highlight the results of published expression profiling studies of ovarian tissues/cell types (follicular tissue and cells, luteal tissue, and oocytes) in livestock species. Emphasis will be on studies utilizing high-throughput approaches for expression profiling (EST sequencing, microarrays, and proteomics). Emphasis in the discussion of such results will be on documented or potential physiological significance of findings.

8.2 Transcriptomics of ovarian tissues: EST sequencing

Generation of cDNA libraries from specific tissues and characterization of ESTs from such libraries are essential to the characterization of transcriptome composition of tissues and species of interest and can hence facilitate the discovery of new genes of functional significance. Such efforts have also accelerated the development of microarray resources in livestock species required for comparative transcriptome profiling. To date, EST sequencing efforts in livestock have been most extensive in cattle and pigs, with over 1.5 million bovine and 1.4 million porcine EST sequences deposited in the Genbank in contrast to the approximately 39,000 equine ESTs present. Despite the fact that ovarian tissues are represented in many of the cDNA libraries from which ESTs were generated, drawing inferences about ovarian gene expression, or ovary-specific gene expression in particular via data mining of these large electronic resources, is compromised because in many instances, mixed tissue libraries were utilized (Smith et al. 2001).

8.2.1 EST sequence analysis of the follicular and luteal transcriptomes

Significant effort has been placed on the characterization of ESTs from cDNA libraries generated from porcine ovarian (follicular and luteal) tissues. Caetano et al. (2003) reported the construction of a normalized cDNA library from swine ovarian follicles ranging in size from 2 to 10 mm and generation of 3479 unique EST sequence clusters from this library. Physiologically relevant information reported was limited because the depth of EST sequencing was limited and because emphasis was on details of library construction and limited gene ontology analysis for sequences obtained.

The most extensive EST sequencing efforts published to date, focused specifically on ovarian tissues in livestock, were conducted by the investigators at the University of

Missouri-Columbia using pigs (Jiang et al. 2004). Eleven porcine cDNA libraries were constructed from whole ovary and (or) various ovarian structures harvested at specific stages of differentiation including fetal, neonatal, and prepubertal ovaries; specific ovarian structures collected on day 0 (follicle), day 5 (follicle and CL), and day 12 of the estrous cycle; and follicles of different size classifications harvested from weaned sows. Approximately 15,613 EST sequences, representing 8507 gene clusters were generated from these libraries. The majority of clusters (68%) had consensus sequences homologous to tentative consensus sequences for mature transcripts represented in The Institute for Genomic Research Porcine Gene Index.

Gene ontology analysis revealed most of the cDNA-encoded proteins that function in binding, catalysis, and transport, and (or) are involved in cell growth and maintenance and (or) metabolism. Since cDNAs were not subtracted or libraries normalized, frequency of the detection of clones of interest across libraries was quantified using "electronic Northern analysis" as a means to approximate expression levels across tissues from which libraries were constructed. For example, cDNA encoding for FTH1, INHA, and AKR1C3 were represented at the highest frequency in the library generated from highly estrogenic 6-mm follicles versus 2-, 4-, or 8-mm follicles. Authors proposed a key role for FTH1 in storage of iron for cytochrome-containing enzymes (Jiang et al. 2004) involved in steroidogenesis.

Comparison of frequency of individual clones sequenced from the day 12 follicle versus the day 12 CL library by virtual Northern analysis revealed 12 differentially expressed genes. For example, frequency of clones for TIMP1 was greater in the day 12 CL library than the day 12 follicle library. Increased TIMP1 expression is characteristic of the extensive extracellular matrix remodeling characteristic of the developmental transition from follicular to luteal tissue in sheep (Smith et al. 1994a), swine (Smith et al. 1994b), and cattle (Smith et al. 1996).

While a greater number of total EST sequences have been reported for the cow versus the pig, ovarian tissue-specific EST projects are more limited in scope in cattle. Casey and colleagues (2004) generated a cDNA library from bovine CL collected on days 6, 8, and 14 after estrus. Three hundred fifty-one unique sequences were obtained from 960 ESTs characterized. Genes encoding for proteins with documented or potential importance to luteal function were represented in ESTs generated, including regulators of steroidogenesis (e.g., STAR), transcription (e.g., CTCF), antioxidants (e.g., CAT), and tissue remodeling factors (e.g., TIMP2). Temporal expression of select genes of interest (identified from EST sequencing) in CL on days 6, 8, 14, and 17 of the estrous cycle was examined using Northern blot analysis. Messenger RNA (mRNA) for SQLE (squalene epoxidase), which catalyzes a rate-limiting step in cholesterol synthesis was decreased sevenfold in regressed CL collected on day 17, coinciding with the characteristic decrease in systemic progesterone. Results suggest that reduced cholesterol synthesis may accompany the decrease in progesterone production characteristic of luteal regression.

8.2.2 *Oocyte EST sequencing projects*

The oocyte is a key regulator of ovarian follicular development and early embryogenesis. A developmental program intrinsic to the oocyte controls the rate of follicular development (Eppig et al. 2002), and bidirectional communication between the oocyte and nearby cumulus and granulosa cells is critical for the normal growth of ovarian follicles (Eppig 2001; Matzuk et al. 2002). Furthermore, during the initial cleavage divisions after fertilization, embryonic development is supported by the products of maternal effect genes (mRNAs and proteins) synthesized and stored in oocytes during oogenesis (Telford et al. 1990; De Sousa et al. 1998). Such products of maternal effect genes are critical for the interval between fertilization and the maternal-to-embryonic transition when transcriptional activity of the embryonic genome becomes robust, and regulation of embryogenesis is shifted from control by maternal (oocyte-derived) mRNA and proteins to control by products of the embryonic genome (Bettegowda et al. 2008a). Despite the important regulatory role of the oocyte in the control of follicular development and early embryogenesis, composition of the oocyte transcriptome and identities and functions of key oocyte-specific genes involved in the above processes are relatively unknown in livestock species, and the products of important oocyte-expressed genes were likely underrepresented in the results of ovarian EST projects and hence unidentified.

To gain a better understanding of the composition of the transcriptome of bovine oocytes, we constructed a cDNA library from a pool of 200 immature, germinal vesicle (GV) and mature, metaphase II (MII) stage oocytes and sequenced a limited number of ESTs (Yao et al. 2004). Our initial analysis of 230 EST sequences from the bovine oocyte cDNA library (unnormalized) revealed novel information about oocyte-expressed genes. The 230 ESTs represented 102 unique sequences of which 46 displayed significant similarities to sequences encoding for known genes present in the Genbank database. While some ESTs correspond to housekeeping genes critical to the function of all cell types (e.g., RPL15) and some represent genes previously known to be expressed both in oocytes and other tissues (e.g., CKS1B), the majority of the ESTs encoded either for genes whose expression in mammalian oocytes to our knowledge had not previously been reported (e.g., CART) or for genes of unknown function (Yao et al. 2004). Results of this small-scale EST sequencing project stimulated a series of functional studies (described below) that enhanced understanding of oocyte regulation of follicular development and early embryogenesis in cattle (Bettegowda et al. 2007).

8.2.3 Oocyte regulation of follicular development and early embryogenesis: The story of JY-1

Among the oocyte EST described above (Yao et al. 2004) encoding for genes of unknown function, one EST sequence (represented by 14 fully sequenced clones with two different sizes: 455 and 355 bp) was selected for further analysis because it was completely novel and did not show significant homology to sequences of any known genes or ESTs deposited in the Genbank. The name JY-1 was assigned to the putative novel gene encoding for this transcript. Identification of this novel EST was significant because at the time of its discovery, there were 4.9 million human, 3.7 million mice, and approximately 228,000 bovine EST sequences deposited in the Genbank. Since this time, additional JY-1 EST obtained from two-cell embryos as part of the MU bovine genome project (genome.rnet.missouri.edu/bovine) have been deposited in the Genbank. Based on these results, we hypothesized that the expression of JY-1 is oocyte-specific and of potential functional significance to female fertility, and conducted a series of experiments to further characterize JY-1 expression and actions relevant to follicular development and early embryogenesis.

Our published results (Bettegowda et al. 2007) established that the JY-1 gene encodes for a species-specific secreted protein belonging to a novel protein family. JY-1 mRNA and protein are expressed in an ovary-specific fashion, present throughout follicular development in primordial through antral follicles and restricted exclusively to the oocyte. The oocyte-specific expression of JY-1 supports a potential specific role for JY-1 in the regulation of follicular development.

Recombinant JY-1 protein (rJY-1) is biologically active, and treatment with rJY-1 in combination with follicle-stimulating hormone (FSH) reduces granulosa cell estradiol production but stimulates progesterone production (Figure 8.1). Biologic actions of JY-1 on bovine granulosa cells are novel and do not mimic the reported effects of other well-known oocyte-specific growth factors (*GDF9* and *BMP15*) on bovine granulosa cells (McNatty et al. 2005; Spicer et al. 2006). Results demonstrate pronounced effects of rJY-1 on granulosa cell function and mimic those initiated in preovulatory granulosa cells during the luteinization process.

Our results also support an important regulatory role for JY-1 in bovine early embryonic development. Abundance of JY-1 mRNA is temporally regulated during early embryonic development in a manner characteristic of maternal effect genes, and the JY-1 gene is not transcribed in early embryos. JY-1 small interfering RNA (siRNA) injection into zygotes revealed that oocyte-derived JY-1 is required for embryos to reach the blastocyst stage, and most JY-1 siRNA-injected embryos do not progress past the 8- to 16-cell stage (Figure 8.2). The mechanisms involved in JY-1 regulation of early embryonic development are currently under investigation.

Given the fact that originally identified JY-1 sequences in bovine oocyte cDNA library were totally novel, we investigated the presence of JY-1 orthologs in other species using available genome sequence

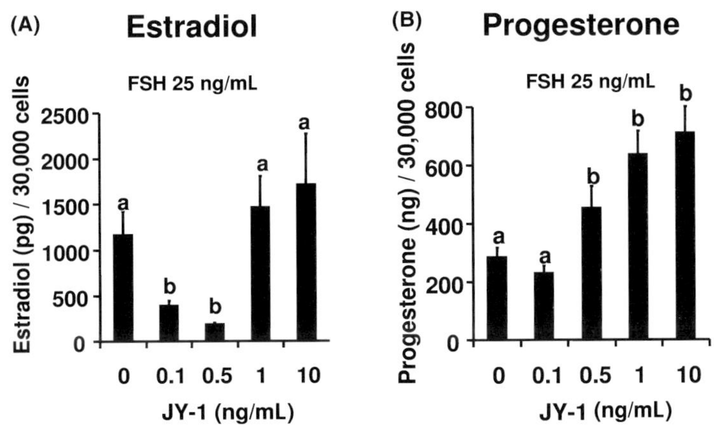

Figure 8.1 Effect of recombinant JY-1 protein (rJY-1) on granulosa cell estradiol and progesterone production. (A) Effect of rJY-1 on FSH-stimulated estradiol production by bovine granulosa cells. (B) Effect of rJY-1 on progesterone production by FSH-treated bovine granulosa cells. Concentrations of estradiol and progesterone were normalized to 30,000 cells. Data are depicted as mean ± SEM (a,b; $P < 0.05$). Copyright 2007 National Academy of Sciences, U.S.A.

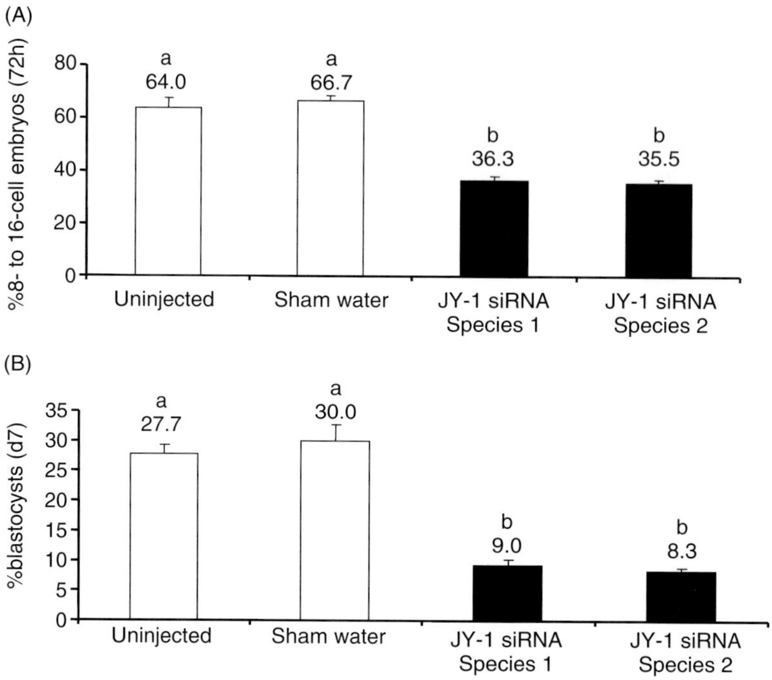

Figure 8.2 Effect of microinjection of individual JY-1 siRNA species on the development of IVF embryos. Presumptive one-cell IVF embryos were subjected to one of the following microinjection treatments: (1) uninjected control, (2) sham water injection, (3) JY-1 siRNA 1 (25 μM), or (4) JY-1 siRNA 2 (25 μM). Microinjected embryos were cultured *in vitro* for 7 days with rate of development to 8- to 16-cell stage denoted on day 3 and rate of development to blastocyst stage denoted on day 7. (A) Effect of microinjection of individual JY-1 siRNA species on percentage of embryos reaching the 8- to 16-cell stages. (B) Effect of microinjection of individual JY-1 siRNA species on percentage of embryos reaching the blastocyst stage. Data are shown as mean ± SEM (a,b; $P < 0.0001$). Copyright 2007 National Academy of Sciences, U.S.A.

resources. JY-1-like sequences are present at chromosomal locations of other vertebrate species (e.g., mice, rats, humans) syntenic to the JY-1 locus on bovine chromosome 29, but lack exons 1 and 2 and do not encode for a functional protein (Figure 8.3). Results support species specificity in evolution and function of this novel oocyte-specific gene initially identified through oocyte EST sequencing efforts.

8.2.4 Oocyte EST sequencing in swine

Sequencing of GV oocyte ESTs has also been performed in swine as part of a pig embryogenesis EST sequencing project (Whitworth et al. 2004). Approximately 1668 oocyte EST sequences were reported, including those encoding known oocyte-specific genes such as ZP1 and ZP3 and many unique sequences. Virtual Northern blot analysis revealed 37 EST clusters whose representation in the GV oocyte versus four cell embryo cDNA libraries differed significantly. Many of such clusters correspond to new products of the embryonic genome in swine or maternal transcripts that were depleted or degraded during the maternal-to-embryonic transition whose functional significance has not been investigated.

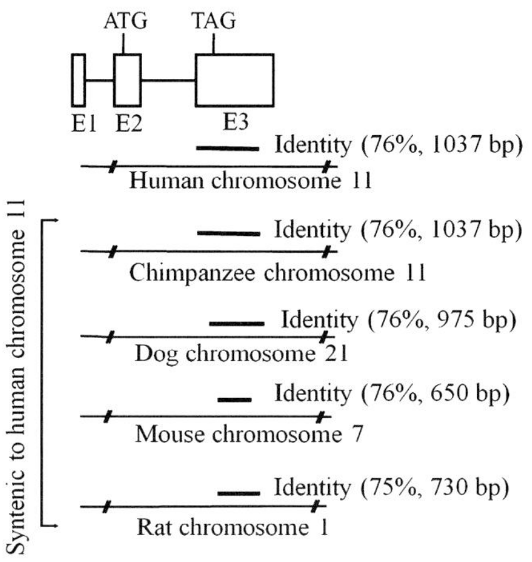

Figure 8.3 JY-1 gene structure and potential orthologs in other species. Genomic DNA databases at NCBI for bovine, human, chimpanzee, dog, mouse, and rat were also searched with the nucleotide sequence of the 1.5-kb bovine JY-1 cDNA. Structure of JY-1 gene on bovine chromosome 29 is denoted. The JY-1 gene has three exons (E1, E2, and E3) separated by two introns. The start (ATG) and stop (TAG) codons of the open reading frame are indicated within the exons. JY-1-like sequences corresponding to exon 3 (length and sequence identity noted) were identified on syntenic chromosomes in human (chromosome 11), chimpanzee (chromosome 11), dog (chromosome 21), mouse (chromosome 7), and rat (chromosome 1), but lack exons 1 and 2 and do not encode for a functional protein. Copyright 2007 National Academy of Sciences, U.S.A.

8.3 Transcriptomics of ovarian tissues: Microarray studies

The recent development of tools for expression profiling in livestock has provided reproductive biologists new opportunities to examine global changes in gene expression in ovarian tissues during key developmental time points, in response to hormonal treatments, and as a tool for phenotyping or predicting reproductive potential. Several years ago, the availability of appropriate homologous platforms for gene expression profiling in livestock species was a significant impediment to the application of microarray technology to the study of ovarian gene expression. The above-described EST sequencing efforts have facilitated advances in the application of microarray technology to the study of gene expression in ovarian tissues, either through the use of custom cDNA microarrays generated by individual investigators or via the use of high-density arrays commercially available for cattle and swine. The following section summarizes prominent microarray studies of gene expression in ovarian tissues (follicle, CL, oocyte) in livestock species with emphasis on studies where physiological insight was gained.

8.3.1 Follicular growth and development

Antral follicle growth in cattle is gonadotropin (FSH and luteinizing hormone [LH]) dependent and occurs in a characteristic wave-like pattern (Ireland et al. 2000; Fortune et al. 2001). Two or three waves of follicular growth are typical of the bovine estrous cycle. Follicular waves are generally characterized by growth of a cohort of multiple small antral follicles and subsequent atresia of all follicles (termed subordinate follicles) but one that continues to grow (termed the dominant follicle). During each wave, a dominant follicle grows to ovulatory size and either undergoes atresia or ovulates if present during the follicular phase of the estrous cycle.

Evans et al. (2004) published the first microarray study of mechanisms regulating follicular development in cattle. Granulosa and theca cells were collected from individual dominant and the largest subordinate follicles on day 3 after the initiation of a follicular wave. A bovine cDNA microarray (BOTL-4) representing over 1300 genes, including receptors, signaling proteins, transcription factors, and apoptosis regulators, was utilized. Approximately 261 differentially expressed genes in granulosa and (or) theca cells of dominant versus subordinate follicles were detected by microarray analysis. Authors hypothesized that granulosa and theca cells from growing dominant follicles would have greater expression of genes linked to inhibition of apoptosis and lower expression of proapoptotic genes than cells from subordinate follicles. Differential expression of 11 genes linked to the regulation of apoptosis in granulosa and theca cells of dominant versus subordinate follicles was detected by microarray analysis and confirmed by quantitative real time PCR. In the granulosa layer, mRNA for DICE1 and MCL1 was greater in granulosa cells of dominant follicles, whereas mRNA for PTGS1, TNF, CAD, and DRAK2 was greater in the granulosa cells of subordinate follicles. Within the theca layer, mRNA abundance for CASP13, P58IPK, APAF1, BTG3, and TSBCLL was greater in subordinate versus dominant follicles. Authors also noted increased expression of the inhibin co-receptor betaglycan (TGFBR3) in granulosa and theca cells of subordinate versus dominant follicles. Given the marked enhanced production of estradiol and the presence of estradiol receptors in dominant versus subordinate follicles early in development, it was suggested that the genes above may in fact represent estradiol target genes functionally associated with the selection of a dominant follicle during follicular waves in cattle (Evans et al. 2004). However, the regulation and functional significance of most of the genes above to follicular development in cattle, with the exception of TGFBR3 (described below) has not yet been determined.

From the same microarray study described above, 83 genes encoding for signal transduction proteins were found to be differentially expressed in granulosa (45 genes) and theca cells (38 genes) of dominant versus subordinate follicles (Forde et al. 2008). Differential expression of a subset of such genes was examined using Q-RT-PCR. Expression of mRNA for BCAR1 was greater in the granulosa cells of dominant follicles, whereas mRNA for TGFBR3, FIBP, SIPA1, PPID, and RANGAP1 was higher in the granulosa cells of subordinate follicles. In contrast, mRNA for FRAP1, GNAI3, CAMK1, FIBP, STX5, WNT2B, DGCR2, and FMNL3 was higher in the theca of dominant versus subordinate follicles, and EPHA4 expression was increased in the theca layer of subordinate versus dominant follicles. Expression of mRNA for the genes above was further examined in granulosa and (or) theca cells of follicles collected at specific stages of a follicular wave representing wave emergence, selection, and dominance. Expression profiles supported important roles for CAMK1 and the receptor tyrosine kinase EphA4 in theca cells and BCAR1 in granulosa cells for the development of dominant follicles. The authors also utilized siRNA-based RNA knockdown procedures to investigate the contribution of TGFBR3 and FIBP to granulosa cell estradiol production. Transfection of FIBP siRNA into granulosa cells, and accompanying knockdown of FIBP mRNA led to a significant increase in estradiol production and the ratio of estradiol to progesterone produced by cultured bovine granulosa cells. Knockdown of TGFBR3 RNA using similar procedures also significantly increased the estradiol to progesterone ratio. Although accompanying demonstration of knockdown of TGFBR3 and FIBP protein following transfection of their respective siRNAs would further increase confidence in the results presented, described studies (Forde et al. 2008) support a functional role for TGFBR3 and FIBP3 in negative regulation of estradiol production and illustrate how results of microarray studies can be utilized as a platform to formulate novel, specific hypotheses and generate accompanying functional data that can enhance understanding of regulation of follicular growth and development.

To investigate the molecular mechanisms controlling FSH and LH signaling in growing dominant follicles, Mihm et al. (2006) conducted microarray studies of dominant follicles collected from cows at 2, 2.5, 3, 3.5, 4, 5, and 5.5 days after the emergence of the first follicular wave. An expanded version of the above-described cDNA microarray, referred to as BOTL-5, was used to screen for genes that were differentially regulated in dominant follicles across different days of development. Alterations in mRNA abundance for 60 genes in the granulosa layer were observed from day 2 to day 5.5 of the follicular wave. From the microarray results, Q-RT-PCR analysis was performed for genes from above linked to apoptosis regulation, cell proliferation, steroidogenesis, growth factor signaling, and for known FSH-, LH-, and estradiol-regulated genes.

A decrease in the follicular fluid estradiol to progesterone ratio, characteristic of a decrease in estradiol-producing capacity, was coincident with a specific gene expression profile in granulosa cells during dominant follicle growth. Concentrations of mRNA encoding for FSHR, ESR2, INHA, ACVR1, CCND2, and the proapoptotic gene SIVA decreased in granulosa cells, while mRNA for LHCGR increased from day 2 to day 5.5 of the first follicular wave. The observed reduction in mRNA for FSHR and known genes positively regulated by FSH (ESR2, ACVR1, INHA, CCND2), coupled with an observed increase in granulosa cell LHCGR mRNA, provided new and convincing molecular evidence supporting a shift in control of dominant follicle development from FSH to LH dependence in cattle (Mihm et al. 2006).

Gene expression profiling was also used to characterize differences in ovary and ovarian follicle transcriptome composition between a randomly selected control line and a selected index line of pigs that ovulates approximately 6.7 more oocytes per cycle than the control line (Caetano et al. 2004). A porcine follicle microarray spotted with cDNAs representing 3636 unique clones was interrogated with labeled cDNA derived from RNA isolated from ovarian tissue and isolated follicles (2–7 mm) from the control and index lines. Differential expression of 71 genes in ovarian tissue and 59 additional genes in follicular samples across days of the follicular phase was noted. The design of this study precludes definitive conclusions about the physiological significance of differential gene expression observed relative to the ovulation rate phenotype of interest. However, 12 of the differentially expressed genes above were subsequently mapped to quantitative trait loci associated with the ovulation rate in swine (Caetano et al. 2005), which illustrates the utility expression profiling data holds for molecular genetics applications relevant to reproductive traits of interest.

8.3.2 Luteinization of the dominant follicle

Using the porcine ovarian ESTs (Jiang et al. 2004) described above, Agca et al. (2006) generated a porcine ovary cDNA array representing 8009 unique genes and used such array to investigate changes in transcriptome composition associated with luteinization of preovulatory follicles in the pig. RNA samples isolated from whole preovulatory porcine follicles >6 mm in diameter obtained before estrus versus similar sized luteinized follicles collected approximately 1 day after the preovulatory LH surge (day 2 post estrus) were utilized to interrogate the array. There were 107 and 43 genes with mRNA identified as decreased or increased in luteinized follicles, respectively (Agca et al. 2006). Pronounced changes in mRNA abundance for specific genes functionally associated with the characteristic shift in follicular steroidogenic and proliferative capacity associated with the luteinization process were detected. Steroidogenic factors with greater mRNA concentrations in the estrogenic follicles collected prior to the LH surge included STAR, CYP17A1, POR (donates electrons to P450 complexes), and HSD3B1. INHBB, PTGES, and AKR1C3 mRNA concentrations were also greater in estrogenic versus luteinized follicles. Furthermore, dimeric dihydrodiol dehydrogenase (SUS2DD), which converts progesterone to its inactive form, was more highly expressed in estrogenic follicles collected prior to the LH surge, which indicates that progesterone may be metabolized by estrogenic preovulatory follicles. Greater mRNA abundance for CCND2 and members of the Wnt/β-catenin pathways were also observed in preovulatory estrogenic follicles relative to luteinized follicles collected after the LH surge, reflective of differences in proliferative status.

Following the LH surge, luteinized follicles had greater expression of cell adhesion- and migration-related mRNAs and mRNA for factors regulating blood flow. These included mRNA for cell surface antigens CD9, CD24, and LGALS3. Six mRNAs encoding for proteins with growth inhibitory functions, including PPP2CB and PDCD4, were upregulated in luteinized follicles collected after the LH surge. Messenger RNAs for vasodilatory proteins upregulated in luteinized follicles included MAOB and LAP3.

The porcine ovarian microarray study described above (Agca et al. 2006) identified distinct differences in the transcriptome of preovulatory estrogenic versus luteinized porcine follicles. Previously undocumented changes in gene expression associated with the luteinization process were identified. It is worth noting that there were commonalities in changes in gene expression (CYP19A1, LRP8, CJA1, LHCGR, JAK3) accompanying luteinization in the study above and in a previously published study in the bovine using suppressive subtractive hybridization (SSH) (Ndiaye et al. 2005), providing further confidence in microarray results and mechanistic changes in gene expression accompanying the luteinization process across species.

8.3.3 CL regression

The principle endocrine factors governing luteal development and regression have been well characterized, but the intracellular mechanisms by which these processes occur are not thoroughly understood. Casey et al. (2005) used the above-described 351 CL ESTs (Casey et al. 2004) combined with 83 ESTs from an ovarian cortex library to generate a 434 gene cDNA array. Corpora lutea were obtained from heifers on days 16 through 19 following estrus, then classified as regressing or non-regressing based on systemic progesterone concentrations and evidence of apoptosis (oligonucleosome formation) in luteal tissue. Results of array analysis indicated that mRNA for steroidogenic factors including HSD3B1, STAR, SCARB1, and CYP11A1 were downregulated in regressing as compared with non-regressing CL, as was GSTA1, which functions to protect lipids from oxidative stress. In contrast, mRNA for CLU, a factor linked to mammary tissue involution and tissue structural/remodeling components (COLA1 and MGP) was increased in regressing corpora lutea. Such results (Casey et al. 2005) illustrate the utility of array approaches for the characterization of temporal changes in mRNA abundance during luteal regression, but alone, provide limited new biologic insight into mechanisms involved in luteal regression due to lack of precisely timed samples relative to initiation of luteolysis, limited number of genes represented on the array, and absence of complementary hormonal regulation or functional studies.

8.3.4 Oocyte maturation

One of the first array-based attempts to characterize the oocyte transcriptome and changes in RNA transcript profiles associated with bovine oocyte maturation was reported by Dalbies-Tran and Mermillod (2003). In this study, mRNA was collected from bovine oocytes before and after *in vitro* maturation. ^{32}P-labeled probes generated from cDNA amplified from the two oocyte populations were hybridized to Atlas human arrays consisting of cDNA fragments representing 1176 known genes spotted on nylon membranes. Positive signals for 300 genes were detected, and mRNA abundance for 37 and 33 genes was decreased and increased, respectively, within *in vitro* matured MII versus immature GV oocytes harvested prior to maturation. At the time of this study, platforms for gene expression profiling in livestock species were limited. Hence, conclusions about lack of expression for specific genes could be attributed to insufficient sequence homology between human cDNAs spotted on the array and mRNA sequences for corresponding bovine genes.

In contrast, in a recent microarray study of changes in the oocyte transcriptome associated with *in vitro* oocyte maturation, Fair et al. (2007) reported that approximately 54% of the 23,000 transcripts represented on the bovine Affymetrix array were detected. Abundance of 209 transcripts was increased and 612 transcripts decreased in MII oocytes following *in vitro* maturation relative to their immature GV stage counterparts. Gene ontology classification revealed prominent changes in genes involved in the regulation of mitogen-activated protein (MAP) kinase activity, translation initiation, and transcription accompanying meiotic maturation. While descriptive, this data set will provide a foundation for future studies focused on the identification of components of the maternal mRNA pool and associated pathways aberrantly regulated during meiotic maturation *in vitro* versus *in vivo*. It is well established that the maternal pool of mRNA and proteins is critical to early embryonic development until control is transferred to products of the embryonic genome (Bettegowda et al. 2008a). Given that rates of embryonic development to blastocyst stage *in vitro* are 2.5- to 3-fold greater when *in vivo* matured oocytes are fertilized and cultured versus *in vitro* matured counterparts (Rizos et al. 2002), it appears likely that alterations in the maternal RNA pool associated with *in vitro* oocyte maturation are functionally associated with poor oocyte developmental competence and reduced rates of embryonic development.

8.3.5 Oocyte competence

Oocyte competence, defined as the ability of an oocyte to be fertilized and develop to the blastocyst stage is progressively acquired during the period of oocyte growth accompanying follicular development (Eppig et al.

2002; Matzuk et al. 2002). Pituitary gonadotropins and bidirectional local communication between the oocyte and adjacent cumulus cells are critical for both nuclear and cytoplasmic maturation (acquisition of the ability to complete meiosis, to ensure monospermic fertilization, and to undergo preimplantation development) (Eppig et al. 2002; Gosden 2002). Competencies acquired by the nuclear and cytoplasmic compartments during the final stages of maturation support the notion that oocyte quality depends on a multitude of factors, many of which can be assessed only at the molecular level.

We have utilized functional genomics approaches to identify the differences in RNA transcript profiles of both the oocyte and adjacent cumulus cells associated with poor developmental competence of bovine oocytes and established a series of markers predictive of oocyte competence (Patel et al. 2007; Bettegowda et al. 2008b). The prepubertal calf model of poor oocyte competence has been foundational to such studies (Revel et al. 1995; Damiani et al. 1996). Oocyte RNA transcript profiling experiments using a bovine cDNA array with approximately 15,000 genes represented (Suchyta et al. 2003) revealed a total of 193 genes coding for transcripts displaying greater mRNA abundance in adult oocytes and 223 genes coding for transcripts displaying greater mRNA abundance in compromised prepubertal oocytes (Patel et al. 2007). Such results formed the foundation for subsequent studies focused on elucidation of the functional significance of the markers identified. Of particular interest from the oocyte microarray studies were genes in the regulation of hormone secretion ontology category (FST, INHBA, INHBB), which were overrepresented in the adult oocyte samples. Subsequent studies revealed a positive association of mRNA abundance for FST in oocytes with oocyte competence and a potential functional role for FST in bovine early embryonic development (Patel et al. 2007). Real-time reverse transcription–polymerase chain reaction (RT-PCR) analysis using a set of samples distinct from those used in microarray experiments confirmed lower amounts of mRNA for FST in oocytes collected from prepubertal (low quality oocytes) versus adult animals and in late cleaving versus early cleaving two-cell stage bovine embryos (Figure 8.4). Early cleaving two-cell stage embryos develop to the blastocyst stage at an approximately fourfold greater rate than their late cleaving counterparts. Given that activation of the embryonic genome and transition from oocyte to embryonic control of development occurs later (eight-cell stage), differences in transcript abundance between early and late cleaving bovine embryos are reflective of differences in competence of oocytes from which they were derived. Furthermore, preliminary results indicate that follistatin supplementation during the initial stages of *in vitro* embryo culture (prior to embryonic genome activation) can enhance rates of blastocyst development (Lee et al. 2007). Results support a positive association of FST mRNA abundance with oocyte quality and suggest the potential for a functional role for FST in bovine early embryonic development.

Available evidence (Hagemann 1999) indicates that competence of oocytes for *in vitro* embryo production in cattle is influenced by stage of the follicular wave during which oocytes were collected, with developmental competence of oocytes from small follicles greater when recovered during the growth phase (early in a wave) than the dominance phase (when a dominant follicle is present). Using a bovine cDNA array containing 2304 bovine oocyte/embryo-enriched ESTS and

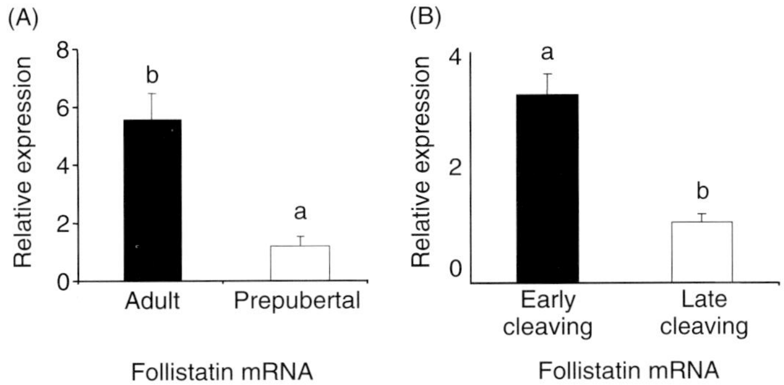

Figure 8.4 Positive relationship between follistatin mRNA abundance and oocyte competence in two distinct models of poor oocyte competence. (A) Quantitative real-time RT-PCR analysis of follistatin mRNA abundance in oocytes collected from adult (black bars) and prepubertal (white bars) animals (model of poor oocyte competence). (B) Quantitative real-time RT-PCR analysis of follistatin mRNA abundance in two-cell stage bovine embryos (collected before embryonic genome activation) that cleaved early (≤30h post fertilization) versus those that cleaved late (30–36h post fertilization; model of poor oocyte competence). Approximately fourfold greater development to blastocyst stage was observed for embryos cleaving early versus late. Data were normalized relative to abundance of RPS18 (endogenous control) and are shown as mean ± SEM (a,b; $P < 0.05$).

various control genes (Sirard et al. 2005), Ghanem et al. (2007) identified 51 transcripts differentially expressed between oocytes collected from small follicles in the growth (good quality oocytes) versus dominance phase (poor quality oocytes). Greater mRNA abundance for ANXA2, S100A10, PP, and RPL24 in the oocytes of small follicles collected during the growth phase and for MSX1 and BMP15 in oocytes of small follicles collected during the dominance phase was confirmed by Q-RT-PCR. The authors also investigated the relationship between G6PDH activity in oocytes collected from an abattoir (determined by brilliant cresyl blue [BCB] staining) and mRNA abundance for five genes associated with oocyte competence in the above model. Several reports indicate that the activity of G6PDH is negatively associated with bovine oocyte competence (Pujol et al. 2004; Alm et al. 2005). Using BCB staining as the criteria for the classification of oocyte competence, a similar relationship between mRNA abundance and oocyte competence was observed for PTTG1, MSX1, PP, and RPL24 as was observed when stage of the follicular wave at collection was used as criteria for the classification of oocyte competence, providing evidence of a relationship between oocyte mRNA abundance for the genes above and developmental competence in a second model predictive of oocyte quality. The functional contribution of observed differences in mRNA abundance of PTTG1, MSX1, PP, and RPL24 to oocyte competence remains to be elucidated.

We (Bettegowda et al. 2008b) have utilized a similar approach to identify cumulus cell markers associated with poor quality oocytes in the prepubertal model system. Approximately 110 genes encoding for transcripts displaying greater mRNA abundance in cumulus cells surrounding GV oocytes collected from adult animals and 45 genes encoding for transcripts displaying greater mRNA abundance in cumulus cells surrounding compromised prepubertal oocytes. Genes in the cysteine-type endopeptidase (cathepsin) activity category (CTSB, CTSS, CTSZ) were overrepresented in the cumulus cell samples harvested from oocytes of prepubertal animals. Q-RT-PCR analysis confirmed that higher amounts of mRNA for the cathepsins above are present in cumulus cells surrounding poor quality oocytes harvested from the ovaries of prepubertal versus adult animals. We also established a negative relationship between cumulus cell cathepsin B, S, and Z expression and embryonic phenotype (blastocyst development) for oocytes collected from adult animals. Furthermore, addition of a cathepsin inhibitor during *in vitro* oocyte maturation can enhance subsequent rates of blastocyst development following parthenogenetic activation or *in vitro* fertilization (IVF) and is associated with reduced rates of cumulus cell apoptosis. Collectively, results support a potential functional relationship between cumulus cell cathepsin expression and oocyte competence in cattle and suggest that cumulus cell cathepsin expression is predictive of an oocyte's embryo development potential.

The relationship between granulosa cell gene expression and oocyte competence has also been investigated in the bovine model using SSH and differential display RT-PCR (ddRT-PCR) procedures (Robert et al. 2001). Cumulus oocyte complexes were aspirated from small (<4 mm) or large (>5 mm) follicles of heifers treated with FSH and pooled in groups of five. Granulosa cells from individual pools of follicles were harvested and stored for subsequent analysis and pools of oocytes subjected to *in vitro* maturation, IVF, and embryo culture to assess developmental competence. Granulosa cells from follicle pools, where 100% versus 0% competence were obtained, were then subjected to RNA isolation. Using ddRT-PCR and SSH, 5 and 18 potential transcripts, respectively, were identified as potentially differentially expressed in the granulosa cells of follicles bearing oocytes that did or did not develop into blastocysts following IVF. While more research is needed, results of this study and the cumulus cell transcript profiling study described above (Bettegowda et al. 2008b) illustrate the potential diagnostic applicability of ovarian cumulus and granulosa cells as indicators of oocyte competence.

8.4 Proteomics of ovarian tissues

While a vast amount of novel information about the biology of the oocyte has been obtained using RNA transcript profiling approaches, oocytes display pronounced posttranscriptional regulatory mechanisms that control RNA translation and stability (Bettegowda and Smith 2007). Hence, it is the protein products that modulate such processes as nuclear and cytoplasmic maturation and development through the maternal-to-embryonic transition following fertilization. Ellederova et al. (2004) used two-dimensional gel electrophoresis and matrix-assisted laser desorption/ionization (MALDI) mass spectrometry to characterize composition of the porcine oocyte proteome. From 350 spots on the gel, proteins in 35 spots were identified by mass spectrometry with 18 spots representing individual proteins. Proteins in greatest abundance (equal to or greater than β-actin) in porcine oocytes included peroxiredoxins, spermine synthase, and ubiquitin carboxyl-terminal hydrolase isozyme L1. When examining changes in the oocyte proteome at different stages of porcine *in vitro* maturation, intensity of all but six spots remained stable, most of which did not contain a single protein. The exception was antiquitin, a member of the aldehyde dehydrogenase family whose abundance increased during meiotic maturation.

In a subsequent study from the same group, Susor et al. (2007) characterized changes in the proteome of porcine oocytes during *in vitro* maturation, but used ^{35}S incorporation to identify changes in synthesis of specific proteins during *in vitro* maturation. Intensity of 16 protein spots changed from GV to MII stages, with intensity of four spots increased. The identity of proteins in one such spot that increased in intensity was determined to be ubiquitin C-terminal hyhydrolase-L1 (UCHL1) by mass spectrometry. Addition of a specific inhibitor of UCHL1 during *in vitro* maturation significantly blocked the progression of porcine oocytes to MII, with the majority of oocytes arrested at metaphase I (MI) in response to treatment with the highest concentration of inhibitor. These biologically relevant results obtained using proteomics approaches suggest an important role for UCHL1 in completion of the first meiosis and transition to anaphase in porcine oocytes.

Bhojwani et al. (2006) used two-dimensional gel electrophoresis, MALDI-time of flight (TOF) mass spectrometry, and Pro-Q diamond phosphoprotein staining to assess changes in the bovine oocyte phosphoproteome (phosphorylated proteins) during *in vitro* maturation. Oocytes were cultured for 0, 10, and 24 h to represent GV, MI, and MII stages of meiotic maturation. Approximately 550 spots were detected for total proteins at each stage of maturation and identity of proteins in 40 spots obtained, four of which differed in abundance at different stages of *in vitro* maturation. Approximately 190, 270, and 250 spots representing phosphorylated proteins were detected at GV, MI, and MII stages, respectively, which indicate that overall number of phosphorylated proteins increases in response to maturation signals. Analysis of two-dimensional gels revealed that proteins in seven spots were shown to be differentially phosphorylated at the above stages of meiotic maturation, including cyclin E2 and a truncated form of cyclin E2, protein disulfide isomerase 3 ER 60 precursor, peroxiredoxin 2, β-actin, aldose reductase, and uridine monophosphate (UMP) synthase. However, the importance of changes in abundance and (or) phosphorylation of the above proteins to progression of meiotic maturation (nuclear and cytoplasmic) remains to be elucidated.

The most comprehensive study on characteristics of the bovine oocyte and cumulus cell proteome was reported by Memili et al. (2007). In this study, 5253 and 1950 proteins were identified in cumulus cells and oocytes, respectively, using differential detergent fractionation two-dimensional liquid chromatography followed by electrospray ionization tandem mass spectrometry. Approximately 12% of proteins were common to the oocyte and cumulus cells. The importance of this study as a resource for future investigations relevant to oocyte and cumulus cell function is highly significant. For example, proteins representing 338 transcription factors and 241 receptor-ligand pathways present in the cumulus cells and oocyte were detected, including 18 pathways for growth factors. Such information provides a tremendous foundation for formulation of specific hypotheses and future studies of differential expression and potential function.

8.5 Future research directions

The above-described results illustrate the contribution of EST sequencing at a species level to gene discovery and the value of EST projects for specific cell types (e.g., oocyte) likely underrepresented in libraries used to generate the majority of currently reported EST sequences. While gene predictions from available genome sequence hold value, confirmation of expression is still important. Results also illustrate how gene discovery from EST sequencing projects can be linked to functional studies of biologic significance. While virtual Northern analysis procedures described above do hold potential for characterization of differences in transcriptome composition, such approaches are biased toward the detection of differences in frequency for abundant transcripts. Such efforts can only be undertaken using clone frequency data obtained from libraries that were not subjected to normalization and are limited in depth by the number of clones sequenced from individual libraries. Results of biologic interest for future study should be confirmed using appropriate quantitative procedures, such as real-time RT-PCR (Q-RT-PCR). *In silico* analysis of EST frequency cannot replace the scope and power afforded by microarray approaches for global investigation of tissue-specific gene expression, but such questions remain relatively unexplored in livestock species. Further improvements in parallel platforms for EST sequencing at greatly reduced costs may ultimately result in greater application of EST sequencing as a means for transcriptome characterization between specific cell types of interest, stages of development, and so on, but cannot currently replace the power and practicality of microarray technology for addressing such questions.

Widespread availability of platforms for performing RNA transcript profiling experiments in important livestock species (cattle and swine) has advanced incorporation of such technology into studies of ovarian function of farm animals, and such approaches may soon become routine. Development of technology for high-density gene expression profiling in other farm species is certainly not far behind. While the wet lab component of microarray experiments is not technically demanding, data analysis and interpretation can present significant obstacles due to the sheer volume of data generated and accompanying statistical challenges (e.g., multiple testing). Hence, ovarian biologists applying such technologies face an oncoming explosion of information and potential data overload. To emphasize results of functional and biologic significance, a logical and systematic approach grounded in sound experimental design with sufficient biologic replication, appropriate statistical analysis incorporating control of false discovery rate, and utilization of available tools to facilitate interpretation of biologic themes can relieve the enormity of such experiments and facilitate the generation of new biologically significant data relevant to an individual's model system of interest (Smith and Rosa 2007). Although challenging due to less extensive annotation/ontology classification for genes in livestock species, functional categories of co-regulated genes and gene pathways can be mined, and hypotheses about common regulatory elements can be formulated and investigated, given the availability of genome sequence information. The application of such data mining approaches will move end points of experiments beyond the simplicity of solely reporting lists of upregulated and downregulated transcripts and also form a foundation for delineation of results of greatest potential biologic relevance and subsequent testing of significance in functional studies. Advances in technologies for testing gene function (e.g., siRNA, morpholinos) and appropriate methods of delivery into primary cultures of ovarian cells now make such studies not only possible but also necessary to confirm physiological significance of microarray data.

Documented use of proteomics approaches to investigate the regulation of ovarian function in farm animals is much more limited than other approaches described. Nevertheless, the challenges associated with the application of such technologies and the potential rewards are illustrated in the above studies. The most up-to-date and highly sensitive procedures for protein identification must be utilized to stretch sensitivity from mere detection of the most highly abundant proteins to that necessary to obtain a "snapshot" reflective of the proteome composition of cell/tissue types of interest. Sophisticated separation procedures are also required to facilitate the identification of individual proteins at a high frequency. While cost and logistics of technologies are not conducive to high-throughput analysis of multiple samples in most settings, the value of information on proteome composition for ovarian tissues/cell types of interest cannot be underestimated, and the potential for subsequent functional studies derived from results of proteomics investigations is evident. Major investment in categorizing the proteome composition of ovarian tissues/cell types at key stages of development and archiving of such results in a searchable, categorized publicly accessible database would greatly accelerate rate of advancement for reproductive biologists in unlocking the secrets of ovarian function beyond that possible using traditional functional genomics approaches.

References

Agca, C., Ries, J.E., Kolath, S.J., Kim, J.H., Forrester, L.J., Antoniou, E., Whitworth, K.M., Mathialagan, N., Springer, G.K., Prather, R.S., and Lucy, M.C. 2006. Luteinization of porcine preovulatory follicles leads to systematic changes in follicular gene expression. *Reproduction* **132**: 133–145.

Alm, H., Torner, H., Lohrke, B., Viergutz, T., Ghoneim, I.M., and Kanitz, W. 2005. Bovine blastocyst development rate in vitro is influenced by selection of oocytes by brilliant cresyl blue staining before IVM as indicator for glucose-6-phosphate dehydrogenase activity. *Theriogenology* **63**: 2194–2205.

Bettegowda, A., Lee, K.B., and Smith, G.W. 2008a. Cytoplasmic and nuclear determinants of the maternal-to-embryonic transition. *Reproduction, Fertility, and Development* **20**: 45–53.

Bettegowda, A., Patel, O.V., Lee, K.B., Park, K., Ireland, J.J., and Smith, G.W. 2008b. Identification of novel cumulus cell molecular markers predictive of oocyte competence: Functional and diagnostic implications. *Biology of Reproduction* **79**: 301–309.

Bettegowda, A. and Smith, G.W. 2007. Mechanisms of maternal mRNA regulation: Implications for mammalian early embryonic development. *Frontiers in Bioscience* **12**: 3713–3726.

Bettegowda, A., Yao, J., Sen, A., Li, Q., Lee, K.B., Kobayashi, Y., Patel, O.V., Coussens, P.M., Ireland, J.J., and Smith, G.W. 2007. JY-1, an oocyte-specific gene, regulates granulosa cell function and early embryonic development in cattle. *Proceedings of the National Academy of Sciences of the United States of America* **104**: 17602–17607.

Bhojwani, M., Rudolph, E., Kanitz, W., Zuehlke, H., Schneider, F., and Tomek, W. 2006. Molecular analysis of maturation processes by protein and phosphoprotein profiling during in vitro maturation of bovine oocytes: A proteomic approach. *Cloning and Stem Cells* **8**: 259–274.

Caetano, A.R., Edeal, J.B., Burns, K., Johnson, R.K., Tuggle, C.K., and Pomp, D. 2005. Physical mapping of genes in the porcine ovarian transcriptome. *Animal Genetics* **36**: 322–330.

Caetano, A.R., Johnson, R.K., Ford, J.J., and Pomp, D. 2004. Microarray profiling for differential gene expression in ovaries and ovarian follicles of pigs selected for increased ovulation rate. *Genetics* **168**: 1529–1537.

Caetano, A.R., Johnson, R.K., and Pomp, D. 2003. Generation and sequence characterization of a normalized cDNA library from swine ovarian follicles. *Mammalian Genome* **14**: 65–70.

Casey, O.M., Fitzpatrick, R., McInerney, J.O., Morris, D.G., Powell, R., and Sreenan, J.M. 2004. Analysis of gene expression in the bovine corpus luteum through generation and characterisation of 960 ESTs. *Biochimica et Biophysica Acta* **1679**: 10–17.

Casey, O.M., Morris, D.G., Powell, R., Sreenan, J.M., and Fitzpatrick, R. 2005. Analysis of gene expression in non-regressed and regressed bovine corpus luteum tissue using a customized ovarian cDNA array. *Theriogenology* **64**: 1963–1976.

Dalbies-Tran, R. and Mermillod, P. 2003. Use of heterologous complementary DNA array screening to analyze bovine oocyte transcriptome and its evolution during in vitro maturation. *Biology of Reproduction* **68**: 252–261.

Damiani, P., Fissore, R.A., Cibelli, J.B., Long, C.R., Balise, J.J., Robl, J.M., and Duby, R.T. 1996. Evaluation of developmental competence, nuclear and ooplasmic maturation of calf oocytes. *Molecular Reproduction and Development* **45**: 521–534.

De Sousa, P.A., Watson, A.J., Schultz, G.A., and Bilodeau Goeseels, S. 1998. Oogenetic and zygotic gene expression directing early bovine embryogenesis: A review. *Molecular Reproduction and Development* **51**: 112–121.

Ellederova, Z., Halada, P., Man, P., Kubelka, M., Motlik, J., and Kovarova, H. 2004. Protein patterns of pig oocytes during in vitro maturation. *Biology of Reproduction* **71**: 1533–1539.

Eppig, J.J. 2001. Oocyte control of ovarian follicular development and function in mammals. *Reproduction* **122**: 829–838.

Eppig, J.J., Wigglesworth, K., and Pendola, F.L. 2002. The mammalian oocyte orchestrates the rate of ovarian follicular development. *Proceedings of the National Academy of Sciences of the United States of America* **99**: 2890–2894.

Evans, A.C., Ireland, J.L., Winn, M.E., Lonergan, P., Smith, G.W., Coussens, P.M., and Ireland, J.J. 2004. Identification of genes involved in apoptosis and dominant follicle development during follicular waves in cattle. *Biology of Reproduction* **70**: 1475–1484.

Fair, T., Carter, F., Park, S., Evans, A.C., and Lonergan, P. 2007. Global gene expression analysis during bovine oocyte in vitro maturation. *Theriogenology* **68**(Supplement 1): S91–S97.

Forde, N., Mihm, M., Canty, M.J., Zielak, A.E., Baker, P.J., Park, S.D., Lonergan, P., Smith, G.W., Coussens, P.M., Ireland, J.J., and Evans, A.C. 2008. Differential expression of signal transduction factors in ovarian follicle development; a role for betaglycan and FIBP in granulosa cells in cattle. *Physiological Genomics*. **33**: 193–204.

Fortune, J.E., Rivera, G.M., Evans, A.C., and Turzillo, A.M. 2001. Differentiation of dominant versus subordinate follicles in cattle. *Biology of Reproduction* **65**: 648–654.

Ghanem, N., Holker, M., Rings, F., Jennen, D., Tholen, E., Sirard, M.A., Torner, H., Kanitz, W., Schellander, K., and Tesfaye, D. 2007. Alterations in transcript abundance of bovine oocytes recovered at growth and dominance phases of the first follicular wave. *BMC Developmental Biology* **7**: 90.

Gosden, R.G. 2002. Oogenesis as a foundation for embryogenesis. *Molecular and Cellular Endocrinology* **186**: 149–153.

Hagemann, L.J. 1999. Influence of the dominant follicle on oocytes from subordinate follicles. *Theriogenology* **51**: 449–459.

Ireland, J.J., Mihm, M., Austin, E., Diskin, M.G., and Roche, J.F. 2000. Historical perspective of turnover of dominant follicles during the bovine estrous cycle: Key concepts, studies, advancements, and terms. *Journal of Dairy Science* **83**: 1648–1658.

Jiang, H., Whitworth, K.M., Bivens, N.J., Ries, J.E., Woods, R.J., Forrester, L.J., Springer, G.K., Mathialagan, N., Agca, C., Prather, R.S., and Lucy, M.C. 2004. Large-scale generation and analysis of expressed sequence tags from porcine ovary. *Biology of Reproduction* **71**: 1991–2002.

Lee, K.B., Bettegowda, A., Ireland, J.J., and Smith, G.W. 2007. Effect of follistatin treatment post fertilization on time to first cleavage, development to the blastocyst stage and cell allocation of in vitro produced bovine embryos. *Reproduction, Fertility, and Development* **19**: 191.

Matzuk, M.M., Burns, K.H., Viveiros, M.M., and Eppig, J.J. 2002. Intercellular communication in the mammalian ovary: Oocytes carry the conversation. *Science* **296**: 2178–2180.

McNatty, K.P., Juengel, J.L., Reader, K.L., Lun, S., Myllymaa, S., Lawrence, S.B., Western, A., Meerasahib, M.F., Mottershead, D.G., Groome, N.P., Ritvos, O., and Laitinen, M.P. 2005. Bone morphogenetic protein 15 and growth differentiation factor 9 co-operate to regulate granulosa cell function in ruminants. *Reproduction* **129**: 481–487.

Memili, E., Peddinti, D., Shack, L.A., Nanduri, B., McCarthy, F., Sagirkaya, H., and Burgess, S.C. 2007. Bovine germinal vesicle oocyte and cumulus cell proteomics. *Reproduction* **133**: 1107–1120.

Mihm, M., Baker, P.J., Ireland, J.L., Smith, G.W., Coussens, P.M., Evans, A.C., and Ireland, J.J. 2006. Molecular evidence that growth of dominant follicles involves a reduction in follicle-stimulating hormone dependence and an increase in luteinizing hormone dependence in cattle. *Biology of Reproduction* **74**: 1051–1059.

Ndiaye, K., Fayad, T., Silversides, D.W., Sirois, J., and Lussier, J.G. 2005. Identification of downregulated messenger RNAs in bovine granulosa cells of dominant follicles following stimulation with human chorionic gonadotropin. *Biology of Reproduction* **73**: 324–333.

Patel, O.V., Bettegowda, A., Ireland, J.J., Coussens, P.M., Lonergan, P., and Smith, G.W. 2007. Functional genomics studies of oocyte competence: Evidence that reduced transcript abundance for follistatin is associated with poor developmental competence of bovine oocytes. *Reproduction* **133**: 95–106.

Pujol, M., Lopez-Bejar, M., and Paramio, M.T. 2004. Developmental competence of heifer oocytes selected using the brilliant cresyl blue (BCB) test. *Theriogenology* **61**: 735–744.

Revel, F., Mermillod, P., Peynot, N., Renard, J.P., and Heyman, Y. 1995. Low developmental capacity of in vitro matured and fertilized oocytes from calves compared with that of cows. *Journal of Reproduction and Fertility* **103**: 115–120.

Rizos, D., Ward, F., Duffy, P., Boland, M.P., and Lonergan, P. 2002. Consequences of bovine oocyte maturation, fertilization or early embryo development in vitro versus in vivo: Implications for blastocyst yield and blastocyst quality. *Molecular Reproduction and Development* **61**: 234–248.

Robert, C., Gagne, D., Bousquet, D., Barnes, F.L., and Sirard, M.A. 2001. Differential display and suppressive subtractive hybridization used to identify granulosa cell messenger RNA associated with bovine oocyte developmental competence. *Biology of Reproduction* **64**: 1812–1820.

Sirard, M.A., Dufort, I., Vallee, M., Massicotte, L., Gravel, C., Reghenas, H., Watson, A.J., King, W.A., and Robert, C. 2005. Potential and limitations of bovine-specific arrays for the analysis of mRNA levels in early development: Preliminary analysis using a bovine embryonic array. *Reproduction, Fertility, and Development* **17**: 47–57.

Smith, G.W., Goetz, T.L., Anthony, R.V., and Smith, M.F. 1994a. Molecular cloning of an ovine ovarian tissue inhibitor of metalloproteinases: Ontogeny of messenger ribonucleic acid expression and in situ localization within preovulatory follicles and luteal tissue. *Endocrinology* **134**: 344–352.

Smith, G.W., Juengel, J.L., McIntush, E.W., Youngquist, R.S., Garverick, H.A., and Smith, M.F. 1996. Ontogenies of messenger RNA encoding tissue inhibitor of metalloproteinases 1 and 2 within bovine periovulatory follicles and luteal tissue. *Domestic Animal Endocrinology* **13**: 151–160.

Smith, G.W. and Rosa, G.J. 2007. Interpretation of microarray data: Trudging out of the abyss towards elucidation of biological significance. *Journal of Animal Science* **85**: E20–E23.

Smith, M.F., Kemper, C.N., Smith, G.W., Goetz, T.L., and Jarrell, V.L. 1994b. Production of tissue inhibitor of metalloproteinases-1 by porcine follicular and luteal cells. *Journal of Animal Science* **72**: 1004–1012.

Smith, T.P., Grosse, W.M., Freking, B.A., Roberts, A.J., Stone, R.T., Casas, E., Wray, J.E., White, J., Cho, J., Fahrenkrug, S.C., Bennett, G.L., Heaton, M.P., Laegreid, W.W., Rohrer, G.A., Chitko McKown, C.G., Pertea, G., Holt, I., Karamycheva, S., Liang, F., Quackenbush, J., and Keele, J.W. 2001. Sequence evaluation of four pooled-tissue normalized bovine cDNA libraries and construction of a gene index for cattle. *Genome Research* **11**: 626–630.

Spicer, L.J., Aad, P.Y., Allen, D., Mazerbourg, S., and Hsueh, A.J. 2006. Growth differentiation factor-9 has divergent effects on proliferation and steroidogenesis of bovine granulosa cells. *The Journal of Endocrinology* **189**: 329–339.

Suchyta, S.P., Sipkovsky, S., Kruska, R., Jeffers, A., McNulty, A., Coussens, M.J., Tempelman, R.J., Halgren, R.G., Saama, P.M., Bauman, D.E., Boisclair, Y.R., Burton, J.L., Collier, R.J., DePeters, E.J., Ferris, T.A., Lucy, M.C., McGuire, M.A., Medrano, J.F., Overton, T.R., Smith, T.P., Smith, G.W., Sonstegard, T.S., Spain, J.N., Spiers, D.E., Yao, J., and Coussens, P.M. 2003. Development and testing of a high-density cDNA microarray resource for cattle. *Physiological Genomics* **15**: 158–164.

Susor, A., Ellederova, Z., Jelinkova, L., Halada, P., Kavan, D., Kubelka, M., and Kovarova, H. 2007. Proteomic analysis of porcine oocytes during in vitro maturation reveals essential role for the ubiquitin C-terminal hydrolase-L1. *Reproduction* **134**: 559–568.

Telford, N.A., Watson, A.J., and Schultz, G.A. 1990. Transition from maternal to embryonic control in early mammalian development: A comparison of several species. *Molecular Reproduction and Development* **26**: 90–100.

Whitworth, K., Springer, G.K., Forrester, L.J., Spollen, W.G., Ries, J., Lamberson, W.R., Bivens, N., Murphy, C.N., Mathialagan, N., Green, J.A., and Prather, R.S. 2004. Developmental expression of 2489 gene clusters during pig embryogenesis: An expressed sequence tag project. *Biology of Reproduction* **71**: 1230–1243.

Yao, J., Ren, X., Ireland, J.J., Coussens, P.M., Smith, T.P., and Smith, G.W. 2004. Generation of a bovine oocyte cDNA library and microarray: Resources for identification of genes important for follicular development and early embryogenesis. *Physiological Genomics* **19**: 84–92.

9

Physiological Genomics of Preimplantation Embryo Development in Production Animals

Luc J. Peelman

9.1 Introduction

A mammalian genome contains on average between 20,000 and 25,000 protein coding genes. With this relatively small amount, an organism needs to produce all the different metabolites necessary for its development and function. To be able to do this, several layers of regulation and modification exist. It has become clear that much more of the genome is transcribed than was previously anticipated. A new box of RNAs has been opened, the regulation and function of which remains, for the most part, an enigma. It has been estimated that in mouse preimplantation development around 15,500 genes are expressed at one time or another and that this number is probably similar to that for other mammals (Stanton et al. 2003).

In the long run toward adulthood, many molecular hurdles have to be overcome by an organism, which is equipped with a unique genome resulting from the fusion of sperm and oocyte, each contributing unique genetic material and molecular toolbox, and nutrients supplied by the mother and the environment in general. Each of these factors puts constraints on the possibilities of the developing organism. In order to cope with this enormous variation, the organism and its individual cells have developed an admirable flexibility. However, there are limits to this flexibility, and many developing embryos do not make it to adulthood. A newly fertilized oocyte soon encounters the first formidable hurdle (Figure 9.1).

To survive to the blastocyst stage, a fertilized oocyte must switch on its own genome after several rounds of cell division and stop using the material that was stored in the oocyte. Not much later, during the early morula stage, cell differentiation commences. The morula develops into a blastocyst with an inner cell mass (ICM) containing

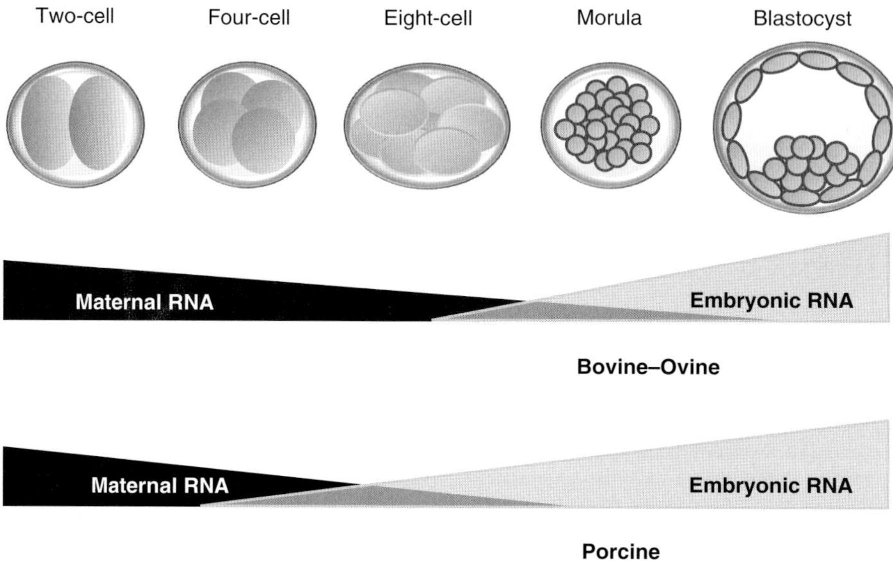

Figure 9.1 Schematic representation of preimplantation embryo development from the 2-cell stage with indication of maternal RNA degradation and embryo genome activation.

the embryonic stem cells and an outer trophoblast layer that will form the extra embryonic tissue. The blastocyst must then hatch from the zona pellucida before being able to implant. All of these transitions need well-orchestrated sweeps of gene expression. The study of gene expression during preimplantation embryo development is a real challenge mainly due to the innate complexity of expression profiling, the individual embryo variation in expression patterns, the difficulty of obtaining *in vivo* embryos, and the small amount of starting material with which to work. Many of the studies are performed using *in vitro*–produced embryos, introducing an extra level of variation due to the differences in media and culture conditions used. The small amount of starting material necessitates an extra amplification step or pooling of embryos for several applications like the use of microarrays for transcription profiling.

9.2 Preimplantation developmental stages and transcriptomics

9.2.1 *Timing of the first cleavage division—Developmental competence*

Developmental competence, which is the ability of an oocyte to proceed through maturation, fertilization, and embryo development, is largely determined by the quality of the oocyte. Oocytes competent to complete nuclear maturation, meaning to be able to progress through meiosis, do not all show the same capacity to reach the blastocyst stage. This difference has been linked to variations in cytoplasmic maturation, including differences in oocyte activation, pronucleus formation, and preimplantation development. Molecular mechanisms governing these processes are only barely understood. Oocytes of lower quality have a delayed first cell cycle, slower cleavage rate, and lower blastocyst yield (Lonergan et al. 1999).

Leoni et al. (2007) used ovine oocytes from prepubertal animals (30–40 days old) as a model of low-quality oocytes and compared their developmental capacity and gene transcription profile with oocytes from adult animals. They observed similar maturation and fertilization rates but a delayed first cell cycle, slower cleavage rate, and lower blastocyst yield in the prepubertal oocyte group. Of the 11 genes studied, seven (activin, $p34^{cdc2}$, glucose-transporter 1, Na^+K^+ATPase, E-cadherin, zona occludens protein 2, and poly(A)polymerase) showed a significant reduction in relative mRNA abundance, indicating that the lower developmental competence in prepubertal oocytes is associated with deficiencies in the mRNA storage of germinal vesicle oocytes.

Several genes associated with the timing of the first cell division and developmental competence have been identified. Comparison of mRNA contents of slow and fast cleaving bovine embryos revealed differences in histone 3 (H3A), preimplantation embryo development (Ped; Fair et al. 2004a,b), *HPRT*, *G6PD*, *IGF-I* and *IGF-IR* (Lonergan et al. 2000), *GLUT-5*, sarcosine (*SOX*), Mn-superoxide dismutase (*MnSOD*), *Cx43*, *IFN-τ*, *IGF-II*, *BAX* (Gutierrez-Adan et al. 2004), histone 2 (H2A), isocitrate dehydrogenase (*IDH*), and YY1- and E4TF1-associated factor 1 (*YEAF1*) genes in cattle (Dode et al. 2006).

The Ped gene was identified in the mouse as the Qa-2 antigen gene located in the Q region of the mouse major histocompatibility complex (MHC; Warner et al. 1987). The mouse Ped gene has two alleles indicated as fast and slow in reference to their effect on the preimplantation developmental growth rate. The fast allele gives a higher expression of the Qa-2 antigen in early embryos and is correlated with a faster cleavage rate and development. Ped homologs have been identified in humans as the *HLA-G* gene (Juriscova et al. 1996), and in bovine as the MHC I4221.1 gene (Fair et al. 2004a). In this study, the fast allele of the bovine Ped gene had a threefold higher transcription in fast-cleaving two-cell embryos than in slow-cleaving ones. Fair et al. also found lower transcription in *in vitro*-produced embryos from different stages up to the blastocyst compared to *in vivo* embryos, with the largest difference starting at the 16-cell stage, which coincides with the onset of a major activation stage of the embryonic expression.

Fair et al. (2004b) performed a limited suppression subtractive hybridization (SSH) experiment between early and late-cleaving ttwo-cell bovine embryos to identify genes associated with developmental competence. Between 30 and 40 clones from each library were sequenced. Of these clones, three (H3A, cyclin B1, and BMP15) were chosen for further analysis using real-time polymerase chain reaction (PCR). H3A was found to be more abundant in early-cleaving embryos, whereas the transcription of the two other genes was variable. In another study, the transcription levels of H2A were also higher in fast-cleaving bovine embryos than in slow-cleaving embryos (Dode et al. 2006), and histone methyltransferase G9a, which regulates lysine 9-acetylated histone H3 methylation, was found to be essential for early embryo development (Tachibana et al. 2002). Taken together, these results indicate that early embryos must have enough histone mRNAs in store for normal development, and the abundance can eventually be used as a marker for competence. Histone genes are often used as reference genes in real-time reverse transcriptase (RT)-PCR experiments. However, as these and other studies have shown, expression of histones is variable during preimplantation development and more than one reference gene should be used for accurate normalization (Goossens et al. 2005).

In the study of Dode et al. (2006), two other genes were found to be upregulated in early-cleaving embryos. *YEAF1* is involved in transcription. Its role in early embryo development is not yet known. *IDH* catalyzes a key regulatory step in the tricarboxylic acid (TCA) cycle, which potentially plays an important role in bovine oocyte maturation (Cetica et al. 2003), and *IDH* also has an important role in protection against oxidative stress (Lee et al. 2002).

Abundance of mRNA for insulin like growth factor I (IGF-I) may be an indicator of embryonic developmental competence, at least in cattle. It was found in all early-cleaving bovine two-cell embryos, but in none of the late-cleaving embryos. IGF-I receptor (IGF-IR) mRNA was found in all two-cell embryos. IGF-I and IGF-IR mRNA were found in all blastocysts, regardless of cleavage status (Lonergan et al. 2003).

It has been observed that male bovine *in vitro*–produced embryos have higher developmental rates than female (Hasler et al. 1995; Massip et al. 1996). Male embryos have a higher competency, develop faster to the blastocyst stage, and reach the expanded blastocyst stage faster. It was postulated that this difference is due to a gene located on the Y chromosome or a consequence of differences in expression between male and female X chromosome-linked genes, with females having a higher expression because they possess two gene copies before X inactivation (Gutierrez-Adan et al. 1996). Oxygen radicals are necessary for normal embryonic development. Regulation of the amount of oxygen radicals is influenced by, among others, two X chromosome-linked genes, glucose-6-phosphate dehydrogenase (*G6PD*) and hypoxanthine phosphoribosyl transferase (*HPRT*), two genes that also play an important role in energy metabolism. Differential expression of both genes was found between male and female blastocysts (Gutierrez-Adan et al. 2000), and a correlation with cleavage rate was found in another study (Lonergan et al. 2000). No difference in expression was found in two-cell embryos, but blastocysts derived from fast-cleaving embryos had a higher expression of *G6PD* and *HPRT* than blastocysts from slow-cleaving embryos. In the mouse, adenylation of the poly(A)-tail of *HPRT* occurred during oocyte maturation and was associated with an increase in its translation (Paynton and Bachvarova 1994). Low oocyte competence, indeed, seems to be associated with altered poly-adenylation patterns and differences in maternal RNA degradation (Brevini et al. 2002). It has been shown that a shorter poly(A)-tail is correlated with low developmental competence (Brevini-Gandolfi et al. 1999, 2000). In conjunction with this, the relative abundance of the poly(A) polymerase (*PAP*) mRNA was found to be lower in ovine prepubertal oocytes than in oocytes derived from adult ewes (Leoni et al. 2007). It has been shown that maternal mRNAs stored in the cytoplasm of the oocyte have short poly(A)-tails and become translationally active only after lengthening of the poly(A)-tail by *PAP* and associated factors. Regulation of this process is an important step in early embryonic development (Gandolfi and Gandolfi, 2001).

Genes that may be stress induced, such as *SOX*, *MnSOD*, *BAX*, interferon tau (*IFN-τ*), and *G6PD*, were found to be expressed in greater amounts in slow-cleaving embryos and in *in vitro*-produced embryos than in fast-cleaving embryos and in *in vivo* embryos, respectively, whereas genes functioning in metabolism, growth, and differentiation such as *GLUT-5*, *Cx43*, *IGF-II*, and *IGF-IIR* had higher mRNA concentrations in fast-cleaving embryos and in *in vivo* embryos (Gutierrez-Adan et al. 2004). The two patterns may be a reflection of the health status of the embryo. However, some contradicting results exist for the influence of *IFN-τ*. In some studies, higher transcription of *IFN-τ* was related to reduced competence (Kubisch et al. 1998; Wrenzycki et al. 2001b), whereas in other studies higher transcription of *IFN-τ* was linked to higher quality (Hernandez-Ledezma et al. 1993; Russell et al. 2006).

9.2.2. Major onset of embryonic expression

Probably the most crucial period of preimplantation development is the activation of the embryonic genome. The embryo starts development under the control of maternal RNAs and proteins and has to switch on its own genome for further development. Failing to do this adequately will lead to the death of the embryo. Minor genome activation starts in most mammalian species around the first cell division. Timing of the major embryonic genome activation (EGA) is somewhat different from species to species. In mice, EGA occurs rapidly (in late one-cell embryos), whereas in bovine and ovine EGA is more delayed (eight- to 16-cell stage). In pigs and humans, EGA occurs around the two-cell stage (Telford et al. 1990). An important aspect of the maternal to zygotic transition (MZT) next to EGA is the degradation of maternal, oocyte-specific transcripts. Given the importance of EGA, it has been studied extensively. Several research groups compared the transcription profile of the embryo, before and after EGA, with gene expression profiles. In cattle, the major onset of embryonic expression is relatively late, starting around the eight- to 16-cell stage (48 to 64 h after fertilization). Due to this late onset, EGA in cattle is somewhat easier to study than in other species where EGA starts earlier (see also Figure 9.1).

Generally, two approaches have been used: (1) comparing transcript profiles of oocytes and two- to four-cell stage embryos with blastocyst profiles and (2) blocking transcription by addition of α-amanitin to the culture medium. Alpha-amanitin is a specific inhibitor of RNA polymerase II and blocks *de novo* mRNA synthesis. The last approach was used by Misirlioglu and coworkers (2006) in combination with microarray analysis using the Affymetrix GeneChip Bovine Genome Array (Affymetric Inc., Santa Clara, CA). They found 258 genes that were at least twofold increased in *in vitro* eight-cell embryos compared with *in vitro* matured oocytes (MII). Gene ontology analysis was carried out using NetAffx Analysis Center and Cowbase (http://www.agbase.msstate.edu). Ontology analysis identified regulators of transcription (*NFYA*, *USF2*), cell adhesion (*DSC2*, *COL12A1*), signal transduction (*PTGER4*, *ADRBK1*), transporters (*CRABP1*), metabolism-related genes, and immune response-related genes. On the other hand, 124 genes were found to be increased in MII compared with eight-cell stage embryos. Transcriptome comparison of eight-cell stage embryos with α-amanitin-treated eight-cell stage embryos revealed 233 genes with a twofold or more increase. Among these genes were *DSC2* and *CRABP1*, which were also found in the comparison between eight-cell embryos and MII. Another gene, purine nucleoside phorphorylase (*NP*), was 149-fold higher in eight-cell embryos and not detectable in the α-amanitin-treated eight-cell embryos. Maternal degraded transcripts were identified by comparing MII with α-amaniti-treated eight-cell embryos. A total of 147 transcripts were at least twofold increased. Among these were several genes involved in DNA methylation and metabolism (Misirlioglu et al. 2006).

The expression of some important nucleolar proteins was studied using the α-amanitin block in pig preimplantation embryos. It was shown that RNA polymerase I and RNA Pol I-associated factor PAF53 are transcribed *de novo* from the embryonic genome and that activation of these genes is delayed in *in vitro* embryos compared with *in vivo* embryos (Bjerregaard et al. 2004).

Using a microarray with 15,529 human cDNAs, Adjaye et al. (2007) found 164 transcripts that were bovine oocyte specific compared with blastocyst, and 1324 that were blastocyst specific compared with oocyte. Oocyte and blastocyst had 419 transcripts in common. Pathway analysis revealed differential expression of genes involved in 107 distinct signaling and metabolic pathways. A comparable but small-scale study in which the transcription profile of oocytes was compared with that of blastocysts was performed by Mamo et al. (2006). They used a combination of cDNA array analysis and real-time RT-PCR to study the transcription of 82 selected genes. Of these genes, 35 were found to be differently regulated.

9.2.3 Compaction

The first cell differentiation processes start as a consequence of EGA and involve compaction and cell allocation. The timing of the first cell differentiation is, just as for EGA, different from species to species. It is earliest (eight-cell) in mouse embryos (Reeve 1981) and latest in pig embryos, where the first cell differentiation starts only shortly before formation of the blastocyst (Reima et al. 1993). In bovine, the first cell differentiation occurs at the 16- to 32-cell stage, and in rabbit it starts at the 32- to 64-cell stage (Koyama et al. 1994). The cell differentiation is accompanied by clear morphological changes in the embryo. The spherical blastomeres flatten onto each other, forming a morula. Cell-to-cell adhesion occurs by formation of epithelial zonula adherens (ZA) of which the transmembrane E-cadherin protein is an important constituent. E-cadherin binds homotypically extracellularly and with catenin cytoplasmically (Figure 9.2).

The catenins are linked to the actin cytoskeleton (Aberle et al. 1996). Following formation of ZA, tight junctions (TJs) are formed. TJs are ring-like structures around a cell which are responsible for sealing cells together and allowing the establishment of apical and basal polarity. They consist of integral membrane proteins (occludin [OCLN], claudins, and junction adhesion molecule [JAM]) that are linked to the actin cytoskeleton by a number of cytoplasmic plaque proteins (including different isoforms of zona occludens 1-3 [ZO-1-3]; (Stevenson and Keon 1998). Miller and coworkers (2003) studied six bovine TJ genes (*JAM*, *OCLN*, Pan ZO-1, Pan ZO-2, ZO-1α+, and ZO-2β+) by semiquantitative RT-PCR. They found a dramatic increase in total TJ transcripts during transition from morula to blastocyst stage just before cavitation, which indicates stage-dependent rather than time-dependent control.

Further aspects of cell polarization include distribution of microvilli, restriction of plasma membrane components to the apical surface, and cytoplasmic polarization. The developmental fate of blastomeres is dependent on their location in the embryo, with inner cells becoming part of the ICM, containing the embryonic stem cells, and the outer cells forming the trophectoderm that will develop into extra embryonic membranes.

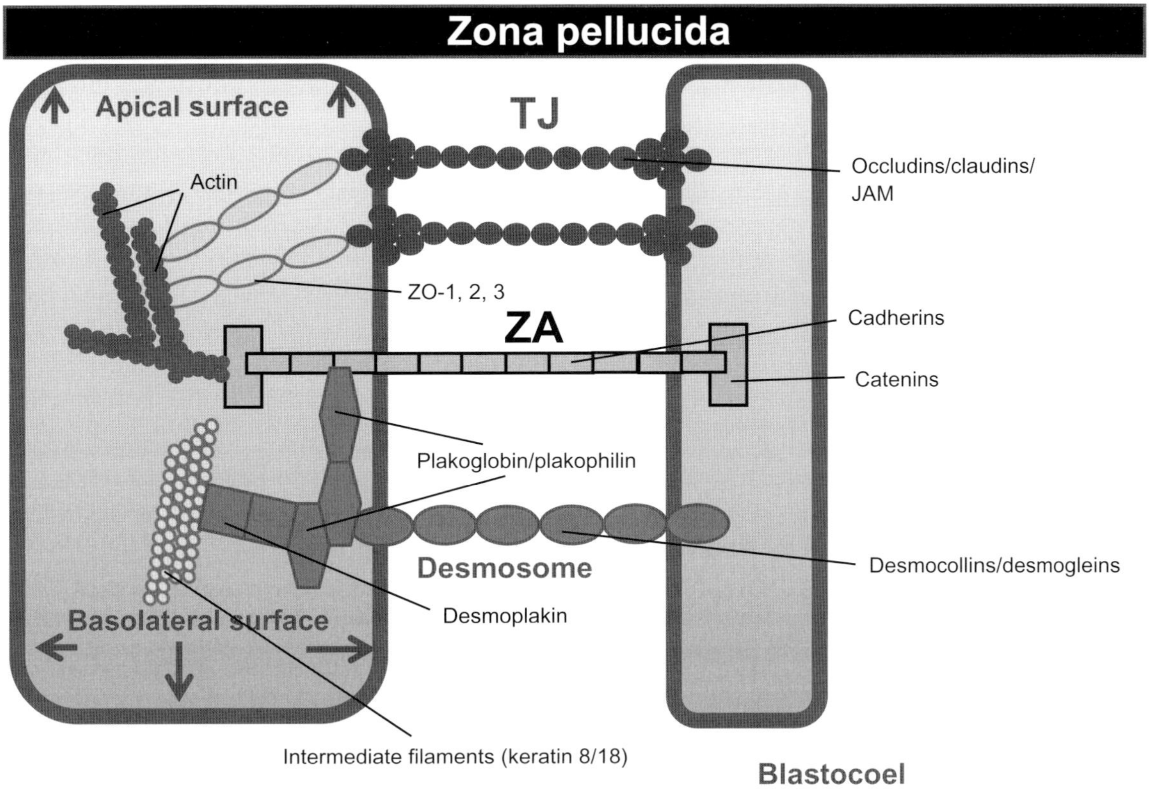

Figure 9.2 Schematic representation of the cell–cell junctions in trophectoderm cells of the morula and blastocyst.

9.2.4 *Blastocyst formation*

After formation of the morula, the outer cells start pumping sodium into the interstitial spaces and, due to osmosis, water follows, expanding the spaces to a fluid-filled cavity, the so-called blastocoel. After differentiation of the trophectoderm cells and cavitation, the embryo is called a blastocyst and contains around 32 to 64 cells.

Important gene families controlling the different facets of blastocyst formation have been previously mentioned: E-cadherin–catenin cell adhesion family, the TJ gene family, the Na/K-ATPase gene family, and the aquaporin gene family (Watson and Barcroft 2001). Also important are other genes involved in cell–cell contact and TJ formation.

Transcription in the blastocyst is compared with that of previous stages in different studies using techniques such as differential display RT-PCR, SSH, and, more recently, microarray analysis, mostly in combination with real-time RT-PCR for verification of results. Subtraction between bovine morula and blastocyst was used to establish a cDNA library for studying expression of the genes during blastocyst formation. Seventy-one clones representing 33 different expressed sequence tags (ESTs) were generated (Ponsuksili et al. 2002). Of these, 19 were verified by real-time RT-PCR and 84% (16/19) followed the SSH pattern

(El-Halawany et al. 2004). Goossens et al. (2007a) used SSH between two- to eight-cell embryos and blastocysts. Sixty-five clones representing 36 known genes, five sequences homologous to genomic sequences, and two sequences with no match in the database were sequenced. Twelve genes were verified by real-time RT-PCR and 75% (9/12) were found to be in agreement with results from SSH. Of the non-ribosomal and mitochondrial genes commonly found in these types of experiments, the following genes were detected in both studies: keratin 18 (*KRT18*), fibronectin (*FN1*), adenine nucleotide translocator 2, and elongation factor 1 alpha. One of these genes, *KRT18*, is not expressed in two- to eight-cell embryos, *in vitro* and *in vivo*, and becomes relatively abundant in blastocysts (El-Halawany et al. 2004; Goossens et al. 2007a). Immunofluorescent staining showed the KRT18 protein is localized at cell–cell contact sites of the trophectoderm, but not at those of the ICM. Hence, KRT18 is a potential marker for trophoblast differentiation (Goossens et al. 2007a). Knockout of KRT18 in the mouse leads to trophoblast fragility and early embryonic lethality (Hesse et al. 2000). A similar mRNA pattern was observed for keratin 8 (*KRT8*; El-Halawany et al. 2004). KRT8, together with KRT18, comprises the intermediate filaments and influences the three-dimensional formation of cell–cell contacts in embryonic visceral endoderm (Jackson et al. 1980).

Na/K-ATPase is confined to the basolateral membrane domain of the trophectoderm, and enzyme activity increases just prior to blastocyst formation in the mouse (Watson and Barcroft 2001). Na/K-ATPase consists of a catalytic α subunit and a noncatalytic, glycosylated β subunit, which are both encoded by different genes. Isoforms $\alpha1$, $\alpha2$, and $\alpha3$ were found in all stages of bovine preimplantation, from zygote to blastocyst, as detected by conventional RT-PCR. In the same study, isoform $\beta1$ was detected only in the morula and blastocyst stages, and isoform $\beta2$ was detected from the eight-cell stage on (Betts et al. 1997). Isoforms $\alpha1$ and $\alpha3$ were localized by immunofluorescence to encircle the entire margin of each blastomere in bovine embryos from zygote up to morula. In blastocysts, isoform $\alpha1$ was confined to the basolateral membranes of trophectoderm cells and to the periphery of ICM cells. Isoform $\alpha3$ was confined to apical cell surfaces of the trophectoderm and was not detected in ICM, indicating an involvement in blastocyst formation (Betts et al. 1998). The presence of isoform $\alpha1$ in trophectoderm and ICM of bovine blastocysts was later confirmed (Wrenzycki et al. 2003). The mRNA amount of isoform $\beta3$ (ATP1B3) was found to be significantly lower in *in vitro* two- to eight-cell embryos compared with *in vitro* blastocysts, but no significant difference was found between *in vivo* two- to eight-cell embryos and blastocysts (Goossens et al. 2007a). It was shown that *in vivo*-produced morulas had a more firm and more prolonged compaction and that they started blastulation at a later embryonic age and cell number (Van Soom et al. 1997), which might be related to a slower increase in *ATP1B3* expression.

E-cadherin and beta-catenin transcripts are present in all stages throughout (*in vitro*) bovine preimplantation development. The proteins were detected by immunocytochemistry to encircle the cell margins of all blastomeres up to the eight-cell stage. From the morula on, protein distribution became similar to that of Na/K-ATPase $\alpha1$ (Barcroft et al. 1998). In the same study, the expression of zonula occludens protein 1 (*ZO-1*) was not detected before the morula stage, where it appeared as punctae between the outer cells. In the blastocyst, the protein was confined to continuous rings at the apical points of the trophectoderm cell contact. The protein was not found in the ICM; however, *ZO-1* mRNA was found in ICM cells (Wrenzycki et al. 2003). Expression of E-cadherin was suppressed in one of the first RNA interference studies performed in bovine preimplantation embryos (Nganvongpanit et al. 2006). The E-cadherin mRNA in morula stage embryos was reduced by 80% compared with the controls, and the number of embryos reaching the blastocyst stage was reduced from around 40% to 22%.

Another type of junction, called desmosomes, is formed in the trophectoderm from the time of cavitation on (see Figure 9.2). Desmosomes are spot-like junctions that maintain the integrity of the epithelium during blastocyst expansion. The extracellular domain is formed by several proteins such as desmogleins and desmocollins. The intracellular domain comprises plakoglobin and plakophilin, which link desmoplakin (DSP) to E-cadherin. DSP also attaches to the intermediate filaments (see keratin 8 and 18) of the cell. No transcripts of desmoglein 1 (*DG1*) and desmocollin I (*DSC1*) were found in bovine preimplantation embryos. *DSC2* and *DSC3* transcripts were found in two- to four-cell embryos up to hatched blastocysts. *DSC2* was predominantly found in trophectoderm cells as is the mRNA of plakophilin (Plako;Wrenzycki et al. 1998; Wrenzycki et al. 2003). No difference in transcription was found between *in vitro* and *in vivo* embryos.

Another structural gene involved in blastocyst formation, cell proliferation, cell adhesion, and cell mobility is *FN1*. Expression of FN1 mRNA and protein was found to be significantly higher in blastocysts than in earlier stages, and it was also differently expressed between *in vitro*-produced embryos and *in vivo*-produced embryos (Ponsuksili et al. 2002; El-Halawany et al. 2004; Mohan et al. 2004; Goossens et al. 2007a). The FN1 protein was predominantly expressed in the ICM and formed filamentous structures between the ICM and the trophectoderm (Goossens et al. 2007a). FN1 interacts with several types of ligands, such as heparin, fibrin, immunoglobulins, and DNA, and acts as a bridge between the collagen matrix and integrins at the cytotrophoblast surface (Aplin et al. 1999). *FN1* knockout mice die shortly after gastrulation (George et al. 1993). A new *FN1* splice variant specific for bovine blastocysts was detected, bringing the total of *FN1* splice variants in bovine to nine. This splice variant was different from the one present in cumulus cells surrounding the oocyte (Goossens et al. 2007b).

Another gene involved in cytoskeletal organization that is significantly upregulated in blastocysts is *MYL6*, which is the smooth muscle isoform of myosin light chain. The protein was found mainly around the blastocoel cavity, and no difference in expression was found between *in vivo* and *in vitro* embryos (Goossens et al. 2007a).

Bovine embryos become dependent on aerobic metabolism from the blastocyst stage on. One aspect of this change is that the embryos start utilizing glucose instead of pyruvate and lactate, as can be seen in the expression profile of several glucose transporters (Wrenzycki et al. 2003). Transcription profiles of *GLUT-1*, *GLUT-2*, *GLUT-3*, *GLUT-4*, and *GLUT-8* have been studied in bovine preimplantation development. *GLUT-2* is not transcribed in preimplantation embryos up to the blastocyst stage (Augustin et al. 2001; Lazzari et al. 2002). *GLUT-1* and *GLUT-4* have higher mRNA content in trophectoderm cells compared with ICM cells, whereas mRNA content is similar for *GLUT-3* (Wrenzycki et al. 2003).

Expression of *GLUT* genes is influenced by environmental conditions (Lazzari et al. 2002), except for *GLUT-1*, and is species-specific (Wrenzycki et al. 2003). *GLUT-3*, *GLUT-4*, and *GLUT-8* mRNA abundance is higher in *in vitro*-produced embryos compared with *in vivo* blastocysts, and these genes are possibly involved in the development of Large Offspring Syndrome (Lazzari et al. 2002; Knijn et al. 2005).

OCT-4, a member of the POU transcription factor family, is involved in transcriptional regulation during early development and cell differentiation and is often used as a marker for totipotency as it acts as a transcription factor for many genes specifically expressed in pluripotent cells in the mouse (Yeom et al. 1996). However, *OCT-4* expression was observed in bovine and porcine ICM and trophectoderm cells from *in vivo*– and *in vitro*–produced embryos, which is in contrast with the mouse, indicating that in pig and cattle, *OCT-4* is also expressed in non-pluripotent cells (Kirchhof et al. 2000; Kurosaka et al. 2004). *OCT-4* transcription is also influenced by the culture conditions. Transcription was lower in blastocysts derived from oocytes matured in TCM-199 supplemented with bovine serum albumin (BSA) or 10% serum. This could indicate that these blastocysts would reach the differentiated stage earlier (Russell et al. 2006).

9.2.5 Hatching

During the development of the embryo up to the blastocyst stage, it is protected by a glycoprotein membrane, called the zona pellucida (ZP), which surrounds the plasma membrane. Before the blastocyst can expand and implant, it needs to hatch from the ZP. Hatching involves the embryonic production of proteases that will digest the ZP. The blastocyst normally hatches out at the pole opposite the ICM, and some specialized trophectoderm cells seem to be involved (Sathananthan et al. 2003). However, relatively little is known about the genes involved in the process. A lower caspase activity was found in expanded blastocysts and also in the non-expanded blastocysts with a higher hatching rate (Jousan et al. 2008), but this is more an indication of the developmental capacity of the embryo and not the hatching capabilities per se.

9.3 Preimplantation developmental systems and transcriptomics

9.3.1 In vivo *preimplantation development*

Systematic functional genomic analyses of *in vivo* preimplantation embryo development are very scarce in production animals. Several studies have been performed comparing transcription profiles of *in vivo* with *in vitro* and/or somatic cell nuclear transfer (SCNT) embryos of certain stages, but a transcription profile of all important developmental stages from oocyte to implantation has not yet been made. The main reason for this is that it is difficult and expensive to obtain enough *in vivo* embryos for making large, representative, stage-specific cDNA libraries and ESTs or large-scale microarray studies. The situation is improving with new technologies for linear amplification of the RNA before hybridization with the arrays. A large-scale study was published by Adjaye et al. (2007) comparing the transcriptomes of bovine oocytes and blastocyst against 15,529 human cDNAs.

One of the more extensive studies of transcription in *in vivo* pig embryos was performed by Whitworth and coworkers (2004). They made cDNA libraries from germinal vesicle-stage oocytes and *in vivo*- and *in vitro*-produced two-cell and blastocyst-stage embryos and sequenced the 3′ ends. In this way, the expression of 2489 gene clusters was scored and virtual Northern blotting was used to compare the expression of *in vivo* and *in vitro* porcine embryos. Thirty-eight clusters were found to be different between *in vitro* and *in vivo* two-cell stage embryos and thirty-seven between *in vitro* and *in vivo* blastocysts. In a follow-up study, Whitworth et al. (2005) used a 15-K microarray and found 1409 and 1696 differentially detected cDNAs between *in vitro* and *in vivo* two-cell and blastocysts, respectively.

The ewe oviduct is sometimes used as a surrogate *in vivo* system for the production of bovine preimplantation embryos. Approximately 100 zygotes can be transferred and cultured in the ligated ewe oviduct. The quality of the blastocysts obtained is similar to that of *in vivo*-produced blastocysts (Galli and Lazzari 1996; Enright et al. 2000), and it has been reported that the transcription pattern of some developmentally important genes is similar to that of *in vivo*-derived bovine embryos (Lazzari et al. 2002). However, deviations for some genes may be observed. It is also of note that there exist considerable transcription differences and thus gene regulation between sheep and cow. Rizos et al. (2004) compared the transcription of eight genes between ovine and bovine embryos cultured under the same *in vitro* conditions, and found significantly higher mRNA abundance for *MnSOD*, survivin, and *GLUT-5* in ovine blastocysts, whereas that of connexin 31 (*Cx31*), *IFN-τ*, and sarcosine (SOX) was higher in bovine blastocysts. The two other genes investigated (E-cad and Na/K ATPase) showed no difference. The differences were thought to be related to species-specific differences in the adaptability of embryos to culture conditions.

9.3.2 *The effect of* in vitro *production (IVP)*

It is generally acknowledged that *in vivo*-derived embryos are of superior quality in comparison with *in vitro*-produced embryos. Many differences between *in vitro* and *in vivo* embryos have been observed: cytoplasm color and density (Pollard and Leibo 1994), general morphology (Van Soom et al. 1997), metabolism (Khurana and Niemann 2000), tolerance of lower temperature (Leibo and Loskutoff 1993), developmental capacity (van Wagtendonk-de Leeuw et al. 2000), incidence of chromosomal abnormalities (Viuff et al. 1999), pregnancy rate, and frequency of heavier fetuses after transfer (Hasler 2000).

Several research groups compared the transcriptome of *in vivo* with *in vitro*-produced embryos starting from oocytes up to hatching blastocysts. The first group of those studies focused on candidate genes chosen because of their known functions in early development from different species (Wrenzycki et al. 1998; Eckert and Niemann 1998; Lequarré et al. 2001; Knijn et al. 2002; Lazzari et al. 2002; El-Halawany et al. 2004) or identified by means of SSH (Tesfaye et al. 2004; Goossens et al. 2007a). It was shown that the relative abundance of the transcripts studied varied through the preimplantation period and that changes in transcript abundance at the blastocyst stage was, in many cases, a consequence of perturbation in an earlier stage (Lonergan et al. 2003).

Most studies comparing *in vivo* with *in vitro*-produced embryos focus on the blastocyst stage. Corcoran and coworkers (2006) used the bovine BOLT5 microarray with 3888 spots representing 932 bovine EST clones from a bovine total leukocyte cDNA library and complemented this with 459 additional amplicons to compare *in vitro* and *in vivo* (ewe oviduct) transcriptomes. In total, 384 genes were found to be differently transcribed, with 85% downregulated *in vitro* compared with *in vivo*. Many of these genes are involved in the regulation of transcription and translation and point to a deregulation of these processes in *in vitro*-produced embryos. Miles and coworkers (2006) used small amplified RNA (SAR)-SAGE to compare the transcription profile of porcine in *in vitro*- and *in vivo*-produced embryos. A total of 20,029 and 23,453 unique putative transcripts were detected in *in vitro* and *in vivo* porcine blastocysts, respectively. Of these, around 900 were differentially expressed. The *in vitro* blastocysts showed reduced transcription in biological processes such as cellular metabolism, organization, and response to stress.

Mohan et al. (2004) used SSH to compare transcription in *in vitro* and *in vivo* bovine blastocysts. Both categories of embryos yielded 32 ESTs. These corresponded to 32 and 22 known genes for *in vivo*- and *in vitro*-produced embryos, respectively. Only three of the genes (galectin-1, *FN1*, and filamin A) were tested using real-time RT-PCR. No difference was found for filamin A. Galectin-1 mRNA was about three times more abundant in *in vivo* blastocysts than in *in vitro* blastocysts. Galectin-1 is involved in cell–cell and cell–matrix interactions and is a regulator of cell transformation (Liu et al. 2002). Transcription of another galectin, galectin-3, was found to be three times higher in blastocysts than in morulas (Ponsuksili et al. 2002).

9.3.3 *Epigenetic modifications and SCNT*

In vitro culture systems and somatic cell (SCNT) procedures can have profound effects on gene expression in preimplantation embryos through epigenetic modifications. Epigenetic modifications in mammals primarily result from changes in the methylation/demethylation of DNA, mostly Cs in CpG dinucleotide motifs, and alterations of histones (ref). DNA methylation is, in general, an expression-repressive mechanism that possibly developed to protect the genome against the uncontrolled action of transposons (Yoder et al. 1997). Two main waves of DNA methylation reprogramming take place, one during germ cell development and one during preimplantation development (Reik and Walter 2001). Highly methylated primordial germ cells undergo rapid genome-wide demethylation and parent-of-origin methylation of certain genes. The second wave of DNA methylation reprogramming, between fertilization and blastocyst formation, starts with a rapid active paternal-specific demethylation independent of replication, which is then followed by a stepwise maternal methylation decline up to the morula stage. This passive decline is linked to the absence of DNMT1, the primary DNA methyltransferase. Active paternal-specific demethylation has been found in pig, cattle, and, to a lesser degree, sheep (Dean et al. 2003). Methylation starts again during the first cell differentiation steps in the blastocyst, leading to hypermethylation of ICM cells and undermethylation of trophectoderm cells (Dean et al. 2001).

Transcription of DNA methyltransferases has been studied in bovine preimplantation embryos. Transcription of *DNMT1*, *DNMT2*, *DNMT3a*, and *DNMT3b* was found by Golding and Westhusin (2003) in all stages, from two-cell up to blastocyst. The same authors found a new *DNMT2* splice variant in the embryos, and the embryo-specific variant, Dnmt1o, found in mice, was not detected, indicating differences in regulation between species. Different transcription between male and female blastocysts was reported for *DNMT3a* and *DNMT3b*, along with hnRNP methyltransferase-like 1 (*HMT1*) and interleukin enhancer binding factor 3 (*ILF3*), pointing to epigenetic differences between *in vitro* male and female bovine blastocysts (Bermejo-Alvarez et al. 2008). Significantly higher mRNA amounts of the histone methyltransferases *SUV39H1* and *G9A* and of the heterochromatin protein 1 (*HP1*) were found in blastocysts derived from male donor cells when compared with *in vivo* blastocysts (Nowak-Imialek et al. 2007). A difference was also observed between blastocysts derived from male and female fibroblasts, indicating that the epigenetic modifications are influenced by donor cell line. Differences in the transcription patterns of other genes involved in chromatin remodeling and histone acetylation/deacetylation during bovine preimplantation development have been described (McGraw et al. 2007).

Cloning of livestock animals through nuclear transfer (NT) has been hindered by low efficiency. Early gestational losses of NT embryos are often associated with aberrant placental development linked to deregulation of gene expression, mainly by epigenetic modifications (Wells et al. 2004). Aberrant transcription of some genes (acrogranin, estrogen receptor-like 2 [*ERR2*], and caudal-related homeobox gene 2 [*CDX2*]) involved in preimplantation and early placental development has been reported in cloned bovine blastocysts (Hall et al. 2005).

The epigenetic status of the donor cell has to be efficiently erased and reprogrammed after transfer for SCNT to be successful. Chromatin remodeling is often incomplete, as is reflected in aberrant DNA methylation and histone modification in bovine embryo clones (Santos et al. 2003). It was shown by using differential methylation hybridization (DMH) and bisulfite sequencing that methylation remodeling in *in vitro*-produced porcine blastocysts produced porcine blastocysts that deviated considerably from *in vivo* blastocysts, whereas remodeling in parthenogenetic and SCNT-derived blastocysts was more similar to *in vivo* blastocysts (Bonk et al. 2008).

The effect of SCNT on gene expression in preimplantation embryos was studied by RT-PCR for individual bovine genes (Daniels et al. 2000, 2001; Wrenzycki et al. 2001a, 2004; Park et al. 2003; Camargo et al. 2005; Jang et al. 2005; Sawai et al. 2005) and porcine genes (McElroy et al. 2008). A whole genome approach has been applied using microarray platforms (Pfister-Genskow et al. 2005; Smith et al. 2005; Somers et al. 2006). All of these studies reported significant differences in transcription between SCNT and *in vivo* or *in vitro*-produced embryos. Beyhan et al. (2007) also took into account the source of the donor nuclei, implying that the source of the donor nuclei can have important consequences for embryonic development. A striking result from comparing the four microarray studies is that the genes found differently expressed between SCNT and *in vitro* embryos are mostly different in all of the studies. The reasons for this observation can be many, ranging from differences in culture conditions to donor cells and microarray platforms used. Another important aspect was that in all four studies less than 1% of the genes were differently transcribed between SCNT and in vitro fertilized (IVF) blastocysts, indicating that reprogramming fails only for a limited number of genes. However, as mentioned earlier, the transcription profile of IVF blastocysts can also be called into question. Also, there are indications that the failure to reprogram is not gene specific but rather is a random process. This may also account for (some) differences found between the studies.

9.3.4 *Effect of culture medium on early embryo development*

Several studies have been performed to gauge the effect of the environment on gene expression on preimplantation development. Until now, most of these studies looked at the transcription of individual genes, often focusing on stress, imprinting, and apoptosis-related genes. In one such study, McElroy et al. (2008) studied the effect of culture conditions and SCNT on the expression of *HSP70.2*, integrin beta 1 (*ITGB1*), phosphoglycerate kinase 1 (*PGK1*), *BAX*, and *IGF2R* in porcine preimplantation embryos. The amount of *BAX* mRNA was higher in *in vitro*-produced blastocysts and SCNT blastocysts cultured in a medium with addition of 10% fetal bovine serum (FBS) on day 4 compared with *in vivo* blastocysts, whereas the mRNA content was lower for *HSP70.2*, *IGF2R*, and *ITGB1* in *in vitro* than in *in vivo* blastocysts. BAX is a pro-apoptotic molecule. Lower developmental capacity of *in* vitro-produced embryos has been linked to aberrant apoptosis. Therefore, it can be expected that the expression of pro-apoptotic genes is higher in *in vitro* and SCNT produced embryos than in *in vivo* embryos. However, it has been shown that the mRNA content of a certain gene can change quickly over time and that for several genes involved in apoptosis, such as *BAX*, *BCL-2*, caspase 3, and caspase 7, the mRNA content bears no relation to active protein content; hence, the mRNA content of these genes cannot be used as a reliable marker for apoptosis in preimplantation development (Vandaele et al. 2008). In agreement with this, Knijn et al. (2005) found no difference in transcription of *BAX* and *BCL-2* under different culture systems. They, however, found differences for *XIAP*, an X chromosome-linked inhibitor of apoptosis. Also, no difference in transcription of *BAX* and *BCL-2* could be found in porcine blastocysts after addition of melatonin (Rodriguez-Osorio et al. 2007). On the other hand, a higher *BAX* mRNA amount was found in blastocysts produced in synthetic oviductal fluid (SOF) medium compared with *in vivo* blastocysts (Rizos et al. 2002), which may be due to the presence of calf serum. It has been shown that calf serum has an influence on the expression of several genes (Rizos et al. 2003). Addition of leptin during bovine oocyte maturation had no effect on cleavage rate, but increased number developed to blastocysts and the proportion of apoptotic cells was reduced. Transcription of the leptin receptor (*LEPR*), signal transducer and activator of transcription 3 (*STAT3*), and baculoviral inhibitor of apoptosis protein repeat-containing 4 (*BIRC4*) was increased, whereas that of *BAX* was reduced, indicating a positive effect of leptin on preimplantion embryo development (Boelhauve et al. 2005).

Given that *HSP70.2* is a molecular chaperone that is generally upregulated in response to stress, it is surprising that the transcription of *HSP70.2* was lower in *in vitro*-produced embryos than in *in vivo* porcine blastocysts (McElroy et al. 2008). It is also contrary to the transcription pattern found for the *HSC70* gene (Bernardini et al. 2003). However, regulation of expression of heat shock proteins is complex and sensitive to minor changes in environment and manipulation. Some reports mention no difference in *HSP70.1* transcription between *in vitro*-produced embryos and *in vivo* bovine blastocysts (Wrenzycki et al. 2001b; Lazzari et al. 2002), whereas others do (Knijn et al. 2005). Also, addition of serum, BSA, or polyvinyl alcohol (PVA) influences transcription of *HSP70.1* (Wrenzycki et al. 1999; Lazzari et al. 2002). A higher relative mRNA abundance was observed in preimplantation embryos cultured in a medium with addition of serum compared with addition of BSA or PVA. On the other hand, no difference in transcription for *HSP70.1* was found under different culture conditions by de Oliveira et al. (2006).

Addition of BSA or PVA was found to have a beneficial effect on the incidence of Large Offspring Syndrome (Thompson et al. 1995; van Wagtendonk-de Leeuw et al. 2000). It was also shown that the transcription profiles of some developmentally important genes were more similar to that in *in vivo* bovine embryos in PVA embryos compared with embryos grown in medium supplemented with BSA or serum (Wrenzycki et al. 1999; Wrenzycki et al. 2001b). Several indications exist that *in vitro* culture systems put a considerable amount of oxidative stress on embryos. This can be seen in a significant upregulation of antioxidative enzymes such as copper–zinc containing superoxide dismutase (Cu/Zn-SOD; Lazzari et al. 2002).

Salazar et al. (2007) used differential hybridization to identify genes differently expressed in porcine morula after *in vitro* culture with malathion, a widely used organophosphate insecticide, added to the medium. Nine genes, of which three were unknown, were found to be downregulated by malathion. These include cytochrome c oxidase I and III and MHC I. Malathion may interfere with mitochondrial electron transport. This is in agreement with the finding that genes involved in mitochondrial biogenesis such as cytochrome oxidase I (*COXI*) and nuclear respiratory factor I (*NRFI*) and mitochondrial transcription factor A (*mtTFA*) have an influence on developmental capacity in bovine preimplantation development (May-Panloup et al. 2005).

An interesting experiment to study the relationship between transcription profile and *in vitro* bovine embryo production success was performed by El-Sayed et al. (2006). These authors took biopsies of bovine blastocysts (day 7), and the remaining embryo (60%–70%) was transferred to recipients after re-expansion. Based on the success of the pregnancy, the embryos were divided in three groups and the biopsies were used in microarray analysis. A homemade array containing 219 clones and the BlueChip bovine cDNA microarray containing around 2000 clones (Sirard et al. 2005) were used to profile the transcription differences between embryos giving no pregnancy (G1) versus embryos leading to calf delivery (G3), or embryos that were resorbed (G2) versus embryos leading to calf delivery (G3). Fifty-two and fifty-eight genes were differently transcribed between G1 and G3, and between G2 and G3, respectively. Biopsies from G3 embryos had higher transcription of genes involved in implantation (*COX2*, *CDX2*), carbohydrate metabolism (*ALOX15*), growth (BMP15), and signal transduction (*PLAU*), whereas those from embryos resulting in no pregnancy were enriched for transcripts from genes involved in inflammation (*TNF-α*), transcription (*MSX1*, *PTTG1*), glucose metabolism (*PGK1*, *AKR1B1*), and implantation inhibition (*CD9*).

9.4 Future research directions

Functional genomics of preimplantation development is a thriving discipline that has a promising future in bridging basic science and practical applications. As in many related disciplines, data are gathered at an astounding rate using the new technologies exploring whole genomes. However, the data flow too often stops at presenting lists of differently transcribed genes. What is definitely needed is a follow-up of these studies with research on protein expression, and functional analysis. Integration of transcriptomics and proteomics in biological systems is necessary to make meaning of all the data.

Acknowledgments

The author wishes to thank Karen Goossens for critically reading the manuscript, corrections, and suggestions.

References

Aberle, H., Schwartz, H., and Kemler R. 1996. Cadherin-catenin complex: Protein interactions and their implications for cadherin function. *Journal of Cellular Biochemistry* **61**: 514–523.

Adjaye, J., Herwig, R., Brink, T.C., Herrmann, D, Greber, B, Sudheer, S, Groth, D, Carnwath J.W., Lehrach, H., and Niemann, H. 2007. Conserved molecular portraits of bovine and human blastocysts as a consequence of the transition from maternal to embryonic control of gene expression. *Physiological Genomics* **31**: 315–327.

Aplin, J.D., Haigh, T., Jones, C.J., Church, H.J., and Vicovac, L. 1999. Development of cytotrophoblast columns from explanted first-trimester human placental villi: Role of fibronectin and integrins alpha5beta1. *Biology of Reproduction* **60**: 828–838.

Augustin, R,. Pocar, P., Navarrete-Santos, A., Wrenzycki, C., Gandolfi, F., Niemann, H., and Fischer, B. 2001. Glucose transporter expression is developmentally regulated in in vitro derived bovine preimplantation embryos. *Molecular Reproduction and Development* **60**: 370–376.

Barcroft, L.C., Hay-Schmidt, A., Caveney, A., Golfoyle, E., Overstrom, E.W., Hyttel, P., and Watson, A.J. 1998. Trophectoderm differentiation in the bovine embryo: Characterization of a polarized epithelium. *Journal of Reproduction and Fertility* **114**: 327–329.

Bernardini, C., Fantinati, P., Castellani, G., Forni, M., Zannoni, A., Seren, E., and Bacci, M.L. 2003. Alteration of constitutive heat shock protein 70 (HSC70) production by in vitro culture of porcine preimplantation embryos. *Veterinary Research Communications* **27S**: 575–578.

Bermejo-Alvarez, P., Rizos, D., Rath, D., Lonergan, P., and Gutierrez-Adan, A. 2008. Epigenetic differences between male and female bovine blastocysts produced in vitro. *Physiological Genomics* **32**: 264–272.

Betts, D.H., Barcroft, L.C., and Watson, A.J. 1998. Na/K ATPase-mediated $^{86}RB^{+}$ uptake and asymmetrical trophectoderm localization of α1 and α3 Na/K-ATPase isoforms during bovine preattachment development. *Developmental Biology* **197**: 77–92.

Betts, D.H., MacPhee, D.J., Kidder, G.M., and Watson, A.J. 1997. Ouabain sensitivity and expression of Na/K-ATPase α- and β-subunit isoform genes during bovine early development. *Molecular Reproduction and Development* **46**: 114–126.

Beyhan, Z., Ross, P.J., Iager, A.E., Kocabas, A.M., Cunniff, K., Rosa G.J., and Cibelli, J.B. 2007. Transcriptional reprogramming of somatic cell nuclei during preimplantation development of cloned bovine embryos. *Developmental Biology* **305**: 637–649.

Bjerregaard, B., Wrenzycki, C., Strejcek, F., Laurincik, J., Holm, P., Ochs, R.L.,

Rosenkranz C., Callesen, H., Rath, D., Niemann, H., and Maddox-Hyttel, P. 2004. Expression of nucleolar-related proteins in porcine preimplantation embryos produced in vivo and in vitro. *Biology of Reproduction* **70**: 867–876.

Boelhauve, M., Sinowatz, F., Wolf, E., and Paula-Lopes, F.F. 2005. Maturation of bovine oocytes in the presence of leptin improves development and reduces apoptosis of in vitro-produced blastocysts. *Biology of Reproduction* **73**: 737–744.

Bonk, A.J., Li, R., Lai, L., Hao, Y., Liu, Z., Samuel, M., Fergason, E.A., Whitworth, K.M., Murphy, C.N., Antoniou, E., and Prather, R.S. 2008. Aberrant DNA methylation in porcine in vitro-, parthenogenetic-, and somatic cell nuclear transfer-produced blastocysts. *Molecular Reproduction and Development* **75**: 250–264.

Brevini-Gandolfi, T.A.L, Favetta, L.A., Lonergan, P., and Gandolfi, F. 2000. The mechanism regulating maternal mRNA stability and translation is affected in bovine embryos with low developmental competence. *Theriogenology* **53**: 268.

Brevini-Gandolfi, T.A.L., Favetta, L.A., Mauri, L., Luciano, A.M., Cillo, F., and Gandolfi, F. 1999. Changes in poly(A) tail length of maternal transcripts during in vitro maturation of bovine oocytes and their relation with developmental competence. *Molecular Reproduction and Development* **52**: 427–433.

Brevini, T.A.L., Lonergan, P., Cillo, F., Favetta, L.A., Fair, T., and Gandolfi, F. 2002. Evolution of mRNA polyadenylation between oocyte maturation and first embryonic cleavage in cattle and its relation with developmental competence. *Molecular Reproduction and Development* **63**: 510–517.

Camargo, L.S., Viana, J.H., Sá, W.F., Ferreira, A.M., Vale Filho, V.R. 2005. Developmental competence of oocytes from prepubertal Bos indicus crossbred cattle. *Animal Reproduction Science* **85**: 53–59.

Cetica, P., Pintos, L., Dalvit, G., and Beconi, M. 2003. Involvement of enzymes of amino acid metabolism and tricarboxylic acid cycle in bovine oocyte maturation in vitro. *Reproduction* **126**: 753–763.

Corcoran, D., Fair, T., Park, S., Rizos, D., Patel, O.V., Smith, G.W., Coussens, P.M., Ireland, J.J., Boland, M.P., Evans, A.C., Lonergan, P. 2006. Suppressed expression of genes involved in transcription and translation in in vitro compared with in vivo cultured bovine embryos. *Reproduction* **131**: 651–660.

Daniels, R., Hall, V.J., French, A.J., Korfiatis, N.A., and Trounson, A.O. 2001. Comparison of gene transcription in cloned bovine embryos produced by different nuclear transfer techniques. *Molecular Reproduction and Development* **60**: 281–288.

Daniels, R., Hall, V., and Trounson, A.O. 2000. Analysis of gene transcription in bovine nuclear transfer embryos reconstructed with granulosa cell nuclei. *Biology of Reproduction* **63**: 1034–1040.

De Oliveira, A.T.D., Lopes, R.F.F., and Rodrigues, J.L. 2006. Gene expression and developmental competence of bovine embryos produced in vitro with different serum concentrations. *Reproduction in Domestic Animals* **41**: 129–136.

Dean, W., Santos, F., and Reik, W. 2003. Epigenetic reprogramming in early mammalian development and following somatic nuclear transfer. *Seminars in Cell and Developmental Biology* **14**: 93–100.

Dean, W., Santos, F., Stojkovic, M., Zakhartchenko, V., Walter, J., Wolf, E., and Reik, W. 2001. Conservation of methylation reprogramming in mammalian development: Aberrant reprogramming in cloned embryos. *Proceedings of the National Academy of Sciences of the United States of America* **98**: 13734–13738.

Dode, M.A.N., Dufort, I., Massicotte, L., and Sirard, M. 2006. Quantitative expression of candidate genes for developmental competence in bovine two-cell embryos. *Molecular Reproduction and Development* **73**: 288–297.

Eckert, J. and Niemann, H. 1998. mRNA expression of leukaemia inhibitory factor (LIF) and its receptor subunits glycoprotein 130 and LIF-receptor-β in bovine embryos derived in vitro or in vivo. *Molecular Human Reproduction* **4**: 957–965.

El-Halawany, N., Ponsuksili, S., Wimmers, K., Gilles, M., Tesfaye, D., and Schellander, K. 2004. Quantitative expression analysis of blastocyst-derived gene transcripts in preimplantation developmental stages of in vitro produced bovine embryos using real-time polymerase chain reaction technology. *Reproduction, Fertility and Development* **16**: 753–762.

El-Sayed, A., Hoelker, M., Rings, F., Salilew, D., Jennen, D., Tholen, E., Sirard, M.A., Schellander, K., Tesfaye, D. 2006. Large-scale transcriptional analysis of bovine embryo biopsies in relation to pregnancy success after transfer to recipients. *Physiological Genomics* **28**: 84–96.

Enright, B.P., Lonergan, P., Dinnyes, A., Fair, T., Ward, F.A., Yang, X., and Boland, M.P. 2000. Culture of in vitro produced bovine zygotes in vitro vs in vivo: Implications for early embryo development and quality. *Theriogenology* **54**: 659–673.

Fair, T., Gutierrez-Adan, A., Murphy, M., Rizos, D., Martin, F., Boland, M.P., and Lonergan, P. 2004a. Search for the bovine homolog of the murine Ped gene and characterization of its messenger RNA expression during bovine preimplantation development. *Biology of Reproduction* **70**: 488–494.

Fair, T., Murphy, M., Rizos, D., Moss, C., Martin, F., Boland, M.P., and Lonergan, P. 2004b. Analysis of differential maternal mRNA expression in developmentally competent and incompetent bovine two-cell embryos. *Molecular Reproduction and Development* **67**: 136–144.

Galli, C. and Lazzari, G. 1996. Practical aspects of IVM/IVF in cattle. *Animal Reproduction Science* **42**: 371–379.

Gandolfi, T.A. and Gandolfi, F. 2001. The maternal legacy to the embryo: Cytoplasmic components and their effects on early development. *Theriogenology* **55**: 1255–1276.

George, E.L., Georges-Laboueusse, E.N., Patel-King, R.S., Rayburn, H., and Hynes, R.O. 1993. Defects in mesoderm, neural tube and vascular development in mouse embryos lacking fibronectin. *Development* **119**: 1079–1091.

Golding, M.C. and Westhusin, M.E. 2003. Analysis of DNA (cytosine 5) methyltransferase mRNA sequence and expression in bovine preimplantation embryos, fetal and adult tissues. *Gene Expression Patterns* **3**: 551–558.

Goossens, K., Van Poucke, M., Van Soom A., Vandesompele, J., Van Zeveren, A., and Peelman, L.J. 2005. Selection of reference genes for quantitative real-time PCR in bovine preimplantation embryos. *BMC Developmental Biology* **5**: 27.

Goossens, K., Van Soom, A., Van Poucke, M., Vandaele, L., Vandesompele, J., Van Zeveren, A., and Peelman, L.J. 2007a. Identification and expression analysis of genes associated with bovine blastocyst formation. *BMC Developmental Biology* **7**: 64.

Goossens, K., Van Soom, A., Van Zeveren, A., and Peelman, L.J. 2007b. A new bovine embryo specific fibronectin splice variant. *Proceedings of 2nd International Meeting on Mammalian Embryogenomics, Paris, France*, p. 144.

Gutierrez-Adan, A., Behboodi, E., Anderson, G.B., Medrano, J.F., and Murray, J.D. 1996. Relationship between stage of development and sex of bovine IVM-IVF embryos cultured in vitro versus in the sheep oviduct. *Theriogenology* **46**: 515–525.

Gutierrez-Adan, A., Oter, M., Martinez-Madrid, B., and De la Fuente, J. 2000. Differential expression of two genes located on the X chromosome between male and female in vitro-produced bovine embryos at the blastocyst stage. *Molecular Reproduction and Development* **55**: 146–151.

Gutierrez-Adan, A., Rizos, D., Fair, T., Moreira, P.N., Pintabo, B., De la Fuente, J., Boland, M.P., and Lonergan, P. 2004. Effect of speed of development on mRNA expression pattern in early bovine embryos cultured in vivo or in vitro. *Molecular Reproduction and Development* **68**: 441–448.

Hall, V.J., Ruddock, N.T., and French, A.J. 2005. Expression profiling of genes crucial for placental and preimplantation development in bovine in vivo, in vitro, and nuclear transfer blastocysts. *Molecular Reproduction and Development* **72**: 16–24.

Hasler, J.F. 2000. In vitro production of cattle embryos: Problems with pregnancies and parturition. *Human Reproduction* **15**: 47–58.

Hasler, J.F., Henderson, W.B., Hurtgen, P.J., Jin, Z.Q., McCauley, A.D., Mower, S.A., Neely, B., Shuey, L.S., Stokes, J.E., and Trimmer, S.A. 1995. Production, freezing and transfer of bovine IVF embryos and subsequent calving results. *Theriogenology* **43**: 141–152.

Hernandez-Ledezma, J.J., Mathialagan, N., Villanueva, C., Sikes, J.D., and Roberts, R.M. 1993. Expression of bovine trophoblast interferons by in vitro-derived blastocysts is correlated with their morphological quality and stage development. *Molecular Reproduction and Development* **36**: 1–6.

Hesse, M., Franz, T., Tamai, Y., Taketo, M.M., and Magin, T.M. 2000. Targeted deletion of keratins 18 and 19 leads to trophoblast fragility and early embryonic lethality. *EMBO Journal* **19**: 5060–5070.

Jackson, B.W., Grund, C., Schmid, E., Burki, K., Franke, W.W., and Illmensee, J. 1980. Formation of cytoskeletal elements during mouse embryogenesis. Intermediate filaments of the cytokeratin type and desmosomes in preimplantation embryos. *Differentiation* **17**: 161–179.

Jang, G., Jeon, H.Y., Ko, K.H., Park, H.J., Kang, S.K., Lee, B.C., and Hwang, W.S. 2005. Developmental competence and gene expression in preimplantation bovine embryos derived from somatic cell nuclear transfer using different donor cells. *Zygote* **13**: 187–195.

Jousan, F.D., De Castro e Paula, L.A., Brad, A.M., Roth, Z., and Hansen, P.J. 2008. Relationship between group II caspase activity of bovine preimplantation embryos and capacity for hatching. *Journal of Reproduction and Development*. February 14 [Epub ahead of print].

Juriscova, A., Casper, R.F., MacLusky, N.J., Mills, G.B., and Librach, C.L. 1996. HLA-G expression during preimplantation embryo development. *Proceedings of the National Academy of Sciences of the United States of America* **93**: 161–165.

Khurana, N.K. and Niemann, H. 2000. Energy metabolism in preimplantation bovine embryos derived in vitro or in vivo. *Biology of Reproduction* **62**: 847–856.

Kirchhof, N., Carnwath, J.W., Lemme, E., Anastassiadis, K., Schöler, H., and Niemann, H. 2000. Expression pattern of Oct-4 in preimplantation embryos of different species. *Biology of Reproduction* **63**: 1698–1705.

Knijn, H.M., Wrenzycki, C., Hendriksen, P.J.M., Vos, P.L.A.M., Herrmann, D., van der Weijden, G.C., Niemann, H., and Dieleman, S.J. 2002. Effects of oocyte maturation regime on the relative abundance of gene transcripts in bovine blastocysts derived in vitro or in vivo. *Reproduction* **124**: 365–375.

Knijn, H.M., Wrenzycki, C., Hendriksen, P.J.M., Vos. P.L.A.M., Zeinstra, E.C., van der Weijden, G.C., Niemann, H., and Dieleman, S.J. 2005. In vitro and in vivo culture effects on mRNA expression of genes involved in metabolism and apoptosis in bovine embryos. *Reproduction, Fertility and Development* **17**: 775–784.

Koyama, H., Suzuki, H., Yang, X., Jiang, S., and Foote, R.H. 1994. Analysis of polarity of bovine and rabbit embryos by scanning electron microscopy. *Biology of Reproduction* **50**: 163–170.

Kubisch, H.M., Larson, M.A., and Roberts, R.M. 1998. Relationship between age of blastocyst formation and interferon-tau secretion by in vitro-derived bovine embryos. *Molecular Reproduction and Development* **49**: 254–260.

Kurosaka, S., Eckardt, S., and McLaughlin, K.J. 2004. Pluripotent lineage definition in bovine embryos by Oct4 transcript localization. *Biology of Reproduction* **71**: 1578–1582.

Lazzari, G., Wrenzycki, C., Herrmann, D., Duchi, R., Kruip, T., Niemann, H., and Galli, C. 2002. Cellular and molecular deviations in bovine in vitro-produced embryos are related to the large offspring syndrome. *Biology of Reproduction* **67**: 767–775.

Lee, S.M., Koh, H.J., Park, D.C., Song, B.J., Huh, T.L., and Park, J.W. 2002. Cytosolic NADP(+)dependent isocitrate dehydrogenase status modulates oxidative damage to cells. *Free Radical Biology and Medicine* **32**: 1185–1196.

Leibo, S.P. and Loskutoff, N.M. 1993. Cryobiology of in vitro-derived bovine embryos. *Theriogenology* **39**: 81–94.

Leoni, G.G., Bebbere, D., Succu, S., Berlinguer, F., Mossa, F., Galioto, M., Bogliolo, L., Ledda, S., and Naitana, S. 2007. Relations between relative mRNA abundance and developmental competence of ovine oocytes. *Molecular Reproduction and Development* **74**: 249–257.

Lequarré, A.S., Feugang, J.M., Malhomme, O., Donnay, I., Massip, A., Dessy, F., and Van Langendonckt, A. 2001. Expression of Cu/Zn and Mn superoxide dismutases during bovine embryo development: influence of in vitro culture. *Molecular Reproduction and Development* **58**: 45–53.

Liu, F.T., Patterson, R.J., and Wang, J.L. 2002. Intracellular functions of galectins. *Biochimica Biophysica Acta* **1572**: 263–273.

Lonergan, P., Khatir, H., Piumi, F., Rieger, D., Humblot, P., and Boland, M.P. 1999. Effect of time interval from insemination to first cleavage on the developmental characteristics, sex and pregnancy rates following transfer of bovine preimplantation embryos. *Journal of Reproduction and Fertility* **117**: 159–167.

Lonergan, P., Gutierrez-Adan, A., Pintado, B., Fair, T., Ward, F., De la Fuente, J., and Boland, M. 2000. Relationship between time of first cleavage and the expression of IGF-I growth factor, its receptor, and two housekeeping genes in bovine two-cell embryos and blastocysts produced in vitro. *Molecular Reproduction and Development* **57**: 146–152.

Lonergan, P., Rizos, D., Gutiérrez-Adan, A., Moreira, P.M., Pintado, B., de la Fuente, J., and Boland, M.P. 2003. Temporal divergence in the pattern of messenger RNA expression in bovine embryos cultured from the zygote to blastocyst stage in vitro or in vivo. *Biology of Reproduction* **69**: 1424–1431.

Mamo, S., Sargent, C.A., Affara, N.A., Tesfaye, D., El-Halawany, N., Wimmers, K., Gilles, M., Schellander, K., and Ponsuksili, S. 2006. Transcript profiles of some developmentally important genes detected in bovine oocytes and in vitro-produced blastocysts using RNA amplification and cDNA microarrays. *Reproduction in Domestic Animals* **41**: 527–534.

Massip, A., Mermillod, P., Van Langendonck, A., Reichenbach, H.D., Lonergan, P., Berg, U., Carolan, C., De Roover, R., and Brem, C. 1996. Calving outcome following transfer of embryos produced in vitro in different conditions. *Animal Reproduction Science* **44**: 1–10.

May-Panloup, P., Vignon, X., Chrétien, M., Heyman, Y., Tamassia, M., Malthièry, Y., and Reynier, P. 2005. Increase of mitochondrial DNA content and transcripts in early bovine embryogenesis associated with upregulation of mtTFA and NRF I transcription factors. *Reproductive Biology and Endocrinology* **3**: 65.

McElroy, S.L., Kim, J.H., Jeong, Y.W., Lee, E.G., Park, S.M., Hossein, M.S., Koo, O.J., Abul Hashem, M.D., Jang, G., Kang, S.K., Lee, B.C., and Hwang, W.S. 2008. Effects of culture conditions and nuclear transfer protocols on blastocyst formation and mRNA expression in pre-implantation porcine embryos. *Theriogenology* **69**: 416–425.

McGraw, S., Vigneault, C., and Sirard, M. 2007. Temporal expression of factors involved in chromatin remodeling and in gene regulation during early bovine in vitro embryo development. *Reproduction* **133**: 597–608.

Miles, J.R., Blomberg, L.A., Krisher, R.L., Everts, R.E., Sonstegard, T.S., Van Tassell, C.P., and Zuelke, K.A. 2006. *Comparison of gene expression from day 6 in vivo- and in vitro-produced porcine blastocysts*. Entrez GEO datasets series GSE5571.

Miller, D.J., Eckert, J.J., Lazzari, G., Duranthon-Richoux, V., Sreenan, J., Morris, D., Galli, C., Renard, J., and Fleming, T.P. 2003. Tight junction messenger RNA expression levels in bovine embryos are dependent upon the ability to compact and in vitro culture methods. *Biology of Reproduction* **38**: 1394–1402.

Misirlioglu, M., Page, G.P., Sagirkaya, H., Kaya, A., Parrish, J.J., First, N.L., Memili, E. 2006. *Proceedings of the National Academy of Sciences of the United States of America* **103**: 18905–18910.

Mohan, M., Hurst, A.G., and Malayer, J.R. 2004. Global gene expression analysis comparing bovine blastocysts flushed on day 7 or produced in vitro. *Molecular Reproduction and Development* **62**: 288–298.

Nganvongpanit, K., Müller, H., Rings, F., Gilles, M., Jennen, D., Hölker, M., Tholen, E., Schellander, K., and Tesfaye, D. 2006. Targeted suppression of E-cadherin gene expression in bovine preimplantation embryo by RNA interference technology using double-stranded RNA. *Molecular Reproduction and Development* **73**: 153–163.

Nowak-Imialek, M., Wrenzycki, C., Herrmann, D., Lucas-Hahn, A., Lagutina, I., Lemme, E., Lazzari, G., Galli, C., and Niemann, H. 2007. Messenger RNA expression patterns of histone-associated genes in bovine preimplantation embryos derived from different origins. *Molecular Reproduction and Development* **75**: 731–743.

Park, S.H., Park, S.B., and Kim, N.H. 2003. Expression of early development-related genes in bovine nuclear transferred and fertilized embryos. *Zygote* **11**: 355–360.

Paynton, B. and Bachvarova, R. 1994. Polyadenylation and deadenylation of maternal mRNAs during oocyte growth and maturation in the mouse. *Molecular Reproduction and Development* **37**: 172–188.

Pfister-Genskow, M., Myers, C., Childs, L.A., Lacson, J.C., Patterson, T., Betthauser, J.M., Goueleke, P.J., Koppang, R.W., Lange, G., Fisher, P., Watt, S.R., Forsberg, E.J., Zheng, Y., Leno G.H., Schultz, R.M., Liu, B., Chetia, C., Yang, X., Hoeschele, I., and Eilertsen, K.J. 2005. Identification of differentially expressed genes in individual produced bovine preimplantation embryos produced by nuclear transfer: improper reprogramming of genes required for development. *Biology of Reproduction* **72**: 546–555.

Pollard, J.W. and Leibo, S.P. 1994. Chilling sensitivity of mammalian embryos. *Theriogenology* **41**: 101–106.

Ponsuksili, S., Tesfaye, D., El-Halawany, N., Schellander, K., and Wimmers, K. 2002. Stage-specific expressed sequence tags obtained during bovine development by differential display RT-PCR and suppression subtractive hybridization. *Prenatal Diagnosis* **22**: 1135–1142.

Reeve, W.J.D. 1981. Cytoplasmic polarity develops at compaction in rat and mouse embryos. *Journal of Embryology and Experimental Morphology* **62**: 351–367.

Reik, W. and Walter, J. 2001. Evolution of imprinting mechanisms: The battle of the sexes begins in the zygote. *Nature Genetics* **27**: 255–256.

Reima, I., Lehtonen, E., Virtanen, I., and Flechon, J.E. 1993. The cytoskeleton and associated proteins during cleavage, compaction and blastocyst differentiation in the pig. *Differentiation* **54**: 34–45.

Rizos, D., Gutierrez-Adan, A., Moreira, P., O'Meara, C., Fair, T., Evans, A.C.O., Boland, M.P., Lonergan, P. 2004. *Molecular Reproduction and Development* **69**: 381–386.

Rizos, D., Gutierrez-Adan, A., Perez-Garnelo, S., de La Fuente, J., Boland, M.P., and Lonergan, P. 2003. Bovine embryo culture in the presence or absence of serum: Implications for blastocyst development, cryotolerance, and messenger RNA expression. *Biology of Reproduction* **68**: 236–243.

Rizos, D., Lonergan, P., Boland, M.P., Arroyo-Garcia, R., Pintado, B., de La Fuente, J., and Gutierrez-Adan, A. 2002. Analysis of differential messenger RNA expression between bovine blastocysts produced in different culture systems: Implications for blastocyst quality. *Biology of Reproduction* **66**: 589–595.

Rodriguez-Osorio, N., Kim, I.J., Wang, H., Kaya, A., and Memili, E. 2007. Melatonin increases cleavage rate of porcine preimplantation embryos in vitro. *Journal of Pineal Research* **43**: 283–288.

Russell, D.F., Baqir, S., Bordignon, J., and Betts, D.H. 2006. The impact of oocyte maturation media on early bovine embryonic development. *Molecular Reproduction and Development* **73**: 1255–1270.

Salazar, Z., Ducolomb, Y., Betancourt, M., Bonilla, E., Cortés, L., Hernandez-Hernandez, F., and Gonzalez-Marquez, H. 2007. Gene expression analysis on the early development of pig embryos exposed to malathion. *International Journal of Toxicology* **26**: 143–149.

Santos, F., Zakhartchenko, V., Stojkovic, M., Peters, A., Jenuwein, T., Wolf, E., Reik, W., and Dean, W. 2003. Epigenetic marking correlates with developmental potential in cloned bovine preimplantation embryos. *Current Biology* **13**: 1116–1121.

Sathananthan, H., Menezes, J., and Gunasheela, S. 2003. Mechanics of human blastocyst hatching in vitro. *Reproductive Biomedicine Online* 7: 228–234.

Sawai, K., Kageyama, S., Moriyasu, S., Hirayama, H., Minamihashi, A., and Onoe, S. 2005. Analysis of mRNA transcripts for insulin-like growth factor receptors and binding proteins in bovine embryos derived from somatic cell nuclear transfer. *Cloning Stem Cells* 7: 189–198.

Sirard, M., Dufort, I., Vallée, M., Massicotte, L., Gravel, C., Reghenas, H., Watson, A.J., King, W.A., and Robert, C. 2005. Potential and limitations of bovine-specific arrays for the analysis of mRNA levels in early development: Preliminary analysis using a bovine embryonic array. *Reproduction, Fertility and Development* **17**: 47–57.

Smith, S.L., Everts, R.E., Tian, X.C., Du, F., Sung, L., Rodriguez-Zas, S.L., Jeong, B., Renard, J., Lewin, H.A., and Yang, X. 2005. Global gene expression profiles reveal significant nuclear reprogramming by the blastocyst stage after cloning. *Proceedings of the National Academy of Sciences of the United States of America* **102**: 17582–17587.

Somers, J., Smith, C., Donnison, M., Wells, D.N., Henderson, H., McLeay, L., and Pfeffer, P.L. 2006. Gene expression profiling of individual bovine nuclear transfer blastocysts. *Reproduction* **131**: 1073–1084.

Stanton, J.A., Macgregor, A.B., and Green, D.P. 2003. Gene expression in the mouse preimplantation embryo. *Reproduction* **125**: 457–468.

Stevenson, B.R. and Keon, B.H. 1998. The tight junction: Morphology to molecules. *Annual Review of Cell and Developmental Biology* **14**: 89–110.

Tachibana, M., Sugimoto, K., Nozaki, M., Ueda, J., Ohta, T., Ohki, M., Fukuda, M., Takeda, N., Niida, H., Kato, H., and Shinkai, Y. 2002. G9a histone methyltransferase plays a dominant role in euchromatic histone H3 lysine 9 methylation and is essential for early embryogenesis. *Genes and Development* **16**: 1779–1791.

Telford, N.A., Watson, A.J., and Schultz, G.A. 1990. Transition from maternal to embryonic control in early development: A comparison of several species. *Molecular Reproduction and Development* **26**: 90–100.

Tesfaye, D., Ponsuksili, S., Wimmers, K., Gilles, M., and Schellander, K. 2004. A comparative expression analysis of gene transcripts in post-fertilization developmental stages of bovine embryos produced in vitro or in vivo. *Reproduction in Domestic Animals* **39**: 396–404.

Thompson, J.G., Gardner, D.K., Pugh, P.A., McMillan, W.H., and Tervit, H.R. 1995. Lamb birth weight is affected by culture system utilized during in vitro pre-elongation development of ovine embryos. *Biology of Reproduction* **53**: 1385–1391.

Van Soom, A., Boerjan, M.L., Bols, P.E.J., Vanroose, G., Lein, A., Coryn, M., and de Kruif, A. 1997. Timing of compaction and inner cell allocation in bovine embryos produced in vivo after superovulation. *Biology of Reproduction* **57**: 1041–1049.

Van Wagtendonk-de Leeuw, A.M., Mullaart, E., de Roos, A.P., Merton, J.S., den Daas, J.H., Kemp, B., and de Ruigh, L. 2000. Effects of different reproduction techniques: AI MOET or IVP, on health and welfare of bovine offspring. *Theriogenology* **53**: 575–597.

Vandaele, L., Goossens, K., Peelman, L., and Van Soom, A. 2008. mRNA expression of BCL-2, BAX, caspase-3 and -7 cannot be used as a marker for apoptosis in bovine blastocysts. *Animal Reproduction Science* **106**: 167–173.

Viuff, D., Rickords, L., Offenberg, H., Hyttel, P., Avery, B., Greve, T., Olsaker, I., Williams, J.L., Callesen, H., and Thomsen, P.D. 1999. A high proportion of bovine blastocysts produced in vitro are mixoploid. *Biology of Reproduction* **60**: 1273–1278.

Warner, C.M., Gollnick, S.O., and Goldbard, S.B. 1987. Linkage of the preimplantation-embryo-development (Ped) gene to the mouse major histocompatibility complex (MHC). *Biology of Reproduction* **36**: 606–610.

Watson, A.J. and Barcroft, L.C. 2001. Regulation of blastocyst formation. *Frontiers in Bioscience* **6**: D708–D730.

Wells, D.N., Forsyth, J.T., McMillan, V., and Oback, B. 2004. The health of somatic cell cloned cattle and their offspring. *Cloning Stem Cells* **6**: 101–110.

Whitworth, K., Agca, C., Kim, J., Patel, R.V., Springer, G.K., Bivens, N.J., Forrester, L.J., Mathialagan, N., Green, J.A., and Prather, R.S. 2005. Transcriptional profiling of pig embryogenesis by using a 15-K member unigene set specific for pig reproductive tissues and embryos. *Biology of Reproduction* **72**: 1437–1451.

Whitworth, K., Springer, G.K., Forrester, L.J., Spollen, W.G., Ries, J., Lamberson, W.R., Bivens, N., Murphy, C.N., Mathialigan, N., Green, J.A., and Prather, R.S. 2004. Developmental expression of 2489 gene clusters during pig embryogenesis: An expressed sequence tag project. *Biology of Reproduction* **71**: 1230–1243.

Wrenzycki, C., Herrman, D., Carnwath, J.W., and Niemann, H. 1998. Expression of RNA from developmentally important genes in preimplantation bovine embryos produced in TCM supplemented with BSA. *Journal of Reproduction and Fertility* **112**: 387–398.

Wrenzycki, C., Herrman, D., Carnwath, J.W., and Niemann, H. 1999. Alterations in the relative abundance of gene transcripts in preimplantation bovine embryos cultures in medium supplemented with either serum or PVA. *Molecular Reproduction and Development* **53**: 8–18.

Wrenzycki, C., Herrmann, D., Keskintepe, L., Martins, A., Sirisathien, S., Brackett, B., and Niemann, H. 2001b. Effects of culture system and protein supplementation on mRNA expression in preimplantation bovine embryos. *Human Reproduction* **16**: 893–901.

Wrenzycki, C., Herrmann, D., Lucas-Hahn, A., Lemme, E., Korsawe, K., and Niemann, H. 2004. Gene expression patterns in in vitro-produced and somatic nuclear transfer-derived preimplantation bovine embryos: Relationship to the large offspring syndrome? *Animal Reproduction Science* **82–83**: 593–603.

Wrenzycki, C., Herrmann, D., and Niemann, H. 2003. Timing of blastocyst expansion affects spatial messenger RNA expression patterns of genes in bovine blastocysts produced in vitro. *Biology of Reproduction* **68**: 2073–2080.

Wrenzycki, C., Wells, D., Herrmann, D., Miller, A., Oliver, J., Tervit, R., and Niemann, H. 2001a. Nuclear transfer protocol affects messenger RNA expression patterns in cloned bovine blastocysts. *Biology of Reproduction* **65**: 309–317.

Yeom, Y.I., Fuhrmann, G., Ovitt, C.E., Brehm, A., Ohbo, K., Gross, M., Hübner, K., and Schöler, H.R. 1996. Germline regulatory element of Oct-4 specific for the totipotent cycle of embryonal cells. *Development* **122**: 881–894.

Yoder, J.A., Walsh, C.P., and Bestor, T.H. 1997. Cytosine methylation and the ecology of intragenomic parasites. *Trends in Genetics* **13**: 335–340.

10

Physiological Genomics of Conceptus–Endometrial Interactions Mediating Corpus Luteum Rescue

Troy L. Ott and Thomas E. Spencer

10.1 Introduction

Placental mammals require luteal progesterone for part or all of gestation. The domestic animals covered in this chapter utilize prostaglandin $F_{2\alpha}$ (PGF) of uterine origin for mediating luteal regression. In this regard, pregnancy establishment has necessitated evolution of conceptus (embryo/fetus and associated extraembryonic membranes) strategies to alter uterine PGF production so that it no longer induces luteal regression. Early studies examining the effects of the conceptus on uterine PGF production revealed complexity both on the part of regulation of endometrial PGF production and release and on the part of conceptus signaling designed to abrogate the luteolytic production of PGF. Although we are learning more and more about the complex biochemical dialogue that is initiated at maternal recognition of pregnancy signaling, this review will focus only on those studies using genomic and proteomic approaches to unravel the key signaling events designed to inhibit regression of the corpus luteum (CL).

Common domestic animals belong to the Perissodactyla (equidae) and Artiodactyla (bovidae, ovidae, caprinae, suidae) orders, which are both members of the superorder Laurasiatheria. Horses, swine, and domestic ruminants exhibit uterine-dependent ovarian cycles. Placental mammals have evolved reproductive strategies that rely on extended periods of intrauterine development followed by birth of live offspring that exhibit a broad range of development and mobility at birth, including some that are substantially mobile within minutes to hours of birth. Without exception, Eutherian domesticated farm animals require conceptus-mediated rescue of CL function and luteal progesterone production for part or all of gestation (Bazer et al. 2008). In contrast, the dog and cat exhibit an extended luteal phase in the absence of conceptus signaling. It is thought that a reproductive strategy that requires CL rescue ensures repeated opportunities for mating and pregnancy establishment at shortened intervals. This evolutionary strategy is taken to the extreme in rodents, which exhibit 4- to 5-day estrous cycles and do not form a functional CL unless mated.

10.2 Physiological genomics of luteal regression

All domestic farm species covered in this chapter have uteri that produce PGF, which acts on the CL to initiate its functional and structural regression. Recently, genomics techniques have been utilized to examine the transcriptomes of CL at various stages of development and regression. Casey and coworkers (2005) conducted one of the earliest studies of differentially expressed genes in the functional and regressing CL of cattle using a custom ovarian cDNA microarray. This analysis yielded 15 differentially expressed genes, of which seven increased and eight decreased in regressed CL compared with non-regressed CL. These genes fell broadly into the following categories: extracellular matrix (ECM), cell structure, oxygen metabolism, apoptosis, steroid biosynthesis, and metabolism. In general, genes in the ECM category increased during luteolysis, with decorin (*DCN*) showing the greatest increase (Casey et al. 2005). *DCN* is a small proteoglycan associated with endothelial cell angiogenesis, particularly that associated with inflammation (Nelimarkka et al. 2001), and collagen fibril assembly (Casey et al. 2005). There was also upregulation of collagen genes associated with formation of type I collagen. As expected, all the genes in the steroid biosynthesis category were reduced in regressing CL compared with functional CL. In addition, there was a large decrease in the insulin-like growth factor 2 receptor (*IGF2R*) and an increase in expression of the apoptosis-related gene, clusterin (*CLU*), in regressing CL. Together these changes in gene expression paint a fairly intuitive picture of CL regression involving decreased steroid biosynthesis and metabolic activity associated with functional regression of the CL, along with increases in ECM remodeling and apoptosis associated with structural regression of the CL.

Recently, Goravanahally and coworkers (2009) compared gene expression profiles in bovine CL before (day 4) and after (day 10) responsiveness to PGF develops. Because CL of both statuses contained receptors for PGF, the differences in patterns of gene expression might reveal critical signaling pathways mediating PGF-induced CL regression. Of the 167 differentially expressed genes detected, the majority were divided equally (~18% each) between genes involved in protein synthesis and modification and genes involved in transcriptional regulation and DNA synthesis. Slightly lower percentages were involved in cell signaling (~12%) and steroidogenesis and metabolism (~10%). In response to PGF, expression of both calcium/calmodulin-dependent protein kinase kinase 2, beta (*CAMKK2*) and guanine nucleotide binding protein (G protein), beta polypeptide 1 (*GNB1*) were increased in the CL of cattle. The authors suggested that the combined increase in expression of *CAMKK2* due to the developmental transition (day 4 to day 10) and PGF treatment may have a critical role in increased luteolytic sensitivity to PGF in cattle. This increase in *CAMKK2* occurred at a developmental stage (day 10), when PGF had an increased ability to elicit a rise in intracellular calcium concentrations compared with those in day 4 CL. *CAMKK2* is thought to increase intracellular calcium via phosphorylation of calcium/calmodulin-dependent protein kinases such as *CAMK1* and *CAMK4*. Once phosphorylated, kinase activity increased 10- to 20-fold. Furthermore, CAMKs can activate mitogen-activated protein kinase 1 (*MAPK1*) and *MAPK3* in several ligand-stimulated pathways.

Although there are a large number of candidate genes identified and characterized in CL of farm species, there are few studies that have attempted to characterize the transcriptomes of the CL at various stages of function or regression. In this regard, the rodent provides little help as a comparative model. Rodents exhibit 4- to 5-day estrous cycles but lack a true luteal phase. In the absence of mating, the CL does not become fully functional and produces scant progesterone for about 2 days and increased activity of aldo-keto reductase family 1, member C1 (dihydrodiol dehydrogenase 1; 20-alpha [3-alpha]-hydroxysteroid dehydrogenase [*AKR1C1*]) producing 20α-hydroxyprogesterone, which does not support pregnancy or the decidual reaction (see Bachelot and Binart 2005). Mating-induced surges in prolactin (PRL) are the signal, which is initially responsible for establishing a fully functional CL that will support gestation in rodents.

Stocco and coworkers (2001) conducted cDNA expression array experiments to examine the effects of PRL and PGF on the transcriptome of the rat CL. In response to mating, the rodents produce diurnal surges of PRL from the anterior pituitary which support CL function for the first half of gestation (Soares et al. 2007). As in the farm species, PGF is responsible for luteal regression in rodents, but its origin is the ovary and not the uterus. Stocco and coworkers (2001) hypothesized that PRL effects were mediated in part by antagonizing the effects of PGF, including decreasing expression of the PGF receptor (*PTGFR*) and phospholipase C, gamma 1 (*PLCG1*), which mediates PFG signaling. In addition, they showed that PRL inhibited expression of transforming growth factor beta 1 (*TGFB1*), a pro-apoptotic cytokine (Stocco et al. 2001). Consistent with this hypothesis, PRL induced expression of a number of genes involved in steroid biosynthesis, whereas PGF inhibited these same genes and reduced expression of the luteinizing hormone (LH) receptor.

Foyouzi and coworkers (2005) found that genes involved in steroidogenesis and in maintaining the antioxidant status of the CL were regulated by PGF-induced luteolysis in the day 19 mouse CL. Using microarray analysis, the same authors found that *AKR1C1* expression increased in response to exogenous PGF, which would result in the conversion of progesterone to its inactive metabolite 20α-hydroxyprogesterone. They further showed that CL undergoing induced regression expressed higher concentrations of cytochrome P450, family 17, subfamily A, polypeptide 1 (*CYP17A1*), which would result in increased production of androstenedione. Androgens have previously been shown to induce abortion in mice (Sridaran et al. 1991). Interestingly, Sidran et al. (1991) also showed that *CYP19A1* and *HSD17B7* (hydroxysteroid (17-beta) dehydrogenase 7) expression decreased in response to PGF. CYP19A1 converts androstenedione to estrone, and HSD17B7 converts estrone to the more biologically active 17β-estradiol. These results suggest that PGF action on the rodent CL involves increased metabolism of progesterone to an inactive metabolite and a reduction in the ability of the CL to produce estrogens, perhaps resulting in further increases in androstenedione.

Another family of genes involved in luteal function is the superoxide dismutase (*SOD*) family. These genes are involved in converting the superoxide radical to hydrogen peroxide. Foyouzi and coworkers (2005) found a decrease in *SOD2*, *SOD3*, and copper chaperone of *SOD1* (*CCS*) expression at the end of pregnancy in mice, which may reduce the capacity of luteal cells to cope with superoxide accumulation. In that study, other members of the free radical scavenging family were high in functioning CL and reduced in response to PGF. Oxidative stress enzymes in the glutathione S-transferase (*GST*) family were also shown to be increased by PRL and decreased by PGF in the rat CL (Stocco et al. 2001). Thus, it is likely that responses to PGF include an increase in free radicals that coincides with luteal regression. Whether this is actually inducing regression or merely a result of CL regression remains to be determined.

In a recently published study using the bonnet monkey (*Macaca radiata*), PGF and chorionic gonadotropin (CG), the luteotropin in humans and subhuman primates, were found to have opposite effects on a number of genes that are thought to be critical for CL function (Priyanka et al. 2009). Not surprisingly, genes associated with steroid biosynthesis, including steroidogenic acute regulatory protein (*STAR*), *CYP11A1*, hydroxy-delta-5-steroid dehydrogenase, 3 beta- and steroid delta-isomerase 1 (*HSD3B1*), and *CYP19A1* were increased by CG and decreased by PGF. This analysis also showed that the monkey CL contained all the necessary enzymes for *de novo* cholesterol biosynthesis and that these enzymes were increased by CG (Priyankda et al. 2009). The authors suggested that the effect of PGF was to interrupt LH signaling downstream of receptor binding. This is supported by the fact that CL regression in monkeys occurs without noticeable changes in LH secretion. In addition, PGF treatment also reduced expression of the LH/choriogonadotropin receptor (*LHCGR*). Bogan and coworkers (2008) used genomic approaches to identify differentially expressed genes at different stages of CL development in the rhesus macaque CL. They identified changes in gene families associated with immune function; hormone and growth factor signaling; steroidogenesis; and prostaglandin biosynthesis, metabolism, and signaling. They divided CL developmental stages in early (3–5), mid (7–8), mid-late (10–12), late (14–16), and very late (18–19) after the mid-cycle LH surge. Using an Affymetrix® (Affymetrix, Santa Clara, CA) rhesus macaque genome DNA microarray, they identified 3234 differentially expressed genes (1418 up and 1816 down; Bogan et al. 2008). Early in CL development, most transcripts were increased, whereas later in development the converse was true. One of the strengths of this study was their extensive validation of microarray results using both quantitative polymerase chain reaction (qPCR) and protein quantification. Not surprisingly, genes involved in immune function were expressed at highest concentrations in the late and very late stage CL, supporting a role for the immune system in CL regression. Bishop and coworkers (2009) used the macaque array to examine the effects of LH and progesterone on CL gene expression. They identified nearly 1500 LH-regulated genes in the macaque CL, with less than one-third of these being affected by steroid ablation and progestin replacement. Nevertheless, several genes in the canonical steroid biosynthetic pathway were affected similarly by LH or steroid withdrawal and replacement including *STAR*, sterol-C4-methyl oxidase-like (*SC4MOL*), and *CYP19A1* (Bishop et al. 2009). Results from these genomic studies fit with the concept that the primate conceptus rescues the CL using a different mechanism from that of the farm species. Namely, CG mimics LH function and abrogates the luteolytic effects of PGF produced in the ovary. In addition, these studies provide further support for the concept that intraluteal prostaglandin E (PGE) and PGF are important regulators of CL function in primates.

10.3 Physiological genomics of blocking luteal regression

10.3.1 Conceptus signals

In the early part of the 20th century, Loeb was the first to suggest and later demonstrate that the uterus could influence CL function in guinea pigs (Short 1967). Later research by Moor and Rowson in the 1960s showed that products of the ovine conceptus were responsible for blocking luteal regression (Short 1967). This set off a flurry of work to identify factors produced by the conceptus that could extend the lifespan of the CL (see McCracken et al. 1999). This early work ruled out a systemic effect of the conceptus on CL lifespan and found that the conceptus must be present in the uterus by day 12 to extend CL function in sheep (Bazer and Roberts 1983; Spencer and Bazer 2004). This was followed by studies in cattle that showed that the critical period was 2 to 3 days later. Similar work in pigs established that conceptus signaling commenced around days 11 to 12. These studies were important for two reasons. First, they narrowed the search for the conceptus signal responsible for blocking luteal regression to a defined window of early pregnancy. Second, they demonstrated that there were no conceptus signals prior to this period that were absolutely required for pregnancy establishment. This time of early pregnancy, when the conceptus must signal the dam to extend CL function, was termed the period of maternal recognition of pregnancy by R.V. Short (1967), a designation that persists today. Work conducted during the late 1960s and early 1970s by a number of laboratories identified the luteolytic signal as PGF. A number of studies, most notably those from the McCracken laboratory, showed that the uterus of sheep and cattle produced pulses of PGF around the time of luteal regression and that these pulses were absent in early pregnant animals (McCracken et al. 1999).

Once the luteolytic signal and the critical window for the conceptus to block this signal were defined, work began to determine the nature of the conceptus factor(s) responsible. Early work on effects of the conceptus on endometrial function utilized conceptus and endometrial tissue cultured in the presence of radiolabeled amino acids to identify proteins that were produced during the period of maternal recognition of pregnancy (see Bazer and Spencer 2006). This approach allowed identification of *de novo* synthesized proteins following two-dimensional gel electrophoresis and Western blotting, which are referred to today as proteomic techniques. These studies revealed a group of low-molecular-weight proteins produced by the conceptus that, when introduced into the uterine lumen in their purified form, extended CL lifespan in cattle, sheep, and goats (Bazer 1992). These proteins were later cloned and determined to be members of the interferon (*IFN*) family of genes that were most closely related to the Type I IFN omega family (see Roberts et al. 2008). Conceptus IFNT was shown to alter the pattern of PGF release and abolish the large pulses of PGF that were, by then, known to mediate CL regression. Other studies also suggested that there was an increase in the PGE : PGF ratio that might also contribute to the maintenance of CL function (Arosh et al.

2009), although the significance of PGE in luteal maintenance is not clear because hysterectomy (and removal of any uterine PGE) results in extension of CL lifespan in sheep (Kiracofe and Spies 1966). In either case, there is now evidence for significant increases in endometrial and luteal PGE production during early pregnancy in ruminants (Asselin and Fortier 2000) and swine (Waclawik and Ziecik 2007). Part of this increase in PGE is due to the activity of 9 keto reductase (20-beta hydroxysteroid dehydrognease or carbonyl reductase [*CBR*]), which can convert PGF to PGE (Waclawik and Ziecik 2007).

The conceptus signals mediating luteal maintenance in pigs differ from those in ruminants in two important ways. The first was the effect of the conceptus on uterine PGF production. Bazer and Thatcher (1977) first proposed the endocrine–exocrine theory for CL rescue in swine, which established that in cyclic pigs PGF was released toward the uterine vasculature (endocrine secretion) and that the conceptus changed the direction of secretion toward the uterine lumen (exocrine secretion). This hypothesis was built upon Bazer and Thatcher's observations that a purple iron transport protein, uteroferrin, was secreted into the uterine lumen during early pregnancy but toward the uterine vasculature in cyclic pigs. Discussion of this phenomenon with a colleague and eminent fetal physiologist, Dr. Donald Barron, formed the initial concept of the endocrine–exocrine theory (F.W. Bazer, personal communication). The second major difference was that the conceptus signal mediating this effect was the steroid estrogen and not an IFN as in ruminants. Work from the Bazer lab was the first to show that there were two periods of conceptus estrogen production, around day 12 and again at day 15 after mating, which were required to extend CL function for a period close to the length of gestation in swine (114 days; Geisert et al. 1982a).

Our understanding of the effects of the conceptus on luteal maintenance in horses lags well behind that for ruminants and swine. Attempts to establish roles for equine conceptus steroid or protein hormones have not been met with much success (Sharp et al. 1989). Although the horse conceptus does produce steroid and protein hormones, these have not been shown to be responsible for alterations in uterine prostaglandin production in horses (Bettridge 2007). Interestingly, recent evidence suggests that the horse conceptus produces a unique Type I IFN, IFN delta (*IFND*), similar to that produced by the pig conceptus (Cochet et al. 2009). Whether this IFN induces changes in the endometrium associated with maintaining CL function remains to be determined. The most compelling evidence for a signal mediating CL rescue in horses comes from studies showing that rapid and sustained conceptus migration between uterine horns, which occurs between days 12 and 17 after mating, was critical for CL rescue. If the conceptus was confined to the tip of one uterine horn, it could not rescue CL function (Sharp et al. 1989). Furthermore, introduction of a glass ball (~30 mm) into the uterine lumen of cyclic mares inhibited estrous behavior and extended progesterone production in about 40% of mares (Nie et al. 2001), suggesting that physical contact of the endometrium had the ability to alter the luteolytic mechanism.

One can summarize from the studies just mentioned that ruminants utilize IFNT and swine estrogen as the initial conceptus signals responsible for blocking luteal regression. Both of these conceptus signals alter PGF release by the endometrium to maintain CL function. In addition, introduction of these hormones into the uterus at the appropriate times will extend CL function. Horse conceptus signaling involves embryo migration between uterine horns, and, undoubtedly, some conceptus produced factors that have yet to be determined.

10.3.2 Uterine responses to conceptus signals

Ruminants

A number of genomic studies of the physiological responses to conceptus signaling have been conducted in recent years (reviewed by Spencer et al. 2008). These include studies examining the effects of pregnancy and progesterone on endometrial gene expression in sheep (Spencer et al. 1999; Gray et al. 2006; Satterfield et al. 2010); pregnancy on endometrial gene expression in cattle (Klein et al. 2006; Bauersachs et al. 2006, 2007, 2008, 2009; Forde et al. 2009); pregnancy on gene expression profiles in the caruncular and intercaruncular endometrium of cattle (Mansouri-Attia et al. 2009); and pregnancy on gene expression in porcine endometrium (Ka et al. 2009). In addition, Chen and coworkers (2007) examined the effects of IFNT on gene expression in an immortalized ovine luminal epithelial cell line, and Kim and coworkers (2003), in human cell lines. Of these, the work by Gray and coworkers (2006), Klein and coworkers (2006), Forde and coworkers (2009), and Ka and coworkers (2009) are most relevant to the present discussion on physiological genomics of CL rescue because endometrial tissues analyzed in these studies were from the early period of the window of pregnancy recognition signaling (e.g., day 14 for sheep, day 18 for cattle, days 13 and 16 for cattle, and day 12 for swine, respectively). The work by Mansouri-Attia and coworkers (2009) focused on day 20 of pregnancy in cattle, approximately 5 days after pregnancy signaling commences. In addition, there are a large number of additional studies that have been conducted using candidate gene approaches that grew out of discoveries using proteomic analysis of endometrial- and conceptus-conditioned culture media (Wolf et al. 2003; Spencer and Bazer 2004). Of course, early studies in sheep and cattle focused on known IFN-stimulated genes (ISGs), and later work focused on the interaction between progesterone and IFNT in regulating uterine gene expression. This work was recently reviewed by Spencer and coworkers (2007, 2008) and Bazer and coworkers (2008) and summarized in Figure 10.1.

The difficulty with global gene expression profiling studies is teasing out endometrial responses to the conceptus that are involved in CL rescue from the responses mediating conceptus growth, attachment, elongation, and placentation, which all occur in response to continued conceptus signaling and maintenance of progesterone production. Clearly, there are dramatic changes in gene expression associated with induction of uterine secretion, remodeling, and immune accommodation at the fetal–maternal interface. Relevant to the present discussion, however, oxytocin receptor (*OXTR*) and estrogen receptor alpha (*ESR1*) expression are both reduced in pregnant or IFNT-treated endometrium compared with cyclic endometrium. In the absence of *OXTR* and *ESR1*, the endometrium does not respond to oxytocin (OT) from the CL and posterior pituitary and cannot produce the pulsatile pattern of PGF that induces luteal regression (Spencer and Bazer 2004). Spencer and Bazer (2004) developed a model for conceptus-mediated CL rescue in ruminants. In this model IFNT, binds the Type I IFN receptor on the endometrial luminal epithelium and activates a signaling pathway resulting in

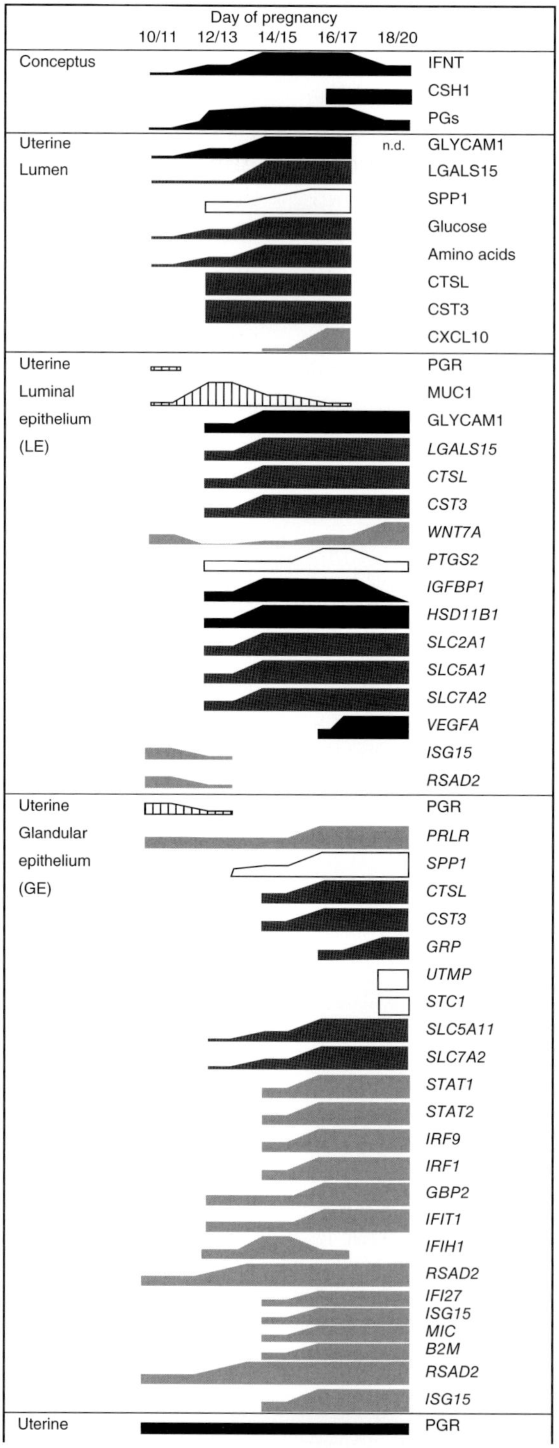

Figure 10.1 Temporal and spatial roadmap of progesterone and IFNT stimulated genes in the uterus during establishment of pregnancy in sheep. The relative abundance of mRNA or protein in the uterus across days of pregnancy is indicated as well as regulation of genes by progesterone and/or IFNT. ▥ = downregulated by progesterone; □ = induced or stimulated by progesterone; ■ =induced or stimulated by IFNT; ■ = induced by progesterone and stimulated by IFNT; n.d. = not determined due to inability to flush intact conceptuses from the uterine lumen on days 18–20 of pregnancy.

inhibition of *ESR1* gene transcription, which abrogates *OXTR* expression and thus production of luteolytic pulses of PGF. Available data support the idea that IFN regulatory factor two (*IRF2*), a potent transcriptional repressor, is involved in IFNT inhibition of *ESR1* gene transcription. The luminal epithelia of the endometrium is responsible for production of the bulk of PGF during luteal regression. Interestingly, PTGS2 (prostaglandin-endoperoxide synthase 2 [prostaglandin G/H synthase and cyclooxygenase]), the enzyme responsible for producing the precursor to PGF, PGG2, is not suppressed by the conceptus or IFNT. *PTGS2* is expressed during early pregnancy in the ovine endometrium as well as conceptus trophectoderm and is postulated to mediate production of prostaglandins that are critical for uterine receptivity and conceptus survival in rodents, including PGE2 and PGI2 (Wang and Dey 2005).

In the sheep endometrium, the transcriptional repressor *IRF2* is specifically and constitutively expressed in the endometrial luminal epithelia and increases during early pregnancy. IRF2 is a potent repressor of IFN-stimulated gene transcription (see Spencer et al. 2007; Bazer et al. 2009). Therefore, the signaling components that mediate induction of classical ISGs are apparently suppressed in the endometrial luminal epithelium but remain functional in the glandular epithelium and stroma. Although the receptors for IFNT (interferon alpha subunit receptors one and two [IFNAR1 and IFNAR2]) are most abundant on the endometrial epithelia, the transcription factors that govern classical Type I IFN responses in many different cell types (signal transducer and activator of transcription one and two [STAT1, STAT2] and IFN regulatory factor nine [IRF9], which forms the ISGF3 complex) are not present, most probably due to the potent IRF2 repressor that is specifically expressed in the endometrial lumenal and superficial glandular epithelium and increased during early pregnancy. Thus, the canonical JAK-STAT pathway mediating the effects of IFNT in the stroma and glandular epithelium is not active in the endometrial lumenal epithelium of pregnancy. Consequently, most classical ISGs (interferon-stimulated gene 15 [*ISG15*], beta-2-microglobulin [*B2M*], radical S-adenosyl-containing domain two [*RSAD2*], etc.) are not expressed or induced by IFNT in the endometrial lumenal epithelium of early pregnancy, which may have implications in maternal tolerance of the conceptus allograft (Johnson et al. 2002; Choi et al. 2003; Song et al. 2007). There is at least one classical ISG, *MX1*, that does not comply with this model and is induced the endometrial lumenal epithelium of pregnant sheep (Ott et al. 1998; Johnson et al. 2002; Hicks et al. 2003) and cattle (Mansouri-Attia et al. 2009). The mechanism of induction of *MX1* (Assiri et al. 2007), as well as many nonclassical IFNT-stimulated genes in the endometrial luminal epithelium, remains to be determined but is postulated to involve a unique, STAT1-independent signaling pathway (Spencer and Bazer 2004; Bazer et al. 2009).

Available data support the idea that ovarian progesterone and conceptus IFNT act synergistically on the endometrial epithelia of the ruminant uterus to regulate genes important for conceptus development and production of IFNT as well as uterine receptivity conceptus implantation. Expression of many endometrial lumenal epithelium genes is initiated between days 10 and 12 post-estrus/mating in both cyclic and pregnant ewes (Figure 10.1). Hormone replacement studies in sheep found that P4 induces the expression of many genes in the endometrial lumenal epithelium that encode adhesion proteins (galectin-15 [*LGALS15*], insulin-like growth factor binding protein one [*IGFBP1*]), enzymes (*PTGS2*), a protease (cathepsin L [*CTSL1*]), a protease inhibitor (cystatin C [*CST3*]), cell proliferation factors (gastrin-releasing peptide [*GRP*]), glucose transporters (*SLC2A1, SLC5A1*), and cationic amino acid (arginine, lysine, and ornithine) transporter (*SLC7A2*). In the glandular epithelium, P4 induces genes that encode a cell proliferation factor (*GRP*), a glucose transporter (*SLC5A11*), adhesion protein (secreted phosphoprotein one [*SPP1*]), regulator of calcium/phosphate homeostasis (stanniocalcin one [*STC1*]), and an immunomodulatory factor (uterine milk protein [*UTMP*]). Of particular note, IFNT stimulates a number of those P4-induced genes that encode secreted proteins (including *CST3, CTSL1, GRP, LGALS15*) as well as transporters for glucose (*SLC2A1, SLC5A11*) and amino acids (*SLC7A1, SLC7A2*). IFNT stimulation of most of these genes requires P4 action. The combinatorial effects of P4 and IFNT on the endometrium are hypothesized to result in specific changes in the intrauterine milieu necessary for conceptus elongation and development (see Spencer et al. 2008).

Swine

Swine conceptuses produce estrogen between days 11 and 15 after mating which is responsible for changing the direction of endometrial PGF secretion towards the uterine lumen and away from the uterine vasculature (Bazer and Thatcher 1977; Bazer et al. 1998). In addition, the conceptus also produces significant amounts of IFN gamma (IFNG; LeFevre et al. 1990; Murphy et al. 2009) and delta (Lefevre and Boulay 1993), as well as interleukin 1, beta (IL1B; Ross et al. 2003), during the initial period of conceptus elongation. For example, there is a several hundred fold increase in *IFNG* mRNA as the conceptus develops from its spherical to its filamentous forms between days 10 and 14 (Ross et al. 2009). It is clear that pig conceptus IFN is affecting endometrial function via induction of *STAT1* (Joyce et al. 2007); however, it does not appear that this IFN plays a role in blocking luteal regression.

Green and coworkers (2006) used a custom cDNA microarray to identify 4827 genes that were differentially expressed in the porcine endometrium across the estrous cycle. This report also presents an excellent overview of the technologies, databases, and challenges associated with transcriptional profiling experiments. Although the experiments examined changes in gene expression at days 0, 3, 6, 10, 12, 14, and 18, days 10 and 12 are most relevant to the present discussion because they represent the transcriptome of the endometrium at the time when conceptus signals are first received (http://genome.rnet.missouri.edu/swine). Interestingly, day 12 endometrium exhibited the highest number of differentially expressed genes (542) compared with the other days. The largest proportion of differentially expressed genes from days 10–14 were associated with signal transduction, particularly those associated with receptor tyrosine kinase activity (Green et al. 2006). This is in contrast to differentially expressed genes from days 0 and 18, which clustered predominantly in the immune function theme. This experiment is an excellent example of how global gene expression profiling can be used to survey the physiological "landscape" at the time when the conceptus is first signaling the endometrium to rescue CL function.

Pregnancy recognition signaling in swine coincides with conceptus and endometrial estrogen production that is responsible for the endocrine–exocrine switch in endometrial PGF secretion (Perry et al. 1973; Bazer and Thatcher 1977; Bazer et al. 1998; Tayade et al. 2007; Franczak and Bogacki 2009). It is now clear that although PGF and other prostaglandins are detrimental to CL function when secreted into the uterine vasculature, they are critical to successful conceptus growth, attachment, and placentation (Ashworth et al. 2006; Waclawik et al. 2006). Although the functioning of the endocrine–exocrine switch has been known for some time (Gross et al. 1988), the actual genes and physiological pathways mediating this effect have not been established. Moreover, few studies have attempted to define the transcriptome of the pregnant endometrium in swine in response to conceptus estrogen (Jiang et al. 2003; Ka et al. 2009).

Candidate gene approaches have defined a critical role for OT in inducing uterine PGF production during luteolysis in swine (Carnahan et al. 1996). Regulation appears to occur both at the level of OXTR number (Ludwig et al. 1998; Franczak et al. 2005; Oponowicz et al. 2006) and OXTR coupling to its second messenger system in the endometrium (Ludwig et al. 1998). These changes occur on a backdrop of elevated PGFS and endometrial capacity to produce PGF. The effects of conceptus estrogen on pulsatile release of endometrial PGF likely involves changes in OXTR numbers that are mediated via the nuclear ESR1 (Franczak and Bogacki 2009). Recently, however, there has been evidence that estrogen may also act via a membrane-bound ESR1 through activation of Akt to alter translation initiation in porcine endometrial cells (Wollenhaupt et al. 2007).

Recently, Ka and coworkers (2009) utilized a technique called annealing control primer-based reverse transcription PCR (ACP RT-PCR; Hwang et al. 2003) to identify differentially expressed genes between endometrium collected from cyclic or pregnant pigs at day 12 after estrus. Day 12 represents the earliest period that conceptus estrogen-mediated responses could be detected and changes in gene expression are likely involved in mediating the endocrine–exocrine switch in pigs. It was interesting that three of the conceptus-induced endometrial genes identified using this approach, *S100A7A* (S100 calcium binding protein A7A), *GSN* (gelsolin [amyloidosis, Finnish type]), and *TRPV6* (transient receptor potential cation channel, subfamily V, member 6), are involved in calcium regulation (Eckert et al. 2004; Sun et al. 1999; Hoenderop et al. 2005). Prior work has shown that calcium increases in uterine secretions at the time of conceptus elongation (Geisert et al. 1982a,b) and that the calcium ionophore A23187 is able to activate the endocrine–exocrine switch (Gross et al. 1990). Whether any of these genes are involved in the endocrine–exocrine switch remains to be determined.

Horse

It has been over two years since the first assembly of the horse genome was published (see Ramery et al. 2009). The horse genome has now been sequenced at close to seven times coverage, which is similar to that available for mice, rats, and dogs. However, the horse genome lags behind these species in terms of annotation. To date, there have been no genomic or proteomic studies focused on the conceptus signals mediating CL rescue in the mare. What is available stems from early "proteomic" studies that examined the array of horse conceptus secretory products and their effects on CL rescue in mares (see Sharp et al. 1989; Allen 2001).

The physiological genomics of CL rescue in the mare remains enigmatic. Neither conceptus estrogen nor proteins have been determined to rescue CL function when introduced into the uterine lumen of the mare. Clearly, however, the equine uterus exhibits reduced responsiveness to OT and lowered PGF production during early pregnancy (Sharp et al. 1989; Starbuck et al. 1998). Most intriguing is the fact that the horse conceptus is propelled rapidly between uterine horns between days 10 and 14 after mating and that this movement is critical for blocking luteolytic production of PGF by the endometrium (Allen 2001). This movement is clearly affected by conceptus production of prostaglandins (Stout et al. 2001; 2002); however, plastic spheres introduced into the uterine lumen are also propelled between uterine horns (although to a lesser extent) and extended CL function in approximately 75% of mares (Rivera del Alamo et al. 2008). The current evidence suggests that either some factor on the surface of the equine capsule or present in low concentrations adjacent to the conceptus mediates this response, or the physical contact of the equine capsule (or glass bead) is responsible for altering uterine PGF production (Rivera del Alamo et al. 2008). Both in the area of conceptus signals and in the area of endometrial responses to those signals, the physiological genomics of CL rescue is certainly ripe for investigation in equids.

10.4 Future research directions

Although global transcriptional profiling has been widely used for less than a decade in livestock species, it has greatly expanded the known universe of genes participating in conceptus–endometrial interactions that meditate rescue of CL function in domestic farm animals. In many cases these studies have confirmed previous work utilizing candidate gene approaches. However, there still remains quite a bit of validation work that must take place to confirm genes identified using global screening methods. In addition, proteomic technologies have matured and been automated to allow cataloging of proteins that regulate critical cellular processes. As the robustness and sensitivity of these techniques improves, they will provide new opportunities for evaluating the transcriptional and translational control of gene expression in the endometrium, conceptus, and CL. With publication of human and animal genomes and their more complete annotation, newer bioinformatic and statistical tools are allowing these genes and their protein products to be organized into families for evaluation of pathways activated in the endometrium and CL by conceptus signals. However, we are still well away from a complete understanding of the process. What these technologies have provided in essence is the cast of characters for a story that is not near fully written. Future advances will rely on the painstaking determination of how these characters interact during the process of establishment of pregnancy. This will allow determination of the etiology of pregnancy failure, and guide attempts to regulate reproductive processes to improve efficiency of animal agriculture. This holistic or systems approach to reproductive biology represents a great advance over the incremental approaches of the past (Hiendleder et al. 2005). Only by taking this approach can the physiological genomics of conceptus–endometrial interactions be understood and, even more importantly, be manipulated to improve animal agriculture.

Acknowledgments

The research reported here was supported in part by National Research Initiative Competitive Grant No. 2005-35203-16252 from the USDA Cooperative State Research, Education, and Extension Service to TES, and National Research Initiative Competitive Grant No. 2000-02398 from the USDA Cooperative State Research, Education, and Extension Service to TLO. Thanks to Ms. Shannon Boone for help preparing this chapter.

References

Allen, W.R. 2001. Fetomaternal interactions and influences during equine pregnancy. *Reproduction* **121**: 513–527.

Arosh, J.A., Banu, S.K., Kimmins, S., Chapdelaine, P.K., MacLaren, L.A., and Fortier, M.A. 2009. Effect of interferon-tau on prostaglandin biosynthesis, transport, and signaling at the time of maternal recognition of pregnancy in cattle: Evidence of polycrine actions on prostaglandin E2. *Endocrinology* **145**: 5280–5293.

Ashworth, M.D., Ross, J.W., Hu, J., White, F., Stein, D.R., DeDilva, U., Johnson, G.A., Spencer, T.E., and Geisert, R.D. 2006. Expression of porcine endometrial prostaglandin synthase during the estrous cycle and early pregnancy, and following endocrine disruption of pregnancy. *Biology of Reproduction* **74**: 1007–1015.

Asselin, E. and Fortier, M.A. 2000. Detection and regulation of the messenger for a putative bovine endometrial 9-keto-prostaglandin E_2 reductase: Effect of oxytocin and interferon-tau. *Biology of Reproduction* **62**: 125–131.

Assiri, A. and Ott, T.L. 2007. Cloning and characterizing of the oMX1 promoter/enhancer region. *Developmental and Comparative Immunology* **31**: 847–857.

Bachelot, A. and Binart, N. 2005. Corpus luteum development: Lessons from genetic models in mice. *Current Topics in Developmental Biology* **68**: 49–85.

Bauersachs, S., Mitko, K., Blum, H., and Wolf, E. 2007. Technical note: Bovine oviduct and endometrium array version 1: A tailored tool for studying bovine endometrium biology and pathophysiology. *Journal of Dairy Science* **90**: 4420–4423.

Bauersachs, S., Mitko, K., Ulbrich, S.E., Blum, H., and Wolf, E. 2008. Transciptome studies of bovine endometrium reveal molecular profiles characteristic for specific stages of estrous cycle and early pregnancy. *Experimental and Clinical Endocrinology and Diabetes* **116**: 371–384.

Bauersachs, S., Ulbrich, S.E., Gross, K., Schmidt, S.E.M., Meyer, H.H.D., Wenigerkind, H., Vermehren, M., Sinowatz, F., Blum, H., and Wolf, E. 2006. Embryo-induced transcriptome changes in bovine endometrium reveal species-specific and common molecular markers of uterine receptivity. *Reproduction* **132**: 319–331.

Bauersachs, S., Ulbrich, S.E., Zakhartchenko, V., Minten, M., Reichenbach, M., Reichenbach, H., Blum, H., Spencer, T.E., Wolf, E. 2009. The endometrium responds differently to cloned versus fertilized embryos. *Proceedings of the National Academy of Sciences of the United States of America* **106**: 5681–5686.

Bazer, F.W. 1992. Mediators of maternal recognition of pregnancy in mammals. *Proceedings of the Society for Experimental Biology and Medicine* **199**: 373–384.

Bazer, F.W., Burghardt, R.C., Johnson, G.A., Spencer, T.E., and Wu, G. 2008. Interferons and progesterone for establishment and maintenance of pregnancy: Interactions among novel cell signaling pathways. *Reproductive Biology* **8**: 179–211.

Bazer, F.W., Ott, T.L., and Spencer, T.E. 1998. Maternal recognition of pregnancy: Comparative aspects. *Trophoblast Research* **12**: 375–386.

Bazer, F.W. and Roberts, R.M. 1983. Biochemical aspects of conceptus–endometrial interactions. *Journal of Experimental Zoology* **228**: 373–383.

Bazer, F.W. and Spencer, T.E. 2006. Methods for studying interferon tau stimulated genes. *Methods in Molecular Medicine* **122**: 367–380.

Bazer, F.W., Spencer, T.E., and Johnson, G.A. 2009. Interferons and uterine receptivity. *Seminars in Reproductive Medicine* **27**: 90–102.

Bazer, F.W., Spencer T.E., Ott, T.L., and Johnson G.A. 2008. Mediators of maternal recognition of pregnancy. In: Aplin, J.D., Fazleabas, A.T., and Glasser, S.R., Giudice, L.C. (eds.), *The Endometrium*, 2nd Edition. London: Informa Healthcare, pp. 268–285.

Bazer, F.W. and Thatcher, W.W. 1977. Theory of maternal recognition of pregnancy in swine based on estrogen controlled endocrine versus exocrine secretion of prostaglandin F2alpha by the uterine endometrium. *Prostaglandins* **14**: 397–400.

Betteridge, K.J. 2007. Equine embryology: An inventory of unanswered questions. *Theriogenology* **68**(Supplement 1): S9–S21.

Bishop, C.V., Hennebold, J.D., Stouffer, and R.L. 2009. The effects of luteinizing hormone ablation/replacement versus steroid ablation/replacement on gene expression in the primate corpus luteum. *Molecular Human Reproduction* **15**: 181–193.

Bogan, R.L., Murphy, M.J., Stouffer, R.L., Hennebold, J.D. 2008. Systematic determination of differential gene expression in the primate corpus luteum during the luteal phase of the menstrual cycle. *Molecular Endocrinology* **22**: 1260–1273.

Carnahan, K.G., Prince, B.C., and Mirando, M.A. 1996. Exogenous oxytocin stimulates uterine secretion of prostaglandin $F_{2\alpha}$ in cyclic and early pregnant swine. *Biology of Reproduction* **55**: 838–843.

Casey, O.M., Morris, D.G., Powell, R., Sreenan, J.M., and Fitzpatrick, R. 2005. Analysis of gene expression in non-regressed and regressed bovine corpus luteum tissue using a customized ovarian cDNA array. *Theriogenology* **64**: 1963–1976.

Chen, Y., Antoniou, E., Liu, Z., Hearne, L.B., and Roberts, R.M. 2007. A microarray analysis for genes regulated by interferon-tau in ovine luminal epithelial cells. *Reproduction* **134**: 123–135.

Choi, Y., Johnson, G.A., Spencer, T.E., and Bazer, F.W. 2003. Pregnancy and interferon tau regulate major histocompatibility complex class I and beta2-microglobulin expression in the ovine uterus. *Biology of Reproduction* **68**: 1703–1710.

Cochet, M., Vaiman, D., and Lefevre, F. 2009. Novel interferon delta genes in mammals: Cloning of one gene from the sheep, two genes expressed by the horse conceptus and discovery of related sequences in several taxa by genomic database screening. *Gene* **433**: 88–99.

Eckert, R.L., Broome, A.M., Ruse, M., Robinson, N., Ryan, D., and Lee K. 2004. S100 proteins in the epidermis. *Journal of Investigational Dermatology* **123**(1): 23–33.

Forde, N., Carter, F., Fair, T.M., Crowe, M.A., Evans, A.C.O., Spencer, T.E., Bazer, F.W., McBride, R., Boland, M.P., O'Goara, P., Lonergan, P., and Roche, J.F. 2009. Progesteron-regulated changes in endometrial gene expression contribute to advanced conceptus development in cattle. *Biology of Reproduction* **81**(4): 784–794.

Foyouzi, N., Cai, Z., Sugimoto, Y., and Stocco, C. 2005. Changes in the expression of steroidogenic and antioxidant genes in the mouse corpus luteum during luteolysis. *Biology of Reproduction* **72**: 1134–1141.

Franczak, A. and Bogacki, M. 2009. Local and systemic effects of embryos on uterine tissues during early pregnancy in pigs. *The Journal of Reproduction and Development* **55**: 262–272.

Franczak, A., Ciereszko, R., and Kotwica, G. 2005. Oxytocin (OT) action in uterine tissues of cyclic and early pregnant gilts: OT receptors concentration, prostaglandin F(2)alpha secretion, and phosphoinositide hydrolysis. *Animal Reproduction Science* **88**(3–4): 325–339.

Geisert, R.D., Renegar, R.H., Thatcher, W.W., Roberts, R.M., and Bazer, F.W. 1982b. Establishment of pregnancy in the pig: I. Interrelationships between preimplantation development of the pig blastocyst and uterine endometrial secretions. *Biology of Reproduction* **27**: 925–39.

Geisert, R.D., Thatcher, W.W., Roberts, R.M., andBazer, F.W. 1982a. Establishment of pregnancy in the pig: III. Endometrial secretory response to estradiol valerate administered on day 11 of the estrous cycle. *Biology of Reproduction* **27**(4): 957–65.

Goravanahally, M.P., Salem, M., Yao, J., Inskeep, E.K., and Flores, J.A. 2009. Differential gene expression in the bovine corpus luteum during transition from early phase to midphase and its potential role in acquisition of luteolytic sensitivity to prostaglandin F2 alpha. *Biology of Reproduction* **80**: 980–988.

Gray, C.A., Abbey, C.A., Beremand, P.D., Choi, Y., Farmer, J.L., Adelson, D.L., Thomas, T.L., Bazer, F.W., and Spencer, T.E. 2006. Identification of endometrial genes regulated by early pregnancy, progesterone, and interferon tau in the ovine uterus. *Biology of Reproduction* **74**: 383–394.

Green, J.A., Kim, J-G., Whitworth, K.M., Agca, C., and Prather, R.S. 2006. The use of microarrays to define functionally-related genes that are differentially expressed in the cycling pig uterus. *Society of Reproduction and Fertility Supplement* **62**: 163–176.

Gross, T.S., Lacroix M.C., Bazer F.W., Thatcher W.W., and Harney J.P. 1988. Prostaglandin secretion by perifused porcine endometrium: Further evidence for an endocrine versus exocrinesecretion of prostaglandins. *Prostaglandins* **35**: 327–41.

Gross, T.S., Mirando, M.A., Young, K.H., Beers, S., Bazer, F.W., and Thatcher, W.W. 1990. Reorientation of prostaglandin F secretion by calcium ionophore, estradiol, and prolactin in perifused porcine endometirum. *Endocrinology* **127**: 637–642.

Hiendleder S., Bauersachs S., Boulesteix A., Blum H., Arnold G.J., Fröhlich T., and Wolf E. 2005. Functional genomics: Tools for improving farm animal health and welfare. *Revue Scientifique et Technique* **24**(1): 355–77.

Hicks, B.A., Etter, S.J., Carnahan, K.G., Joyce, M.M., Assiri, A.A., Carling, S.J., Kodali, K., Johnson, G.A., Hansen, T.R., Mirando, M.A., Woods, G.L., Vanderwall, D.K., and Ott, T.L. 2003. Expression of the uterine Mx protein in cyclic and pregnant cows, gilts, and mares. *Journal of Animal Science* **81**: 1552–1561.

Hoenderop, J.G., Nilius B., Bindels R.J. 2005. Calcium absorption across epithelia. *Physiological Reviews* **85**(1): 373–422.

Hwang, I.T., Kim, Y.J., Kim, S.H., Kwak, C.I., Gu, Y.Y., Chun, J.Y. 2003. Annealing control primer system for improving specificity of PCR amplification. *Biotechniques* **35**(6): 1180–1184.

Jiang, Z., Zhang, M., Wasem, V.D., Michal, J.J., Zhang, H., Wright, W.W. 2003. Census of genes expressed in porcine embryos and reproductive tissues by mining an expressed sequence tag database based on human genes. *Biology of Reproduction* **69**: 1177–1182.

Johnson, G.A., Joyce, M.M., Yankey, S. J., Hansen, T.R., and Ott, T.L. 2002. The interferon stimulated gene (ISG) 17 and Mx have different temporal and spatial expression in the ovine uterus suggesting more complex regulation of the *Mx* gene. *The Journal of Endocinology* **174**: R7–R11.

Joyce, M.M., Burghardt, R.C., Geisert, R.D., Burghardt, J.R., Hooper, R.N., Ross, J.W., Ashworth, M.D., and Johnson, G.A. 2007. Pig conceptuses secrete estrogen and interferons to differentially regulate uterine STAT1 in a temporal and cell type-specific manner. *Endocrinology* **148**: 4420–4431.

Ka, H., Seo, H., Kim, M., Choi, Y., and Lee, C. 2009. Identification of differentially expressed genes in the uterine endometrium on day 12 of the estrous cycle and pregnancy in pigs. *Molecular Reproduction and Development* **76**: 75–84.

Kim, S., Choi, Y., Bazer, F.W., and Spencer, T.E. 2003. Identification of genes in the ovine endometrium regulated by interferon tau independent of signal transducer and activator of transcription 1. *Endocrinology* **144**: 5203–5214.

Kiracofe, G.H. and Spies, H.G. 1966. Length of maintenance of naturally formed and experimentally induced corpora lutea in hysterectomized ewes. *Journal of Reproduction and Fertility* **11**: 275–279.

Klein, C., Bauersachs, S., Ulbrich, S.E., Einspanier, R., Meyer, H.H.D., Schmidt, S.E.M., Reichenbach, H., Vermehren, M., Sinowatz, F., Blum, H., and Wolf, E. 2006. Monozygotic twin model reveals novel embryo-induced transcriptome changes of bovine endometrium in the preattachment period. *Biology of Reproduction* **74**: 253–264.

Lefevre, F. and Boulay, V. 1993. A novel and atypical type one interferon gene expressed by trophoblast during early pregnancy. *The Journal of Biological Chemistry* **268**: 19760–19768.

Lefevre, F., L'Haridon, R., Berras-Cuesta, F., and La bonnardiere, C. 1990. Production, purification and biological properties of an Escherichia coli-derived recombinant porcine alpha interferon. *The Journal of General Virology* **71**: 1057–1063.

Ludwig, T.E., Sun, B.C., Carnahan, K.G., Uzumcu, M., Yelich, J.V., Geisert, R.D., and Mirando M.A. 1998. Endometrial responsiveness to oxytocin during diestrus and early pregnancy in pigs is not controlled solely by changes in oxytocin receptor population density. *Biology of Reproduction* **58**(3): 769–77.

Mansouri-Attia, N., Aubert, J., Reinaud, P., Giraud-Delville, C., Taghouti, G., Galio, L., Everts, R.E., Degrelle, S., Richard, C., Hue, I., Yang, X., Tian, X. C., Lewin, H.A., Renard, J-P., and Sandra, O. 2009. Gene profiling of bovine endometrium at implantation. *Physiological Genomics* **39**(1): 14–27.

McCracken, J.A., Custer, E.E., and Lamsa, J.C. 1999. Luteolysis: A neuroendocrine-mediated event. *Physiological Reviews* **79**: 263–304.

Murphy, S.P., Tayade, C., Ashkar, A.A., Hatta, K., Zhang, J., and Croy, B.A. 2009. Interferon gamma in successful pregnancies. *Biology of Reproduction* **80**: 848–859.

Nelimarkka, L., Salminen, H., Kuopio, T., Nikkari, S., Ekfors, T., Laine, J., Pelliniemi, L., and Jarvelainen, H. 2001. Decorin is produced by capillary endothelial cells in inflammation-associated angiogenesis. *American Journal Pathology* **158**: 345–353.

Nie, G.J., Johnson, K.E., Braden, T.D., and Wenzel, J.G.W. 2001. Use of a glass ball to suppress behavioral estrus in mares. *Proceedings of the Annual Convention of the AEEP* **47**: 246–248.

Oponowicz, A., Franczak, A., Kurowicka, B., and Kotwica, G. 2006. Relative transcript abundance of oxytocin receptor gene in porcine uterus during luteolysis and early pregnancy. *Journal of Applied Genetics* **47**: 345–351.

Ott, T.L., Yin, J., Wiley, A.A., Kim, H.T., Gerami-Naini, B., Spencer, T.E., Bartol, F.F., Burghardt, R.C., and Bazer, F.W. 1998. Effects of the estrous cycle and early pregnancy on uterine expression of Mx protein in sheep (ovis aries). *Biology of Reproduction* **59**: 784–794.

Perry, J.S., Heap, R.B., and Amoroso, E.C. 1973. Steroid hormone production by pig blastocysts. *Nature* **245**(5419):45–47.

Priyanka, S., Jayaram, P., Sridaran, R., and Medhamurthy, R. 2009. Genome-wide gene expression analysis reveals a dynamic interplay between luteotropic and luteolytic factors in the regulation of corpus luteum function in the bonnet monkey (Macaca radiate). *Endocrinology* **150**: 1473–1484.

Ramery, E., Closset, R., Art, T., Bureau, F., and Lekeux, P. 2009. Expression microarrays in equine sciences. *Veterinary Immunology and Immunopathology* **127**: 197–202.

Rivera del Alamo, M.M., Reilas, T., Kindahl, H., and Katila, T. 2008. Mechanisms behind intrauterine device-induced luteal persistence in mares. *Animal Reproduction Science* **107**: 94–106.

Roberts, R.M., Chen, Y., Ezashi, T., and Walker, A.M. 2008. Interferons and the maternal-conceptus dialog in mammals. *Seminars in Cell and Developmental Biology* **19**: 170–177.

Ross, J.W., Malayer, J.R., Ritchey, J.W., and Geisert, R.D. 2003. Characterization of the interleukin-1B system during porcine trophoblastic elongation and early placental attachment. *Biology of Reproduction* **69**: 1251–1259.

Ross, J.W., Ashworth, M.D., Stein, D.R., Couture, O.P., Tuggle, C.K., and Geisert, R.D. 2009. Identification of differential gene expression during porcine conceptus rapid trophoblastic elongation and attachment to uterine luminal epithelium. *Physiological Genomics* **36**: 140–148.

Satterfield, M.C., Gao, H., Li, X., Wu, G., Johnson, G.A., Spencer, T.E., and Bazer, F.W. 2010. Select nutrients and their associated transporters are increased in the ovine uterus following early progesterone administration. *Biology of Reproduction* **82**(1): 224–231.

Sharp, D.C., McDowell, K.J., Weithenauer, J., Thatcher, W.W. 1989. The continuum of events leading to maternal recognition of pregnancy in mares. *Journal of Reproduction and Fertility Supplement* **37**: 101–107.

Short, R.V. 1967. Reproduction. *Annual Review of Physiology* **29**: 373–400.

Soares, M.J., Konno, T., and Alam, S.M.K. 2007. The prolactin family: Effectors of pregnancy-dependent adaptations. *Trends in Endocrinology and Metabolism* **18**: 114–121.

Song, G., Bazer, F.W., and Spencer, T.E. 2007. Pregnancy and interferon tau regulate RSAD2 and IFIH1 expression in the ovine uterus. *Reproduction* **133**: 285–295.

Spencer, T.E. and Bazer, F.W. 2004. Conceptus signals for establishment and maintenance of pregnancy. *Reproductive Biology and Endocrinology* **2**: 1–15.

Spencer, T.E., Johnson, G.A., Bazer, F.W., Burghardt, R.C., and Palmarini, M. 2007. Pregnancy recognition and conceptus implantation in domestic ruminants: Roles of progesterone, interferons and endogenous retroviruses. *Reproduction Fertility and Development* **19**: 65–78.

Spencer, T.E., Sandra, O., and Wolf, E. 2008. Genes involved in conceptus-endometrial interactions in ruminants: Insights from reductionism and thoughts on holistic approaches. *Reproduction* **135**: 165–179.

Spencer, T.E., Stagg, A.G., Ott, T.L., Johnson, G.A., Ramsey, W.S., and Bazer, F.W. 1999. Differential effects of intrauterine and subcutaneous administration of recombinant ovine interferon tau on the endometrium of cyclic ewes. *Biology of Reproduction* **61**: 464–470.

Sridaran, R., Smith, C.J., and Richards, J.S. 1991. Effects of in vivo dihydrotestosterone treatment on changes in nocturnal surge of prolactin, luteal ultrastructure and P-450 mRNA and protein content in pregnant rats. *Molecular and Cellular Endocrinology* 77: 75–83.

Starbuck, G.R., Stout, T.A.E., Lamming, G.E., Allen, W.R., and Flint, A.P.F. 1998. Endometrial oxytocin receptor and uterine prostaglandin secretion in mares during the oestrous cycle and early pregnancy. *Journal of Reproduction and Fertility* **113**: 173–179.

Stocco, C., Callegari, E., and Gibori, G. 2001. Opposite effect of prolactin and prostaglandin F2a on the expression of luteal genes as revealed by rat cDNA expression array. *Endocrinology* **142**: 4158–4161.

Stout, T.A.E. and Allen, W.R. 2001. Role of prostaglandins in intrauterine migration of the equine conceptus. *Reproduction* **121**: 771–775.

Stout, T.A.E. and Allen, W.R. 2002. Prostaglandin E2 and F2a production by equine conceptuses and concentrations in conceptus fluids and uterine flushings recovered from early pregnant and dioestrous mares. *Reproduction* **123**: 261–268.

Sun, H.Q., Yamamoto, M., Mejillano, M., and Yin, H.L. 1999. Gelsolin, a multifunctional actin regulatory protein. *Journal of Biological Chemistry* **274**(47): 33179–33182

Tayade, C., Fang, Y., and Croy, B.A. 2007. A review of gene expression in porcine endometrial lymphocytes, endothelium and trophoblast during pregnancy success and failure. *Journal of Reproduction and Development* **53**: 455–463.

Waclawik, A., Rivero-Muller, A., Blitek, A., Kaczmarek, M.M., Brokken, L.J.S., Watanabe, K., Rahman, N.A., and Ziecik, A.J. 2006. Molecular cloning and spatiotemporal expression of prostaglandin F synthase and microsomal prostaglandin E synthase-1 in porcine endometrium. *Endocrinology* **147**: 210–221.

Waclawik, A. and Ziecik, A.J. 2007. Differential expression of prostaglandin (PG) synthesis enzymes in conceptus during peri-implantation period and endometrial expression of carbonyl reductase/PG 9-ketoreductase in the pig. *Journal of Endocrinology* **194**: 499–510.

Wang, H. and Dey, S.K. 2005. Lipid signaling in embryo implantation. *Prostaglandins and Other Lipid Mediators* **77**(1–4): 84–102.

Wolf, E., Arnold, G.J., Bauersachs, S., Beier, H.M., Blum, H., Einspanier, R., Fohlich, T., Herrler, A., Hiendleder, S., Kolle, S., Prelle, K., Reichenbach, H-D., Stojkovic, M., Wenigerkind, H., and Sinowatz, F. 2003. Embryo-maternal communication in bovine—Strategies for deciphering a complex cross-talk. *Reproduction in Domestic Animals* **38**: 276–289.

Wollenhaupt, K., Brussow, K-P., Tiemann, U., and Tomek, W. 2007. The embryonic pregnancy signal oestradiol influences gene expression at the level of translational initiation in porcine endometrial cells. *Reproduction in Domestic Animals* **42**: 167–175.

11

Physiological Genomics of Placental Growth and Development

Sukanta Mondal

11.1 Introduction

The placenta (Greek, *plakuos*: flat cake) is a functional connection between the embryo and the uterus which meets the significant challenge of accommodating the nutritional and growth regulatory needs of the developing fetus. This feto-maternal organ begins at implantation of the blastocyst and is delivered with the fetus at birth. The placenta plays a critical role in (1) mediating implantation, (2) establishing the interface for nutrient and gas exchange between maternal and fetal circulation, (3) regulating maternal recognition of pregnancy, (4) altering the local immune environment, and (5) stimulating maternal cardiovascular and metabolic functions through production of paracrine and endocrine hormones. It produces a plethora of functional molecules, viz. prolactin (PRL), growth hormone (GH), placental lactogen (PL), insulin-like growth factors (IGFs), prolactin-related proteins (PRPs), and prolactin-like proteins (PLPs), which contribute to successful pregnancy and establishment of placenta.

The recent developments in molecular biology and biotechnology have resulted in unlimited access to the genome and have enhanced the pace and precision of creating gene sequences and functional genomics to meet the challenges of food, agriculture, and animal improvement. The development of new innovative technologies for sequencing of whole genomes and for the mass screening of transcriptomes has revolutionized genetic profiling, mapping as well as our understanding of underlying physiological mechanisms. These molecular technologies have provided new ways of evaluating reproductive potential and the basic physiological mechanisms that limit reproductive performance. These technologies will also provide new tools for managing and monitoring livestock fertility. Therefore, understanding of genetic mechanisms and pathways involved in plancetal growth and function has opened new vistas for improving livestock. Furthermore, these genomic and transcriptomic approaches are helpful in functional genomics and have led to newer ways of evaluating the genetic components of phenotypes in livestock species. Advance biotechnological approaches such as microarray, siRNA, and bioinformatics tools have tremendous importance in understanding the complex mechanisms of poor reproductive efficiency, which could lead to safe and effective gene-based strategies for enhancement of reproduction and production.

11.2 Placental development: Basics

11.2.1 The origin of placenta

After fertilization, the next major event is trophoblast differentiation, which is required for implantation. For the first 4 to 6 days, preimplantation development takes place within the oviduct. During this period the zygote undergoes cleavage division and differentiation of innermost cells into the inner cell mass and the surrounding cells into trophoectoderm, which occurs around the 16-cell stage. The inner cell mass subsequently develops into the fetus and the trophoectoderm gives rise to the placenta. This process differs somewhat between species but also shares substantial similarities across a broad range of species. For example, in humans, after the trophoblast attaches to the endometrium, the embryo invades the endometrium through differentiation of the trophoectoderm into cytotrophoblast (inner layer) and syncytiotrophoblast (outer layer). The syncytiotrophoblast cells migrate and ultimately line the maternal spiral arteries opening them and causing maternal blood to flow across the cytotrophoblast. The cytotrophoblast forms columns of cells (villi) that invade the endometrium and anchor to the maternal decidua. The trophoblast cells produce vascular endothelial growth factor (VEGF), placenta-derived growth factor (PDGF), and fibroblast growth factor (FGF), which augment angiogenesis and placental development.

11.2.2 Primary cell types of placenta

In general, placentas of various species consist of two primary cell populations: an outer epithelial layer derived from the trophoectoderm (trophoblast), and inner vascular and stromal layers derived from the allantois (extra embryonic mesoderm). Trophoblast cells are one of the earliest differentiating cells and show species to species variation in their development and organization into placental structure. The trophoblast layer generates the extensive area for nutrient exchange as well as interacting closely with the uterus to produce a plethora of macromolecules that promote maternal blood flow to the implantation site. The trophoblast layer is covered by extensive microvilli, which spread diffusely across the placenta in pigs and are clustered into microcotyledons in horses or macro-cotyledons in ruminants. Rodents and primates possess a hemochorial placenta in which maternal blood directly bathes fetal chorionic villi cells. In ruminants, the separation between maternal and fetal blood is more extensive. However, ruminants do form transient synepitheliochorial placentas. The maternal endometrial epithelium eventually regrows, and there is minimal invasion and maximum cellular separation between maternal and fetal compartments.

11.2.3 Placenta—An endocrine organ

The placenta is directly responsible for mediating and/or modulating the maternal environment necessary for fetal growth and development. As an active endocrine organ, the placenta is capable of secreting a plethora of hormones, growth factors, cell adhesion molecules, extracellular matrix metalloproteinase, and cytokines, which play crucial roles in implantation and placentation. Although several transcription factors are involved in placental development, the exact role of specific transcription factors is unclear. Considerable evidence indicates that basic helix-loop-helix (bHLH) transcription factors are involved in placental trophoblast cell development. Hand1, a bHLH transcription factor, is essential for the differentiation of trophoblast giant cells. The expression of Hand1 mRNA is not detectable at early postimplantation stages and highly expressed in the trophoblast giant cell layer surrounding the implanted conceptus (Cross et al. 1995; Scott et al. 2000). Firulli et al. (1998) observed that mouse embryos that are homozygous for a Hand1 null mutation did not survive beyond day 8. The outer layer of trophoblast cells in Hand1 knockout mice failed to undergo the characteristic morphological giant cell appearance (Riley et al. 1998). Mash 2, another bHLH gene, is required for the maintenance of giant cell precursors (Guillemot et al. 1994), and its overexpression in Rcho-1 cells prevented giant cell differentiation. In contrast, formation of syncytiotrophoblast cells in mice is controlled by a distinct genetic pathway that is governed by glial cell missing-1 (GCM-1). In humans, GCM-1 was shown to regulate the activity of the syncytiotrophoblast aromatase gene (*CYP19*). Anson-Cartwright et al. (2000) reported that the labyrinth layer of the placenta does not form in GCM-1 null mutants and embryonic death occurs at day 10. The zinc finger proteins GATA-2 and GATA-3 are expressed in trophoblast giant cells, and their disruption caused arrested development at midgestation and depressed mouse placental lactogen-I (Prl3d1) gene expression (Ma et al. 1997). In primates, the *CHS* genes (PL homologs) are derivatives of the ancestral growth hormone gene, whereas in rodents and ruminants, the PL genes are derived from the ancestral prolactin (*PRL*) gene. In rodents, Prl3d1 gene is specifically expressed in trophoblast giant cells and Prl3d1 mRNA is reduced in Hand1 mutants (Firulli et al. 1998). Co-transfection of Hand1 with a Prl3d1 promoter reporter gene construct results in dose-dependent transactivation (Cross et al. 1995). Deletion of an 86-bp region of the promoter (between -274 and -88 relative to the transcription start site) prevents transactivation by Hand1, suggesting that Hand1 regulates Prl3d1 gene promoter activity.

11.3 Placental hormones and peptides

11.3.1. PRL

Prolactin plays a crucial role in placental growth and development, mammary gland development, and immune responses. The members of the PRL family of genes include PLs, PLPs, proliferins (PLF), and PLF-related proteins (PLF-RP). The PRL and GH genes are closely related and evolved from a common ancestral gene. The PRL family genes in human, rat, mouse, and cow are located in chromosomes 6, 17, 13, and 22, respectively. Two exon–intron organizations have been described; (1) five exon–four intron structure for both PRL and other members of PRL families, and (2) a six exon–five intron structure for members of the rodent PLP-C subfamily. The members of the PRL family possess four conserved cysteine residues

(Nicoll et al. 1986). However, placental lactogen-I (PL-I), PL-I variants, (PL-Iv), and prolactin-like protein-A (PLP-A) possess a fifth cysteine (Cohick et al. 1996), and PRL, PLF, and members of the PLP-C family has six cysteine residues (Roby et al. 1993). The PRL family members possess post-translational addition of carbohydrates, except PL-II and PLP-Cv. Glycosylation is an important post-translational modification in eukaryotic cells which influences the structure and biological function of proteins, including effects on protein stability, protein secretion, protein half-life, receptor interaction, and subsequent downstream biological activities. PRL family members expressed by spongiotrophoblast cell possess distinct glycosylation patterns that involve Asn-linked oligosaccharides containing both N-acetyl galactosamine (GalNAC) and sialic acid. There are differences in glycosylation patterns between mouse and rat PRL. There are two putative N-linked glycosylation sites that correspond to two glycoproteins of 29 and 33 kDa in rat, whereas mice possess a single putative N-linked glycosylation site that corresponds to a 29 kDa protein species.

The members of the PRL family are expressed in cell type-, location- and temporal-specific patterns in the uteroplacental compartment and the anterior pituitary. Rodent trophoblast giant cells, spongiotrophoblast cells, and invasive trophoblast cells each produce a unique subset of PRL family members. Transcriptional control of trophoblast giant cell-specific gene expression has been studied using the Rcho-I trophoblast cell line (Lu et al. 1994; Peters et al. 2000). During the last week of gestation, a population of trophoblast cells exits the chorioallantoic placenta and invades the uterine mesometrial compartment. Here they express a subset of PRL family members (Ain et al. 2003; Wiemers et al. 2003). In mice and rats, at least 19 different genes with some similarity to PRL, such as PL, PLPs, PRPs, PLF, and PLF-RP, have been identified. In the ruminant placenta, unlike GH, no PRL activity has been reported. However, PRP genes are expressed in binucleate cells of cow and sheep placentae (Anthony et al. 1995). In cattle and sheep, binucleate trophoblast cells and endometrial heterokaryons (fusion of binucleate trophoblast cells with endometrial epithelial cells) produce PLs. In cattle, PRP-1 is expressed in the placenta during the early peri-implantation period and before bPL can be detected (Yamada et al. 2002).

The members of the PRL family exhibit two types of biological functions: classical and nonclassical. Classical actions involve biological effects mediated through PRL and/or GH receptors, whereas nonclassical actions represent the mechanisms of ligand-mediated biological activities. The cellular targets for nonclassical members of the PRL family include endothelial cells (angiogenesis), erythrocyte and megakaryocyte precursors (erythropoiesis), natural killer cells (immune response), eosinophils, and hepatocytes. The PRL receptor gene, a member of the cytokine receptor superfamily, is comprised of 10 exons and 9 introns. The receptor has two 5′ promoters that direct transcription of a 598 amino acid protein, which is composed of an extracellular domain (ECD), a hydrophobic transmembrane domain, and a cytoplasmic region homologous to GH receptors.

Species differences exist in ligand–receptor interaction. In some species, ligands that are produced at the feto-maternal interface bind with the PRL receptor, whereas in other species hormones/cytokines are produced to activate both PRL and GH receptors. PRL binding to the receptor causes dimerization, which induces protein tyrosine phosphorylation and activation of JAK2 kinase and STATS 1 to 5 (Prigent-Tessier et al. 2001). The auto-paracrine effects of PRL in decidua are mediated by the activation of PRL signal transduction, which involves stimulation of Jak2-STAT5 and activation of phosphatidyl inositol 3 kinase/Akt signaling (Prigent-Tessier et al. 2001).

11.3.2 GH

The GH gene is a member of a multigene family that includes chorionic somatomammotropin and prolactin as well as several other genes, which evolved through a series of gene duplications (Gootwine 2004). GH stimulates cell growth and proliferation either directly or indirectly through insulin-like growth factor I (IGF-I). GH receptors (GHRs) are expressed in the bovine (Scott et al. 1992; Kolle et al. 1997) and sheep placenta (Lacroix et al. 1999). Although the activity of GH is first detected in the fetal pituitary and fetal circulation around days 50 to 60 of pregnancy in ruminants, fetal GH is the main source of GH activity in the feto-placental unit. However, GH concentrations in the maternal circulation are less than that in the fetal umbilical cord during early pregnancy, suggesting that the placenta may be an additional source (Lacroix et al. 1999). Lacroix et al. (1996, 1999) reported expression of GH by the trophoectoderm and syncytial cells of placenta between days 27 and 75 of pregnancy in sheep.

The GH and PRL genes are structurally similar and evolved from a common ancestral precursor. The GH genes are located in chromosomes 17, 10, 11, and 22 in human, rat, mouse, and cow, respectively. Gene duplication is one of the mechanisms that allowed the evolution of placental-specific endocrine activity. Although cattle, sheep, and goats are evolutionarily related, they differ from each other in the ways their placental GH and PRL-like hormones evolved. Humans possess a cluster of five highly related genes. One of these is expressed in the pituitary (GH-N) while four are expressed in the placenta: placental GH variant (GHv) and the chorionic somatomammotropins, CS-A, CS-B, and CS-L (Lacroix et al. 2002).

Placental GH is the product of the GHv gene expressed in the syncytiotrophoblast of human placenta. During pregnancy, GHv expression gradually replaces pituitary GH expression, which becomes undetectable by the end of the gestation (Lacroix et al. 2002). In primates, at least five genes code for GH-like proteins. One is expressed in pituitary and four in the placenta. This cluster of GH-like genes has evolved from duplications of the single GH gene.

Duplication at the GH locus occurred both in sheep and goat (Yamamo et al. 1991), but not in cattle (Woychick et al. 1982). Two types of GH transcripts encoding two GH-like proteins have been detected in sheep (Lacroix et al. 1996). The sequence of one of these is identical to that for the pituitary oGH gene (Orian et al. 1988). However, the other differs from pituitary oGH by the substitution of three amino acids: one in the signal peptide, the second at the border of helix 1 of the GH molecule, and the third in the loop structure at the binding site (de Vos et al. 1992). In sheep, there are two alleles: the GH1 allele contains a single gene copy (GH1), whereas in the GH2 allele, the gene is duplicated (GH2-N and GH2-Z). The sequence of the GH1 allele is identical to that of the pituitary oGH gene (Ofir and Gootwine 1997). The sequence of the GH2-N gene is similar to that of the pituitary GH gene, but the duplicated GH2-Z gene copy of the GH2 allele has a three amino acid substitution similar to GH placental cDNA variant (Lacroix et al. 1996). The GH activity in sheep placenta during early pregnancy evolved to include both extrapituitary expression of the original pituitary GH gene, and through creation of a new GH gene copy by gene duplication that codes for slightly modified protein (Gootwine et al. 1996). Yamano et al. (1991) investigated GH genes in a goat genomic library and found two types of fragments, one containing a single GH gene (gGH1) and the other containing two genes arranged in tandem (gGH2 and gGH3). The tandem arrangement of the gGH2 and gGH3 genes is similar to that seen for ovine GH2-N and GH2-Z. The ovine GH gene is expressed in uninucleate and binucleate trophoblasts and heterokaryons of placenta during days 35 to 70 of pregnancy. Transcripts for members of GH family (GHv, CS-A, CS-B, and CS-L) were detected in the human placenta. Transcription of GH family genes is regulated by a locus control region located 23 kb upstream of the cluster. In primates, four members of the GH family, CS-1, CS-2, CS-3, and GHv, possess extensive homology with human GH family and are expressed in chorio-allantoic placenta (Golos et al. 1993).

11.3.3 PL

Placental lactogen, a member of the GH/PRL gene family, is secreted from the placenta of primate, rodent, and ruminant. Although its function and secretory control are not completely understood, it has myriad effects during pregnancy, such as placental angiogenesis, maternal, and fetal intermediary metabolism, mammary gland growth and development, ovarian and placental steroidogenesis, and luteal function (Corbacho et al. 2002; Gertler and Djiane 2002; Gootwine 2004). The first ruminant PL was detected in goat (Buttle et al. 1972), and subsequently Forsyth (1974) and Kelly et al. (1974) identified PL in sheep placental tissue. PLs are produced by binucleate cells of the conceptus trophoectoderm and are secreted into both the maternal and fetal circulation in ruminants. PL is detectable in trophoblastic tissue by days 16 and 36 in ewe and cow, respectively, and continues to be synthesized throughout pregnancy (Gootwine 2004). It is thought that fully granulated binucleate cells migrate across the microvillar junction in the placenta and fuse with maternal uterine epithelial cells to form either transiently surviving trinucleate cells in cattle or a persistent feto-maternal syncytia in sheep and goat (Soares 2004). In ewes, oPL can be detected in maternal circulation by day 50 with maximum levels between days 120 and 140 and declining thereafter until parturition. The concentration of this hormone is lower in the fetus but shows a similar decline with the advancement of pregnancy (Kappes et al. 1992). The concentrations of PL in bovine maternal and fetal circulation are similar to those in sheep, except that PL levels are much lower than that in sheep (Gootwine 2004). In goat, the concentration starts to increase at approximately days 45 to 60 of the gestation and either peaks or reaches a plateau during the last third of pregnancy. The lower levels in sheep and goat compared with cow suggest that there has been some divergence in the function of PL in these species.

PL arose during mammalian evolution through three independent events. One was a duplication of the GH gene in primate (Chen et al. 1986) and two separate duplications of the PRL gene to give PLs and other prolactin-like placental proteins in rodent (Lin et al. 2000) and ruminant (Anthony et al. 1995). Bovine and ovine PL are structurally more similar to PRL than to GH. Schuler et al. (1988) cloned and characterized the ruminant PL, which is 67% identical with oPL, 51% with bPRL, 30% with bovine prolactin-related cDNAI, 30% with rodent placental hormone, and 20% with human PL and bGH. Both oPL and bPL possess an N-terminal disulfide loop that is characteristic of mammalian prolactin but is not present in somatotropins (Nicoll et al. 1986). Ovine PL is a non-glycosylated, single-chain, 23-kDa polypeptide consisting of 198 amino acids. However, bovine PL is structurally different from oPL, having the apparent molecular weight of 32 to 34 kDa, and is secreted as multiple isoforms due to differential splicing of bPL transcripts and allelic variants of the gene (Kessler and Schuler 1991; Yamakawa et al. 1990). Bovine PL is heavily glycosylated and contains N-linked triantennary oligosaccharides and one or more O-linked carbohydrate chains. However, other members of this gene family such as ovine, porcine, and human PRL are glycosylated, and N-linked carbohydrates are attached at a different portion of the molecule (asparagines 31 in PRL compared with asparagine 53 in bPL). Moreover, the placenta does not secrete non-glycosylated bPL, in contrast to PRL, of which only a portion of the secreted protein is glycosylated (Byatt et al. 1992). Depending on the protein, glycosylation can dramatically affect biological activity. Enzymatic removal of N-linked oligosaccharides increases the affinity of bPL for the bovine somatotropin receptor by twofold (Byatt et al. 1992). However, deglycosylation did not affect activity using a somatotropin bioassay or in a lactogenic bioassay. Therefore, although glycosylation may affect receptor binding affinity, the biological activity of bPL is not dependent on the presence of oligosaccharides.

The members of the PL family exhibit a similar mechanism of action and receptor activation via homo- and hetero-dimerization of the receptor ECD and subsequent trans-phosphorylation of receptor-associated JAK2 or other related kinases (Kelly et al. 1974). Ruminant PLs can bind to both PRL and GH receptors (Anthony et al. 1995), whereas oPL can mimic the action of oPRL (Sakal et al. 1997). In ruminants, PLs signal through PRL-R homodimers and PRL-R/GH-R heterodimers and, in the absence of PRL-R, may act as GH-R antagonist. Some of the GH-like effects of PLs may be mediated through the interaction with PRL-R/GH-R heterodimers. Ruminant PLs have somatogenic activity in a heterologous system, but in a homologous system ruminant PLs antagonize GH activity (Herman et al. 1999; Warren et al. 1999; Gertler and Djiane 2002). Studies on the interaction of oPLs with the ECDs of oGHR and bovine and ovine PRL receptor (PRL-Rs) revealed that oPL can heterodimerize GH-Rs and PRL-Rs (Gertler and Djiane 2002). In ruminants, PLs activate Jak/STAT and mitogen-activated protein kinase signaling pathways (Anthony et al. 1998; Gertler and Djiane 2002). The nature of PL signal transduction differs depending on whether PRL-R forms homodimers or PRL-R/GH-R heterodimers are formed. Heterodimer interaction results in prolonged STAT3 activation leading to distinct cellular responses (Gertler and Djiane 2002).

11.3.4 PRPs

Prolactin-related proteins are nonclassical members of the PRL/GH family that have been found in the cow, sheep, goat, mouse, and rat placentas. They play important roles in the regulation of implantation and placental formation in mammals. In cattle, at least 13 placental PRPs have been identified and are thought to play vital roles in implantation and formation of placentomes in cattle (Kesssler et al. 1991; Yamada et al. 2002). Prolactin-related protein-I belongs to the PRL/GH family and shares a 63% similarity to bovine PRL and 45% to bovine GH (Schuler and Hurley 1987). The N-terminal regions of the bPRP-I and bPRP-VI proteins are rich in hydrophobic amino acids characteristic of a signal peptide (Schuler and Hurley 1987). Bovine PRP-I and PRP-VI mature proteins have three disulfide bonds with six cysteine residues at positions 39, 42, 97, 215, 232, and 238. bPRP-I has three N-glycosylation sites at positions 70-72, 92–94, and 159–161. Recently, bPRP-VII, bPRP-VIII, and bPRP-IX were cloned and characterized from the bovine placenta (Ushizawa et al. 2005a, b). Bovine PRP-VII contains a 929 nucleotide ORF, which encodes a protein of 238 amino acids. The predicted amino acid sequence is 63% homologous to bPRP-I and 70% to bPRP-VI (Ushizawa et al. 2005a). Bovine PRP-VII has eight cysteine residues with four disulfide bonds, which is more than other bPRPs. The bPRP-VIII and bPRP-IX cDNAs consist of 909 and 910 bp ORFs, which correspond to proteins of 236 and 238 amino acids, respectively. The inferred amino acid sequence of bPRP-VIII was 69% identical to bPRP-VI, 66% to bPRP-VII, 61% to bPRP-I and bPRP-III, 58% to bPRP-IV and bPRP-V, 57% to bPRP-IX, and 42% to bPRP-II (Ushizawa et al. 2005b). The deduced amino acid sequence of bPRP-IX protein showed 81% homology to bPRP-IV, 76% to bPRP-I, 70% to bPRP-II, 60% to bPRP-VII, 57% to bPRP-VI and bPRP-VIII, and 53% to bPRP-III and bPRP-V. Phylogenetic analysis revealed that bPRP-VIII, bPRP-III, bPRP-VI, and bPRP-VII comprise one clade, whereas bPRP-IX, bPRP-II, and bPRP-IV comprise another clade (Ushizawa et al. 2005b). The N-terminal regions of bPRP-VIII and bPRP-IX possess two consensus sequences for N-glycosylation and Asn X Ser/Thr at positions 60–62 and 233–235, whereas bPRP-IX had four consensus sequences for N-glycosylation at positions 70–72, 92–94, 146–148, and 160–162. Ushizawa et al. (2007b) cloned and characterized PRP-I and PRP-VI cDNA in goat. The full length cPRP-I and cPRP-VI cDNA contained 717- and 720-bp open reading frames corresponding to proteins of 238 and 239 amino acids, respectively. The inferred amino acid sequence of cPRP-VI is 74% identical to bPRP-VI. The cPRP-I showed 72% homology to bPRP-I, 61% to bRPR-II, 72% to bPRP-IV, 76% to bPRP-IX, and 71% to bPRP-XII (Ushizawa et al. 2007b). Like bPRP-I, cPRP-I contains three disulfide bonds with six cysteine residues. In contrast to bPRP-VI, cPRP-VI has eight cysteine residues with six residues at positions 39, 42, 43, 97,174, and 215, and an extra two cysteines at positions 232 and 239. Caprine PRP-I possesses two consensus sequences for N-glycosylation at positions 70–72 and 92–94 and an atypical N-glycosylation site, Asn X Cys, at position 95–97 (Ushizawa et al. 2007b). Unlike bovine PRP-VI, which has only one consensus sequence, cPRP-VI has three consensus sequences for N-glycosylation at positions 48–50, 60–62, and 70–72, with the atypical glycosylation site (Asn X Cys) at positions 95–97.

Recently, Ushizawa et al. (2007a) cloned two novel ovine PRPs: oPRP-I and oPRP-II. Ovine PRP-II had a typical PRP sequence similar to bovine PRP-I. Ovine PRP-I had a shorter sequence lacking 52 bp from the coding region of other PRP sequences (positions 529–580). Phylogenetic analysis revealed that oPRP-I and bPRP-I, bPRP-II, bPRP-IV, bPRP-IX, bPRP-XII, bPRP-XIV, and cPRP-I are closely related. In contrast, oPRP-II was more distant from bPRP-I, bPRP-II, bPRP-IV, bPRP-IX, bPRP-XII, bPRP-IV, and cPRP-I. Both oPRP-I and oPRP-II are expressed in trophoblast binucleate cells in cattle and goats. Ovine PRP-I expression declined from early to mid-gestation, whereas oPRP-II expression remained constant throughout the gestation period.

11.3.5 PLPs

Prolactin-like proteins belong to the GH/PRL family and have structural similarity to PRL and PL. Lin et al. (1997) cloned, characterized, and expressed three members of mouse PLP, including PRL-like protein A (PLP-A), PLP-B, and decidual/trophoblast PRL-related protein (d/t PRP). Mouse PLP-A is synthesized as a 227 amino acid precursor and is secreted as a glycoprotein of 196 amino acid, which is 78% homologous to rat PLP-A. PLP-B encodes a protein of 230 amino acids consisting of a mature glycoprotein of 201 amino acids which shares 66% identity with rat PLP-B. Decidual/trophoblast PRP encodes a precursor protein of 240 residues and a secreted glycoprotein of 211 amino acids with 64% homology with rat d/t PRP. PLP-A, PLP-B, and d/t PRP are expressed in placenta or decidua. Expression of PLP-A mRNA is maximum on day 12 in rodent trophoblast giant cells, whereas PLP-B mRNA is high on day 10 in decidual cells and on day 12 in spongiotrophoblast (Lin et al. 1997). Decidual/trophoblast PRP mRNA is abundant in the decidual layer on day 8 of gestation (Lin et al. 1997). In rats, nine PLP genes have been identified that are structurally similar to PRL and GH. Iwatsuki et al. (1998) characterized PLP-H which encodes a mature protein of 239 amino acids including a 31 amino acid signal sequence. At the amino acid level, PLP-H shares 78% homology with PLP-C and 67% with PLP-D. PLP-H possesses two putative N-glycosylation sites and eight cysteine residues, of which six are highly conserved in the placental PRL family (Iwatsuki et al. 1998). Similar to PLP-C and PLP-D, PLP-H mRNA first appears on day 14 of pregnancy and expression increases until term. Iwatsuki et al. (1996) cloned a rat PLP-D cDNA which encodes a protein of 240 amino acids including a signal peptide of 29 amino acids. It contains a putative N-glycosylation site and six cysteine residues that are highly conserved in the placental PRL family. The deduced amino acid sequence of PLP-D is 80% homologous to PLP-E and 73% to decidual PRL-related protein. Like PLP-H, PLP-D mRNA is also expressed in spongiotrophoblast and trophoblast giant cells and is first detected at day 14 of pregnancy and increases until term (Iwatsuki et al. 1996). In the bovine placenta, two prolactin-like proteins (bPLP-I and bPLP-II) were identified which resemble bovine prolactin but are different from bovine PL or PRPs (Yamakawa et al. 1990). The inferred amino acid sequences of bPLP-I and bPLP-II share 45–51% identity with bPRL and 23–24% with bGH (Yamakawa et al. 1990). At the nucleotide and amino acid level, bPLP-I and bPLP-II share 62% and 39% homology, respectively. Bovine PLP-I, bPLP-II, PLs, PRLs, and other prolactin-like proteins from cow, mouse, and rat possess seven common amino acid residues: five are conserved among other members of the family, and the other two residues are conserved in bovine, mouse, and rat PLs, PRLs, and PRL-like proteins.

11.3.6 IGFs

Insulin-like growth factors IGF-I and IGF-II play key roles in the regulation of embryonic and fetal growth and development. IGF-I is a single-chain basic protein of 70 amino acids, and IGF-II is a slightly acidic, single-chain, peptide of 67 amino acids (Sara and Hall 1990; Forbes and Westwood 2008 for reviews). IGFs exert their biological effects by binding to cell surface receptors. Two distinct subtypes of receptors for IGFs have been identified. Type I IGF receptor binds IGF-I with equal or greater affinity than IGF-II and also binds insulin with low affinity. In contrast, Type II IGF receptor typically binds IGF-II with greater affinity than IGF-I and does not bind insulin. Human IGF-I and IGF-II are present in the placenta as early as 6 weeks of gestation (Han et al. 1996) and augment the proliferation and survival of placental fibroblast. IGF-I has been found to regulate both the differentiation of cytotrophoblasts into syncytiotrophoblasts (Bhaumick et al. 1992) and into extravillous cells (Lacey et al. 2002). In mice, knockdown of IGF-II in the placenta reduced diffusional exchange surface area and reduced permeability for nutrients. Conversely, in guinea pigs Sferruzzi-Perri et al. (2006) observed that maternal IGF-II increases the total surface area of the placenta for nutrient exchange, whereas IGF-I did not affect the surface area of the placenta but diverted nutrients from mother to fetus.

IGFs circulate in the blood bound to IGF-binding proteins (IGF-BPs). IGF-BPs abolish the acute insulin-like actions, restricts their permeability through capillaries, and inhibits their access to membrane receptors. In humans, six separate IGF-BPs have been described, viz. IGF-BP1, IGFBP-2, IGFBP-3, IGFBP-4, IGFBP-5, and IGFBP-6 (Han et al. 1996). IGF-BP3 is primarily responsible for maintaining IGF levels in the blood. Other IGF-BPs found in the blood stream, including IGFBP-1 and IGFBP-2 can cross endothelial barriers and transport IGFs from the circulation to peripheral tissues (Forbes and Westwood 2008). The role of IGFBP-1 in the placenta is controversial. IGFBP-1 has been found to enhance both basal and IGF-II-induced extravillous trophoblast migration (Irving and Lala 1995). IGFBP-3 is predominantly expressed in trophoblast, fibroblasts of the villous stroma, amnion, and chorion of fetal membranes (Han et al. 1996; Rogers et al. 1996). IGFBP-3 has been found to inhibit IGF-stimulated mitogenesis of placental fibroblasts (Rogers et al. 1996).

As mentioned previously, IGFs induce their effects on cellular proliferation, differentiation, and survival by binding to and activating specific receptors. Two forms of IGF receptors have been identified: a heterotetrameric type-1 receptor, which resembles the insulin receptor, and a monomeric type-II receptor, which is structurally different from the insulin or type-1 receptor. IGF-IR is a heterotetrameric glycoprotein consisting of two alpha subunits of 706 amino acids and two transmembrane β subunits of 627 residues (Sara and Hall 1990). The IGF-IIR is a single-chain, membrane-spanning, glycoprotein that is also known as cation-independent mannose-6-phosphate receptor. The IGF-IR is found in trophoblast, villous endothelium, and the mesenchymal core of the placenta. Studies on transgenic mice lacking the IGF-IR revealed that a reduction in the number of placental IGF-IR might be a contributory factor in pregnancies complicated by intrauterine growth restriction (IUGR). Binding of IGF-I to its receptor results in the activation of two signaling cascades: the PI3K pathway or the mitogen-activated protein kinase (MAPK/ERK1/2) pathway. Activation of IGF-IR results in autophosphorylation of tyrosine residues in the intracellular β subunits and subsequent activation of PI3K and MAPK pathways, resulting in the transcription of target genes involved in cellular proliferation and differentiation.

11.4 Transcriptomics of placental development

11.4.1 Assessment of transcriptional regulation of placental genes through microarray

Evaluation of gene expression is an effective way of identifying genes important in the regulation of traits that are of economic importance in livestock production. Precise knowledge of gene expression profiles is necessary to improve the reproductive efficiency of mammals. Using microarrays as tools for screening for expression of thousands or tens of thousands of genes has been a revolutionary breakthrough in identifying candidate genes that are critical during early pregnancy. The detailed gene expression profiles in the preimplantation embryo and placenta provide insights into the molecular mechanisms that are vital to furthering our knowledge of embryogenesis, implantation, and placental development. Ushizawa et al. (2007c) evaluated global gene expression in the placenta and classified them into 10 clusters. Increased expression was found for PL, pregnancy-associated glycoprotein-1 (PAG-1), and the sulfotransferase family member estrogen preferring member I (SULTIEI) gene. Expression of transcription factor AP-2 alpha (TFAP2A) was high, whereas that of transcription factor AP-2 beta (TFAP2B) was low and was intermediate to that of transcription factor AP-2 gamma (TFAP2C). *In situ* hybridization revealed that TFAP2A, TFAP2B, and TFAP2C mRNA were localized in different sets of trophoblast cells (Ushizawa et al. 2007c). In cow placenta, TFAP2A was expressed in cotyledonary epithelial cells including binucleate cells; TFAP2B was specifically expressed in binucleate cells; and TFAP2C was expressed in mononucleate cells. Recently, Ushizawa et al. (2007c) evaluated gestational stage-specific gene expression profiles in bovine placentomes using microarray and *in silico* analysis. They suggested that the genes TFAP2A, TFAP2B, and TFAP2C may have different roles in the differentiation and proliferation of trophoblasts. Ishida et al. (2007) analyzed cDNA from mid- to late-stage mouse placenta to understand the molecular basis of placental development and function, using microarray. They reported that the expression patterns of apolipoprotein A-II (Apoa 2), apolipoprotein C-II (Apoc 2), CEA-related cell adhesion molecule 14 (Ceacam 14), cell repressor of E1A-stimulated genes (Creg 1), flavin-containing monooxygenase 1 (Fmo1), insulin-like growth factor-II (IGF-II), serine protease inhibitor, Kazal type 3 (Spink 3), serine protease inhibitor 1-1(Spi 1-1), and trophoblast-specific protein alpha (Tpbpa) were similar to mouse PL.

11.4.2 Genomic imprinting

The underlying genetic mechanisms that control interactions between different cell types within the feto-maternal interface and the relative combinations of the maternal and zygotic genes are poorly understood. Genomic imprinting is an epigenetic phenomenon that results in the differentiated expression of a gene or chromosomal region according to the parental origin of inheritance (Joyce and Ferguson-Smith 1999). Imprinted genes play a crucial role in feto-placental development by affecting the growth and nutrient transfer capacity of the placenta in mammals. Georgiades et al. (2000) investigated the *in vivo* function of mouse chromosome 12 imprinting by generating conceptuses that inherited both copies of this chromosome from either the father (paternal uniparental disomy for chromosome 12 [pUPD-12]) or the mother (mUPD-12). Maternal UPD-12 animals died perinatally and exhibited embryonic and placental growth retardation. In contrast, pUPD-12 conceptuses died late in gestation and had a variety of defects including placentomegaly. Georgiades et al. (2001) identified a variety of defects in cell function at the feto-maternal interface, such as compromised invasion of the maternal decidualized endometrium and the central maternal artery, abnormalities in the wall of the central maternal artery, and defects within the zygote-derived cellular layer of the labyrinth. Recently, Zhou et al. (2007) investigated the imprinting status of L-arginine: glycine amidinotransferase (GATM) and paternally expressed gene (PEG10) on days 75 and 90 of pregnancy in pig placentas. Biallelic expression of the GATM gene was observed in the placenta of pigs. In contrast, the PEG10 gene was monoallelically expressed in the porcine placenta on days 75 and 90 of gestation. It was observed that deletion of IGF-II gene in the labyrinthine trophoblast of the placenta restricted placental growth by interfering with the permeability of the placenta to nutrients (Hemberger et al. 2002).

11.4.3 Tracking gene expression signatures using bioinformatics tools

Bioinformatics allows the conceptualization of biology in terms of molecules and then applying informatics techniques to understanding and organizing the information associated with these molecules and their expression patterns. It involves developing and applying computational methods for managing and analyzing information about the sequence, structure, and function of biological molecules and systems. It involves the development of new algorithms and statistics to assess the relationships among members of large datasets; the analysis and interpretation of various types of data including nucleotide and amino acid sequences, protein domains, and protein structures; and the development and implementation of tools that enable efficient access and management of different types of information. It involves the creation of extensive electronic databases on genomes and protein sequences, and techniques such as the three-dimensional modeling of biomolecules and biological systems. These revolutionary technologies have provided a new understanding of biology, with widespread applications to medicine, agriculture, and ecology. Large databases of cDNA sequences of tens of thousands of genes in thousands of tissue samples provide the source data for identifying candidate genes that are associated with placental and fetal growth and development.

Ming Wong and Walker (2001) studied the expression patterns of IGF and placental steroid synthesis (PSS) genes in human cDNA libraries. They observed that either IGF/PSS genes, including placental lactogen-4 (PL-4), human growth hormone (hGH), pregnancy-associated plasma protein-A (PAPP-A), eosinophil major basic protein (EMBP), placental alkaline phosphatases (PLAP), placental aromatose P450; cholesterol side chain cleavage enzyme (P450scc), and 3 beta hydroxy steroid dehydrogenase (3 beta-HSD) share a similar expression profile across these libraries. They chose these eight genes as their bait to look for other genes that showed very similar expression. They observed 10 genes that were not previously linked to IGF/PSS that had expression patterns similar to the eight genes. Out of 10 genes, six genes including malignant melanoma metastasis suppressor, placenta specific-1 (PLAC-1), pregnancy specific glycoprotein 10 (PSG-10), pregnancy specific beta 1 glycoprotein (PSG-beta1), serine palmitoyl transferase (SPT), and TONDU are associated with cell growth in fetal and /or cancer tissues. Four are EST sequences, namely, PLAC2, PLAC3, PLAC4, and PLAC5, which occur predominantly in placental/fetal tissue or tumors suggesting the involvement of PLAC genes in tissue growth. Other known IGF/PSS genes such as metalloprotease ADAM 12, early placenta insulin like peptide (EPIL), IGF binding proteins, and placental growth factor are also found to be co-expressed less consistently with PLAC2, PLAC3, PLAC4 and PLAC5 genes. The genes identified by coexpression analysis are useful candidates for exploring their roles in placental and fetal development. Recently, Jiang et al. (2004) analyzed ESTs and identified 5024 genes in bovine placenta with human orthologs. A total of 24 preferentially expressed genes (PEG) and 39 highly expressed genes (HEG) were found in the placenta. Transcriptional profiles were similar in the placenta, ovary, and mammary gland.

11.5 Future research directions

Exploring how genomes affect reproductive efficiency will undoubtedly lead to the development of tools to optimize reproductive management.

This has been greatly aided by the advent of genomic and bioinformatics technologies. Understanding the mechanisms of preimplantation embryo development and placentation has been a challenge to reproductive and developmental biologists. Recent technological advances in genetic manipulation and expression profiling offer excellent opportunities to elucidate the molecular mechanism controlling embryogenesis and placentation. The current advances in molecular biology and biotechnology, particularly functional genomics (DNAarrays), have allowed the identification of embryonic and maternal genes potentially involved in embryo survival and placental development. Validation of the functional involvement of genes that have been identified requires extensive *in vitro* studies before *in vivo* therapy can be applied. Recently, oligo-based and cDNA microarray technologies made it possible to understand many of the factors controlling the regulation of gene transcription and to globally evaluate gene expression profiles. Thus, a genome-wide screening approach coupled with functional assays will help elucidate the complex embryo–uterine crosstalk. The application of molecular genetic technologies to animal agriculture will definitely bring about exciting changes in livestock production and the tailoring of animals to produce products needed by humans.

Acknowledgments

The author wishes to express his deep sense of gratitude to Director, NIANP, Bangalore, for granting permission to write the chapter. Sincere help rendered by Mrs. Rekha is duly acknowledged. I should not forget to acknowledge Shelby Hayes, Editorial Assistant, Wiley-Blackwell, for prompt and helpful service when required. Finally, thanks are due to my wife, Shrabanti, and daughter, Anoushka, who cheerfully tolerated and supported the many hours of absence involved in writing the chapter.

References

Ain, R., Canham, L.N. and Soares, M.J. 2003. Gestational stage-dependent intrauterine trophoblast cell invasion in the rat and mouse: novel endocrine phenotype and regulation. *Deelopmental Biology* **260**: 176–190.

Anson-Cartwright, L., Dawson, K., Holmyard, D., Fisher, S.J., Lazzarini, R.A., and Cross, J.C. 2000. The glial cell missing-1 protein is essential for branching morphogenesis in the chorioallantoic placenta. *Nature* **25**: 311–314.

Anthony, R.V., Liang, R., Kayl, E.P., and Pratt, S.I. 1995. The growth hormone/prolactin gene family in ruminant placentae. *Journal of Reproduction and Fertility* **49**(Supplement): 83–95.

Anthony, R.V., Limesand, S.W., Fanning, M.D., and Liang, R. 1998. Placental lactogens and growth hormone. In: Bazer, F.W. (ed.), *The Endocrinology of Pregnancy*. Totowa, NJ: Humana Press, pp. 461–490.

Bhaumick, B., George, D., and Bala, R.M. 1992. Potentiation of epidermal growth factor-induced differentiation of cultured human placental cells by insulin-like growth factor-I. *Journal of Clinical Endocrinology and Metabolism* **74**: 1005–1011.

Buttle, H.L., Forsyth, I.A., and Knaggs, G.S. 1972. Plasma prolactin measured by radioimmunoassay and bioassay in pregnant and lactating goats and the occurrence of a placental lactogen. *Journal of Endocrinology* **53**: 483–489.

Byatt, J.C., Warren, W.C., Eppard, P.J., Staten, N.R., Krivi, G.G., and Collier, R.J. 1992. Ruminant placental lactogens: Structure and biology. *Journal of Animal Science* **70**: 2911–2923.

Chen, E.Y., Liao, Y.C., Smith, D.H., Barrera-Saldana, H.A., Gelinas, R.E., and Seeburg, P.H. 1986. The human GH basis: Nucleotide sequence, biology and evolution. *Genetics* **4**: 479–497.

Cohick, C.B., Dai, G., Xu, L., Deb, S., Kami, T., Leavan, G., Sipirer, C., Siprirer, J.K., and Soares, M.J. 1996. Placental lactogen-I variants utilizes the prolactin receptor signaling pathway. *Molecular and Cellular Endocrinology* **116**: 49–58.

Corbacho, A.M., Martinez De la Escalera, G., and Clapp, C. 2002. Roles of prolactin and related members of the prolactin/growth hormone/placental lactogens family in angiogenesis. *Journal of Endocrinology* **173**: 219–238.

Cross, J.C., Flannery, M.L., Blanar, M.A., Steingrimson, E., Jenkins, N.A., Copland, N.G., Rutter, W.J., and Werb, Z. 1995. Hxt encodes a basic halix-loop-helix transcription factor that regulates trophoblast cell development. *Development* **121**: 2513–2523.

de Vos, A.M., Ultsch, M., and Kossiakoff, A.A. 1992. Human growth hormone and extracellular domain of its receptor: crystal structure of the complex. *Science* **255**: 306–312.

Firulli, A.B., McFadden, D.G., Lin, Q., Srivastava, D., and Olson, E.N. 1998. Heart and extraembryonic mesodermal defects in mouse embryos lacking the bHLH transcription factor Hand1. *Nature Genetics* **18**: 266–270.

Forbes, K. and Westwood, M. 2008. The IGF axis and placental function: A mini review. *Hormone Research* **69**: 129–137.

Forsyth, I.A. 1974. The comparative study of placental lactogenic hormones. A review. In: Josimovich, J.B. (ed.), *Lactogenic Hormones, Fetal Nutrition and Lactation*. New York: Wiley and Sons, p. 49.

Georgiades, P., Watkins, M., Burton, G.J., and Ferguson-Smith, A.C. 2001. Roles for genomic imprinting and the zygotic genome in placental development. *Proceedings of National Academy of Sciences* **98**(8): 4522–4527.

Georgiades, P., Watkins, M., Surani, M.A., and Ferguson-Smith, A.C. 2000. Parental origin specific developmental defects in mice with unipaternal disomy for chromosome 12. *Development* **127**: 4719–4728.

Gertler, A. and Djiane, J. 2002. Mechanism of ruminants placental lactogen action: Molecular and in vitro studies. *Molecular Genetics and Metabolism* **75**: 189–201.

Golos, T.G., Durning, M., Fisher, J.M., and Fowler, P.D. 1993. Cloning of four GH/chorionic somatomammotropins-related cDNAs differentially expressed during pregnancy in the rhesus monkey placenta. *Endocrinology* **133**: 1744–1752.

Gootwine, E. 2004. Placental hormones and fetal-placental development. *Animal Reproduction Science* **82–83**: 551–566.

Gootwine, E., Ofir, R., and Yossafi, S. 1996. Charactterization of *PuvII* polymorphisms between the ovine growth hormone GH2-N and GH2-Z gene copies. *Animal Biotechnology* 7: 135–143.

Guillemot, F., Nagy, A., Auerbach, A., Rossant, J. and Joyner, A.L. 1994. Essential role of Mash-2 in extraembryonic development. *Nature* **371**: 333–336.

Han, V.K., Bassett, N., Walton, J. and Challis, J.R. 1996. The expression of insulin like growth factor (IGF) and IGF-binding protein (IGFBP) genes in the human placenta and membranes: Evidence for IGF-IGFBP interactions at the feto-maternal interface. *Journal of Clinical Endocrinology and Metabolism* **81**: 2680–2693.

Hemberger, M.M., Hughes, J., Dean, W., Ferguson-Smith, A., Fundele, R., Stewart, F., Kelsey, G., Fowden, A., Sibley, C., and Reik, W. 2002. Placental specific IGF-II is a major modulator of placental and fetal growth. *Nature* **417**: 945–948.

Herman, A., Helman, D., Livnah, O., and Gertler, A. 1999. Ruminant placental lactogens act as antagonists to homologous growth hormone receptors by ovine placental lactogen. *Journal of Biological Chemistry* **274**: 7631–7639.

Irving, J.A. and Lala, P.K. 1995. Functional role of cell surface integrins on human trophoblast cell migration: Regulation by TGF-beta, IGF-II, and IGFBP-1. *Experiment on Cell Research* **217**: 419–427.

Ishida, M., Ohashi, S., Kizaki, Y., Naito, J., Horiguchi, K., and Hasegawa, T. 2007. Expression profiling of mouse placental lactogen II and its correlative genes using a cDNA microarray analysis in the developmental placenta. *Journal of Reproduction and Development* **53**(1): 69–76.

Iwatsuki, K., Oda, M., Tanak, S., Ogawa, T., and Shiota, K. 1998. Molecular cloning and characterization of a new member of the rat placental prolactin (PRL) family, PRL-like protein H. *Biochemistry* **139**(12): 4976–4983.

Iwatsuki, K., Shinozaki, M., Hattori, N., Hirasawa, K., Itagaki, S., Shiota, K., and Ogawa, T. 1996. Molecular cloning and characterization of a new member of the rat placental prolactin (PRL) family, PRL-like protein D (PLP-D). *Endocrinology* **137**(9): 3849–3855.

Jiang, Z., Wu, X.L., Garcia, M.D., Griffin, K.B., Michal, J.J., Ott, T.L., Gaskins, C.T., and Wright, R.W. Jr. 2004.Comparative gene-based *in silico* analysis of transcriptomes in different bovine tissues and (or) organs. *Genome* **47**: 1164–1172.

Joyce, J.A. and Ferguson-Smith, A.C. 1999. *Development: Genetics, Epigenetics and Environmental Regulation*. Russo, V., Cove, D., Edgar, L., Jaenisch, R., and Salamini, F. (eds.). Berlin: Springer, pp. 421–434.

Kappes, S.M., Warren, W.C., Pratt, S.L., Liang, R., and Anthony, R.V. 1992. Quantification and cellular localization of ovine placental lactogen messenger ribonucleic acid expression during mid- and late gestation. *Endocrinology* **131**: 2829–2838.

Kelly, P.A., Robertson, H.A., and Fiesen, H.G. 1974. Temporal pattern of placental lactogen and progesterone secretion in sheep. *Nature* **248**: 435–437.

Kessler, M.A., Duello, T.M., and Schuler, L.A. 1991. Expression of prolactin related hormones in the early bovine conceptus, and potential for paracrine effect on the endometrium. *Endocrinology* **129**: 1885–1895.

Kessler, M.A. and Schuler, I.A. 1991. Structure of the bovine placental lactogen gene and alternative splicing of transcripts. *DNA Cell Biology* **10**: 93–104.

Kolle, S., Sinowatz, F., Boie, G., Lincoln, D., and Waters, M.J. 1997. Differential expression of the growth hormone receptor and its transcript in bovine uterus and placenta. *Molecular and Cellular Endocrinology* **131**: 127–136.

Lacroix, M.C., Devinoy, E., Servely, J.L., Pusissant, C., and Kann, G. 1996. Expression of the growth hormone gene in ovine placenta: Detection and cellular localization of the protein. *Endocrinology* **137**: 4886–4892.

Lacey, H., Haigh, T., Westwood, M., and Aplin, J.D. 2002. Mesenchymall derived insulin like growth factor-I provides a paracrine stimulus for trophoblast migration. *BMC Development Biology* **2**: 5–11.

Lacroix, M.C., Devinoy, E., Cassy, S., Servely, J.L., Vidaud, M., and Kann, G. 1999. Expression of growth hormone and its receptor in the placental and feto-maternal environment during early pregnancy in sheep. *Endocrinology* **140**: 5587–5597.

Lacroix, M.C., Guibourdenche, J., Frendo, J.L., Muller, F., and Evain-Brion, D. 2002. Human placental growth hormone—A review. *Placenta* **23**(Supplement A): S87–S94.

Lin, J., Poole, J., and Daniel, I.H.L. 1997. Three new members of the mouse prolactin/growth hormone family are homologous to proteins expressed in the rat. *Endocrinology* **138**: 5541–5549.

Lin, J., Poole, J., and Linzer, D.I. 2000a. Three new members of the mouse prolactin/growth hormone family are homologous to proteins expressed in the rat. *Endocrinology* **138**(12): 5541–5549.

Lin, J., Toft, D.J., Bengtson, N.W., and Linzer, D.I. 2000b. Placental prolactins and the physiology of pregnancy. *Recent Progress in Hormone Research* **55**: 37–51.

Lu, X.J., Deb, S., and Soares, M.J. 1994. Spontaneous differentiation of trophoblast cells along the spongiotrophoblast pathway: Expression of the placental prolatin gene family and modulation by retinoic acid. *Development Biology* **163**: 86–97.

Ma, G.T., Roth, M.E., Groskopf, J.C., Tsai, F.Y., Orkin, S.H., Grosveld, F., Engel, J.D., and Linzer, D.Z. 1997. GATA-2 and GATA-3 regulate trophoblast specific gene expression in vivo. *Development* **124**: 907–914.

Ming Wong, S.L. and Walker, M.G. 2001. A bioinformatics approach to identifying fetal development genes. *Gene Function and Disease* **2**(5–6): 221–225.

Nicoll, A.S., Mayer, G.L., and Russel, S.M. 1986. Structural features of prolactin and GHs that can be related to their biological properties. *Endocrine Review* **7**: 169–203.

Ofir, R. and Gootwine, E. 1997. Ovine growth hormone gene duplication—Structural and evolutionary implications. *Mammalian Genome* **8**: 770–772.

Orian, J.M., O'Mahoney, J.V., and Brandon, M.R. 1988. Cloning and sequencing of the ovine growth hormone gene. *Nucleic Acids Research* **16**: 9046–9049.

Peters, T.J., Chapman, B.M., and Soares, M.J. 2000. Trophoblast differentiation: An in vitro model for trophoblast giant cell development. In: Tuan, R.S. and Lo, C.W. (eds.), *Developmental Protocols*. Totowa, NJ: Humana Press, pp. 301–311.

Prigent-Tessier, A., Barkai, U., Tessier, C., Cohen, H., and Gibori, G. 2001. Characterization of a rat uterine cell line, UIII cells: Prolactin (PRL) expression and endogenous regulation of PRL-dependent genes; estrogen receptor β, α2-macroglobulin, and decidual PRL involving the Jak2 and Stat5 pathway. *Endocrinology* **142**: 1242–1250.

Riley, P., Anson-Cartwright, L., Riley, P., Reda, D., and Cross, J.C. 1998. The Hand1 bHLH transcription factor is essential for placentation and cardiac morphogenesis. *Nature Genetics* **18**: 271–275.

Roby, K.E., Deb, S., Gibori, G., Sipires, C., Levan, G., Knok, S.C.M., and Soares, M.J. 1993. Decidual PRL related proteins: Identification, molecular cloning and characterization. *Journal of Biological Chemistry* **268**: 3136–3142.

Rogers, J., Wiltrout, L., Nanu, L., and Fant, M.E. 1996. Developmentally regulated expression of IGF binding protein-3 (IGFBP-3) in human placental fibroblasts: Effect of exogenous IGFBP-3 on IGF-1 action. *Regulatory Peptides* **61**: 189–195.

Sakal, E., Bignon, C., Groselaude, J., Kantor, A., Shapira, R., Leibovich, H., Helman, D., Nespoulous, C., Shamay, A., Rowlinson, S.W., Djiane, J., and Gertler, A. 1997. Large scale preparation and characterization of recombinant ovine placental lactogen. *Journal of Endocrinology* **152**: 317–327.

Sara, V.R. and Hall, K. 1990. Insulin-like growth factors and their binding proteins. *Physiological Reviews* **70**(3): 591–614.

Schuler, L.A. and Hurley, W.L. 1987. Molecular cloning of a prolactin related mRNA expressed in bovine placenta. *Proceedings of National Academy of Sciences of the United States of America* **84**(16): 5650–5654.

Schuler, L.A., Shimonura, K., Kessles, M.A., Fieles, C.G., and Bremel, R.D. 1988. Bovine placental lactogen: Molecular cloning and protein schuture. *Biochemistry* **27**(22): 8443–8448.

Scott, I.C., Anson-Cartwright, L., Riley, P., Reda, D., and Cross, J.C. 2000. The Hand1 basic helix-loop-helix transcription factor regulates trophoblast giant cell differentiation via multiple mechanisms. *Molecular and Cellular Biology* **20**: 530–541.

Scott, P., Kessler, M.A., and Schuler, L.A. 1992. Molecular cloning of the bovine prolactin receptor and distribution of prolactin and growth hormone receptor transcripts in fetal and utero-placental tissues. *Molecular and Cellular Endocrinology* **89**: 47–58.

Sferruzzi-Perri, A.N., Owens, J.A., Pringle, K.G., Robinson, J.S., and Roberts, C.T. 2006. Maternal insulin like growth factors-I and -II act via different pathways to promote fetal growth. *Endocrinology* **147**: 3344–3355.

Soares, M.J. 2004. The prolactin and growth hormone families: Pregnancy-specific hormones/cytokines at the maternal-fetal interface. *Reproductive Biology and Endocrinology* **2**: 51–58.

Ushizawa, K., Kaneyama, K., Tanahashi T., Tokunaga, T., Tsunoda, Y., and Hashizume, K. 2005a. Cloning and expression of a new member of prolactin related protein in bovine placenta: Bovine PRP-VII. *Biochemistry Biophysics Research Communication* **326**(2): 435–441.

Ushizawa, K., Takahashi, T., Hosoe, M., Kaneyama, K., and Hashizume, K. 2005b.

Cloning and expression of two new prolactin related proteins, prolactin related protein-VIII and -IX in bovine placenta. *Reproductive Biology and Endocrinology* **3**: 68–80.

Ushizawa, K., Takahashi, T., Hosoe, M., Ohkoshi, K., and Hashizume, K. 2007a. Expression and characterization of novel ovine orthologs of bovine placental prolactin related proteins. *BMC Molecular Biology and Endocrinology* **8**: 95–101.

Ushizawa, K., Takahashi, T., Hosoe, M., Kizaki, K., Abe, Y., Sasada, H., Sato, E., and Hashizume, K. 2007b. Gene expression profiles of novel caprine placental prolactin related proteins similar to bovine placental prolactin- related protein. *BMC Developmental Biology* **7**: 16–28.

Ushizawa, K., Takahashi, T., Hosoe, M., Ishiwata, H., Kaneyama, K., Kizaki, K. and Hashizume, K. 2007c. Global gene expression analysis and regulation of the principal genes expressed in bovine placenta in relation to the transcription factor AP-2 family. *Reproductive Biology and Endocrinology* **5**: 17.

Warren, W.C., Byatt, J.C., Huynth, M., Paik, K., Pegg, G., and Staten, N.R. 1999. Evaluation of the somatogenic activity of bovine placental lactogen with cell lines transfected with the bovine somatotropin receptor. Life *Science* **65**: 2755–2767.

Wiemers, D.O., Ain, R., Ohboshi, S. and Soares, M.J. 2003. Migratory trophoblast cells express a newly identified member of the prolactin gene family. Journal of *Endocrinology* **179**: 335–346.

Woychick, R.P., Camper, S.A., Lyons, R.H., Horowitz, S., Goodwin, E.C., and Rottman, F.M. 1982. Cloning and nucleotide sequencing of the bovine growth hormone gene. *Nucleic Acids Research* **10**: 7197–7210.

Yamakawa, M., Tanaka, M., Koyama, M., Kagesato, Y., Watahiki, M., Yamamoto, M., and Nakashima, K. 1990. Expression of new members of the prolactin growth hormone gene family in bovine placenta: Isolation and characterization of two prolactin-like cDNA clones. *Journal of Biological Chemistry* **265**: 8915–8920.

Yamada, O., Todoroki, J., Kizaki, D., Takahashi, T., Imai, K., Patel, O.V., Schuler, L.A., and Hashizume, K. 2002. Expression of prolactin-related protein I at the fetomaternal interface during the implantation period in cows. *Reproduction* **124**: 427–437.

Yamano, Y., Abe, M., Mikawa, S., Kioka, N., Manabe, E., Sakai, H., Komano, T., Utsumi, K., and Iritani, A. 1991. Structural analysis of repetitive DNA sequences in the goat growth hormone gene region. *Agriculture and Biological Chemistry* **55**: 633–639.

Zhou, Q.Y., Huang, J.N., Xiong, Y.Z., and Zhav, S.H. 2007. Imprinting analyses of the porcine GATM and PEG10 genes in placental on days 75 and 90 of gestation. *Genes and Genetic System* **82**: 265–269.

12

Cellular, Molecular, and Genomic Mechanisms Regulating Testis Function in Livestock

Kyle Caires, Jon Oatley, and Derek McLean

12.1 Introduction

The production of sperm occurs in the testis and is essential for male fertility and ultimately the production of offspring. Structurally, the testis is organized with seminiferous tubules that produce sperm and the cells in the interstitial space between the tubules. The seminiferous tubules lack blood vessels and include differentiating germ cells and Sertoli cells that are contained within a single cell layer of peritubular myoid cells that form the final cell barrier or outside ring of the tubule. The interstitial space comprises Leydig cells, fibroblasts, some immune cells, and the cells that make up blood vessels. This complex organization presents some challenges when investigating how individual cell types contribute to the overall process of sperm production. For example, spermatogenesis includes mitosis of undifferentiated germ cells, followed by meiosis for chromosomal reduction to produce haploid gametes. Therefore, the testis has a complex set of germ cells that differentiate through a unique cellular process, meiosis, which occurs continually throughout the life of the male. In addition, the germ cells interact with, and are regulated by, somatic cells through intimate contact within the seminiferous tubule. The cells present in the interstitial space also influence germ cell differentiation by providing factors that may directly regulate the germ cells or regulate the somatic cells of the seminiferous tubule.

Testis development and sperm production are controlled during development and in mature animals by the hypothalamus and pituitary gland. Gonadotropin-releasing hormone (GnRH) from the hypothalamus stimulates the pituitary gland to produce the gonadotropin follicle-stimulating hormone (FSH) and luteinizing hormone (LH), which regulate the Sertoli and interstitial Leydig cells, respectively. LH stimulates Leydig cells to produce testosterone. FSH and testosterone act on Sertoli cells to stimulate these cells to proliferate during development and support germ cell differentiation in mature animals. Testosterone and inhibin, a protein hormone produced by Sertoli cells under the regulation of FSH, feed back to the hypothalamus and pituitary gland to suppress the production of GnRH and the gonadotropins. This feedback is critical to controlling the pulsatile release of GnRH to maintain serum hormone concentrations within normal physiological ranges. Disruption of the positive or negative signaling associated with the hypothalamic–pituitary–testis loop results in complete or partial loss of sperm production.

The complex environment of the testis to support germ cell differentiation and sperm production creates challenges for identifying and characterizing the mechanisms that regulate spermatogenesis. The ability to simulate the seminiferous tubule environment to support germ cell differentiation from immature diploid spermatogonia into haploid sperm *in vitro* has not been consistently achieved in mammals. Therefore, the specific signals and genes responsible for each development step must be investigated in combination with all other signals from germ cells and somatic cells at each developmental stage. Primary culture of Sertoli cells and germ cells for short periods of time has been used to investigate aspects of hormone action and signal transduction in these cell types. However, removal of the cells from the physiological environment of the testis eliminates cell interactions and any potential paracrine signaling. Although the results from these projects provide valuable information, they must be interpreted with caution. Fortunately, the development of multiple experimental approaches including transgenic animals, cloning, germ cell transplantation, and ectopic testis grafting enables scientists to investigate physiological mechanisms within the testis. The elucidation of the complete genome sequence from laboratory and livestock species and the ability to isolate the complete set of transcripts from a cell or tissue (transcriptome) has greatly aided the characterization of genes and proteins regulating testicular function. The impact of these research approaches has improved our understanding of testis biology. However, most research in testis biology using genomics-focused techniques has been conducted using mice or rats as model organisms.

The aim of this chapter is to provide information about techniques that are used to investigate spermatogenesis in livestock and the information that has been learned from experimentation. In addition, we will provide background on spermatogenesis and discuss several projects that used genomics-based approaches to determine basic mechanisms that regulate somatic or germ cell development in the testis. Examples of how genomics-based approaches have generated large databases of information regarding gene expression profiles of the testis in rodents will be referenced to provide information and resources for the reader to understand how these datasets can be developed and interpreted to gain insight into testis biology.

12.2 Spermatogenesis

12.2.1 *Germ cell differentiation: Basics*

The entire process of spermatogenesis is dependent on the formation of the testis during embryonic and postnatal development (Cupp and Skinner 2005). During embryonic development the SRY gene is expressed in primitive Sertoli cells, stimulating a cascade of events leading to the formation of sex cords in the embryonic gonad. These cords result from the aggregation of primordial germ cells (PGCs), primitive Sertoli cells, and pre-peritubular cells that have migrated from the mesonephros. These cells go through periods of proliferation or mitotic arrest in the case of germ cells during the remaining time of embryonic development and at birth have formed seminiferous tubules (Cupp and Skinner 2005). PGCs migrate from the yolk sac to the embryonic gonad and proliferate for a period of time. In the embryonic testis, the PGCs differentiate into gonocytes, the most primitive male germ cell, and then stop proliferating.

Germ cells resume mitosis after birth and migrate from the center of the seminiferous tubule to the base of the tubule. The time of this migration and cell division varies between species, starting around postnatal day 2 in mice (Nagano et al. 2000) to several weeks after birth in bull calves (Curtis and Amann 1981). The paracrine or autocrine signals that stimulate this process are not known, but cells that do not migrate to the base of the tubule undergo apoptosis. Migration of germ cells to the basal portion of the tubule and resumption of mitosis during this time is important for the establishment of the spermatogonial stem cell (SSC) population in mice (McLean et al. 2003).

The initial steps of germ cell differentiation have been determined by analysis of histological sections from the testes of many species. Morphological differences in differentiating germ cells were used to distinguish between different germ cell developmental stages. Initiation of sperm production occurs when SSC differentiation results in production of daughter cells, termed A_{paired} (A_{pr}) spermatogonia, which are committed to differentiation rather than self-renewal (de Rooij and Russell 2000). The A_{pr} spermatogonia then undergo a series of mitotic cell divisions, becoming $A_{aligned}$ (A_{al}) spermatogonia, and give rise to differentiating A-type spermatogonia that undergo another series of amplifying mitotic divisions. These differentiating A-type spermatogonia mature into intermediate and B-type spermatogonia, which enter meiosis, becoming primary and secondary spermatocytes, and eventually haploid spermatids are produced, which undergo a transformation into spermatozoa. Collectively, the SSCs (also termed A_s), A_{pr}, and A_{al} germ cells, are referred to as proliferating spermatogonia and all share very similar phenotypic and likely molecular characteristics (Oatley and Brinster 2008). SSCs are rare, estimated to be present in approximately 1 in 3000 cells of the adult mouse testis (Tegelenbosch and de Rooij 1993).

Following the initial steps of germ cell differentiation, germ cells in the testis undergo mitosis as spermatogonia then differentiate into spermatocytes that undergo meiosis. After meiosis, the haploid germ cells are called spermatids, and upon completion of nuclear repackaging and the formation of the axoneme, these cells spermiate into the lumen for transport to the epididymis. In the bull, spermatogonia first appear at around 12–16 weeks of age, and these cells differentiate into spermatocytes at around 24 weeks of age (Curtis and Amann 1981).

12.2.2 Germ cell differentiation: Regulation

The factors that influence germ cell differentiation are likely regulated by the gonadotropins and testosterone. For example, plasma FSH levels are fairly stable from 4 weeks of age through puberty in bull calves (Amann 1983), indicating that the action of this hormone is regulated by FSH receptor expression by Sertoli cells. In contrast, plasma LH levels increase from 8 to 12 weeks of age and then begin to fluctuate during sexual development (Amann 1983). Testosterone, produced by Leydig cells under the influence of LH, has a slight increase at around 4 weeks of age, then declines, and gradually increases from 12 weeks of age to reach maximal levels at 24–28 weeks of age. The peak of serum testosterone concentrations coincide with the appearance of meiotic cells (Amann 1983). Although there is a general understanding of hormone regulation and morphological changes that occur during the establishment of the testis leading to sperm production, a detailed understanding of the factors required to initiate this process is lacking. One approach for identifying the proteins associated with testis development, the establishment of spermatogenesis, and germ cell differentiation is to profile the genes expressed in the testis during development. Several platforms are available for this type of characterization including the production of cDNA libraries from isolated testicular germ and somatic cells. Gene expression profiling with gene microarrays is the most prevalent approach for characterizing the genes involved in development or function of specific cells or tissues.

12.3 Transcriptomics of testis in bulls

12.3.1 Microarray analysis on testis tissue grafts

The use of microarray technology in male reproductive biology has allowed the characterization of large numbers of genes that are important for sperm production and testis development in humans (He et al. 2006) and rodents (McLean et al. 2002; Shima et al. 2004; Small et al. 2005; Johnston et al. 2008). With sequence information obtained from a variety of mammalian genome projects (Lewin 2003; Rothschild 2003; Womack 2005), microarray platforms are available for several domestic livestock species, including bovine. Information from protein–protein interaction and metabolism and cell signaling research has facilitated the development of pathway analysis software on a genome-wide level and extend the usefulness of microarray data to provide additional information regarding bioactive molecules in gene expression studies.

Thus, Schmidt et al. (2007) sought to investigate the factors critical for bovine testis development *in vivo* and determine the mechanisms that may be responsible for donor-age-related differences in the ability of bovine testis tissue grafts to produce elongated spermatids (Schmidt et al. 2006a). To accomplish this objective, testis tissue obtained from 2-, 4-, and 8-week-old bull calves were grafted on immunodeficient mice and removed at several tissue age time points. To determine factors potentially responsible for the age-related differences in sperm production in bovine testis grafts, and therefore testis development, the transcriptomes of donor tissues were assayed using Affymetrix Bovine GeneChips (Santa Clara, CA) and deposited in the NCBI Gene Expression Omnibus (GEO) database (http://www.ncbi.nlm.nih.gov/geo; series GEO series # GSE5970). On average, 56% of transcripts present on the GeneChips were expressed in 2-, 4-, and 8-week-old bull testis (Schmidt et al. 2007).

Data processing identified approximately 200 transcripts for further analysis, and ontological clustering of transcripts (twofold difference in expression) indicated significant differences in expression of genes involved in cell communication, maintenance, and signal transduction between 2 and 8 weeks of bovine testis development *in situ* (Schmidt et al. 2007). Thus, age-related gene expression (and subsequent protein expression) in the donor testis tissue before grafting likely affected the ability of grafted tissues to be accepted by the host environment and/or support germ cell differentiation. Microarray analysis identified angiogenin, early growth response 1, insulin-like growth factor 2, insulin-like growth factor-binding protein 3, transgelin 2, and thrombomodulin as candidate genes responsible for donor-tissue age variation in the production of sperm by testis grafts (Schmidt et al. 2007). The authors confirmed the expression patterns of these genes by quantitative polymerase chain reaction (qPCR) (Schmidt et al. 2007), and these findings underscore the importance of genes involved in cell growth and vascular biology for establishment of spermatogenesis in the bull.

Schmidt et al. (2007) also evaluated the expression of genes previously known to be important for germ and Sertoli cell biology in rodents by qPCR in bovine tissue after grafting. The interaction between KIT ligand (KITL), also known as stem cell factor, produced by the Sertoli cells and its receptor KIT on germ cells is important for germ cell differentiation (Mauduit et al. 1999). KIT expression was significantly lower in bovine testis tissue grafts when compared with bovine testis *in situ*, but transcript abundance of KIT and KITLG increased as grafts developed in the recipient mouse (Schmidt et al. 2007). Other Sertoli cell-expressed genes, including clusterin and GATA4, were also found to increase as grafted tissues developed, and these transcripts were significantly lower in grafted tissues from 8-week-old donors when compared with grafts originating from donors of other ages (Schmidt et al. 2007). Immunohistochemical analysis confirmed expression of clusterin and GATA4 protein in these tissues (Schmidt et al. 2007). The expression of GDNF and FGF2, two proteins important for SSC self-renewal and Sertoli cell function, also decreased with donor age. Abundance of transcripts for both these genes in 8-week-old donor tissues were significantly lower than other donor ages evaluated (Schmidt et al. 2007). Together these findings provide insights into factors important for bovine testis development and thus may provide novel targets for improving fertility in bulls and testis tissue grafts.

12.3.2 Ectopic testis xenografting

Scientists have been experimenting with testis transplantation since the late 1800s for a variety of purposes, including hormone therapy (Turner 1938). Ectopic subcutaneous testicular grafting has been utilized as another technique to investigate spermatogenesis of lab and livestock species *ex situ* (Johnson et al. 1996; Honaramooz et al. 2002; Oatley et al. 2004b, 2005a). Ectopic grafting of testicular tissue is a method in which a portion of testicular parenchyma from a donor animal is placed into a recipient animal, usually under the skin on the back of the animal. This recipient animal is usually an immunodeficient nude male mouse (nu+/nu+). The nu/nu mouse strain lacks a thymus and hence does not have the characteristic immune system T cells that could mount an immune response and ultimately reject donor-derived testis tissue. Remarkably, spermatogenesis will initiate in the grafted tissue and elongated sperm are produced (ref). Offspring have been generated by intracytoplasmic sperm injection (ICSI) into oocytes using elongated spermatids recovered from mice testis tissue grafted onto mice (Schlatt et al. 2003). To date, no offspring generated using sperm from cross-species testis grafts (e.g., pig grafted on mouse) have been reported. Some of the potential applications of testis tissue grafting include male germline preservation including the conservation of endangered species, investigation of the effects of toxicants on spermatogenesis, investigation of endocrine regulation of spermatogenesis, and the production of transgenic spermatozoa following the genetic manipulation of testis tissue before grafting. Although testis tissue grafting does produce sperm, it is by no means normal spermatogenesis when considering sperm production and sperm transport. However, this technique provides a novel tool for investigating the fundamental aspects of testis function for large, agriculturally significant species.

Ectopic tissue xenografting provides a useful means for researchers to investigate factors and mechanisms important to germ cell differentiation, Sertoli cell populations, and production of sperm in various livestock species without maintaining those bulls and boars, respectively. For example, in bovine testis tissue grafts, only about 10% of the seminiferous tubules are capable of undergoing complete spermatogenesis after a 24-week grafting period (Oatley et al. 2004a,b). As a result of this relatively low efficiency, factors that upregulate or downregulate Sertoli cell proliferation and germ cell differentiation in the testis tissue grafts can be more easily observed.

The differentiation of testis tissue following grafting varies depending on the age of the donor animal (Oatley et al. 2004a; Schmidt et al. 2007; Caires et al. 2008). During development, testicular cells are exposed to increasing concentrations of FSH and LH. FSH is critical for the establishment of the Sertoli cell population, and LH stimulates developing Leydig cells to produce testosterone. Both of these processes are essential for germ cell differentiation. In adults, negative feedback from testosterone production suppresses FSH and LH production. Therefore, grafting tissue from neonatal animals onto intact, adult animals does not provide a similar endocrine environment that is present in the donor animal. It appears likely that removing the gonad of the recipient mouse prior to testis transplantation results in an ideal environment for cell differentiation, the initiation of spermatogenesis, and hormone production in the grafted testis tissue, resulting in sperm production. However, the high gonadotropin concentrations present in castrated mice may influence cell differentiation. For example, pig testis tissue that was grafted on nude mice showed complete spermatogenesis earlier than in a normal pig testis (Honaramooz et al. 2002). This means that germ cell differentiation was accelerated in grafted pig testis tissue compared with the normal time required for germ cell differentiation in an intact pig. Similarly, many seminiferous tubules in grafted tissue have larger diameters than tubules in testes that are still attached to the animal (Oatley et al. 2004b; Schmidt et al. 2006a; Caires et al. 2008). This may be a result of high FSH concentrations stimulating Sertoli cell proliferation without the normal feedback mechanisms regulating the arrest of Sertoli cell mitosis just before puberty. Although a doubling of the Sertoli cell number increases sperm output in rats, hyperproliferation of Sertoli cells in ectopic grafted testis tissue may result in seminiferous tubules that are too large to support germ cell differentiation.

Variation in spermatid production in testis grafts following the grafting period represents a unique method for evaluating the timing necessary for germ cells to differentiate in different species. For example, bovine testis tissue from 2-week-old calves grafted onto mice and removed 24 weeks later has few seminiferous tubules with elongating spermatids. However, if the grafting period is extended to 36 weeks, the percent of seminiferous tubules supporting germ cell differentiation is significantly higher (Schmidt et al. 2006a). These results suggest that, to some extent, the intrinsic mechanisms regulating germ cell differentiation do not change when the tissue is grafted onto mice or exposed to a different endocrine environment. Similarly, manipulation of the endocrine environment of the host mouse may represent a novel approach for investigating how systemic factors regulate testis development and germ cell differentiation.

12.3.3 Manipulation of testis tissue before xenografting

The success of bovine testis grafts depends on many factors including: donor age, endocrine environment of the recipient, and endogenous treatments of testis tissue prior to (or during) the grafting period with factors that may potentially regulate testis function. Schmidt et al. (2006b) treated testis tissue at the time of grafting with 1 μg/ml of vascular endothelial growth factor (VEGF), a potent angiogenic factor. The hypothesis was that testis tissue treated with VEGF before grafting would significantly increase angiogenesis in the grafts, leading to improved graft survival. Results showed that VEGF treatment increased graft weight and spermatogenesis in grafted tissue but did not increase blood vessel numbers in grafted tissue. VEGF is produced in the testis and gene expression is induced by human chorionic gonadotropin (hCG) treatment (Haggstrom Rudolfsson et al. 2003). In the human testis, VEGF and its receptors, VEGFR-1 and VEGFR-2, are localized to both the Sertoli and Leydig cells. Additionally, VEGFR-1 and VEGFR-2 are found on the testicular capillary endothelial cells (Ergun et al. 1997) and germ cells (Korpelainen et al. 1998). VEGF receptors are differentially expressed on developing germ cells. VEGFR-2 is present on spermatogonia, whereas VEGFR-1 is present on spermatids (Nalbandian et al. 2003). These results indicate that VEGF may have non-endothelial cell targets in the testis. This is interesting because VEGF appears to have a positive effect on Sertoli and germ cell differentiation in the testis.

The action of VEGF on endothelial cells is an active area of research, and much is known about its mechanisms of action in this cell type. VEGF receptors are receptor tyrosine kinases (RTK), which are enzymes that can transfer a phosphate group (via autophosphorylation) to a tyrosine residue in a protein (on the c-terminus end of a receptor) following ligand binding and dimerization. Receptor protein tyrosine kinases (PTKs) possess an extracellular ligand-binding domain, a transmembrane domain, and an intracellular catalytic domain. The transmembrane domain anchors the receptor in the plasma membrane, while the extracellular domains bind growth factors. Characteristically, the extracellular domains of VEGF comprise immunoglobulin-like domain structural motifs (Shibuya and Claesson-Welsh 2006). Phosphorylation is an important function in signal transduction to regulate enzyme activity.

VEGFR-1, which is present in transmembrane and soluble forms, inhibits angiogenesis during early embryogenesis, but it also stimulates angiogenesis and inflammatory responses in postnatal life, playing a role in several human diseases such as rheumatoid arthritis and cancer. The soluble VEGFR-1 is overexpressed in placental trophoblast cells (Shibuya and Claesson-Welsh 2006). VEGFR-2 has critical functions in physiological and pathological angiogenesis through distinct signal transduction pathways regulating the proliferation and migration of endothelial cells (Shibuya and Claesson-Welsh 2006). The downstream targets of both receptors include activation of many signaling pathways (PI3K/Akt, Ras/Raf-MEK/Erk, eNOS/NO, and IP3/Ca2+, PKC, PKA) that lead to changes in gene transcription responsible for their functions as endothelial cell mitogens and vascular permeability factors (Namiecinska et al. 2005). We demonstrated that VEGF treatment supports spermatogonial survival in the bovine testis by blocking apoptosis pathways (Caires et al. 2009). Intracellular signaling cross-talk between VEGF and glial cell line-derived neurotrophic factor (GDNF) occurs in neuronal cells. As will be described in the SSC section of this chapter, GDNF regulates SSC proliferation and self-renewal. Therefore, interaction between GDNF and VEGF in cattle and possibly other mammalian testis may be an important mechanism for establishing the spermatogonia population during testis development.

Ectopic testis tissue grafting may provide an effective method for genetically manipulating male germ cells before differentiation. This approach would generate a large number of genetically modified sperm for the production of transgenic animals. Several techniques can be used to genetically modify the undifferentiated germ cells in the testis tissue, including lipofection, electroporation, or virus-mediated methods. Bovine testis tissue was electroporated with a β-galactosidase expression vector prior to grafting the tissue on mice (Oatley et al. 2004b). The grafts were removed 24 weeks later, stained for β-galactosidase activity, and evaluated for germ cell differentiation and transgene expression. Histological analysis showed that transgene expression was present in both Sertoli and differentiated germ cells but not in interstitial cells, suggesting that SSCs or spermatogonia can be genetically manipulated by electroporation and that the germ cells survive this treatment. Thus, nontargeted introduction of genes is possible with ectopic testis grafting. However, targeted gene deletion has not been attempted, and more precise methods for accomplishing this goal need to be developed.

Manipulation of testis tissue before grafting can also improve our understanding of the factors that regulate spermatogenesis. Testis tissue maintained in culture for 5–7 days before grafting is capable of producing elongating spermatids after the grafting period (Schmidt et al. 2006b). Similarly, testis tissue cryopreserved before grafting can be grafted following thaw, and produce sperm (Caires et al. 2008). As a result, several powerful applications of the technique can be employed, mainly pertaining to investigating the molecular mechanism regulating Sertoli and germ cell proliferation and differentiation in the testis. For example, culturing tissue provides a useful means of directly assessing the effects of growth factors on germ cell and Sertoli cell survival in ectopic testis tissue grafts. Also, because testis tissue can be cryopreserved before grafting and still achieve successful spermatogenesis, male germ-line preservation is another potential use of this technique.

12.3.4 SSC transplantation

Spermatogenesis is the process by which millions of sperm are produced daily within the testis. Spermatogenesis commences at puberty and continues throughout the life of the male. At the foundation of this process are the SSCs, which undergo both self-renewal and differentiation. SSC transplantation experiments pioneered by Brinster and Zimmermann (1994) provided the first and only functional assay for SSCs. This procedure involves injecting a suspension of testicular cells into the seminiferous tubules of an infertile recipient. The SSCs translocate to the basement membrane of the seminiferous tubule and colonize the recipient testis. SSC transplantation has enabled scientists to characterize the biological activity of SSCs, generate transgenic mice, and assess factors important in the culture of SSCs (Jeong et al. 2003; Oatley et al. 2007; McLean 2008). Transplantation of testicular cells from many species into the seminiferous tubules of immunodeficient mice has been used to determine if SSCs are present in a cell population (Dobrinski et al. 1999; Dobrinski et al. 2000; Oatley et al. 2002). Transplantation of SSCs from species that are closely related to mice (i.e., rats and hamsters) will result in complete germ cell differentiation and sperm formation from the donor SSCs (Clouthier et al. 1996). However, transplantation of SSCs from species that are more divergent from mice does not result in germ cell differentiation. Interestingly, the SSCs in these experiments survived but did not differentiate beyond the undifferentiated spermatogonia cell type. However, there are examples of transplantation of donor germ cells from livestock species into the testes of the same species, resulting in donor-derived spermatogenesis (Honaramooz et al. 2003).

Research focused on SSCs in mice and livestock have provided valuable information about the factors that regulate the initiation of germ cell differentiation leading to the production of sperm. Enrichment and culture of primary SSCs is the most direct way of determining the mechanisms regulating SSC self-renewal and proliferation. Rarity of SSCs in the testis poses a challenge for establishing long-term cultures of SSCs when total testis cell populations are utilized. Thus, it is essential that a cell fraction enriched for SSCs be utilized. Because there are currently no known specific morphological or phenotypic markers for SSCs, they cannot be isolated as a pure cell population from the testis. Selection strategies rely on collection of cell fractions enriched for SSCs, which are effective because nearly all somatic cells (e.g., testicular fibroblasts, myoid, Leydig, and Sertoli cells) and more mature germ cells (e.g., differentiating spermatogonia, spermatocytes, and spermatids) are removed. Currently, techniques for isolating SSC-enriched cell fractions from total testis cell populations are available for mice (Kubota et al. 2004a), rats (Hamra et al. 2004; Ryu et al. 2004), and nonhuman primates (Müller et al. 2008). In rodents and primates, isolation of cells expressing specific surface molecules has provided the most efficient means for collecting SSC-enriched fractions. In the adult mouse, isolation of testis cells that express the surface marker Thy1 results in 300-fold enrichment for SSCs compared with total testis cell populations (Kubota et al. 2004a). In the rat, selection of epithelia cell adhesion molecule (Ep-CAM)-positive testis cells results in 120-fold enrichment of SSCs (Ryu et al. 2004). Similarly, isolation of testis cell populations that preferentially bind to laminin also results in enrichment of SSCs from both mouse and rat testes (Shinohara et al. 2003; Hamra et al. 2004). Laminin is a major component of the seminiferous tubular basement membrane of most mammals including livestock. Thus, it is reasonable to hypothesize that laminin-binding cells from testes of livestock species will also be enriched for SSCs; however, this possibility has not been tested. In addition, isolation of SSC-enriched fractions from testes of any livestock species based on expression of the surface molecules Thy1 or Ep-CAM has not been reported. Studies by Aponte et al. (2006) and Izadyar et al. (2002, 2003) used gravity sedimentation through bovine serum albumin (BSA) gradients to isolate spermatogonia-enriched fractions from bull testes; however, the SSC content of these cell populations was not determined.

Maintenance of SSCs *in vitro* for extended periods, in conditions that support their self-renewal, allows for expansion of SSC numbers. Currently, techniques for long-term culture of SSCs are only available for mice (Kubota et al. 2004b; Oatley and Brinster 2008), rats (Ryu et al. 2005), and hamsters (Kanatsu-Shinohara et al. 2008). Previous studies have resulted in short-term proliferation of bovine SSCs, but long-term maintenance has not been achieved (Oatley et al. 2004a,c; Aponte et al. 2008). Additionally, culture of SSCs from other livestock species has not been reported. In rodents, long-term maintenance of SSCs requires culture on mitotically inactive feeder cell monolayers in optimized serum-free media with specific nutrient and growth factor supplementations. These types of conditions have not been evaluated for SSCs of any livestock animal. With rodents, feeder cells derived from mouse embryos are effective at supporting long-term self-renewing SSC expansion. Immortalized STO feeder cell monolayers were shown to support both mouse (Kubota et al. 2004b) and rat (Ryu et al. 2005) SSC expansion for greater than 5 months in culture. With bulls, primary bovine embryonic fibroblasts (BEF) have been used as feeders to support short-term expansion of bovine SSCs (Oatley et al. 2004a). In those studies, SSC numbers increased over a 7-day period but rapidly declined after 14 days, suggesting that long-term self-renewal cannot be supported.

Effective expansion of rodent SSCs *in vitro* has relied on the use of serum-free conditions (Kubota et al. 2004b; Ryu et al. 2005). In contrast, several cell types, including embryonic stem cells, require the addition of fetal bovine serum (FBS) in basal media to support growth. The richness of nutrients in FBS preferentially supports proliferation of rapidly dividing cells. Because SSCs divide relatively slowly (Kubota et al. 2004b), other rapidly dividing cell types such as testicular fibroblasts outgrow SSCs when cultured in serum-containing media, resulting in loss of SSCs over time. Also, FBS appears to have toxic effects on mouse and rat SSCs in culture (Kubota et al. 2004a; Ryu et al. 2005). Previous attempts at culturing bovine SSCs have included FBS in basal media (Dobrinski et al. 2000; Oatley et al. 2002; Oatley et al. 2004a,c; Aponte et al. 2006). In those studies, short-term expansion of bovine SSCs was observed, followed by a rapid decline of SSC numbers at which time fibroblast takeover was observed, which likely impaired SSC proliferation and survival (Dobrinski et al. 2000; Oatley et al. 2002; Oatley et al. 2004c). To date, attempts to maintain SSCs of any livestock species in serum-free media conditions have not been reported.

SSC self-renewing proliferation in serum-free conditions is limited without the addition of specific growth factors. Inclusion of the growth factor GDNF is essential for expansion of mouse, rat, and hamster SSCs when cultured in defined conditions (Kubota et al. 2004b; Ryu et al. 2005; Kanatsu-Shinohara et al. 2008). Additionally, preliminary bovine studies showed that addition of GDNF into cultures of bovine germ cells in nonoptimized medium enhanced short-term expansion over a 14-day period (Oatley et al. 2004c). These results suggest that there is conservation among mammalian species for specific growth factors that influence SSC self-renewal. Additional mouse studies have shown that insulin-like growth factor 1 (IGF-1), epidermal growth factor (EGF), basic fibroblast growth factor (bFGF), and leukemia inhibitory factor (LIF) also influence SSC proliferation in serum-free conditions (Kubota et al. 2004a,b; Kanatsu-Shinohara et al. 2008). To date, influences of these factors on the self-renewal of SSCs *in vitro* from any livestock species has not been evaluated. The entire milieu of growth factors that control SSC self-renewal has yet to be discovered, and SSCs of each species may require specific combinations to promote self-renewing proliferation.

12.4 Reproductive genomics in boars

12.4.1 Boar testis development

Efficient sperm production in boars is dependent on germ and somatic cell maturation during neonatal and prepubertal development. As with all mammals, during this time, somatic cells undergo critical proliferation and differentiation events that ultimately determine the baseline for future sperm production in the adult. However, the mechanisms responsible for these biological processes in the developing boar testis remain unclear. The majority of published work regarding male reproductive biology and testis development has been conducted in rodent models, and thus a better understanding of factors regulating the onset and maintenance of spermatogenesis in boars is lacking.

Somatic cells, including Sertoli cells and Leydig cells, account for the majority of estrogen and testosterone synthesis in the male but also produce nutrients and growth factors essential for regulating germ cell differentiation. Although in theory the boar can generate millions of sperm daily, the number of Sertoli cells contained within the testes ultimately determines the capacity for sperm production. The timeline for Sertoli cell proliferation and differentiation events are highly species dependent, occur before puberty, and are essential for fertility. In rodents, it is established that during prenatal and postnatal testis development, FSH stimulates Sertoli cells to proliferate (Griswold 1993). In contrast, Sertoli cells halt proliferation and initiate terminal differentiation in response to thyroid hormones, testosterone, and retinoic acid signaling (Orth 1982; Buzzard et al. 2003; Holsberger and Cooke 2005).

The majority of research in porcine testis development has focused on biological events occurring between 1 month after birth and pubertal age (approximately 5 months of age in European breeds), and it is known that Sertoli cell expansion in the boar testis occurs between birth and approximately 20 weeks of age (Erickson 1964; Putra and Blackshaw 1985). In mice, ablating the biological activity of thyroid hormones during postnatal life results in an extended period of Sertoli cell proliferation, larger testes, and an increase in spermatogenic capacity (Joyce et al. 1993). A similar study evaluating the effect of postnatal hypothyroidism was conducted in 3-week-old boars, and the authors concluded that no increase in Sertoli cell proliferation occurred following treatment (Klobucar et al. 2003), in striking contrast to findings in rodents.

12.4.2 The Meishan model

The Meishan breed presents a unique model for investigating the endocrine regulation of spermatogenesis in swine. The Meishan is a slow-growing breed of swine originating from the Taihu Lake region outside of Shanghai, China. Evidence suggests that divergence between European breeds and the

Meishan occurred roughly between 2000 and 500,000 years ago (Paszek et al. 1998; Giuffra et al. 2000) while domestication of wild boars took place approximately 9000 years ago (Bokonyi 1974).

When compared with boars originating from conventional breeds of swine, Meishan boars experience hastened puberty and significantly elevated (2- to 10-fold) levels of gonadotropins and testosterone during establishment of spermatogenesis (Lunstra et al. 1997). These endocrine differences are maintained in mature animals due to a larger population of pituitary gonadotrophs and increased expression of genes encoding the subunits of FSH and LH (Li et al. 1997). Furthermore, Meishan boars have a unique testicular composition that is characterized by a twofold increase in the proportion of interstitial tissue compared with total testicular volume. In addition, mean Leydig cell size is threefold larger when compared with boars of European descent (Okwun et al. 1996a,b; McCoard et al. 2003a). In contrast, both testis size and the number of Sertoli cells are significantly reduced in Meishan boars when compared with boars of European descent (McCoard et al. 2003a,b). Interestingly, sperm production is not adversely affected by this phenomenon; in fact, the efficiency of spermatogenesis (daily sperm production per Sertoli cell) is twofold greater in Meishan boars (Okwun et al. 1996a,b). Moreover, Sertoli cell volume is significantly larger in the Meishan boar testes (McCoard et al. 2003a). Thus, the unique physiology of the Meishan boar provides an excellent model for the investigation of factors regulating testis development and affords the opportunity to better understand mechanisms governing germ and somatic cell biology in the male.

Hemicastration of males before puberty (i.e., removal of one testis while leaving the contralateral testis in place) results in compensatory testicular hypertrophy in mammals as a result of increased numbers of germ and Sertoli cells (Brown and Chakraborty 1991; Orth 1993). Thus, prepubertal hemicastration is a useful model for studying factors governing germ and somatic cell proliferation in the mammalian testis. A study evaluated the proliferative response following hemicastration in crossbred Meishan × White Composite boars by dividing animals into large (Lg) and small (Sm) testis groups. Hemicastration stimulated Leydig and Sertoli cell proliferation in Lg testis boars (Lunstra et al. 2003). In contrast, populations of Leydig and Sertoli cells in Sm testis boars expanded due to increases in size, but not number (Lunstra et al. 2003). The number of Sertoli cells per testis was maximal 56 days after birth in Sm testis boars. However, a longer duration of Sertoli cell proliferation was observed in Lg testis boars as Sertoli cell number per testis reached a maximum of 112 days after birth (Lunstra et al. 2003). The mechanisms responsible for differences in somatic cell growth and differentiation are unknown but could provide insight into methods for increasing sperm production in males.

12.4.3 Candidate genes and quantitative trait loci (QTLs) for boar phenotypes

In order to provide an effective means for understanding the genetic mechanisms responsible for the reproductive phenotypes in the Meishan boars, a resource population was developed by scientists from the US Meat Animal Research Center's (MARC) Swine Resource Population. Reciprocal matings of purebred Meishan (Ms) and White Composite (WC; Chester White, Landrace, Large White, and Yorkshire) were conducted to produce F1 and F3 generations, respectively. A genomic scan identified a centrometric region (~80 cM) of the *Sus scrofa* X chromosome that was associated with pubertal gonadotropin concentrations and testis size in boars (Ford et al. 2001; Rohrer et al. 2001). The gene encoding thyroxine-binding globulin (TBG), the primary transporter and regulator of thyroid hormone availability in serum (Bartalena and Robbins 1993), resides within QTL for testis size and plasma FSH as demonstrated by using a comparative mapping approach of the porcine X chromosome (McCoard et al. 2002).

As discussed previously, thyroid hormones regulate Sertoli cell proliferation in rodents (Cooke et al. 1994) and bulls (Majdic et al. 1998), and thus TBG represents a candidate gene influencing testis development in swine. In a follow-up study, Nonneman et al. (2005) developed the hypothesis that TBG is responsible for causing decreased circulating levels of triiodothyronine (T3) and thyroxine (T4), resulting in an extended period of Sertoli cell proliferation responsible for the increased testis size characteristic of European breeds (WC) when compared with Asiatic breeds like the Meishan (Ms). To test this hypothesis, germplasm (F8 and F10) from the original research population consisting of 3/4 WC X 1/4 Ms was used to identify positional single nucleotide polymorphisms (SNPs) in TBG that affect endocrine parameters and testis growth in developing boars (Nonneman et al. 2005). The porcine TBG gene was sequenced and a nonconservative adenine to cytosine polymorphism (codon 226) in exon 2 was identified, which resulted in a consensus change of a histidine (Ms) to an asparagine (WC) in the ligand-binding domain of the mature TBG protein (Nonneman et al. 2005). The consensus C allele was found to be Meishan-specific and is associated with reduced testis size. The A allele was associated with increased affinity of mature TBG for thyroid hormones, a reduction in circulating free T4 and T3, and a significant increase in testis size in WC boars (Nonneman et al. 2005). Similar biochemical activities were observed in a variant form of human TBG *in vitro* (Bertenshaw et al. 1992). The frequency of the A and C alleles in the boar population was 0.71 and 0.29, respectively (Nonneman et al. 2005). The His226Asn SNP also resides within the QTL for increased circulating FSH in boars. Thus, variation in testis size of boars is due, in part, to the effects of thyroid hormones despite conflicting observations in swine (Nonneman et al. 2005).

Another unique breed of swine with atypical testicular characteristics is the Piau, which originated in Brazil (França et al. 2000). Piau swine are similar to the Meishan in growth rate and testis size, but the adult Piau boars contain a significantly greater proportion of seminiferous tubules relative to total testicular volume when compared with Meishan boars (França et al. 2000). Thus, longitudinal studies with Piau, Meishan, and European breeds provide valuable insight into testis biology and how selection pressure for growth and carcass traits affect reproductive performance in domestic swine. Two distinct phases of Sertoli cell proliferation occur during testis development in Piau boars (França et al. 2000). The first phase of Sertoli cell proliferation is responsible for a sixfold increase in cell numbers between birth and 30 days post partum (dpp), similar to observations and proposed mechanism in mice (Vergouwen et al. 1991; Joyce et al. 1993) and rats (Orth 1982). However, a second period of Sertoli cell proliferation occurred in Piau boars between 90- and 120-dpp, as evidenced by a twofold increase in Sertoli cell number per testis during this time (França et al. 2000), in disagreement with rodent and pig studies (Gondos and Berndston 1993). Thus, two distinct populations of Sertoli cells exist in the boar testis. Both periods of Sertoli cell proliferation were associated with high FSH plasma levels, but it is important to note that significant increases in serum testosterone also occurred during these phases. This supports findings in nonhuman primates (Arslan et al. 1993) but is in striking contrast to rodent studies that suggest testosterone is a negative regulator of Sertoli cell proliferation and/or expansion (Buzzard et al. 2003). Therefore, FSH and testosterone may function independently to regulate the mitogenic activity of Sertoli cells in the developing boar testis.

12.4.4 Spermatogenesis in pigs: the critical first 14 days of life

Several reports suggest the importance of Sertoli cell proliferation in the boar testis during the first 2 weeks of life (McCoard et al. 2001, 2003b), and it is known that postnatal hypothyroidism initiated at 21-dpp does not affect testis development in boars. Thus, we hypothesized that during the first 14 days of life critical events occur in Sertoli cell homeostasis that govern the future spermatogenic capacity of mature boars. To test this possibility, we obtained testis tissue from 3-, 5-, 7-, and 14-dpp whiteline composite boars from the Washington State University swine population and evaluated them using two approaches. The first approach was to evaluate testicular growth and markers of Sertoli cell maturation during this period. The second approach utilized the ectopic testis tissue grafting bioassay to evaluate the response of germ and somatic cell populations when supported on a castrated, immunodeficient, nude mouse. In this model the recipients' endocrine environment includes high FSH and low androgen concentrations at the time of grafting, thus supporting testis tissue growth and establishment of germ and somatic cell populations (Oatley et al. 2005a). Grafts synthesize and secrete steroids and other factors that regulate pituitary function (in recipient mice), but more importantly support production of sperm at a tissue age consistent with age of puberty (20–22 weeks) in the donor species. Thus, grafting is a useful biological assay for understanding mechanisms regulating somatic cell proliferation and differentiation events associated with the establishment and maintenance of spermatogenesis.

Our results indicate that testis size in neonatal boars (3-dpp) increases by 1.5-, 2,- 6,- and 12-fold by ages 5-, 7-, 14-, and 21-dpp, respectively, and this was concomitant with an increase in Sertoli cell numbers (Caires et al. 2008). Immunohistochemical analysis indicated that Sertoli cells in the postnatal boar testis maintain a primitive phenotype until at least 3 weeks of age (Caires et al. 2008), as defined by the expression patterns of two well-characterized protein markers of Sertoli cell maturation: cytokeratin 18 (Stosiek et al. 1990) and GATA-1 (Yomogida et al. 1994; Bartůnek et al. 2003). At the time of graft removal, the effect of donor age was evaluated on testis tissue growth and androgen biosynthesis. Testis tissue weight and seminiferous tubule cross-section numbers were significantly greater (twofold) in grafts originating from 3-dpp donors, when compared with all other ages (Caires et al. 2008). In contrast, no differences in either parameters of growth were detected in grafts from 5-, 7-, and 14-dpp donors. Donor age had no effect on androgen production by grafted tissues as no differences in serum testosterone were detected by radioimmunoassay (RIA), or biological assay of vesicular gland weights in recipient mice. Recipient mice supporting testis grafts from 5-, 7-, and 14-dpp donors had FSH concentrations in serum similar to normal physiological concentrations in age-matched, intact nude mice. However, serum FSH concentrations were significantly lower than normal in recipient mice supporting testis grafts from 3-dpp donors (Caires et al. 2008).

Together these results indicate a donor age effect on the ability of testis grafts to grow and exert negative feedback on pituitary FSH. This effect was independent of testosterone and likely due to increased inhibin production from a larger population of Sertoli cells in 3-dpp donor grafts. Porcine testis tissue obtained from 3-, 5-, 7,- and 14-day-old neonatal boars were all capable of producing round and elongated spermatids after grafting. However, spermatid production was significantly greater (eightfold) in testis grafts from 14-day-old donors when compared with all other donor ages (Caires et al. 2008). No differences in the establishment of spermatogenesis were detected in grafts originating from 3-, 5-, and 7-dpp neonatal boars. Thus, we observed intrinsic differences in the biological activity of Sertoli and germ cell populations during neonatal boar testis development associated with the establishment of spermatogenesis. Interestingly, gonocytes and Sertoli cells were also immune-positive for androgen receptor protein during the first 3 weeks of life (Caires et al. 2008), and activity linked to these receptors suggests a functional role in regulating Sertoli cell hypertrophy and germ cell maturation in the neonatal boar testis. The effect of estrogenic compounds on the mitotic activity in germ and somatic cells in the developing boar testis must also be considered. Male pig fetuses secrete significant amounts of estrogens (Haeussler et al. 2007) and exhibitaromatase expression in germ and somatic cells during fetal life (Haeussler et al. 2007; Choi et al. 2009). Furthermore, inhibiting endogenous estrogen synthesis delays puberty, allowing for an extended period of proliferation and expansion of Sertoli and Leydig cells in the developing boar testis (At-Taras et al. 2008; Berger et al. 2008).

These results and unpublished observations from our lab regarding gene expression in the developing boar testis highlight the importance of FSH and thyroid hormones in regulating the initial wave of Sertoli cell proliferation following birth. We propose that this population is critical for supporting the establishment of spermatogenesis in boars. Thus, we postulate a developmental switch in which androgens, potentially in cooperation with FSH and estrogens, promote the subsequent growth and expansion of Sertoli cells in the postnatal testis from 14- to 120-dpp in commercial breeds of swine. The gene coding for androgen receptor is also located close to a QTL region on the *Sus scrofa* X chromosome, affecting testis size and FSH concentration (Nonneman et al. 2005), and thus represents a physiological candidate gene regulating testis development in boars, and should be evaluated in future genomic studies.

12.5 Future research directions

Genomic and functional investigation of the genes and proteins that are important for testis development and spermatogenesis has the potential to impact multiple aspects of reproductive physiology in the male. Improved understanding of the basic mechanisms of testis development and function could lead to modifications of sperm cryopreservation protocols, increased embryo survival, and the ability to screen males for genetic or reproductive potential at younger ages than currently possible. These changes have the potential to increase efficiency and profitability of animal agriculture. In addition, knowledge gained from these studies increases the basic foundation of information about testis biology and could be translated to human health.

References

Amann, R.P. 1983. Endocrine changes associated with onset of spermatogenesis in Holstein bulls. *Journal of Dairy Science* **66**(12): 2606–2622.

Aponte, P.M., Soda, T., van de Kant, H.J., and de Rooij, D.G. 2006. Basic features of bovine spermatogonial culture and effects of glial cell line-derived neurotrophic factor. *Theriogenology* **65**(9): 1828–1847.

Aponte, P.M., Soda, T., Teerds, K.J., Mizrak, S.C., van de Kant, H.J., and de Rooij, D.G. 2008. Propagation of bovine spermatogonial stem cells in vitro. *Reproduction* **136**(5): 543–557.

Arslan, M., Weinbauer, G.F., Schlatt, S., Shahab, M., and Nieschlag, E. 1993. FSH and testosterone, alone or in combination, initiate testicular growth and increase the number of spermatogonia and Sertoli cells in a juvenile non-human primate (*Macaca mulatta*). *The Journal of Endocrinology* **136**(2): 235–243.

At-Taras, E.E., Kim, I.C., Berger, T., Conley, A., and Roser, J.F. 2008. Reducing endogenous estrogen during development alters hormone production by porcine Leydig cells and seminiferous tubules. *Domestic Animal Endocrinology* **34**(1): 100–108.

Bartalena, L. and Robbins, J. 1993. Thyroid hormone transport proteins. *Clinics in Laboratory Medicine* **13**(3): 583–598.

Bartůnek, P., Králová, J., Blendinger, G., Dvorák, M., and Zenke, M. 2003. GATA-1 and c-myb crosstalk during red blood cell differentiation through GATA-1 binding sites in the c-myb promoter. *Oncogene* **22**(13): 1927–1935.

Berger, T., McCarthy, M., Pearl, C.A., At-Taras, E., Roser, J.F., and Conley, A. 2008. Reducing endogenous estrogens during the neonatal and juvenile periods affects reproductive tract development and sperm production in postpuberal boars. *Animal Reproduction Science* **109**(1–4): 218–235.

Bertenshaw, R., Sarne, D., Tornari, J., Weinberg, M., and Refetoff, S. 1992. Sequencing of the variant thyroxine-binding globulin (TBG)-San Diego reveals two nucleotide substitutions. *Biochimica et Biophysica Acta* **1139**: 307–310.

Bokonyi, S. 1974. *History of Domestic Mammals in Central and Eastern Europe*. Budapest: Akademiai Kiado.

Brinster, R.L. and Zimmermann, J.W. 1994. Spermatogenesis following male germ-cell transplantation. *Proceedings of the National Academy of Sciences of the United States of America* **91**: 11298–11302.

Brown, J.L. and Chakraborty, P.K. 1991. Comparison of compensatory pituitary and testicular responses to hemicastration between prepubertal and mature rats. *Journal of Andrology* **12**: 119–125.

Buzzard, J.J., Wreford, N.G., and Morrison, J.R. 2003. Thyroid hormone, retinoic acid, and testosterone suppress proliferation and induce markers of differentiation in cultured rat Sertoli cells. *Endocrinology* **144**(9): 3722–3731.

Caires, K.C., Schmidt, J.A., Oliver, A.P., de Avila, J., and McLean, D.J. 2008. Endocrine regulation of the establishment of spermatogenesis in pigs. *Reproduction in Domestic Animals = Zuchthygiene* **43**(Supplement 2): 280–287.

Caires, K.C., de Avila, J.M., and McLean, D.J. 2009. Vascular endothelial growth factor A (VEGF) regulates germ cell survival during the establishment of spermatogenesis in the bovine testis. *Reproduction* **138**(4): 667–677.

Choi, I., Kim, J.Y., Lee, E.J., Kim, Y.Y., Chung, C.S., Chang, J., Choi, N.J., Chung, H.J., and Lee, K.H. 2009. Ontogeny of expression and localization of steroidogenic enzymes in the neonatal and prepubertal pig testes. *Journal of Andrology* **30**(1): 57–74

Clouthier, D.E., Avarbock, M.R., Maika, S.D., Hammer, R.E., and Brinster, R.L. 1996. Rat spermatogenesis in mouse testis. *Nature* **381**(6581): 418–421.

Cooke, P.S., Zhao, Y.D., and Bunick, D. 1994. Triiodothyronine inhibits proliferation and stimulates differentiation of cultured neonatal Sertoli cells: Possible mechanism for increased adult testis weight and sperm production induced by neonatal goitrogen treatment. *Biology of Reproduction* **51**: 1000–1005. [Published correction appears in *Biology of Reproduction* 1998; 59: 216].

Cupp, A.S. and Skinner, M.K. 2005. Embryonic Sertoli cell differentiation. In: Skinner, M.K. and Griswold, M.D. (eds.), *Sertoli Cell Biology*. Amsterdam: Elsevier, pp. 43–70.

Curtis, S.K. and Amann, R.P. 1981. Testicular development and establishment of spermatogenesis in Holstein bulls. *Journal of Animal Science* **53**: 1645–1657.

de Rooij, D.G. and Russell, L.D. 2000. All you wanted to know about spermatogonia but were afraid to ask. *Journal of Andrology* **21**: 776–798.

Dobrinski, I., Avarbock, M.R., and Brinster, R.L. 1999. Transplantation of germ cells from rabbits and dogs into mouse testes. *Biology of Reproduction* **61**: 1331–1339.

Dobrinski, I., Avarbock, M.R., and Brinster, R.L. 2000. Germ cell transplantation from large domestic animals into mouse testes. *Molecular Reproduction and Development* **57**: 270–279.

Erickson, B.H. 1964. Effects of neonatal gamma irradiation on hormone production and spermatogenesis in the testis of the adult pig. *Journal of Reproduction and Fertility* **8**: 91–100.

Ergun, S., Kilie, N., Fiedler, W., and Mukhopadhyay, A.K. 1997. Vascular endothelial growth factor and its receptors in normal human testicular tissue. *Molecular and Cellular Endocrinology* **131**: 9–20.

Ford, J.J., Wise, T.H., Lunstra, D.D., and Rohrer, G.A. 2001. Interrelationships of porcine X and Y chromosomes with pituitary gonadotropins and testicular size. *Biology of Reproduction* **65**(3): 906–912.

França, L.R., Silva, V.A. Jr., Chiarini-Garcia, H., Garcia, S.K., and Debeljuk, L. 2000. Cell proliferation and hormonal changes during postnatal development of the testis in the pig. *Biology of Reproduction* **63**(6): 1629–1636.

Giuffra, E., Kijas, J.M.H., Amarger, V., Carlborg, O., Jeon, J.T., and Andersson, L. 2000. The origin of the domestic pig: Independent domestication and subsequent introgression. *Genetics* **154**: 1758–1791.

Gondos, B. and Berndston, W.E. 1993. Postnatal and pubertal development. In: Russell, L.D. and Griswold, M.D. (eds.), *The Sertoli Cell*, 1st Edition. Clearwater, FL: Cache River Press, pp. 115–154.

Griswold, M.D. 1993. Actions of FSH on mammalian Sertoli cells. In: Russell, L.D. and Griswold, M.D. (eds.), *The Sertoli Cell*, 1st Edition. Clearwater, FL: Cache River Press, pp. 493–508.

Haeussler, S., Wagner, A., Welter, H., and Claus, R. 2007. Changes of testicular aromatase expression during fetal development in male pigs (*Sus scrofa*). *Reproduction* **133**(1): 323–330.

Haggstrom Rudolfsson, S., Johansson, A., Franck Lissbrant, I., Wikstrom, P., and Bergh, A. 2003. Localized expression of angiopoietin 1 and 2 may explain unique characteristics of the rat testicular microvasculature. *Biology of Reproduction* **69**(4): 1231–1237.

Hamra, F.K., Schultz, N., Chapman, K.M., Grellhesl, D.M., Cronkhite, J.T., Hammer, R.E., and Garbers, D.L. 2004. Defining the spermatogonial stem cell. *Developmental Biology* **269**(2): 393–410.

He, Z., Chan, W.Y., and Dym, M. 2006. Microarray technology offers a novel tool for the diagnosis and identification of therapeutic targets for male infertility. *Reproduction* **132**(1): 11–19.

Holsberger, D.R. and Cooke, P.S. 2005. Understanding the role of thyroid hormone in Sertoli cell development: A mechanistic hypothesis. *Cell and Tissue Research* **322**(1): 133–140.

Honaramooz, A., Behboodi, E., Megee, S.O., Overton, S.A., Galantino-Homer, H., Echelard, Y., and Dobrinski, I. 2003. Fertility and germline transmission of donor haplotype following germ cell transplantation in immunocompetent goats. *Biology of Reproduction* **69**(4): 1260–1264.

Honaramooz, A., Snedaker, A., Bolani, M., Scholer, H., Dobrinski, I., and Schlatt, S. 2002. Sperm from neonatal mammalian testes grafted in mice. *Nature* **418**: 778–781.

Izadyar, F., Spierenberg, G.T., Creemers, L.B., den Ouden, K., and de Rooij, D.G. 2002. Isolation and purification of type A spermatogonia from the bovine testis. *Reproduction* **124**(1): 85–94.

Izadyar, F., Den Ouden, K., Creemers, L.B., Posthuma, G., Parvinen, M., and de Rooij, D.G. 2003. Proliferation and differentiation of bovine type A spermatogonia during long-term culture. *Biology of Reproduction* **68**(1): 272–281.

Jeong, D.K., McLean, D.J., and Griswold, M.D. 2003. Long-term culture and transplantation of murine testicular germ cells. *Journal of Andrology* **24**(5): 661–669.

Johnson, L., Suggs, L.C., Norton, Y.M., Welsh, T.H. Jr., and Wilker, C.E. 1996. Effect of hypophysectomy, sex of host, and/or number of transplanted testes on Sertoli cell number and testicular size of syngeneic testicular grafts in Fischer rats. *Biology of Reproduction* 1996 **54**(5): 960–969.

Johnston, D.S., Wright, W.W., Dicandeloro, P., Wilson, E., Kopf, G.S., and Jelinsky, S.A. 2008. Stage-specific gene expression is a fundamental characteristic of rat spermatogenic cells and Sertoli cells. *Proceedings of the National Academy of Sciences of the United States of America* **105**(24): 8315–8320.

Joyce, K.L., Porcelli, J., and Cooke, P.S. 1993. Neonatal goitrogen treatment increased adult testis size and sperm production in the mouse. *Journal of Andrology* **14**: 448–455.

Kanatsu-Shinohara, M., Muneto, T., Lee, J., Takenaka, M., Chuma, S., Nakatsuji, N., Horiuchi, T., and Shinohara, T. 2008. Long-term culture of male germline stem cells from hamster testes. *Biology of Reproduction* **78**(4): 611–617.

Klobucar, I., Kosec, M., Cebulj-Kadunc, N., and Majdic, G. 2003. Postnatal hypothyroidism does not affect prepubertal testis development in boars. *Reproduction in Domestic Animals = Zuchthygiene* **38**(3): 193–198.

Korpelainen, E.I., Karkkainen, M.J., Tehunen, A., Lakso, M., Rauvala, H., Vierula, M., Parvinen, M., and Alitalo, K. 1998. Overexpression of VEGF in the testis and epididymis causes infertility in transgenic mice: Evidence for nonendothelial targets for VEGF. *The Journal of Cell Biology* **143**: 1705–1712.

Kubota, H., Avarbock, M.R., and Brinster, R.L. 2004a. Growth factors essential for self-renewal and expansion of mouse spermatogonial stem cells. *Proceedings of the National Academy of Sciences of the United States of America* **101**: 16489–16494.

Kubota, H., Avarbock, M.R., and Brinster, R.L. 2004b. Culture conditions and single growth factors affect fate determination of mouse spermatogonial stem cells. *Biology of Reproduction* **71**(3): 722–731.

Lewin, H.A. 2003. The future of cattle genome research: The beef is here. *Cytogenetic and Genome Research* **102**: 10–15.

Li, M.D., MacDonald, G.J., and Ford, J.J. 1997. Breed differences in expression of inhibin/activin subunits in porcine anterior pituitary glands. *Endocrinology* **138**(2): 712–718.

Lunstra, D.D., Ford, J.J., Klindt, J., and Wise, T.H. 1997. Physiology of the Meishan boar. *Journal of Reproduction and Fertility Supplement* **52**: 181–193.

Lunstra, D.D., Wise, T.H., and Ford, J.J. 2003. Sertoli cells in the boar testis: Changes during development and compensatory hypertrophy after hemicastration at different ages. *Biology of Reproduction* **68**: 140–150.

Majdic, G., Snoj, T., Horvat, A., Mrkun, J., Kosec, M., and Cestnik, V. 1998. Higher thyroid hormone levels in neonatal life result in reduced testis volume in postpubertal bulls. *International Journal of Andrology* **21**(6): 352–357.

Mauduit, C., Hamamah, S., and Benahmed, M. 1999. Stem cell factor/c-kit system in spermatogenesis. *Human Reproduction Update* **5**: 535–545.

McCoard, S.A., Fahrenkrug, S.C., Alexander, L.J., Freking, B.A., Rohrer, G.A., Wise, T.H., and Ford, J.J. 2002. An integrated comparative map of the porcine X chromosome. *Animal Genetics* **33**: 178–185.

McCoard, S.A., Lunstra, D.D., Wise, T.H., and Ford, J.J. 2001. Specific staining of Sertoli cell nuclei and evaluation of Sertoli cell number and proliferative activity in Meishan and white composite boars during the neonatal period. *Biology of Reproduction* **64**(2): 689–695.

McCoard, S.A., Wise, T.H., and Ford, J.J. 2003a. Endocrine and molecular influences on testicular development in Meishan and white composite boars. *The Journal of Endocrinology* **178**(3): 405–416.

McCoard, S.A., Wise, T.H., Lunstra, D.D., and Ford, J.J. 2003b. Stereological evaluation of Sertoli cell ontogeny during fetal and neonatal life in two diverse breeds of swine. *The Journal of Endocrinology* **178**(3): 395–403.

McLean, D.J. 2008. Spermatogonial stem cell transplantation and testicular function. In: Hou, S. (ed.), vol. **450**, *Methods in Molecular Biology: Germline Stem Cells.* Totowa, NJ: Humana Press, pp. 149–162.

McLean, D.J., Friel, P.J., Johnston, D.S., and Griswold, M.D. 2003. Characterization of spermatogonial stem cell maturation and differentiation in neonatal mice. *Biology of Reproduction* **69**: 2085–2091.

McLean, D.J., Friel, P.J., Pouchnik, D., and Griswold, M.D. 2002. Oligonucleotide microarray analysis of gene expression in follicle-stimulating hormone-treated rat

Sertoli cells. *Molecular Endocrinology* **16**(12): 2780–2792.

Müller, T., Eildermann, K., Dhir, R., Schlatt, S., and Behr, R. 2008 Glycan stem-cell markers are specifically expressed by spermatogonia in the adult non-human primate testis. *Human Reproduction* **23**(10): 2292–2298.

Nagano, R., Tabata, S., Nakanishi, Y., Ohsako, S., Kurohmaru, M., and Hayashi, Y. 2000. Reproliferation and relocation of mouse malegerm cells (gonocytes) during prespermatogenesis. *The Anatomical Record* **258**: 210–220.

Nalbandian, A., Dettin, L., Dym, M., and Ravindranath, N. 2003. Expression of vascular endothelial growth factor receptors during male germ cell differentiation in the mouse. *Biology of Reproduction* **69**: 985–994.

Namiecinska, M., Marciniak, K., and Nowak, J.Z. 2005. VEGF as an angiogenic, neurotrophic, and neuroprotective factor. *Postepy higieny i medycyny doswiadczalnej* (Online) **59**: 573–583.

Nonneman, D., Rohrer, G.A., Wise, T.H., Lunstra, D.D., and Ford, J.J. 2005. A variant of porcine thyroxine-binding globulin has reduced affinity for thyroxine and is associated with testis size. *Biology of Reproduction* **72**(1): 214–220.

Oatley, J.M., Avarbock, M.R., and Brinster, R L. 2007 Glial cell line-derived neurotrophic factor regulation of genes essential for self-renewal of mouse spermatogonial stem cells is dependent on Src family kinase signaling. *The Journal of Biological Chemistry* **282**: 25842–25851.

Oatley, J.M., Avarbock, M.R., Telaranta, A.I., Fearon, D.T., and Brinster, R.L. 2006. Identifying genes important for spermatogonial stem cell self-renewal and survival. *Proceedings of the National Academy of Sciences of the United States of America* **103**: 9524–9529.

Oatley, J.M. and Brinster, R.L. 2008. Regulation of spermatogonial stem cell self-renewal in mammals. *Annual Review of Cell and Developmental Biology* **24**: 263–286.

Oatley, J.M., de Avila, D.M., McLean, D.J., Griswold, M.D., and Reeves, J.J. 2002. Transplantation of bovine germinal cells into mouse testes. *Journal of Animal Science* **80**: 1925–1931.

Oatley, J.M., de Avila, D.M., Reeves, J.J., and McLean, D.J. 2004a. Testis tissue explant culture supports survival and proliferation of bovine spermatogonial stem cells. *Biology of Reproduction* **70**: 625–631.

Oatley, J.M., de Avila, D.M., Reeves, J.J., and McLean, D.J. 2004b. Spermatogenesis and germ cell transgene expression in xenografted bovine testicular tissue. *Biology of Reproduction* **71**: 494–501.

Oatley, J.M., Reeves, J.J., and McLean, D.J. 2004c. Biological activity of cryopreserved bovine spermatogonial stem cells during in vitro culture. *Biology of Reproduction* **71**: 942–947.

Oatley, J.M., Reeves, J.J., and McLean, D.J. 2005a. Establishment of spermatogenesis in neonatal bovine testicular tissue following ectopic xenografting varies with donor age. *Biology of Reproduction* **72**: 358–364.

Okwun, O.E., Igboeli, G., Ford, J.J., Lunstra, D.D., and Johnson, L. 1996a. Number and function of Sertoli cells, number and yield of spermatogonia, and daily sperm production in three breeds of boar. *Journal of Reproduction and Fertility* **107**(1): 137–149.

Okwun, O.E., Igboeli, G., Lunstra, D.D., Ford, J.J., and Johnson, L. 1996b. Testicular composition, number of A spermatogonia, germ cell ratios, and number of spermatids in three different breeds of boars. *Journal of Andrology* **17**(3): 301–309.

Orth, J.M. 1982. Proliferation of Sertoli cells in fetal and postnatal rats: A quantitative autoradiographic study. *The Anatomical Record* **203**: 485–492.

Orth, J.M. 1993. Cell biology of testicular development in fetus and neonate. In: Desjardins, C. and Ewing, L.L. (eds.), *Cell and Molecular Biology of the Testis*, 1st Edition. New York: Oxford University Press, pp. 3–42.

Paszek, A.A., Flickinger, G.H., Fontanesi, L., Rohrer, G.A., Alexander, L., Beattie, C.W., and Schook, L.B. 1998. Livestock variation of linked microsatellite markers in diverse swine breeds. *Animal Biotechnology* **9**(1): 55–66.

Putra, D.K.H. and Blackshaw, A.W. 1985. Quantitative studies of compensatory testicular hypertrophy following unilateral castration in the boar. *Australian Journal of Biological Sciences* **38**: 429–434.

Rohrer, G.A., Wise, T.H., Lunstra, D.D., and Ford, J.J. 2001. Identification of genomic regions controlling plasma FSH concentrations in Meishan-white composite boars. *Physiological Genomics* **6**(3): 145–151.

Rothschild, M.F. 2003. From a sow's ear to a silk purse: Real progress in porcine genomics. *Cytogenetic and Genome Research* **102**: 95–99.

Ryu, B.Y., Kubota, H., Avarbock, M.R., and Brinster, R.L. 2005. Conservation of spermatogonial stem cell self-renewal signaling between mouse and rat. *Proceedings of the National Academy of Sciences of the United States of America* **102**(40): 14302–14307.

Ryu, B.Y., Orwig, K.E., Kubota, H., Avarbock, M.R., and Brinster, R.L. 2004. Phenotypic and functional characteristics of spermatogonial stem cells in rats. *Developmental Biology* **274**(1): 158–170.

Schlatt, S., Honaramooz, A., Boiani, M., Schöler, H.R., and Dobrinski, I. 2003. Progeny from sperm obtained after ectopic grafting of neonatal mouse testes. *Biology of Reproduction* **68**: 2331–2335.

Schmidt, J.A., de Avila, J.M., and McLean, D.J. 2006a. Grafting period and donor age affect the potential for spermatogenesis in bovine ectopic testis xenografts. *Biology of Reproduction* **75**(2): 160–166.

Schmidt, J.A., de Avila, J.M., and McLean, D.J. 2006b. Effect of vascular endothelial growth factor and testis tissue culture on spermatogenesis in bovine ectopic testis xenografts. *Biology of Reproduction* **75**(2): 167–175.

Schmidt, J.A., de Avila, J.M., and McLean, D.J. 2007. Analysis of gene expression in bovine testis tissue prior to ectopic testis tissue xenografting and during the grafting period. *Biology of Reproduction* **76**(6): 1071–1080.

Shibuya, M. and Claesson-Welsh, L. 2006. Signal transduction by VEGF receptors in regulation of angiogenesis and lymphangiogenesis. *Experimental Cell Research* **312**(5): 549–560.

Shima, J.E., McLean, D.J., McCarrey, J.R., and Griswold, M.D. 2004. The murine testicular transcriptome: Characterizing gene expression in the testis during the progression of spermatogenesis. *Biology of Reproduction* **71**: 319–330.

Shinohara, T., Orwig, K.E., Avarbock, M.R., and Brinster, R.L. 2003. Restoration of spermatogenesis in infertile mice by Sertoli cell transplantation. *Biology of Reproduction* **68**(3): 1064–1071.

Small, C.L., Shima, J.E., Uzumcu, M., Skinner, M.K., and Griswold, M.D. 2005. Profiling gene expression during the differentiation and development of the murine embryonic gonad. *Biology of Reproduction* **72**: 492–501.

Stosiek, P., Kasper, M., and Karsten, U. 1990. Expression of cytokeratin 8 and 18 in human Sertoli cells of immature and atrophic seminiferous tubules. *Differentiation; Research in Biological Diversity* **43**: 66–70.

Tegelenbosch, R.A. and de Rooij, D.G. 1993. A quantitative study of spermatogonial multiplication and stem cell renewal in the C3H/101 F1 hybrid mouse. *Mutation Research* **290**: 193–200.

Turner, C.D. 1938. Intra-ocular homotransplantation of prepuberal testes in the rat. *The American Journal of Anatomy* **63**(1): 101–159.

Vergouwen, R.P., Jacobs, S.G.P.M., Huiskamp, R., Davids, J.A., and de Rooij, D.G. 1991. Proliferative activity of gonocytes, Sertoli cells and interstitial cells during testicular development in mice. *Journal of Reproduction and Fertility* **93**: 233–243.

Womack, J.E. 2005. Advances in livestock genomics: Opening the barn door. *Genome Research* **15**(12): 1699–1705.

Yomogida, K., Ohtani, H., Harigae, H., Ito, E., Nishimune, Y., Engel, J.D., and Yamamoto, M. 1994. Developmental stage- and spermatogenic cycle-specific expression of transcription factor GATA-1 in mouse Sertoli cells. *Development (Cambridge, England)* **120**: 1759–1766.

Part III

Genomics and Reproductive Biotechnology

13

The Epigenome and Its Relevance to Somatic Cell Nuclear Transfer and Nuclear Reprogramming

Jorge A. Piedrahita, Steve Bischoff, and Shengdar Tsai

13.1 Introduction

In this chapter, we will discuss the importance of nuclear reprogramming during somatic cell nuclear transfer (SCNT) and its implications for normal fetal and placental development. Such a topic requires an overview of several interrelated fields. First, the epigenome must be properly defined and its importance to the nuclear reprogramming process described. Second, the relationship of a crucial gene family for placental development and function of imprinted genes must be covered as they are particularly susceptible to epigenetic influences. In the final topic, we cover the effects of SCNT in relation to the epigenome, both the changes that occur and/or fail to occur during nuclear reprogramming, as well as their developmental consequences, in particular, in relation to placentation and fetal growth.

13.2 The epigenome

The late developmental biologist Conrad Waddington described the "epigenetic landscape" as a metaphor for how gene regulation occurs during development. The picture is of a marble rolling down a steep valley with a series of peaks and troughs. These dips in the landscape represent various cell fate decisions, and its initial conception represents the largely irreversible differentiation or lineage commitment as cells progress from a totipotent one-cell zygote to one of the many, diverse cell types that form an adult organism. More recently, epigenetics has been defined as the phenomenon that changes the outcome or phenotype without a change in genotype or underlying DNA sequence. One of the simplest examples of epigenetic regulation at work is the process of cell fate specification. The genomic DNA of all the cells that comprise the body or *soma* is equivalent (with the exception of a few cell types such as B and T cells, which undergo somatic rearrangements), as proven by the cloning of fully differentiated cells to produce a full adult animal (Humpherys et al. 2002). Given this state of "genomic equivalence" on the sequence level, what is responsible for the differences between the varying cell types? The answer is that the differences are encoded in various epigenetic modifications to the chromatin state, including DNA methylation and histone (the proteins around which DNA is wrapped) modifications.

13.2.1 DNA methylation

DNA methylation is a chemical modification of DNA that occurs in mammals only at the cytosine residues of CpG dinucleotides (a methyl group is added at the 5′ position of cytosine to make 5-methyl-cytosine). There exist CpG-rich areas of the genome known as "CpG islands" as well as CpG poor areas referred to as "CpG deserts." Two types of enzymes that actively methylate DNA in mammals have been identified: maintenance methyltransferases methylate DNA that is already methylated on one strand (hemimethylated DNA) and *de novo* methyltransferase that methylate unmethylated DNA. Dnmt1 is an abundant active DNA maintenance methyltransferase with a preference for hemimethylated DNA (Svedruzic 2008) and is responsible for maintaining DNA methylation marks during replication. Sequestering of Dnmt1 in the early stages of mammalian development is responsible for the passive demethylation of the maternal zygotic genome. Dnmt3a and Dnmt3b are *de novo* methyltransferases, which can methylate unmethylated DNA (Okano et al. 1998; Xie et al. 1999). While Dnmt3l does not exhibit intrinsic DNA methylases activity, it connects a specific histone state (unmethylated lysine 4 of histone H3) with the *de novo* methylation of DNA by Dnmt3b and is an intriguing example of the epigenetic cross talk between DNA methylation and histone modifications. Dnmt2, which was thought to be a DNA methyltransferase due to strong sequence homology, has been renamed to *TRDTM1* (tRNA aspartic acid methyltransferase 1) as it turned out not to methylate DNA at all, but instead methylate aspartic acid tRNA. DNA methyltransferases Dnmt1, Dnmt3a, and Dnmt3b have been shown to be essential for development as targeted homozygous mice at these loci do not survive (Li et al. 1992). On the other hand, targeted Dnmt1$^{-/-}$ embryonic stem (ES) cells are viable, but proliferation of these null mutant cells is limited after differentiation (Lei et al. 1996).

13.2.2 The controversy over active DNA demethylation

In comparison to active DNA methylases, the subject of active DNA demethylation is far more controversial. Evidence suggesting that an active demethylase exists, stems from observations that at an early stage in embryonic development, paternal DNA becomes actively demethylated. Because the activation energy to break the covalent bond of 5-methyl-cytosine is high, however, the mechanism of DNA methylation has been suggested to more likely progress via a base-excision repair mechanism (Ooi and Bestor 2008). Mbd2 (methyl-binding domain protein 2), a protein that shows methylation-dependent binding to DNA, was initially reported to be a DNA demethylase (Bhattacharya et al. 1999). However, these results could not be independently reproduced, and Mbd2-deficient mice exhibited normal patterns of DNA methylation. Gadd45a was also recently reported to be an active DNA demethylase that operated via a base-excision repair pathway (Barreto et al. 2007); however, a follow-up study by an independent group found that Gadd45a did not promote DNA demethylation (Jin et al. 2008). Thus, at this point, it is unclear how DNA is demethylated, but what is undisputed is the importance of DNA methylation in affecting gene expression.

Table 13.1 Selected histone 3 modifications and their associated function.

Histone 3 modification	Associated function/ marking
Lysine 4 unmodified	Recruitment of *de novo methylation*
Lysine 4 trimethylation	Active transcription
Lysine 4 and 9, overlapping	Imprinting control region
Lysine 9 trimethylation	Heterochromatin, binds HP1
Lysine 27 trimethylation	Transcriptional repression

13.2.3 Chromatin modifications

The classical view of DNA methylation is that it is a modification to DNA that represses transcription. An example of this would be the methylation of a CpG-rich promoter that prevents transcription factor binding and activation at that locus. However, this view is overly simplistic, as recent findings from genome-wide epigenetic profiles of DNA methylation have revealed a more complex relationship between DNA methylation and transcription. Yet methylation of DNA can have a drastic effect on chromatin structure, initiated by methyl-binding proteins that recognize the methylated DNA and attract additional proteins to the methylated area, resulting in chromatin configuration changes. In addition to the methylation of DNA, there are several other modifications targeted toward histone proteins that participate in modification of chromatin on a regional level. A number of posttranslational modifications can be made to these histone tails, including acetylation, phosphorylation, ubiquitination, and methylation. Table 13.1 summarizes key histone 3 modifications and their overall effects on chromatin structure and gene expression. Overall, the epigenetic language that defines chromatin structure is now only beginning to be understood. However, the current state of knowledge is that this information is likely to be encoded in these modifications to the N-terminal tails of the core histones in what is sometimes referred to as the "histone code." One can imagine the dramatic effect higher-order chromatin structure could have on transcriptional activity; in highly condensed chromatin (heterochromatin), DNA is highly inaccessible to the transcriptional machinery, and transcription is shut down. On the other hand, in uncondensed chromatin (euchromatin), the DNA is highly accessible, and high transcriptional activity is possible.

The function of DNA methylation in the context of regulation of transcription has been controversial; in particular, questions have arisen as to whether DNA methylation is a marker or a regulator of gene expression. The recent observation of a correlation between gene body methylation (methylation in the center of a transcript, as opposed to the promoter) and gene expression in both plants and animals (Hellman and Chess 2007) points to a more complex role for DNA methylation than previously suspected. The classical model for transcriptional control by DNA methylation is that of a methylated promoter, in which demethylation of the promoter permits transcription factor binding, and subsequent transcription initiation to occur. However, this view of methylation-induced transcriptional repression is being challenged by a number of recent studies. Genome-wide profiling of human DNA promoter methylation by Weber et al. (2007) suggested that although DNA methylation is sufficient to inactivate CpG island promoters, in most cases, inactive CpG island promoters remain unmethylated. In other words, DNA methylation is not required for the majority of observed tissue-specific transcriptional repression. Their work instead points to chromatin structure, via histone 3 lysine 4 dimethylation (H3K4me2), as a way to protect these CpG islands from *de novo* methylation, and also as a mechanism of transcriptional repression.

13.2.4 The relationship between chromatin marks and developmental potential

If we suppose that transcription can affect DNA methylation, what then regulates the transcriptional changes that are so precisely modulated during the course of mammalian embryonic development? The work of Bernstein et al. (2006) provided an intriguing clue with the discovery of bivalent chromatin domains with both active and repressive histone modifications (H3K4me3 and H3K27me3). These bivalent markings suggest that certain genes are repressed but poised for transcription in mouse embryonic stem cells. To assess the degree to which these bivalent chromatin domains were present in different cell types, deep sequencing technology from Illumina/Solexa was used to generate comprehensive, high-resolution chromatin maps of pluripotent and lineage-committed mammalian cells after chromatin immunoprecipitation with anti-H3K4me3 and anti-H3K4me27 antibodies (Mikkelsen et al. 2007). A wealth of data was generated by this approach—most interesting of which was that many of the genes marked by bivalent chromatin domains are involved in the regulation of development. These data strongly suggested that cell commitments and developmental potential are represented by histone 3 and lysine 4 and 27 trimethylation.

Another discovery that changed our perception of the transcriptional landscape was recently reported by Richard Young's group at MIT (Guenther et al. 2007). Using ChIP-chip methods, they observed that transcription initiation occurs at the majority of human promoters in all types of cells examined. Only at slightly more than half of the genes where transcription initiates will transcriptional elongation continue. Thus, it is not as simple as which genes are being transcribed and which are not, but a more complex regulation of the completion of elongation. In sum, the transcriptional machinery is poised at most promoters, ready to begin transcription. This near ubiquitous promoter occupancy is likely to have a regulatory role; RNA Pol II occupancy of promoters is tightly correlated with H3K4me3 (Guenther et al. 2007). It could also have a significant implication for nuclear reprogramming after SCNT from two aspects: (1) cell lines with a high proportion of bivalent chromatin domains are likely to be more easily reprogrammed as they are poised to received an inductive signal; and (2) if promoter occupancy is ubiquitous, SCNT reprogramming must likely function at the elongation step, not at the initiation step.

Other modifications such as histone acetylation also play a role in gene regulation: An inverse relationship between DNA methylation and histone acetylation is observed locally by polymerase chain reaction (PCR)-based methods, and globally using a high-resolution oligonucleotide tiling array (Hayashi et al. 2007; Wu et al. 2007). This points to a connection between DNA methylation, histone actetylation, and chromatin structure. The most direct evidence that histone modifications regulate methylation has come from Ooi et al. (2007) who demonstrated that *de novo* methylation is in fact connected to unmethylated histone H3 lysine 4 (H3K4) by physical interactions with Dnmt3l, an enzymatically inactive regulatory factor required for the establishment of DNA methylation patterns in germ cells (Nimura et al. 2006). Ooi et al. (2007) used mass spectrometry to identify the main proteins that interact with Dnmt3l: Dnmt3a2, Dnmt3b, and the four core histones. Further, they demonstrated with peptide interaction assays that Dnmt3l interacts specifically with an H3 tail that is unmodified at lysine 4; this interaction is abolished when lysine 4 is methylated. Taken together, the body of evidence from these recent publications implicates chromatin structure as a driving force behind transcriptional regulation, with DNA methylation serving as a mechanism to reinforce and stabilize transcriptional repression. (We do not, however, exclude the alternative possibility that there is a bidirectional dialogue between methylation and chromatin state.) With such a complex mode of regulation of chromatin structure, how then can one cell type be reprogrammed to behave like another one?

13.3 Epigenetic reprogramming

By reasoning that if all cells share the same genomic DNA on a sequence level, one might come to the conclusion that perhaps by simply changing their epigenetic state (epigenetic reprogramming), it might be possible to alter cellular identity. Currently, four methods have been used for the epigenetic reprogramming of cells: (1) SCNT, (2) fusion of terminally differentiated cells with ES cells, (3) exposure of cells of one type to nuclear extracts from a different cell type (transdifferentiation), and (4) cellular reprogramming by introduction of selected transcription factors using retroviral transgenesis. In all cases, the idea being that factors present in one cell type can reprogram another cell type.

13.3.1 SCNT

The first successful mammalian cloning by SCNT was reported by Campbell and Wilmut in 1996 and provided conclusive evidence that a somatic cell could be reprogrammed back to a totipotent embryonic state, capable of developing into another genetically identical animal (Campbell et al. 1996). In this method, a somatic cell is introduced into an enucleated mature metaphase II oocyte, where factors present in the oocyte "reprogram" the incoming nuclei, eventually giving rise to a developing embryo. ES cells can reprogram somatic cells by cell–cell fusion to form a tetraploid cell with ES cell-like properties (Do and Scholer 2004; Cowan et al. 2005). These reprogrammed cells end up with a transcriptional profile similar to ES cells and have the ability to differentiate into multiple cell types. However, the obvious limitation here is that the end result is a tetraploid cell, effectively excluding it from any therapeutic application. In addition, there have been experiments involving the use of cell extracts from oocytes or pluripotent cells to reprogram somatic nuclei (Collas 2003; Collas and Hakelien 2003; Collas and Taranger 2006). These experiments have demonstrated demethylation of pluripotency-associated promoter regions from a population of cells; however, the method suffers from a lack of rigorous evidence because no report to date shows reactivation of a pluripotency genetic reporter (Pou5f1-eGFP, Nanog-eGFP). It is therefore impossible to exclude the possibility that the reprogrammed cells are in fact derived from the cell extracts themselves. In spite of their individual limitations, taken together, these experiments strongly suggested the existence of reprogramming factors that can act to transform the epigenetic state of somatic cell nuclei. Yet in all cases, these reprogramming factors were not defined as they involved complex mixtures of factors. Thus, no information was provided as to what factors in the cell extracts/nuclei were actually the ones responsible for the reprogramming process.

13.3.2 Yamanaka four-factor experiment

In a seminal experiment from the group of Shinya Yamanaka, the induction of pluripotent stem cells from mouse fibroblasts using four defined factors ("Yamanaka factors") was demonstrated. By introducing these factors, they were able to convert fibroblasts into cells that had many of the characteristics of ES cells, including morphology and differentiation potential (Takahashi and Yamanaka 2006). Further refinements of this approach by Yamanaka and two other groups have resulted in the generation of induced pluripotent (iPs) cells that were able to contribute to germline chimeras (Maherali et al. 2007; Okita et al. 2007; Wernig et al. 2007). Human iPS cells have now also been generated by the direct reprogramming of fetal and adult fibroblasts, using the same defined factor approach as in mice (Takahashi et al. 2007; Yu et al. 2007; Lowry et al. 2008; Park et al. 2008).

The significance of the Yamanaka experiment cannot be downplayed. It demonstrated, irrefutably, that somatic cells could be reprogrammed by defined factors, and provides a starting point for dissecting the complex, stochastic process of nuclear reprogramming. While Dolly also demonstrated conclusively the ability of cells to be reprogrammed, it was done in a global manner and by factors that we still do not know much about. In contrast, the seminal work of Yamanaka showed that it takes a few select proteins to completely reprogram a differentiated cell. This drastically changed our view of nuclear reprogramming from one of a global event requiring a multitude of players, to a more targeted and select system. So what does it take to make an ES cell? The four factors used by Takahashi and Yamanaka (2006) to induce pluripotency are *Oct-3/4*, *Sox2*, *c-Myc*, and *Klf4* (see Figure 13.1 for the description of reprogramming using defined factors). *Oct4* is a transcription factor expressed specifically in embryonal carcinoma cells, early embryos, germ cells, and embryonic stem cells (Okamoto et al. 1990; Scholer et al. 1990). A precise level of *Oct4* expression is required for the maintenance of developmental potency; less than a twofold increase results in differentiation to endoderm or mesoderm, while repression triggers differentiation to trophectoderm (Niwa et al. 2000). The data available show that *Oct4* and *Nanog* together orchestrate the transcription of an interconnected, autoregulatory network of genes responsible for maintaining pluripotency (Wang et al. 2006). *Sox2* is a member of a family of SOX proteins that all recognize a similar binding motif, and it plays a key role in ES cell establishment as supported by the observation that ES cells cannot be established from null Sox2 mice (Avilion et al. 2003). *c-Myc* can recruit a number of histone acetyl

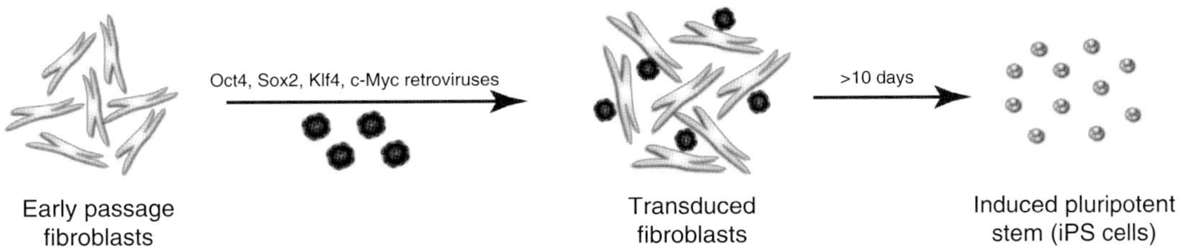

Figure 13.1 Overview of epigenetic reprogramming by retroviral transduction. Early passage fibroblasts are transduced with viruses encoding Oct4, Sox2, Klf4, and c-Myc. Transduced fibroblasts are reprogrammed over the course of >12 days into induced pluripotent stem cells. This experiment was the first example of direct reprogramming with known factors. The reprogrammed cells, however, are not suitable for clinical use as they contain random retroviral integrations (Yamanaka 2007).

transferases, including *GCN5*, *CBP*, and *p300* (Adhikary and Eilers 2005). The oncogenic properties of *c-Myc* are not surprising, given its well-characterized role as a proto-oncogene (Dalla-Favera et al. 1982; Hooker and Hurlin 2006). *c-Myc* probably contributes to the phenotype of self-renewal of iPs cells, as well as their open and active chromatin structure (Yamanaka 2007). *Klf4* is a Kruppel-like factor, zinc finger protein postulated to be required to suppress the apoptotic inducing effects of *c-Myc* (Yamanaka 2007). More recent work has indicated that *c-Myc* is not required for the generation of iPS cells from fibroblasts, and while the number of colonies obtained was reduced, conversely, the specificity of induction was increased (Nakagawa et al. 2008). While none of the factors are expressed at a high level in fibroblasts, by carefully picking a cell type with high endogenous expression of a factor, that factor could also potentially be omitted. This was found indeed to be the case with *Sox2*, which is expressed at a high level in neural stem cells, which when transduced with retroviruses with Oct4 and Klf4 were successfully reprogrammed to iPS cells (Kim et al. 2008). Finally, utilizing a combined chemical and genetic approach, the Oct4 retrovirus itself was replaced with a small molecule BIX-01294, an inhibitor of the G9A histone methyltransferase, by transducing with the factors *Klf4*, *Sox2*, and *c-Myc* in the presence of BIX-01294 (Shi et al. 2008a). This is exciting not only as it is a first step toward replacing viral factors inappropriate for clinical use with small molecules, but also for providing insights into the mechanism of transcription factor-based reprogramming.

13.3.3 Molecular changes during reprogramming

In a bid to understand how the defined factors might reprogram somatic cells on a mechanistic level, the role of DNA methylation in the dynamic regulation of transcription throughout development is of particular interest. During the course of a relatively long reprogramming period (12–40 days post infection), Oct4 promoter demethylation occurs and Oct4 transcription is reactivated as observed by a targeted fluorescent reporter. Interestingly, Southern blot analysis of Oct4 eGFP positive (reprogrammed) and negative (not reprogrammed) populations of a transduced subclone demonstrated that they are clonally derived from the same parent based on similar patterns of proviral integration

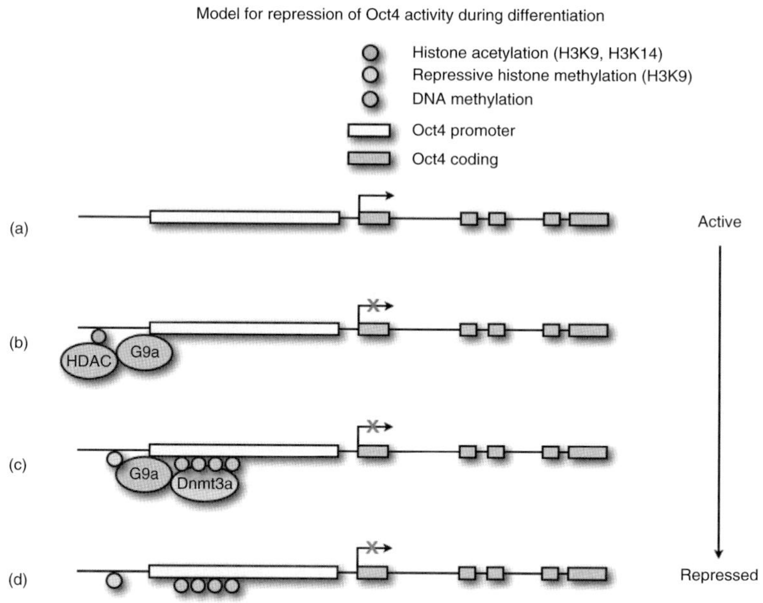

Figure 13.2 Sequential model of Oct4 in mouse ES cells during retinoic acid-induced differentiation. (a) Oct4 promoter is active in uninduced ES cells. (b) Transient transcriptional repression induced by retinoic acid recruits histone deacetylase mediated by G9a. Once histone deacetylation occurs, transcription stops. (c) G9a methylates histone 3 lysine 9 and recruits Dnmt3a/3b for *de novo* methylation. (d) The Oct4 promoter is now repressed by chromatin structure and stabilized by DNA methylation. Note that transcription stops prior to methylation.

and that reprogramming depends on stochastic epigenetic events over the course of extended cell proliferation. Fully reprogrammed clones, as characterized by Oct4 reporter reactivation and positive staining for alkaline phosphatase (AP), SSEA1, Nanog, and Oct4, exhibit complete demethylation at the Oct4 promoter (Meissner et al. 2007). The importance of epigenetic modifications in the reprogramming step is reinforced by the replacement of the Oct factor by BIX-01294, a G9a histone methyltransferase inhibitor. In this context, there are a number of experiments that follow the inactivation of genes downregulated during differentiation. In one study, Feldman et al. 2006 carefully examined the sequence of events that lead to the repression and demethylation of Oct4, one of the most well-known pluripotency-determining genes, during retinoic acid-induced differentiation. They observed that levels of *Oct4* protein dropped 24 h before any change was detected in its promoter methylation, suggesting that *de novo* methylation was secondary to transcriptional and chromatin changes. In subsequent experiments, Oct4 was shown to still undergo transcriptional repression in mutant Dmnt3a/3b–/– ES cells treated with retinoic acid in the complete absence of *de novo* methylation, via repressive histone 3 lysine 9 trimethylation (H3K9me3) by G9a (Figure 13.2). It should be noted that the steps in this sequence repression appeared to be reversible, up until the point of *de novo* methylation, suggesting an essential role for DNA methylation to stabilize the epigenetic change.

13.3.4 Efficiency of reprogramming is improved with chemical inhibitors

The effectiveness of the G9a inhibitor in promoting reprogramming to a pluripotent state suggests that the repressive histone modifications laid down by G9A exist in a dynamic equilibrium, which is subsequently shifted from a repressive to an active ground state in the presence of BIX-01294. However, the G9a inhibitor BIX-01294 would not be predicted to have a direct effect on DNA demethylation, which is consistent with the requirement for the remaining factors *Klf4* and *Sox2*. As *Sox2* has binding sites on the Oct4 promoter, perhaps the demethylation at the locus occurs via a passive demethylation mechanism during replication mediated by *Sox2* binding and exclusion of maintenance methyltransferase Dnmt1. The kinetics of the reprogramming process did not significantly change with BIX-01294, which fits with a model where the rate-limiting step is the change in DNA methylation state. Additional evidence for the importance of chromatin state on reprogramming via defined factors comes from the observation that partially reprogrammed cells are hypomethylated at pluripotency-related genes. This led Mikkelsen et al. (2008) to try treatment with the DNA methyltransferase inhibitor 5-azacytidine (5-AZA) to enhance the efficiency of reprogramming. They found that AZA increased the frequency of appearance of GFP-positive cells in a Nanog-GFP reporter system from 0.25% to 7.5%. Similarly, Huangfu et al. (2008) found that both DNA methyltransferase and histone deacetylase inhibitors improve reprogramming efficiency, and that the small-molecule valproic acid is particularly effective.

Taken together, all of this work is exciting not only for its practical implications in human and veterinary medicine, but also because it provides the framework for an eventual understanding of how a few selective transcription factors can interact in a fascinating epigenetic cross talk between histone modifications, transcription, and DNA methylation to ultimately specify cell fate. In short, nuclear reprogramming, be it by defined factors, cell fusion, or SCNT, involves significant changes to the chromatin state. Yet, in spite of the complexity of the changes involved, the work of Yamanaka showed that only a few factors are required to initiate a complex series of events leading to stable changes in cell fate, that is, nuclear reprogramming. But, in all cases, they involve changes to the chromatin state including DNA methylation and histone modifications.

13.4 Genomic imprinting

While in theory all genes in the genome can be affected by their epigenetic state, the imprinted gene family is particularly relevant as it plays a major role in placental and fetal development and function, and has a unique mode of regulation that is heavily dependent on modification of the epigenome. In the previous section, mechanisms of epigenetic regulation were discussed. This section aims to discuss placental physiology and its control in part by genomic imprinting, an epigenetic phenomenon that results in monoallelic expression of a subset of genes based on parent-of-origin inheritance. This silencing of one parental strand involves epigenetic markings by allele-specific DNA methylation and/or histone modifications (Lewis et al. 2004). We conclude with a discussion on growth phenotypes of imprinted genes.

13.4.1 Genomic nonequivalence

In the early 1980s, scientists were interested in discovering why some vertebrates including fish, lizards, and rarely some birds could develop to term without the contribution of sperm, a process called parthenogenesis, and why mammalian parthenotes succumbed to developmental arrest around the time of implantation. Two outstanding, contradictory hypotheses existed: (1) Hoppe and Illmensee (1982) proposed that the low success rate of parthenogenetic embryos was due to homozygous recessive lethal alleles, yet (2) pronuclei microsurgery experiments by McGrath and Solter (1984) supported the framework that parental genomes were marked (i.e., imprinted) differently and, thus, required both female and male haploid gametic genomes to complete normal development. To test these hypotheses, maternally or paternally derived pronuclei, prior to their fusion, were swapped between one-cell zygotes by microsurgery using pronuclear transfer techniques (Lyle 1997). The newly reconstituted diploid zygotes contained either two maternal, two paternal, or control (one maternal, one paternal) haploid nuclei and were transferred to pseudopregnant dams to develop. Viable pups were obtained only from control pregnancies, while uniparental fetuses always resulted in developmental failure, with parthenogenetic/gynogenetic embryos arresting at midgestion prior to implantation. Striking phenotypic differences were observed between biparental or uniparental pregnancies: litters generated from (1) gynogenotes or parthenotes (two maternal genomes) yielded intrauterine growth-restricted conceptuses, hypovascular placentae, and a reduction in total mass of extraembryonic tissues; and (2) androgenotes (two paternal haploid genomes) in contrast resulted in hydatidiform moles, a trophoblastic neoplasia with a very small embryo component and a very large abnormal placenta. Surprisingly, a sole report of full-term parthenogenetic mice was reported by Illmensee and Hoppe, yet neither the authors nor others could reproduce their results (Marx 1983). Independent work directly confirmed that male and female parental genomes direct fundamentally different developmental programs in mammalian embryos and were necessary for full-term development (Surani et al. 1984). This work eventually led to the discovery and characterization of a set of imprinted genes differentially regulated depending on their parent of origin.

13.4.2 Uniparental models

The initial work on imprinted genes was based on the identification of regions housing these unique genes. To accomplish this, mice with maternal or paternal chromosomal disomies (uniparental disomies) were crossed to test for non-complementation of parental alleles, and this resulted in the initial mapping of imprinted regions to mouse chromosomes 2, 8, and 17 as well as other loci (Cattanach and Kirk 1985). By using these chromosomal rearrangements, investigators succeeded at identifying the first reciprocal set of imprinted genes: maternal expression of IGF2R, and H19 on chromosome 17 by Barlow et al. (1991), and paternal expression of IGF2 on chromosome 7 (DeChiara et al. 1991; Ferguson-Smith et al. 1991). At present, more than 90 imprinted genes have been cataloged, and these represent broad gene class assignments, including coding and noncoding RNAs, small nucleuolar RNAs, microRNAs, and retrogenes (igc.otago.ac.nz/home.html) (Morison et al. 2005). What is striking is how such a small gene number, composed of less than 0.5% of known genes, can have such a drastic influence in placental and fetal development.

13.4.3 Localized imprinting control regions

The bulk of imprinted genes reside in clusters and share regional control mechanisms (Edwards and Ferguson-Smith 2007). These imprinting regional control centers (ICRs) carry a germline imprint in the form of differentially methylated DNA regions (DMRs) that serve to modulate gene activity by *cis*-acting mechanisms. Although both gametic imprints use differential methylation to establish ICRs, parent-of-origin control is not well conserved. Broadly, patterns of ICR control can be summarized as follows: maternal germline methylation shuts down promoters of paternally expressed antisense RNAs and inhibits productive extension of paternally expressed genes (AIR represses IGF2R). Alternatively, paternal germline imprints may serve as insulators between genes and recruit CTCF to shield downstream enhancers to protect gene activity (H19/IGF2 locus) (Edwards and Ferguson-Smith 2007). A summary of all known ICRs is beyond the scope of this review; for additional information, please see Edwards and Ferguson-Smith (2007) and Thorvaldsen and Bartolomei (2007).

13.4.4 Genomic imprinting in evolutionary context

What then is the function of this unique gene family? Comparative imprinting studies among mammalian clades dates the emergence of both placentation and the phenomenon of genomic imprinting to 180–210 million years ago (Hore et al. 2007). The lack of genomic imprinting in avian species, combined with the emergence of parental epigenetic asymmetry, with rudimentary imprinting controls in marsupials and further enhancements of complex imprinting mechanisms in true placental mammals, suggests that the epigenetic phenomenon arose in a stepwise, adaptive manner (Edwards et al. 2007). Figure 13.3 illustrates the changes in placentation, parallel changes in the evolution of the imprinted gene family, and their increasing importance as the placenta develops and becomes more complex (for additional information on the fascinating topic of the molecular evolution of imprinted genes, please see Smits et al. 2008).

13.4.5 The parental conflict hypothesis

Debate over the evolutionary significance of imprinting in mammals has led to the parental conflict hypothesis (Moore and Haig 1991), which predicts that paternally expressed genes act on the placenta to promote extraction of resources from the mother to enhance fetal growth, while maternally expressed genes act to restrain fetal growth to conserve maternal resources for long-term reproductive fitness of the mother. A way of functionally describing imprinted genes is as rheostats controlling the flow of nutrients from the mother to the fetus. In eutherian mammals, the fetus is dependent solely on its mother for its nourishment, provided through the placenta. To ensure normal fetal growth, the flux of nutrients across the placenta must meet developmental energy demand of the growing fetus. The identification of molecular cross talk from fetus to placenta (and vice versa) is central to understanding of the balance between nutrient supply and demand. Additionally, the study of animal models with aberrant fetal growth, such as intrauterine growth restriction (IUGR) or large offspring syndrome

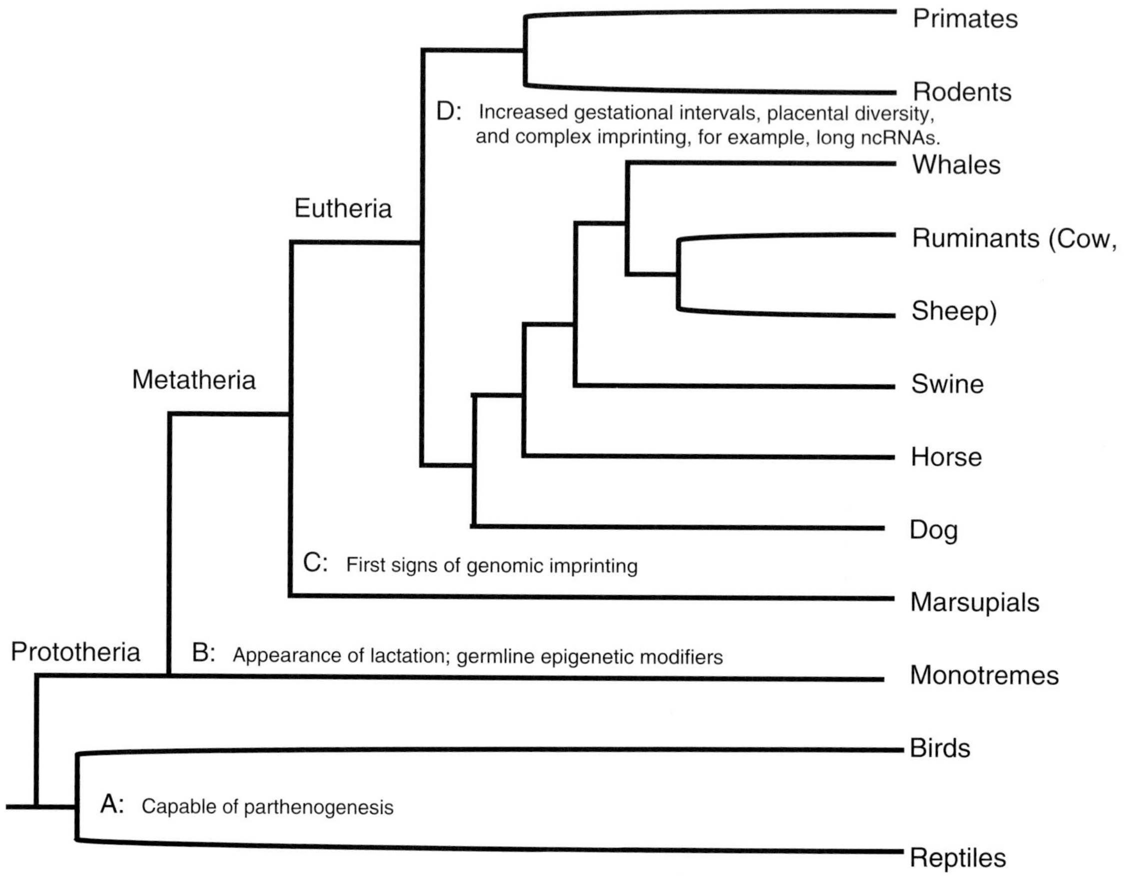

Figure 13.3 Epigenetic and evolutionary events as related to genomic imprinting and mammalian diversification. Mammals are incapable of spontaneous parthenogenesis, as seen in reptiles and rarely in birds. Monotremes nurse their young from simple mammary glands, in which milk drips down a hair shaft as they have no nipples. Genomic imprinting is absent in monotremes, for example, platypus, but coevolved with placentation and viviparity, as first seen in marsupials. Eutherian mammals have different modes of placentation, including (1) diffuse, epitheliochorial (pigs, horses); (2) cotyledonary, epitheliochorial as in ruminants (cattle, sheep); (3) zonary, endotheliochorialas as in dogs; and (4) discoid, hemochorial (primates and rodents). Long gestation intervals enhance the parental asymmetry of resources and are believed to contemporaneously increase complex imprinting mechanisms.

(LOS), provides an invaluable experimental tool to identify supply and demand signals from genetic or epigenetic contributions.

13.4.6 Imprinted genes in fetal placental function

So what is the direct molecular evidence for imprinted genes being involved in the fetal placental function? Disregulation of allelic dosage in specific imprinted genes has profound effects on fetal viability and placental metabolism (see Table 13.2 for summary). Additionally, while the uniparental models represent an exaggerated model of unbalanced imprinting, the resulting fetuses support the prediction of the parental conflict theory with smaller

Table 13.2 Effects on placental physiology by imprinted gene expression (adapted from Angiolini et al. 2006).

Allele	Gene	Gene product	Impact on placental efficiency	Knockout phenotype
Paternally expressed	*Igf2*	Growth factor (placenta and fetus)	↑ Surface area; ↓ thickness of exchange barrier; ↑ fetal demand	Placental and fetal growth restriction; ↓*Slc38a2* expression at E19
	Igf2P0	Growth factor (placenta only)	↑ Surface area; ↓ thickness of exchange barrier	Early placenta growth restriction; late fetal growth restriction; passive diffusion defect; ↑*Slc38a4* expression at E16
	Mest	α/β Hydrolase (placenta and fetus)	↑ Surface area? ↑ angiogenesis, blood flow? fetal demand?	Placental and fetal growth restriction
	Peg 3	Zinc finger transcription factor (placenta and fetus)	↑ Surface area? others?	Placental and fetal growth restriction
	Slc38a4	System A amino acid transporter (placenta and fetus)	↑ Surface area? ↑ amino acid transport; others?	Placental and fetal growth restriction
	Plagl1	Zinc finger transcription factor (placenta and fetus)	↑ Transport? ↑ surface area	Skeletal defects, neonatal lethality, IUGR, disrupted transactivation of *Igf2* and *H19* promoters, dyspnea (Varrault et al. 2006)
	Peg10	Ty3/gypsy retrotransposon-derived gene	↑ Transport? ↑ surface area	Severe growth retardation, absence of spongiotrophoblast layer, embryonic lethality, proto-oncogene agonist of SIAH1 (Ono et al. 2006)
	Rtl1	Sushi-like retroelement; intronless gene	No effect on system A transporters. ↑ Passive diffusion.	Fetal–maternal interface defects. Starvation of trophoblast cells. Placentomegaly (maternal KO); IUGR, late-fetal or neonatal lethality (Sekita et al. 2008)
Maternally expressed	*H19*	Noncoding RNA; effects mediated by *Igf2* (placenta and fetus)	↓ Surface area; ↓ fetal demand	Placental and fetal overgrowth; disproportionate overgrowth of the placenta
	Igf2r	IGF-II clearance receptor (placenta and fetus)	↓ Surface area? others? ↓ fetal demand?	Placental and fetal overgrowth
	Phlda2	Cytoplasmic protein with pleckstrin homology domain (placenta and fetal liver)	↓ Surface area	Placental overgrowth; fetal growth remains unchanged; disproportionate growth of placental layers
	Grb10	Adaptor protein (placenta and fetus)	↓ Surface area? ↓ fetal demand?	Placental and fetal overgrowth; increased insulin signaling protein kinase phosphorylation (Wang et al. 2007)
	Cdkn1c	Cyclin-dependent kinase inhibitor (placenta and fetus)	↓ Surface area? others? ↓ fetal demand?	Placental overgrowth

fetuses and placentas in the pathenotes/gynogenotes and a large placenta in the androgenotes. At the molecular level, the reciprocal imprinting of Igf2 and Igf2r provides an example of parental conflict theory. In mice, the insulin-like growth factor Igf2 is paternally expressed and increases placental and fetal weights as well as nutrient flow, while its receptor is expressed maternally and sequesters the function of Igf2 by binding and, subsequently, trafficking to the lysosome (Barlow et al. 1991). A recent study in murine placentas with promoter-specific deletions in Igf2 evidenced a nexus between fetal nutrient demand and upregulation of imprinted amino acid transporter Slc38a4 (Angiolini et al. 2006). In this study, a gene knockout approach was utilized to clarify the role of placental-specific Igf2 transcripts (mRNAs originating from promoter P0 Monk et al. 2006) on fetal nutrient demand. *Igf2 P0*$^{+/-}$ fetuses exhibited a reduction in placental weight in comparison to wild-type littermates. However, mutant pups were able to support normal fetal growth until term despite reduced placental mass, owing to a compensatory increase in system A amino acid transporters. Additional studies also suggest that aberrant regulation of imprinted genes and overexpression of genes responsible for fetal growth like *Plac1* in mouse reconstructed embryos may lead to placentomegaly (Suemizu et al. 2003). Similarly, inactivation of *Peg10* (an epigenetically regulated gene) further supports the view that imprinted genes are coincident with placental development because Peg10 null mice fail to develop to term and have abnormal placentas (Ono et al. 2006). A more comprehensive summary of imprinted genes that have been directly implicated in fetal/placental development and function is presented in Table 13.2. Note that, in general, the phenotypes of the transgenic mice support the parental conflict hypothesis with paternally imprinted genes having and opposite phenotype to maternally imprinted genes.

However, while there are a limited number of imprinted genes with known functions in placental development in mice and humans, there are almost no studies of their role in domestic species. Among the 90 or so imprinted genes identified to date in mice or humans, the imprinting status is known for only 17 genes in cattle, 15 in sheep, and 11 genes in swine (Otago database; igc.otago.ac.nz/Summary-table.pdf). Of these genes, only two reports exist about their relevance to placental physiology or broader impact on fetal growth. The *callipyge* mutation, a muscle hypertrophy condition in sheep, which has been extensively studied and results from enhanced protein expression of the imprinted DLK1 protein (Charlier et al. 2001). It affects muscle growth and energy utilization. Disregulation of the IGF2 DMR has been correlated with abnormal offspring syndrome (AOS), LOS, a likely consequence of assisted reproductive technique (ART) procedures, and embryo manipulation, where suboptimal embryonic growth results from serum addition or incomplete reprogramming through SCNT (Farin et al. 2006). These two genes, DLK1 and IGF2, are the only published works that specifically address the function of the gene and its effect on growth and fetal and placental development. Other work presents only gene expression changes under a variety of circumstances, but no functional analysis accompanies those observations. Equally disappointing, there are over 60 imprinted genes for which there is no information on their role in the development and function of the placenta in any species. In the majority of cases, it is not even known which cell type expresses these genes and whether there are stage-specific changes or species-specific patterns. Considering that imprinted genes and placentas coevolved; that several studies have described peturbations of imprinted gene expression caused by ARTs (Moore and Haig 1991) including extended *in vitro* embryo culture, microsurgery, intracytoplasmic sperm injection (ICSI), and SCNT; and that placentas within eutherian mammals differ drastically in morphology, it is disappointing as to the lack of research efforts in this area.

Additionally, understanding the biologic merit of epigenetic factors for the design of optimum, long-term selection schemes in livestock is growing in importance, because recombination mapping experiments to identify quantitative trait loci (QTL) in outbred swine populations have pinpointed imprinting as a causal mechanism for phenotypic variation (Knott et al. 1998). Work from previous investigators demonstrated that polar overdominant inheritance of an imprinted gene DLK1 polymorphism is associated with growth and fat deposition in pigs (Kim et al. 2004). And there are over 40 QTL in swine alone that have been associated with parent-of-origin effects suggesting the existing of an imprinted gene in the QTL region (see igc.otago.ac.nz/home.html for complete information on QTL associated with imprinting in swine).

13.5 SCNT and epigenetic abnormalities

The process of SCNT has been described and reviewed recently by others (Keefer 2008; Kishigami et al. 2008). In simple terms, however, there are two main events taking place during SCNT that differ with a normal fertilization event. First, there is a transcription signature in the donor nuclei that is not compatible with embryogenesis. This gene expression signature must then be switched or reprogrammed into one of an early embryo. This is done by transcription factors present in the oocyte and requires the removal of the factors present in the nuclei and their replacement with oocyte-derived factors. For this exchange, the chromatin needs to be easily accessible. The second factor is the chromatin configuration of the somatic cell nuclei compared with that or a sperm/oocyte. In the sperm/oocyte case, the male and female chromatins are packed differently, with the sperm DNA bound by protamines that quickly decondense after fertilization. The oocyte DNA, in contrast, is bound by maternal histones. In a normal fertilization event, this difference in chromatin packing is mirrored by the speed and degree of methylation changes that are seen after fertilization. Thus, in all species examined to date, the paternal DNA is more rapidly demethylated than the oocyte DNA. What we do not know yet is how critical is that difference between the maternal and the paternal demethylation dynamics. In other words, what is the functional importance of early demethylation of the paternally derived DNA. What we do know is that in SCNT, such a difference does not exist, as both the maternal and the paternal chromatins are indistinguishable from a chromatin configuration perspective.

13.5.1 Chemical methods to improve the efficiency of SCNT

The issue of chromatin accessibility in the context of SCNT has been investigated to some extent. It has been suggested that treatment of nuclear donors with chromatin modifies such as trichostatin A (TSA; a specific and potent inhibitor of class I and II mammalian HDAC I) and 5-AZA (an inhibitor of DNMT1), compounds known to "open up" chromatin, facilitate reprogramming after SCNT. What is interesting is that there appear to be species differences in response to chromatin remodeling methods. In mice, multiple groups have reported beneficial effects of TSA/5-AZA treatment but only when using differentiated cells (Kishigami et al. 2006; Shi et al. 2008b). In contrast, treatment of ES cells had no beneficial results. More importantly, pups derived from TSA-treated cells had reduced incidence of AOS (Kishigami et al. 2006). This suggests that ES cells are already in an open chromatin configuration, and additional "relaxation" of chromatin is not beneficial. In swine and cattle, the results are not as striking. There have been some reported benefits on *in vitro* development, but the development to term or the effects on placental abnormalities are not well documented (Li et al. 2008). Additionally, there have been reports on gene expression changes as a result of TSA treatment (Iager et al. 2008). Those changes have been interpreted as being beneficial as they resemble expression signatures of control embryos. However, without data on term viability and lack of information on the effects of the treatment on placental development and function, it is premature to evaluate what will be the effects of chromatin remodeling methods on cloning efficiencies in domestic animals. In addition to the lack of detailed information on chromatin remodeling modifications on SCNT, there is a conceptual problem that will be more difficult to overcome. While TSA and/or 5-AZA treatments can change chromatin structure, they cannot differentiate between the paternal and maternal chromosomes and thus cannot mimic the normal fertilization event. This would require a different chromatin modification of the paternal versus the maternal DNA, something that is not technically feasible. Thus, from a conceptual standpoint, it will be extremely difficult to mimic the chromatin configuration and dynamics that are seen in a normal fertilization event versus what is seen in SCNT.

13.5.2 Epigenetic abnormalities

So how do we know that the reprogramming that occurs during nuclear transfer is abnormal? Studies in this area range from gene expression profiling to methylation analysis of selected regions. Combined, what these two approaches confirm is that the placenta is the organ that is most affected as it seems to be particularly susceptible to epigenetic pertubations. This includes the number of differentially expressed genes between normal and SCNT placentas, as well as the degree of methylation abnormalities. In contrast, the fetus proper is affected to a much lesser extent. In the bovine, our group determined that in a normal pregnancy, the placentas are hypomethylated compared with the somatic tissues. In SCNT pregnancies, however, there was hypermethylation of the cloned placentas compared with the control, but little to no changes in somatic methylation levels (Dindot et al. 2004). In addition, in female clones, the Xist gene was abnormally regulated (Dindot et al. 2004). This abnormal X-inactivation in cloned cattle have been reported by others (Wrenzycki et al. 2002; Xue et al. 2002) and may explain why there are reports of a higher proportion of male SCNT clones, at least in bovine.

13.5.3 Placental abnormalities

Placental abnormalities in SCNT pregnancies have been reported in multiple mammalian species (see the review by Arnold et al. 2008). Yet, the degree of abnormalities seem to differ with some species, such as cattle and sheep, being particularly susceptible to placental defects associated with SCNT, while other species, such as swine, have few reports of abnormal placentation. Whether this is a result of the differences in placental morphologies between species still remains to be determined. These species differences, moreover, are not limited to the placenta. In cattle and sheep, there have been many reports of increase in fetal/offspring weight in response to SCNT. This has been referred as the LOS, although we recently proposed a more accurate classification (AOS) that can reflect the different degrees of severity of the syndrome as well as cover species where the increase in weight is not observed (Farin et al. 2006). For instance, while SCNT results in increases in conceptus weight in cattle and sheep, the opposite is true in swine. In swine, SCNT causes a slight but significant increase in the incidence of IUGR (Estrada et al. 2007). While the phenotypic effect on weight is distinct between species, in all cases reported to date, there have been multiple placental defects identified. In swine, SCNT-derived placentas tend to be hypovascular and show trophoblast hypoplasia and overall terminal villi hypoplasia (Lee et al. 2007). Moreover, a detailed proteomic analysis indicated that proteins involved in apoptosis are disregulated in the SCNT placentas (Lee et al. 2007). More severe defects have been observed in cattle, sheep, and mice (Hill et al. 1999, 2000, 2001, 2002; Ogura et al. 2002; Rhind et al. 2003). As for swine, gene expression profiling indicates that placentas from cloned cattle have significant gene expression changes compared with the controls (Everts et al. 2008). Yet, at this point, there is little known as to what triggers such placental defects. Is it abnormal methylation levels in SCNT placentas tissues that results in global epigenetic abnormalities leading to placental defects, or is it abnormal reprogramming of a selected group of genes involved in differentiation of the placenta with the end results of abnormal placental development?

13.5.4 Angiogenesis

As angiogenesis is a critical component for placental development and function, this system has been well studied in both SCNT and IVF pregnancies in cattle. Recently, it has been reported that the vascular endothelial growth factor A (VEGF-A) system appear to be disregulated in cloned placentas, but no data were presented (Arnold et al. 2008). Previously, it has been reported that the VEGF system was disregulated in placentas from IVF fetuses, so it is likely that the same holds true for SCNT pregnancies. As to what causes VEGF disregulation, this remains to be determined. Arnold et al. (2006) looked at a selected number of candidate genes known to be involved in trophoblast differentiation in cattle and determined that there were expression differences in *Ascl2 (Mash 2)*, *Hand1*, and *PAG9*. Interestingly, there did not seem to be any disruption of imprinting of ASCL2 indicating that the gene expression differences were not due to abnormal imprinting.

In summary, in SCNT we observe both global changes in methylation of the placenta, disregulation of specific genes such as XIST, as well as gene expression changes compared with normal placentas. But what is the evidence that both are connected? That is, that changes in the epigenome of the SCNT placenta are responsible for the defects observed in SCNT. While the direct evidence is lacking, there is a considerable body of knowledge supporting the role of the epigenome in placental development. This includes the observation that overall methylation levels are lower in the placenta than in the somatic tissues and that this difference is present through gestation (Rossant et al. 1986). Treatments known to affect methylation levels such as dietary addition of TSA during pregnancy result in abnormal placentas (Vlahović et al. 1999; Serman et al. 2007). In addition, mutation in several members of the DMNT family, including DMNT1, DMNT3a, DMNT3b, and DMNT3L, all result in placental abnormalities, in addition to other defects.

13.6 Future research directions

So where do we go from here? At this point, we know that the placenta is the organ more susceptible to SCNT; we know that in spite of species differences in fetal outcomes, in all species examined to date, there are placental anomalies associated with SCNT. We also know that the placenta is hypomethylated compared with the somatic tissues and that SCNT results in overall global hypermethylation of the placenta in SCNT pregnancies. Protein and gene expression analysis both confirm that there are significant gene expression differences in the placenta of control and SCNT clones. We also know from candidate gene studies that genes involved in placental development and differentiation are affected by SCNT. What we do not know, however, are the factors responsible for reprogramming an incoming somatic cell nucleus into a totipotent fate. Are there just few as was demonstrated for ES cells, or is this process more complex and requires a larger number of factors? Or is the reprogramming mechanism of SCNT, cell–cell fusion, and direct reprogramming by transcription factors the same but proceeding with different kinetics? If we identify those factors, can we use that information to experimentally modify the incoming nuclei so as to facilitate normal reprogramming? We also do not know to what extent imprinted genes play a role in placental defects associated with SCNT. Clearly, they are not the only genes affected, but could they be the master regulatory genes that then affect other genes, or are they just one more group of genes susceptible to incomplete reprogramming? And while their role in SCNT still remains somewhat tenuous, there is strong evidence that they play a major role in energy flow and in placental development in mammalian species. Why then are they so critically understudied, especially in domestic animals?

References

Adhikary, S. and Eilers, M. 2005. Transcriptional regulation and transformation by Myc proteins. *Nature Reviews. Molecular Cell Biology* **6**: 635–645.

Angiolini, E., Fowden, A., Coan, P., Sandovici, I., Smith, P., Dean, W., Burton, G. et al. 2006. Regulation of placental efficiency for nutrient transport by imprinted genes. *Placenta* **27**(Supplement A): S98–S102.

Arnold, D.R., Bordignon, V., Lefebvre, R., Murphy, B.D., and Smith, L.C. 2006. Somatic cell nuclear transfer alters peri-implantation trophoblast differentiation in bovine embryos. *Reproduction* **132**: 279–290.

Arnold, D.R., Fortier, A.L., Lefebvre, R., Miglino, M.A., Pfarrer, C., and Smith, L.C. 2008. Placental insufficiencies in cloned animals—A workshop report. *Placenta* **29**(Supplement A): S108–S110.

Avilion, A.A., Nicolis, S.K., Pevny, L.H., Perez, L., Vivian, N., and Lovell-Badge, R. 2003. Multipotent cell lineages in early mouse development depend on SOX2 function. *Genes & Development* **17**: 126–140.

Barlow, D.P., Stoger, R., Herrmann, B.G., Saito, K., and Schweifer, N. 1991. The mouse insulin-like growth factor type-2 receptor is imprinted and closely linked to the Tme locus. *Nature* **349**: 84–87.

Barreto, G., Schafer, A., Marhold, J., Stach, D., Swaminathan, S.K., Handa, V., Doderlein, G. et al. 2007. Gadd45a promotes epigenetic gene activation by repair-mediated DNA demethylation. *Nature* **445**: 671–675.

Bernstein, B.E., Mikkelsen, T.S., Xie, X., Kamal, M., Huebert, D.J., Cuff, J., Fry, B. et al. 2006. A bivalent chromatin structure marks key developmental genes in embryonic stem cells. *Cell* **125**: 315–326.

Bhattacharya, S.K., Ramchandani, S., Cervoni, N., and Szyf, M. 1999. A mammalian protein with specific demethylase activity for mCpG DNA. *Nature* **397**: 579–583.

Campbell, K.H., McWhir, J., Ritchie, W.A., and Wilmut, I. 1996. Sheep cloned by nuclear transfer from a cultured cell line. *Nature* **380**: 64–66.

Cattanach, B.M. and Kirk, M. 1985. Differential activity of maternally and paternally derived chromosome regions in mice. *Nature* **315**: 496–498.

Charlier, C., Segers, K., Karim, L., Shay, T., Gyapay, G., Cockett, N., and Georges, M. 2001. The callipyge mutation enhances the expression of coregulated imprinted genes in cis without affecting their imprinting status. *Nature Genetics* **27**: 367–369.

Collas, P. 2003. Nuclear reprogramming in cell-free extracts. *Philosophical Transactions of the Royal Society of London. Series B, Biological Sciences* **358**: 1389–1395.

Collas, P. and Hakelien, A.M. 2003. Teaching cells new tricks. *Trends in Biotechnology* **21**: 354–361.

Collas, P. and Taranger, C.K. 2006. Epigenetic reprogramming of nuclei using cell extracts. *Stem Cell Reviews* **2**: 309–317.

Cowan, C.A., Atienza, J., Melton, D.A., and Eggan, K. 2005. Nuclear reprogramming of somatic cells after fusion with human embryonic stem cells. *Science* **309**: 1369–1373.

Dalla-Favera, R., Bregni, M., Erikson, J., Patterson, D., Gallo, R.C., and Croce, C.M. 1982. Human c-Myc onc gene is located on the region of chromosome 8 that is translocated in Burkitt lymphoma cells. *Proceedings of the National Academy of Sciences of the United States of America* **79**: 7824–7827.

DeChiara, T.M., Robertson, E.J., and Efstratiadis, A. 1991. Parental imprinting of the mouse insulin-like growth factor II gene. *Cell* **64**: 849–859.

Dindot, S.V., Farin, P.W., Farin, C.E., Romano, J., Walker, S., Long, C., and Piedrahita, J.A. 2004. Epigenetic and genomic imprinting analysis in nuclear transfer derived *Bos gaurus/Bos taurus* hybrid fetuses. *Biology of Reproduction* **71**: 470–478.

Do, J.T. and Scholer, H.R. 2004. Nuclei of embryonic stem cells reprogram somatic cells. *Stem Cells* **22**: 941–949.

Edwards, C.A. and Ferguson-Smith, A.C. 2007. Mechanisms regulating imprinted genes in clusters. *Current Opinion in Cell Biology* **19**: 281–289.

Edwards, C.A., Rens, W., Clarke, O., Mungall, A.J., Hore, T., Graves, J.A., Dunham, I., Ferguson-Smith, A.C., and Ferguson-Smith, M.A. 2007. The evolution of imprinting: Chromosomal mapping of orthologues of mammalian imprinted domains in monotreme and marsupial mammals. *BMC Evolutionary Biology* **7**: 157.

Estrada, J., Sommer, J., Collins, B., Mir, B., Martin, A., York, A., Petters, R.M., and Piedrahita, J.A. 2007. Swine generated by somatic cell nuclear transfer have increased incidence of intrauterine growth restriction (IUGR). *Cloning and Stem Cells* **9**: 229–236.

Everts, R.E., Chavatte-Palmer, P., Razzak, A., Hue, I., Green, C.A., Oliveira, R., Vignon, X. et al. 2008. Aberrant gene expression patterns in placentomes are associated with phenotypically normal and abnormal cattle cloned by somatic cell nuclear transfer. *Physiological Genomics* **33**: 65–77.

Farin, P.W., Piedrahita, J.A., and Farin, C.E. 2006. Errors in development of fetuses and placentas from in vitro-produced bovine embryos. *Theriogenology* **65**: 178–191.

Feldman, N., Gerson, A., Fang, J., Li, E., Zhang, Y., Shinkai, Y., Cedar, H., and Bergman, Y. 2006. G9a-mediated irreversible epigenetic inactivation of Oct-3/4 during early embryogenesis. *Nature Cell Biology* **8**: 188–194.

Ferguson-Smith, A.C., Cattanach, B.M., Barton, S.C., Beechey, C.V., and Surani, M.A. 1991. Embryological and molecular investigations of parental imprinting on mouse chromosome 7. *Nature* **351**: 667–670.

Guenther, M.G., Levine, S.S., Boyer, L.A., Jaenisch, R., and Young, R.A. 2007. A chromatin landmark and transcription initiation at most promoters in human cells. *Cell* **130**: 77–88.

Hayashi, H., Nagae, G., Tsutsumi, S., Kaneshiro, K., Kozaki, T., Kaneda, A., Sugisaki, H., and Aburatani, H. 2007. High-resolution mapping of DNA methylation in human genome using oligonucleotide tiling array. *Human Genetics* **120**: 701–711.

Hellman, A. and Chess, A. 2007. Gene body-specific methylation on the active X chromosome. *Science* **315**: 1141–1143.

Hill, J.R., Burghardt, R.C., Jones, K., Long, C.R., Looney, C.R., Shin, T., Spencer, T.E., Thompson, J.A., Winger, Q.A., and Westhusin, M.E. 2000. Evidence for placental abnormality as the major cause of mortality in first-trimester somatic cell cloned bovine fetuses. *Biology of Reproduction* **63**: 1787–1794.

Hill, J.R., Edwards, J.F., Sawyer, N., Blackwell, C., and Cibelli, J.B. 2001. Placental anomalies in a viable cloned calf. *Cloning* **3**: 83–88.

Hill, J.R., Roussel, A.J., Cibelli, J.B., Edwards, J.F., Hooper, N.L., Miller, M.W., Thompson, J.A. et al. 1999. Clinical and pathologic features of cloned transgenic calves and fetuses (13 case studies). *Theriogenology* **51**: 1451–1465.

Hill, J.R., Schlafer, D.H., Fisher, P.J., and Davies, C.J. 2002. Abnormal expression of trophoblast major histocompatibility complex class I antigens in cloned bovine pregnancies is associated with a pronounced endometrial lymphocytic response. *Biology of Reproduction* **67**: 55–63.

Hooker, C.W. and Hurlin, P.J. 2006. Of Myc and Mnt. *Journal of Cell Science* **119**: 208–216.

Hoppe, P.C. and Illmensee, K. 1982. Full-term development after transplantation of parthenogenetic embryonic nuclei into fertilized mouse eggs. *Proceedings of the National Academy of Sciences of the United States of America* **79**: 1912–1916.

Hore, T.A., Rapkins, R.W., and Graves, J.A. 2007. Construction and evolution of imprinted loci in mammals. *Trends in Genetics* **23**: 440–448.

Huangfu, D., Maehr, R., Guo, W., Eijkelenboom, A., Snitow, M., Chen, A.E., and Melton, D.A. 2008. Induction of pluripotent stem cells by defined factors is greatly improved by small-molecule compounds. *Nature Biotechnology* **26**: 795–797.

Humpherys, D., Eggan, K., Akutsu, H., Friedman, A., Hochedlinger, K., Yanagimachi, R., Lander, E.S., Golub, T.R., and Jaenisch, R. 2002. Abnormal gene expression in cloned mice derived from embryonic stem cell and cumulus cell nuclei. *Proceedings of the National Academy of Sciences of the United States of America* **99**: 12889–12894.

Iager, A.E., Ragina, N.P., Ross, P.J., Beyhan, Z., Cunniff, K., Rodriguez, R.M., and Cibelli, J.B. 2008. Trichostatin A improves histone acetylation in bovine somatic cell nuclear transfer early embryos. *Cloning and Stem Cells* **10**: 371–379.

Jin, S.G., Guo, C., and Pfeifer, G.P. 2008. GADD45A does not promote DNA demethylation. *PLoS Genetics* **4**: e1000013.

Keefer, C.L. 2008. Lessons learned from nuclear transfer (cloning). *Theriogenology* **69**: 48–54.

Kim, J., Bergmann, A., Lucas, S., Stone, R., and Stubbs, L. 2004. Lineage-specific imprinting and evolution of the zinc-finger gene ZIM2. *Genomics* **84**: 47–58.

Kim, J.B., Zaehres, H., Wu, G., Gentile, L., Ko, K., Sebastiano, V., Arauzo-Bravo, M.J. et al. 2008. Pluripotent stem cells induced from adult neural stem cells by reprogramming with two factors. *Nature* **454**: 646–650.

Kishigami, S., Mizutani, E., Ohta, H., Hikichi, T., Thuan, N.V., Wakayama, S., Bui, H.T., and Wakayama, T. 2006. Significant improvement of mouse cloning technique by treatment with trichostatin A after somatic nuclear transfer. *Biochemical and Biophysical Research Communications* **340**: 183–189.

Kishigami, S., Wakayama, S., Hosoi, Y., Iritani, A., and Wakayama, T. 2008. Somatic cell nuclear transfer: Infinite reproduction of a unique diploid genome. *Experimental Cell Research* **314**: 1945–1950.

Knott, S.A., Marklund, L., Haley, C.S., Andersson, K., Davies, W., Ellegren, H., Fredholm, M. et al. 1998. Multiple marker mapping of quantitative trait loci in a cross between outbred wild boar and large white pigs. *Genetics* **149**: 1069–1080.

Lee, S.Y., Park, J.Y., Choi, Y.J., Cho, S.K., Ahn, J.D., Kwon, D.N., Hwang, K.C. et al. 2007. Comparative proteomic analysis associated with term placental insufficiency in cloned pig. *Proteomics* **7**: 1303–1315.

Lei, H., Oh, S.P., Okano, M., Juttermann, R., Goss, K.A., Jaenisch, R., and Li, E. 1996. De novo DNA cytosine methyltransferase activities in mouse embryonic stem cells. *Development* **122**: 3195–3205.

Lewis, A., Mitsuya, K., Umlauf, D., Smith, P., Dean, W., Walter, J., Higgins, M., Feil, R., and Reik, W. 2004. Imprinting on distal chromosome 7 in the placenta involves repressive histone methylation independent of DNA methylation. *Nature Genetics* **36**: 1291–1295.

Li, E., Bestor, T.H., and Jaenisch, R. 1992. Targeted mutation of the DNA methyltransferase gene results in embryonic lethality. *Cell* **69**: 915–926.

Li, J., Svarcova, O., Villemoes, K., Kragh, P.M., Schmidt, M., Bøgh, I.B., Zhang, Y. et al. 2008. High in vitro development after somatic cell nuclear transfer and trichostatin A treatment of reconstructed porcine embryos. *Theriogenology* **70**: 800–808.

Lowry, W.E., Richter, L., Yachechko, R., Pyle, A.D., Tchieu, J., Sridharan, R., Clark, A.T., and Plath, K. 2008. Generation of human induced pluripotent stem cells from dermal fibroblasts. *Proceedings of the National Academy of Sciences of the United States of America* **105**: 2883–2888.

Lyle, R. 1997. Gametic imprinting in development and disease. *The Journal of Endocrinology* **155**: 1–12.

Maherali, N.S.R., Xie, W., Utikal, J., Eminli, S., Arnold, K., Stadtfeld, M., Yachechko, R. et al. 2007. Directly reprogrammed fibroblasts show global epigenetic remodeling and widespread tissue contribution. *Cell Stem Cell* **1**: 55–70.

Marx, J.L. 1983. Swiss research questioned. *Science* **220**: 1023.

McGrath, J. and Solter, D. 1984. Completion of mouse embryogenesis requires both the maternal and paternal genomes. *Cell* **37**: 179–183.

Meissner, A., Wernig, M., and Jaenisch, R. 2007. Direct reprogramming of genetically unmodified fibroblasts into pluripotent stem cells. *Nature Biotechnology* **10**: 1177–1181.

Mikkelsen, T.S., Hanna, J., Zhang, X., Ku, M., Wernig, M., Schorderet, P., Bernstein, B.E., Jaenisch, R., Lander, E.S., and Meissner, A. 2008. Dissecting direct reprogramming through integrative genomic analysis. *Nature* **454**: 49–55.

Mikkelsen, T.S., Ku, M., Jaffe, D.B., Issac, B., Lieberman, E., Giannoukos, G., Alvarez, P. et al. 2007. Genome-wide maps of chromatin state in pluripotent and lineage-committed cells. *Nature* **448**: 553–560.

Monk, D., Sanches, R., Arnaud, P., Apostolidou, S., Hills, F.A., Abu-Amero, S., Murrell, A. et al. 2006. Imprinting of IGF2 P0 transcript and novel alternatively spliced INS-IGF2 isoforms show differences between mouse and human. *Human Molecular Genetics* **15**: 1259–1269.

Moore, T. and Haig, D. 1991. Genomic imprinting in mammalian development: A parental tug-of-war. *Trends in Genetics* **7**: 45–49.

Morison, I.M., Ramsay, J.P., and Spencer, H.G. 2005. A census of mammalian imprinting. *Trends in Genetics* **21**: 457–465.

Nakagawa, M., Koyanagi, M., Tanabe, K., Takahashi, K., Ichisaka, T., Aoi, T., Okita, K., Mochiduki, Y., Takizawa, N., and Yamanaka, S. 2008. Generation of induced pluripotent stem cells without Myc from mouse and human fibroblasts. *Nature Biotechnology* **26**: 101–106.

Nimura, K., Ishida, C., Koriyama, H., Hata, K., Yamanaka, S., Li, E., Ura, K., and Kaneda, Y. 2006. Dnmt3a2 targets endogenous Dnmt3L to ES cell chromatin and induces regional DNA methylation. *Genes to Cells* **11**: 1225–1237.

Niwa, H., Miyazaki, J., and Smith, A.G. 2000. Quantitative expression of Oct-3/4 defines differentiation, dedifferentiation or self-renewal of ES cells. *Nature Genetics* **24**: 372–376.

Ogura, A., Inoue, K., Ogonuki, N., Lee, J., Kohda, T., and Ishino, F. 2002. Phenotypic effects of somatic cell cloning in the mouse. *Cloning and Stem Cells* **4**: 397–405.

Okamoto, K., Okazawa, H., Okuda, A., Sakai, M., Muramatsu, M., and Hamada, H. 1990. A novel octamer binding transcription factor is differentially expressed in mouse embryonic cells. *Cell* **60**: 461–472.

Okano, M., Xie, S., and Li, E. 1998. Cloning and characterization of a family of novel mammalian DNA (cytosine-5) methyltransferases. *Nature Genetics* **19**: 219–220.

Okita, K., Ichisaka, T., and Yamanaka, S. 2007. Generation of germline-competent induced pluripotent stem cells. *Nature* **448**: 313–317.

Ono, R., Nakamura, K., Inoue, K., Naruse, M., Usami, T., Wakisaka-Saito, N., Hino, T. et al. 2006. Deletion of Peg10, an imprinted gene acquired from a retrotransposon, causes early embryonic lethality. *Nature Genetics* **38**: 101–106.

Ooi, S.K. and Bestor, T.H. 2008. The colorful history of active DNA demethylation. *Cell* **133**: 1145–1148.

Ooi, S.K., Qiu, C., Bernstein, E., Li, K., Jia, D., Yang, Z., Erdjument-Bromage, H. et al. 2007. DNMT3L connects unmethylated lysine 4 of histone H3 to de novo methylation of DNA. *Nature* **448**: 714–717.

Park, I.H., Zhao, R., West, J.A., Yabuuchi, A., Huo, H., Ince, T.A., Lerou, P.H., Lensch, M.W., and Daley, G.Q. 2008. Reprogramming of human somatic cells to pluripotency with defined factors. *Nature* **451**: 141–146.

Rhind, S.M., King, T.J., Harkness, L.M., Bellamy, C., Wallace, W., DeSousa, P., and Wilmut, I. 2003. Cloned lambs—Lessons from pathology. *Nature Biotechnology* **21**: 744–745.

Rossant, J., Sanford, J.P., Chapman, V.M., and Andrews, G.K. 1986. Undermethylation of structural gene sequences in extraembryonic lineages of the mouse. *Developmental Biology* **117**: 567–573.

Scholer, H.R., Ruppert, S., Suzuki, N., Chowdhury, K., and Gruss, P. 1990. New type of POU domain in germ line-specific protein Oct4. *Nature* **344**: 435–439.

Sekita, Y., Wagatsuma, H., Nakamura, K., Ono, R., Kagami, M., Wakisaka, N., Hino, T. et al. 2008. Role of retrotransposon-derived imprinted gene, Rtl1, in the feto-maternal interface of mouse placenta. *Nature Genetics* **40**(2): 243–248.

Serman, L., Vlahović, M., Sijan, M., Bulić-Jakus, F., Serman, A., Sincić, N., Matijević, R., Jurić-Lekić, G., and Katusić, A. 2007. The impact of 5-azacytidine on placental weight, glycoprotein pattern and proliferating cell nuclear antigen expression in rat placenta. *Placenta* **28**: 803–811.

Shi, L.H., Ai, J.S., Ouyang, Y.C., Huang, J.C., Lei, Z.L., Wang, Q., Yin, S., Han, Z.M., Sun, Q.Y., and Chen, D.Y. 2008. Trichostatin A and nuclear reprogramming of cloned rabbit embryos. *Journal of Animal Science* **86**: 1106–1113.

Shi, Y., Do, J.T., Desponts, C., Hahm, H.S., Scholer, H.R., and Ding, S. 2008. A combined chemical and genetic approach for the generation of induced pluripotent stem cells. *Cell Stem Cell* **2**: 525–528.

Smits, G., Mungall, A.J., Griffiths-Jones, S., Smith, P., Beury, D., Matthews, L., Rogers, J. et al. 2008. Conservation of the H19 noncoding RNA and H19-IGF2 imprinting mechanism in therians. *Nature Genetics* **40**: 971–976.

Suemizu, H., Aiba, K., Yoshikawa, T., Sharov, A.A., Shimozawa, N., Tamaoki, N., and Ko, M.S. 2003. Expression profiling of placentomegaly associated with nuclear transplantation of mouse ES cells. *Developmental Biology* **253**: 36–53.

Surani, M.A., Barton, S.C., and Norris, M.L. 1984. Development of reconstituted mouse eggs suggests imprinting of the genome during gametogenesis. *Nature* **308**: 548–550.

Svedruzic, Z.M. 2008. Mammalian cytosine DNA methyltransferase Dnmt1: Enzymatic mechanism, novel mechanism-based inhibitors, and RNA-directed DNA methylation. *Current Medicinal Chemistry* **15**: 92–106.

Takahashi, K., Tanabe, K., Ohnuki, M., Narita, M., Ichisaka, T., Tomoda, K., and Yamanaka, S. 2007. Induction of pluripotent stem cells from adult human fibroblasts by defined factors. *Cell* **131**: 861–872.

Takahashi, K. and Yamanaka, S. 2006. Induction of pluripotent stem cells from mouse embryonic and adult fibroblast cultures by defined factors. *Cell* **126**: 663–676.

Thorvaldsen, J.L. and Bartolomei, M.S. 2007. SnapShot: Imprinted gene clusters. *Cell* **130**: 958.

Varrault, A., Gueydan, C., Delalbre, A., Bellmann, A., Houssami, S., Aknin, C., Severac, D. et al. 2006. Zac1 regulates an imprinted gene network critically involved in the control of embryonic growth. *Developmental Cell* **11**(5): 711–722.

Vlahović, M., Bulić-Jakus, F., Jurić-Lekić, G., Fucić, A., Marić, S., and Serman, D. 1999. Changes in the placenta and in the rat embryo caused by the demethylating agent 5-azacytidine. *The International Journal of Developmental Biology* **43**: 843–846.

Wang, L., Balas, B., Christ-Roberts, C.Y., Kim, R.Y., Ramos, F.J., Kikani, C.K., Li, C. et al. 2007. Peripheral disruption of the Grb10 gene enhances insulin signaling and sensitivity in vivo. *Molecular and Cellular Biology* **27**(18): 6497–6505.

Wang, J., Rao, S., Chu, J., Shen, X., Levasseur, D.N., Theunissen, T.W., and Orkin, S.H. 2006. A protein interaction network for pluripotency of embryonic stem cells. *Nature* **444**: 364–368.

Weber, M., Hellmann, I., Stadler, M.B., Ramos, L., Paabo, S., Rebhan, M., and Schubeler, D. 2007. Distribution, silencing potential and evolutionary impact of promoter DNA methylation in the human genome. *Nature Genetics* **39**: 457–466.

Wernig, M., Meissner, A., Foreman, R., Brambrink, T., Ku, M., Hochedlinger, K., Bernstein, B.E., and Jaenisch, R. 2007. In vitro reprogramming of fibroblasts into a pluripotent ES-cell-like state. *Nature* **448**: 318–324.

Wrenzycki, C., Lucas-Hahn, A., Herrmann, D., Lemme, E., Korsawe, K., and Niemann, H. 2002. In vitro production and nuclear transfer affect dosage compensation of the X-linked gene transcripts G6PD, PGK, and Xist in preimplantation bovine embryos. *Biology of Reproduction* **66**: 127–134.

Wu, J., Wang, S.H., Potter, D., Liu, J.C., Smith, L.T., Wu, Y.Z., Huang, T.H., and Plass, C. 2007. Diverse histone modifications on histone 3 lysine 9 and their relation to DNA methylation in specifying gene silencing. *BMC Genomics* **8**: 131.

Xie, S., Wang, Z., Okano, M., Nogami, M., Li, Y., He, W.W., Okumura, K., and Li, E. 1999. Cloning, expression and chromosome locations of the human DNMT3 gene family. *Gene* **236**: 87–95.

Xue, F., Tian, X.C., Du, F., Kubota, C., Taneja, M., Dinnyes, A., Dai, Y., Levine, H., Pereira, L.V., and Yang, X. 2002. Aberrant patterns of X chromosome inactivation in bovine clones. *Nature Genetics* **31**: 216–220.

Yamanaka, S. 2007. Strategies and new developments in the generation of patient-specific pluripotent stem cells. *Cell Stem Cell* **1**: 39–49.

Yu, J., Vodyanik, M.A., Smuga-Otto, K., Antosiewicz-Bourget, J., Frane, J.L., Tian, S., Nie, J. et al. 2007. Induced pluripotent stem cell lines derived from human somatic cells. *Science* **318**: 1917–1920.

14

Biotechnology and Fertility Regulation

Valéria Conforti

14.1 Introduction

This chapter describes immunocontraception and immunosterilization as methods of fertility control in animals. Both methods can be broadly defined as biotechnologies that induce an immune response against a molecule whose binding to antibodies impairs reproduction. Theoretically, immunocontraceptive vaccines prevent fertilization of an egg by sperm without interfering with sexual behaviors. The correct use of the term immunocontraception excludes antifertility (AF) vaccines that act by preventing embryo implantation or by inducing abortion. Immunosterilization prevents reproduction and sexual behaviors. Some authors prefer the term immunocastration although it implies irreversibility of the effects. For simplification purposes, vaccines for fertility control will be referred to as AF vaccines herein.

Over the last decades, several types of AF vaccines have been shown to at least temporally suppress reproductive function in either or both sexes. This biotechnology has several advantages over more traditional methods of fertility control, which usually rely on surgery or constant exposure to exogenous steroid hormones. Unlike surgical castration, AF vaccines have the potential to be reversible and are not as invasive. In contrast with animals treated with steroids for fertility control, immunized animals are not exposed to exogenous hormones, which have been associated with deleterious side effects from long-term use.

Some AF vaccines act by stimulating the immune system to respond to a nonself, yet essential, component of fertility. This is the case of active immunization of females against sperm-associated antigens. However, most types of AF vaccines act by inducing an immune response against an endogenous component of the immunized animal, creating an autoimmune condition that is usually transient. Neutralization of an endogenous component is accomplished by introducing an antigen that is structurally similar to the target molecule yet "foreign" enough to elicit an immune response.

14.2 Basic aspects in vaccine development

14.2.1 Safety

Safety is a primary concern in the development of any vaccine. This includes minimization of toxicity and undesirable side effects in the immunized animal, safety of the personnel who administer the vaccine, and environmental safety. The immune response should be specific to the desired molecule so that antibody interaction will be restricted to the type of cell or tissue to be targeted, which helps minimize side effects. If the vaccine is intended for more than one species, one should consider that the immune response to a specific vaccine preparation and potential side effects might differ across species. Environmental safety is particularly important if vaccine preparation includes recombinant DNA technology.

14.2.2 Reversibility and longevity

Depending on the sterilization needs of the target population or individual, a reversible vaccine might be ideal. Pet owners might prefer reversible fertility control for their animals and would probably choose vaccines whose duration of effects could be estimated. Thus, the average duration of fertility control for any vaccine needs to be thoroughly studied before the product becomes commercially available. To date, the literature on the longevity of experimental AF vaccines is still scarce.

In wildlife management, reversible AF vaccines might be desirable because the need for fertility control might change over time depending on fluctuations in population size. For fertility control of feral animals or species considered to be "pests," a potentially irreversible vaccine would be the best choice. In other cases, reversibility might not be an issue as long as the duration of the vaccine prevents reproduction for a sufficient amount of time. For example, feedlot heifers are usually exposed to bulls in open range before going to feedlots. Since pregnancy in feedlot heifers is disadvantageous for numerous reasons (humane, economic, etc.), fertility control is recommended. When in the feedlot, estrus suppression is still desirable as a means to avoid excessive physical activity associated with behavioral estrus. Thus, the vaccine effects would need to last just long enough to suppress estrus during the period of time where heifers may be exposed to bulls, prior to entry into the feedlot, until slaughter.

14.2.3 Frequency of treatments

The number of immunizations necessary for successful fertility control is an important factor to be considered. Producers would benefit from an effective, single-dose vaccine given that management of large herds for each booster injection brings additional costs associated with time, labor, and the cost of each dose itself.

Control of wild or feral populations would be more feasible with a single-dose vaccine, considering the difficulties of recapturing individual animals for booster injections. Ideally, a single-dose contraceptive vaccine would be effective in controlling the size of a population even if each treated individual was immunized only once in its lifetime. Of course, this is assuming that the vaccine could provide long-lasting contraception and that the proportion of individuals captured and released after immunization would be large enough to cause an overall impact on the population size.

The number of injections necessary for the desired vaccine effect and longevity is influenced by the type of delivery system. Single-dose vaccines usually require a delivery system that releases antigen and adjuvant in a slow manner in order to maintain relatively high levels of immunogens in the system for a prolonged period of time. The idea here is that the slow antigenic release would simulate the effects of booster injections. Immunocontraception has been tested using delivery systems, such as polymer microspheres, that slow down antigenic release to control the fertility of species associated with overpopulation problems in North America, including white-tailed deer and horses (Kirkpatrick et al. 1996; Turner et al. 1996).

One of the most frequently used delivery systems for slow, controlled antigenic release is antigen encapsulation in polymers, typically, a mix of lactide and glycolide polymers. The ratio lactide:glycolide dictates the rate at which the antigen is released into the system because each polymer has a certain degradation rate. This delivery system is commonly used in single-dose vaccines.

14.2.4 Route of administration

The route of administration is also an important factor to be considered in the design of a new vaccine as it may affect the intensity of the immune response. However, the ideal route of administration may vary depending on the target population. Wildlife management agents would benefit from the use of an injectable contraceptive vaccine that can be remotely injected intramuscularly using darting rifles or blowguns (Turner et al. 1992). In small animal practice, the route of choice might be subcutaneous injections.

Development of oral AF vaccines still faces several challenges such as the need to protect a protein antigen from degradation by proteases in the digestive tract. Alternatively, expression vectors for gene delivery may allow oral administration of recombinant DNA vaccines (Seleem et al. 2008).

Concerns with potential oral AF vaccines would include difficulty in estimating the dose ingested and the need to protect nontarget populations from accidental ingestion of the vaccine. Attempts to control populations of species considered as "pests," such as sewer rats, would benefit from an oral AF vaccine that could be masked in baits.

14.2.5 Production costs

Besides duration of effectiveness and ease of administration, another major aspect that consumers would consider before choosing between vaccine brands is, obviously, price per dose. Many protocols for preparation of protein antigens used in experimental vaccines include some type of protein purification procedure. Because it is one of the most expensive, time-consuming steps in antigen preparation, protein purification would considerably affect vaccine production costs and could hinder production at a large scale. However, protein purification may reduce deleterious side effects from vaccination that may result from reaction with nontarget tissues. Thus, it is recommended that the pros and cons of protein purification be evaluated thoroughly to determine if it is really needed before marketing the vaccine.

14.2.6 Regulatory requirements for approval of novel vaccines

In the process of developing a new AF vaccine, it is also important to keep in mind that approval for marketing depends on whether the product meets the regulatory requirements of the countries where the vaccine is intended to be marketed. One of the main concerns regarding new biotechnologies in the animal industry is the introduction of chemical compounds into the food chain. According to the United States Department of Agriculture (USDA), animals produced for human consumption cannot be treated with AF vaccines within 90 days before slaughter. Therefore, immunization protocols must be in agreement with mandatory clearance periods.

The ability to produce vaccine batches with consistent chemical structure is also a concern if a vaccine is to be approved for commercialization. In the United States, AF vaccines that alter the physiology of the animal (e.g., suppression of ovarian cyclicity) are considered to be drugs and therefore, are regulated by the Food and Drug Administration (FDA), which requires consistency in chemical structure between batches of the product.

Many AF vaccines that have been used experimentally do not meet this criterion because the antigen is linked to a large protein molecule (carrier protein) by a process called chemical conjugation. In this process, cross-linking agents are typically used to bind to the carrier protein and target antigen through their side chains. Chemical conjugation yields batches of vaccine with inconsistent chemical structure because it is not uncommon to obtain conjugation between the same molecules of either carrier protein or hapten.

14.3 Specific aspects in vaccine development

This section describes the main components of AF vaccines—target antigens and the components used to enhance the immune response to the antigen, namely adjuvants and carrier proteins.

14.3.1 Carrier proteins

The reason for using carrier proteins is that some antigen molecules are too small to be immunogenic and need to be conjugated to larger (carrier) proteins in order to elicit an immune response. This is the case of small peptides (or haptens), which are usually conjugated chemically to a carrier protein in antigen preparation. Typically, the carrier protein is a foreign molecule for immunogenic purposes. One of the most frequently used carrier proteins is keyhole limpet hemocyanin (KLH), a protein obtained from the mollusk giant keyhole limpet. Other commonly used carrier proteins for immunization of mammals are tetanus toxoid (TT), diphtheria toxin (DT), and the bird protein ovalbumin (OVA).

As an alternative to chemical conjugation, some laboratories use recombinant technology to produce molecules of antigen fused to carrier protein, which results in consistent chemical structure. Plasmid vectors encoding sequences of the hapten and carrier protein can be used to transform cells (e.g., *Escherichia coli*) that will then consistently express the fusion protein to be used as an antigen. Other laboratories use DNA constructs as the antigen itself. In either case, the recombinant antigen is designed to be expressed with consistent chemical structure.

14.3.2 Adjuvants

The choice of adjuvant is one of the most crucial factors in the success of any vaccine. Essentially the role of an adjuvant is to augment the immune response to an antigen and in doing so increase the efficacy of the vaccine. There are several mechanisms by which adjuvants work. Basically, an adjuvant may augment the immune response to an antigen by one or more of the following actions: (1) prolonging the time of exposure to the antigen; (2) influencing the distribution or presentation of the antigen; (3) directly activating the immune system, among other actions.

Adjuvants can prolong the exposure to an antigen by entrapping it in some type of reservoir (e.g., water-in-oil emulsions, polymer microspheres). The "depot" effect of antigen entrapment results in the release of low antigen doses for prolonged periods of time, which contributes to selective stimulation of B cells with high-affinity receptors and production of high-affinity antibodies (Siskind and Benacerraf 1969). The oil component of some adjuvants creates the "depot" effect in water-in-oil emulsions and plays a role in the distribution of the antigen through the lymphatic system. Direct activation of the immune system can be done by presenting foreign components that will trigger an immune response, such as bacterial components. The vertebrate immune system will recognize bacterial components (e.g., cell wall, DNA) as nonself components and the immune system will be activated to fight the invading organism.

Cells of the immune system such as phagocytes and dendritic cells can detect the presence of microorganisms by recognition of pathogen-associated molecular patterns (PAMPs) through Toll-like receptors (TLRs). These cells engulf and digest (endophagocytosis) invading microorganisms and then travel to the lymph nodes and the spleen, where presentation of antigenic epitopes to T cells will start a cascade of events that will result in stimulation of antibody production by plasma cells. Basically, the same cascade of events happens upon injection of a vaccine adjuvanted with immunostimulants that are recognized by the immune system as "invading organisms."

Freund's adjuvants have been the adjuvant of choice for experimental immunization for decades (Freund et al. 1937; Broderson 1989; Billiau and Matthys 2001; Stills 2005). Both Freund's complete adjuvant (FCA) and Freund's incomplete adjuvant (FIA) are a combination of 85% light mineral oil and 15% mannide monooleate, which acts as a surfactant. The difference between FCA and FIA is that the former contains heat-killed and dried *Mycobacterial* cells as an immunostimulant. Originally, FCA contained *Mycobacterium tuberculosis* but one disadvantage of using this particular bacterial component was that animals injected with it would test positive in tuberculosis tests. *Mycobacterium avium* and *Mycobacterium butyricum* have been used in more recent preparations to avoid that problem.

Besides their reputation as potent immunostimulators, Freund's adjuvants are also known for their undesirable side effects, including skin lesions (Gendimenico and Mezick 1995), granuloma formation and ulceration (Broderson 1989), arthritis (Haak et al. 1996), and pneumonia (Broderson 1989). These side effects make them unacceptable for therapeutic use in humans and animals. Freund's adjuvants are still commonly used in experimental immunizations and for polyclonal antibody production, but there is a significant concern about pain and distress caused by these adjuvants.

In susceptible strains of rodents, FIA alone (without auto-antigens) can induce arthritis in an acute manner (Holmdahl and Kvick 1992), while FCA (without auto-antigens) can induce chronic arthritis (Pearson 1956). Granuloma formation is more common after injection with FCA than FIA (Billiau and Matthys 2001). A second injection with mycobacteria can cause severe delayed-type hypersensitivity reactions that can be lethal

(Broderson 1989). Thus, Freund's immunization protocols usually have FCA in the first vaccination, followed by booster injections with FIA (Raffel 1948). The undesirable side effects of FCA have led many researchers to seek alternative immunostimulants that could be as potent as mycobacteria but with less toxicity. Alternatively, bacterial components such as muramyl dipeptide (MDP), lipopolysaccharide (LPS), and monophosphoryl lipid A (MPL) have been used as immunostimulants in vaccine preparations. Moreover, synthetic compounds that mimic bacterial components for immunostimulation have shown promising results.

Bacterial DNA is naturally rich in motifs that contain unmethylated cytosines followed by guanines as dinucleotides, flanked by particular base sequences. These CpG motifs are usually differentiated by a hexanucleotide sequence that contains at least one CG dinucleotide. In vertebrates, the frequency of CG dinucleotides in DNA is about three to four times lower compared with bacterial DNA, a phenomenon known as CG suppression (Bird 1986, 1987). Moreover, cytosines are usually methylated on the 5′ position in vertebrate DNA. CpG motifs are recognized by the vertebrate immune system as a "danger signal" announcing the presence of nonself DNA, which elicits an immune response (Bird 1987).

Synthetic CpG oligodeoxynucleotides (ODNs) have been evaluated as immunostimulants to replace bacterial components (Krieg et al. 1995). The initial (innate) response to CpG ODN is fast and not antigen-specific; there is proliferation of B cells (Krieg 1996; Hartmann et al. 2000), activation of natural killer (NK) cells (Ballas et al. 1996), and release of cytokines (Krieg et al. 1999). These cytokines attract additional immune cells that lead to an adaptive (antigen-specific) immune response. Exogenous CpG ODNs can be added to a vaccine preparation; alternatively, genetic engineering may be used to incorporate CpG ODN sequences into constructs containing the antigen sequence (Naz 2006).

More recently, CpG ODN was compared with Freund's adjuvant as an immunostimulant for immunosterilization (Conforti et al. 2007, 2008). All injections in the CpG protocol contained CpG ODN, yet no sign of severe inflammatory reaction was observed in any of the animals in the CpG groups. Thus, CpG ODN was shown to be safe for multiple injections. Additionally, the CpG ODN protocol was reported to be as immunostimulatory as Freund's protocol in those studies.

The search for the ideal adjuvant—one that maximizes the immune response to a specific antigen without causing side effects—is a continuing field of studies.

14.3.3 Antigens

Depending on the type of antigen, an antifertility vaccine may or may not aim at both sexes. Numerous types of molecules could be targeted, including sperm or egg proteins, hormones involved in reproduction, or any other molecules whose neutralization would prevent conception or embryonic development.

Antigens for AF vaccines can be obtained from tissue preparations or produced synthetically. Natural antigens from tissue preparations are usually from a heterologous origin for immunogenic purposes. The following is a discussion on the major types of antigens used experimentally in AF vaccines, the techniques used in their discovery, their effectiveness, and the pros and cons of their use.

The search for new antigens usually starts with *in vivo* injections of potentially antigenic preparations to induce an immune response and to harvest antibodies that could be used in laboratory techniques (e.g., Western blot analysis, sodium dodecyl sulfate-polyacrylamide gel electrophoresis [SDS-PAGE], and immunohistochemistry) for immunochemical characterization of the antigen molecule. Characterization of a cognate protein includes determination of its molecular weight, developmental expression, and tissue specificity, as well as its mechanisms of regulation (e.g., studying the expression of a protein after castration and subsequent hormone supplementation).

Other techniques commonly used in the search for novel antigens are neonatal tolerization and hybridoma technology. Neonatal tolerization is a process that has been shown to effectively raise a specific immune response to molecules whose low immunogenicity would otherwise be masked by more immunogenic molecules present in the same tissue preparation. The protocol includes administration of the immunogenic preparation to which tolerance is to be induced (tolerogen), without adjuvants, to neonatal animals within 24 h of birth. This first tolerogen injection takes advantage of the fact that the neonatal immune system is not able to respond to antigens and will consider foreign molecules as self-components. After some weeks, a second injection of non-adjuvanted tolerogen is given, followed a few days later by administration of an immunosuppressant drug such as cyclophosphamide. Through induced chemical immunosuppression in the tolerized animal any cell population that had escaped the effects of tolerization should be rendered immunologically unresponsive to the tolerogen by the drug. The tolerized/immunosuppressed animal is injected with the immunogenic preparation to which an immune response is desired (immunogen). The resulting immune response will be directed only to the components present in the immunogen preparation but not in the tolerogen preparation. This protocol has been shown to generate antibodies specifically against proteins found in the epididymides, but not in the testes (Joshi et al. 2003b), or against antigens that are found on cauda, but not caput, epididymal sperm membranes (Ensrud and Hamilton 1991).

Following neonatal tolerization, production of monoclonal antibodies (mAbs) against the antigen of interest is important for immunochemical characterization of the antigenic molecule. This can be accomplished by hybridoma technology, which includes harvesting of lymphocytes from the spleen of immunized animals and co-incubation with myeloma cell lines. In the presence of polyethylene glycol, these two types of cell fuse, resulting in continuous production of mAbs.

After identification of a novel antigen, several aspects of the vaccine must be evaluated in order to determine the combination of components that will result in optimal immune response. Trial studies must test different adjuvants as well as different doses of antigen.

14.4 Sperm antigens

Several laboratories have focused on the identification of sperm proteins that could potentially serve as antigens in contraceptive vaccines. Because an immune response against sperm proteins would impair sperm function, whether sperm are in the male or female reproductive tract, sperm vaccines could be used as a contraceptive method for both sexes.

Sperm antigens may be of testicular and/or epididymal origin and may play key roles in fertilization, which makes them potential targets for contraceptive vaccines. Spermatozoa are not capable of fertilizing an egg until they pass through the epididymis, where sperm maturation occurs. The epithelium of the epididymis interacts with spermatozoa by secreting proteins that alter the surface of the sperm cells. Some of these surface proteins are involved in mechanisms that are essential for fertilization such as sperm motility and sperm–egg binding.

Immunization against certain epididymal antigens may interfere with sperm–egg binding. A hamster sperm glycoprotein of epididymal origin and approximate size of 26 kDa (P26h) was used in sperm–egg binding studies. Rabbit antibodies raised against P26h were co-incubated with mature oocytes and spermatozoa to evaluate the effects of anti-P26h antibodies on gamete interaction (Bégin et al. 1995). *In vitro* fertilization systems (oocytes and spermatozoa) from two species (hamster and mouse) were studied separately. Inhibitory effects on sperm–egg binding of co-incubation with anti-P26h IgGs or Fab fragments were compared in both species. In the hamster system, IgGs and Fab fragments had comparable inhibitory effects. In the mouse system, intact IgGs had a reduced inhibitory effect (compared with the hamster system) and Fab fragments did not inhibit gamete interaction. The authors hypothesized that the intact, but not the Fab fragments, anti-P26h IgGs are capable of partial inhibition of sperm–egg interaction in the mouse because they generate steric hindrance.

14.4.1 Identification of epididymis-specific sperm antigens

Testicular antigens are more immunogenic than epididymal antigens. Thus, immunization with whole sperm preparations, and subsequent application of hybridoma technology, favors production of monoclonal antibodies against antigens of testicular origin. Techniques such as tolerization of testicular antigens may be used prior to injection with epididymal preparations in order to induce an immune response specifically against epididymal antigens (Joshi et al. 2003a).

Using neonatal tolerization of testicular antigens followed by immunization with epididymal antigens, Joshi et al. (2003b) identified an epididymis-specific, androgen-regulated protein of ~27 kDa. This protein was found in the epididymal epithelium as well as on spermatozoa from rat epididymides. Antibodies raised against this protein caused agglutination of spermatozoa *in vitro*, suggesting that this protein is a potential antigen for contraceptive vaccines.

14.4.2 Sperm antigens for human contraception

There has been increasing interest in the identification of testis/epididymal antigens that could potentially be used for reversible immunocontraception in humans. A testis/epididymal-specific protein named Eppin has been used as an antigen for contraceptive vaccine trials in nonhuman primates with the ultimate goal of becoming a nonhormonal contraceptive method for men. Immunization with adjuvanted recombinant human Eppin was shown to cause temporary infertility in male bonnet macaques (*Macaca radiate*; O'Rand et al. 2004).

14.4.3 Searching for sperm antigens using the vasectomized model

Another approach to identify sperm antigens relies on the autoimmune effects that follow a vasectomy. In the intact male reproductive tract, sperm proteins that would be interpreted by the immune system as nonself molecules are kept from triggering an autoimmune response by the blood-testis and blood-epididymal barriers. Following vasectomy, obstruction of the reproductive tract and subsequent changes in barrier permeability lead to contact between sperm proteins and blood stream, triggering an autoimmune response. The vasectomized model allows for the harvesting of anti-sperm antibodies (ASA) and identification of sperm-associated auto-antigens that could potentially be used in contraceptive vaccines.

Some auto-antigens of testicular origin identified in the vasectomized model have a potential for immunocontraception because they interfere with sperm motility. Using the vasectomized mice, Wakle et al. (2005) obtained anti-sperm monoclonal antibodies and selected one of them—mAb D5E5—for immunochemical characterization of the cognate sperm-associated auto-antigen. The mAb D5E5 bound to a protein of approximately 70kDa in size that was found to be expressed in both testicular and epididymal sperm, as well as in testicular, but not epididymal, tissue. This protein, named Testis Specific Auto-antigen70 (TSA70), was found in rat, bull, human, and nonhuman primate sperm. *In vitro* co-incubation of mAb D5E5 with mouse sperm was shown to reduce forward progressive motility, making TSA70 a potential target for immunocontraception.

14.4.4 *Female infertility and the discovery of sperm antigens*

Anti-sperm antibodies can also be found in untreated females. In fact, by using anti-sperm antibodies found in the serum of an infertile woman researchers identified a sperm-specific protein in human sperm extract, named 80kDa Human Sperm Antigen (80kDaHSA; Bandivdekar et al. 1991). This protein was found in men on the sperm surface and in testicular and epididymal, but not other, somatic tissues. Synthetic peptides of 80kDaHSA conjugated to KLH have been tested as antigens for contraceptive vaccines in laboratory species. Peptide 1 is one of the peptides of 80kDaHSA obtained by digestion with endoproteinase Lys-C. Immunization with synthetic Peptide 1 caused transient infertility in male rabbits and marmosets—a nonhuman primate model (Khobarekar et al. 2008). No difference was found in sperm count between control and immunized animals; however, agglutination of sperm cells was observed in immunized rabbits, but not marmosets, along with complete loss of progressive motility in both species.

14.4.5 *Sperm antigens for wildlife population control*

Other sperm antigens might be effective for immunocontraception by interfering with mechanisms necessary for fertilization, but not necessarily essential to sperm motility. A marsupial species (*Macropus eugenii*) was used as a model to evaluate a sperm contraceptive vaccine as a potential method of fertility control for free-living overpopulations of kangaroos (Asquith et al. 2006). An anti-sperm immune response was observed in males immunized with homologous whole sperm preparation containing tetanus toxoid, as an immunological marker, and adjuvanted with FCA. Anti-sperm IgGs bound *in vivo* to the acrosome and mid-piece of spermatozoa. Sperm-immunized and control males were allowed to mate with superovulated females. Their results showed that sperm-immunization reduced fertilization rates. The luminal fluid from the rete testis had a higher amount of anti-sperm IgGs compared with the fluid from the cauda epididymis. In contrast, antibody-sperm binding was negligible in the testis, weak in the caput epididymis, but intense in subsequent regions of the epididymis and vas deferens. The authors suggested that *in vivo* systemic anti-sperm antibodies reached spermatozoa in the male reproductive tract through the rete testis. It was speculated that antibody effects were probably not primarily on sperm motility, because sperm from immunized males were found in the female reproductive tract in the mucoid layer surrounding the oocytes.

14.4.6 Sperm antigens in DNA vaccines

DNA vaccines have also shown promising results for immunocontraception using sperm antigens. A recent study using a DNA vaccine reported reduction in fertility rates in female mice after immunization with different vaccine preparations including a construct containing the sequences of a CpG ODN and the antigen of interest (Naz 2006). The DNA construct encoded the sperm-specific dodecamer peptide named YLP_{12}. Treatments consisted of intradermal immunization using gene gun with a preparation containing a YLP_{12} DNA construct with zero, one, or two repeats of a CpG sequence, or the DNA construct plus exogenous synthetic CpG ODN. All YLP_{12} DNA treatments resulted in antibody production, which was detected in both serum and vaginal tract, and reduced fertility. Overall, immunization with YLP_{12} DNA resulted in higher production of IgG2a compared with IgG1, indicating a Th1-biased response. A Th1 immune response is also marked by the secretion of cytokines such as interleukin-2 (IL-2) and interferon-gamma (IFN-γ), while a Th2 response results in expression of IL-4 and IL-10. In that study, expression of both Th1 and Th2 cytokines was detected, with a Th1 bias. Among the YLP_{12} DNA treatment groups no difference was detected in the amount of any class/subclass of antibodies produced; moreover, there was no difference among YLP_{12} DNA treatment groups in the number of animals that had an immune response to the vaccine. However, the vaccine preparations having two CpG repeats or exogenous CpG ODN had the strongest inhibitory effect on fertility rates *in vivo* and on acrosome reaction and sperm–egg binding *in vitro* compared with the other treatment groups. Since antibody titers did not seem to be the reason for the differences observed in the *in vitro* assays, the author speculates that antibody specificity might have played a role in increasing inhibition of acrosome reaction and sperm–egg binding. Additionally, the superiority of these two treatment groups in reducing contraception *in vivo* led the author to speculate that cytokines might have contributed to fertility reduction through sperm- or embryo-toxic effects.

14.5 Zona pellucida antigens

The plasma membrane of the mammalian oocyte is surrounded by a thick outer layer called zona pellucida (ZP). Depending on the species, the ZP is composed of three or four major glycoproteins (ZP1-4). ZP proteins play crucial roles in fertilization for their involvement in oocyte development and sperm binding, among other functions.

14.5.1 ZP vaccines for female contraception

Anti-ZP antibodies can lead to female infertility by preventing sperm binding to receptors or by causing steric hindrance (Liu et al. 1989). Because of these mechanisms of action, some authors believe that ZP vaccines cause infertility without interfering with endocrine function; consequently, ovarian cyclicity as well as estrous and breeding behaviors would be retained. Some authors reported no differences in breeding behavior between ZP immunized and control animals (Kirkpatrick et al. 1990; Turner et al. 1996). Maintenance of these behaviors might be disadvantageous in some cases; for example, if sexually transmitted diseases are a concern, ZP vaccines should not be recommended as a contraceptive method.

14.5.2 ZP immunization and ovarian histopathology

Studies in horses and deer have reported that pZP immunization reduced fertility without altering ovarian histopathology (Liu et al. 1989; McShea et al. 1997). Other studies, however, have reported abnormalities in ovarian histopathology, including alteration in granulosa structure (Skinner et al. 1984), disruption of endocrine ovarian function, and estrous and mating behaviors following ZP immunization (Stoops et al. 2006). Thus, the contraceptive effects of some ZP vaccines might be a result from disruption of ovarian function.

Domestic ewes immunized with a partially purified porcine ZP (pZP) in FCA became infertile but hormone profile and behavioral data suggested that ovarian endocrine function had been compromised (Stoops et al. 2006). Fecal progesterone metabolite profiles revealed lack of estrous cyclicity, which was in agreement with lack of behavioral estrus and mating observed in the pZP/FCA-treated ewes housed with rams. Examination of ovaries revealed aberrant histology including drastic reduction in primordial follicle numbers, absence of follicles at further stages of development, absence of CLs, abnormal granulosa cell organization with missing oocytes, and ZP degeneration. The authors suggested that fertility reduction was primarily related to disruption of ovarian function, and that the purification status of the antigen might have contributed to the ovarian abnormalities observed. Partially purified ZP antigens may contain additional immunogenic proteins that may potentially trigger an immune response against other ovarian components besides the ZP.

In contrast with vaccines made from ovarian preparations, vaccines containing purified recombinant ZP protein antigens might result in a more specific immune response but still may cause alterations in ovarian histology. A study on rabbits immunized with a recombinant myxoma virus-ZP fusion protein evaluated the effects of rabbit ZP2 (rZP2) or rZP3 as antigens (Mackenzie et al. 2006). Immunization with the recombinant fusion protein containing rZP2 resulted in ZP antibody production and binding to the ZP, but no effect on fertility or ovarian histology was observed. However, immunization with rZP3 resulted in antibody production and binding to ZP, transient infertility, and altered ovarian histology. Upon evaluation of the ovaries, corpora lutea were found in both fertile and infertile females, suggesting that immunization against rZP3 did not suppress breeding activity and that ovarian endocrine function was not drastically affected. The abnormalities observed in ovarian follicles of rabbits immunized against rZP3 affected granulosa cells, zona pellucida, and oocytes, but seemed to be temporary.

In summary, the intensity and duration of contraceptive or deleterious effects, if any, from ZP immunization may vary across species, and may depend on factors such as antigen purity. Thus, effects of ZP vaccines on ovarian histology and function should be thoroughly addressed in a species-specific manner. This is particularly important if the contraceptive effects are not intended to be permanent.

14.5.3 ZP immunization during pregnancy

ZP vaccines may not interfere with ongoing pregnancies. ZP-immunized mares that were pregnant at the time of injection had normal pregnancies and gave birth to healthy young offspring (Lyda et al. 2005). This is of interest to wildlife population management since pregnancy diagnosis in free-ranging animals is unfeasible; moreover, the objective of immunocontraception is prevention, not interruption, of pregnancy.

14.5.4 ZP vaccines for wildlife population control

For effective contraception throughout at least one breeding season, most ZP vaccines tested to date would require at least two immunizations (Kirkpatrick et al. 1991; Fayrer-Hosken et al. 2000; Kitchener et al. 2002). However, vaccines that require multiple injections are impractical for wildlife. Thus, single-dose ZP vaccines have been evaluated in wild species and were able to reduce fertility in white-tailed deer, although efficacy was reduced compared with treatment with multiple injections (Turner et al. 1996).

Another study showed promising results after single immunization with pZP in black bears (Lane et al. 2007). Two experiments were conducted: in the first experiment, the first immunization contained pZP in FCA, followed by a second immunization with pZP in FIA. In the second experiment, bears were immunized with a combination of biodegradable polymer pellet (for slow, controlled release) containing pZP and a water-soluble adjuvant called QS-21 plus a liquid emulsion of pZP and QS-21. In both experiments, ZP immunization reduced cub production. Although these two protocols were not compared directly in a single experiment, the single immunization with slow antigenic release seemed to be more efficient than two immunizations in Freund's adjuvants.

14.5.5 The challenge of developing efficient ZP vaccines for cats

ZP vaccines have shown to produce satisfactory antifertility effects in several species; however, development of an effective ZP vaccine for contraception of domestic and exotic felines has been challenging scientists for years (Gorman et al. 2002; Harrenstien et al. 2004; Levy et al. 2005). Anti-ZP antibody production has been reported in female domestic cats after immunization with either heterologous or, to a lesser degree, homologous ZP, yet these ZP vaccines have failed to prevent pregnancies in fertility trials in the species (Levy et al. 2005). This problem has been attributed to lack of cross-reactivity between ZP antibodies and feline ZP *in vivo*.

14.6 LHRH antigens

Luteinizing hormone-releasing hormone (LHRH), a hypothalamic decapeptide, also known as Gonadotropin-releasing hormone (GnRH), regulates reproductive function in both sexes. LHRH is released from hypothalamic neurons that project to the median eminence, where it enters the hypothalamic-hypophyseal portal vessels. Through these vessels LHRH reaches the pituitary, where it binds to receptors on gonadotrophs and stimulates secretion of the gonadotropins luteinizing hormone (LH) and follicle-stimulating hormone (FSH). Through circulation, gonadotropins reach the gonads (ovaries/testes) and stimulate secretion of reproductive hormones—progesterone, estradiol, and testosterone.

14.6.1 Effects of LHRH immunization in males and females

Immunization with synthetic LHRH neutralizes endogenous LHRH, which leads to disruption of the hypothalamic-pituitary-gonadal axis, and consequently impairs steroidogenesis and gametogenesis. LHRH immunization in females causes suppression of estrous cyclicity and behavior; in males, serum testosterone decreases as well as aggression and sex drive. In both sexes, there is gonadal regression.

14.6.2 LHRH immunization and cross-reactivity with isoforms

Since the discovery of isoforms, mammalian LHRH has also been called GnRH-I. There are at least four isoforms found in mammals, namely chicken GnRH-II (GnRH-II), salmon GnRH (GnRH-III), and two forms of lamprey GnRH. Function of these molecules in mammals remain poorly understood but there has been concern that LHRH immunization could generate antibodies that would cross-react with isoforms. Immunization of mice with LHRH produced antibodies that cross-reacted with chicken GnRH-II and lamprey GnRH-III (Khan et al. 2007b). Immunization against LHRH or lamprey GnRH-III appeared to have a stronger inhibitory effect on spermatogenesis compared with GnRH-II, suggesting that, like LHRH, lamprey GnRH-III has a role in spermatogenesis. In another study, manipulation of the LHRH amino acid sequence allowed for production of modified synthetic LHRH peptides that were used for immunization to evaluate cross-reactivity with LHRH isoforms *in vitro*; the resulting antibodies bound to LHRH but showed no cross-reactivity to its isoforms (Turkstra et al. 2005).

14.6.3 Applications of LHRH vaccines

There are numerous applications for LHRH vaccines in animal sciences. Besides immunocontraception, LHRH immunization has been used to control aggressive and sexual behaviors, to decrease steroid-dependent odors such as boar taint, and to improve growth performance in comparison with castrates (reviewed by Bonneau and Enright 1995; Dunshea et al. 2001). Additionally, LHRH immunization has been tested in animal models and humans as a potential tool for research and treatment of steroid-dependent diseases that affect humans, especially prostate cancer (Simms et al. 2000; Hill et al. 2003). Immunization against LHRH is also used to treat intact male dogs diagnosed with benign prostatic hyperplasia, an androgen-dependent condition.

The cattle industry has many potential applications for LHRH vaccines. LHRH immunization for feedlot heifers would suppress estrus, preventing undesirable pregnancies and excessive physical activity associated with estrous behaviors that may cause weight loss or injuries. Currently, AF methods for feedlot heifers are either spaying or daily addition of the synthetic progestin Melengestrol acetate (MGA) to feed. However, sterilization of heifers by spaying requires veterinary services and may result in some death loss. MGA treatment has been associated with interstitial pneumonia in cattle (McAllister et al. 2002). Thus, immunosterilization would be an interesting alternative to traditional methods of fertility control for feedlot heifers. In fact, it has been shown that LHRH immunization has effectively suppressed ovarian function in beef heifers (Stevens et al. 2005; Conforti et al. 2008). However, decreased steroidogenesis following LHRH immunization compromises average daily gain (ADG) in heifers. Therefore, hormone supplementation, usually in the form of an implant, is recommended in order to maintain satisfactory ADG in feedlot heifers immunized against LHRH.

Countries like Brazil, where hormone implants for growth are not allowed, do not castrate bulls until about 2 years of age for maximum growth purposes. However, those bulls are pasture-fattened in large groups, making testosterone-driven aggression a concern because of increased risk of injuries for animals and handlers. Surgical castration at 2 years of age would decrease aggressiveness but risks associated with this practice include complications from screwworm infestation during surgical recovery. Thus, LHRH immunization of 2-year-old bulls presents an attractive, less invasive alternative to castration for reduction of aggressiveness. LHRH immunization has been shown to reduce serum testosterone concentrations of bulls immunized at 2 years of age (Hernandez et al. 2005).

14.6.4 *Recombinant LHRH antigens*

To date, numerous types of LHRH vaccines have been tested; originally, these vaccines had the decapeptide chemically conjugated to a carrier protein. Other studies have used recombinant technology for antigen production with consistent chemical structure. Our laboratory has developed a recombinant fusion protein expressed in *E. coli* cells transformed by a plasmid containing seven LHRH inserts into a fragment of ovalbumin (ova-LHRH; Zhang et al. 1999). The number of LHRH inserts was shown to affect the efficacy of the vaccine; the plasmid containing seven LHRH inserts produced a more immunogenic fusion protein compared with that produced by a plasmid containing only four inserts. The ova-LHRH vaccine has successfully suppressed reproductive function in several species, including mice, rats, sheep, and cattle, which is evidenced by gonadal regression (Zhang et al. 1999; Sosa et al. 2000; Ülker et al. 2005; Conforti et al. 2007; Figure 14.1).

Recombinant technology has also been used to eliminate the need for large carrier proteins in LHRH vaccines. The presence of carrier proteins may cause a phenomenon

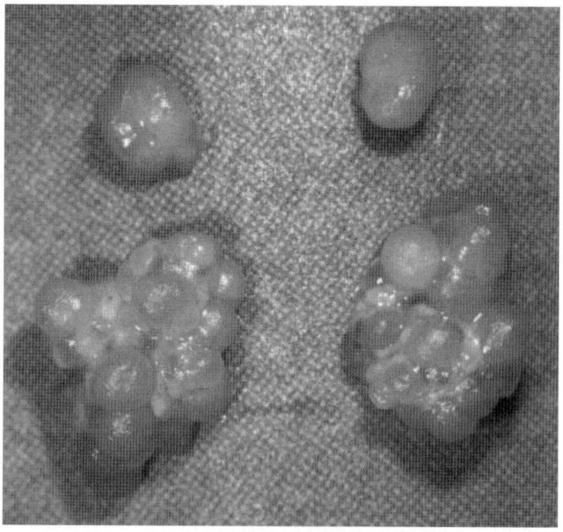

Figure 14.1 Gonadal regression in adult female rats as a consequence of immunization with recombinant ovalbumin-LHRH fusion protein as evidenced by comparison between ovaries from an immunized rat (top pair) and from an untreated control rat (bottom pair). As expected, LHRH immunization caused severe reduction in the number of developing follicles.

known as carrier-induced epitope suppression, which negatively affects the immune response to the hapten, compromising the effectiveness and longevity of a vaccine. In order to avoid epitope suppression, T-helper epitope sequences may be used as a replacement for large carrier proteins in LHRH vaccines. In a recent study, researchers evaluated a vaccine containing plasmid DNA encoding LHRH repeats and selected T-helper epitope nucleotide sequences from viruses and bacteria for enhanced immunogenicity (Khan et al. 2007a). The plasmid DNA was encoated in a virus vector (Hemagglutinating Virus of Japanese Envelop) for gene delivery. Fertility trials showed that immunization of male mice with this vaccine reduced litter numbers.

14.6.5 Suggestions for future studies: Effects of LHRH immunization on pregnancy, on target cells/tissues, and on longevity of vaccine effects

Some aspects of LHRH vaccines still need further investigation. Little is known about the consequences of active LHRH immunization during pregnancy. Most reports on the effect of LHRH antibodies in pregnant females come from passive immunization studies. Administration of LHRH antibodies to pregnant sheep affects the fetal hypothalamic-pituitary-gonadal axis, causing reduction in LH secretion in both male and female fetuses and reduction in fetal FSH secretion in males (Miller et al. 1998). Another study investigated the effects of active and passive immunization against LHRH on early pregnancy in pigs (Tast et al. 2000). Sows received the primary immunization on the day of farrowing and were allowed to mate at the following estrus. A booster injection was given either on day 10 or 20 postmating and resulted in failure to establish pregnancy or abortion, respectively. In another experiment, sows passively immunized against LHRH on day 12 postmating failed to establish pregnancy.

Another aspect of LHRH immunization that needs further investigation is the effect of LHRH antibodies on target cells and tissues. Some authors believe that active LHRH immunization may cause lesions in the median eminence, where terminals of LHRH neurons are not protected by the blood-brain barrier. Active immunization against LHRH in pigs has been reported to cause several signs of inflammation in the median eminence, including fibrosis and tissue disruption by edema (Molenaar et al. 1993). Another author examined the hypothalamus of pigs immunized against LHRH and found no histological changes (Turkstra 2005). Thus, this topic remains controversial. It is worthwhile to note that results of functional atrophy might be mistakenly interpreted as tissue lesions.

Literature on the longevity of LHRH vaccines is still relatively scarce but most authors seem to agree that efficiently immunized animals become temporarily infertile but regain fertility once antibody levels decrease below a certain threshold. However, it has been suggested that long-term effects of LHRH vaccines may depend on the age at which animals are immunized. Neonatal LHRH immunization affected secretion of LHRH in adult sheep (Clarke et al. 1998).

14.6.6 Reviewing the need for purification of certain recombinant LHRH antigens

Clearly, LHRH immunization is an effective method of fertility control, but the future of LHRH vaccines as marketable products will depend on factors such as safety, consistency of results (i.e., percentage of immunized animals that respond effectively to the vaccine for the desired period of time), predictability of chemical structure, and cost per dose. Many experimental LHRH antigens have successfully resulted in immunosterilization but the costs and time required for large scale production would render the vaccine unmarketable. As previously discussed, protein purification would be one of the major obstacles to large scale production of many vaccine preparations. However, protein purification might be eliminated without diminishing the efficacy of LHRH antigens.

In a recent preliminary study, the non-purified form of the recombinant antigen ova-LHRH was tested in female domestic cats. This recombinant fusion protein is found in inclusion bodies in the transformed *E. coli* cells. In previous studies, ova-LHRH vaccines had the antigen purified by nickel chelation chromatography. However, gel analyses indicated that this recombinant fusion protein had basically the same molecular weight before and after purification, suggesting that ova-LHRH is the predominant protein in inclusion bodies of the transformed bacteria. Sera from cats vaccinated with non-purified ova-LHRH had an increasing anti-LHRH antibody activity following a single immunization. The vaccine preparation contained a combination of encapsulated ova-LHRH and CpG ODN for slow release plus the same antigen and adjuvant in a water-in-oil emulsion, the idea being that the nonencapsulated antigen/adjuvant mix would work as a primary injection, while the encapsulated counterpart would mimic booster injections. The immune response observed approximately 8 months after a single injection of non-purified ova-LHRH is comparable with the results obtained in other species after multiple injections of purified ova-LHRH. Moreover, no severe inflammatory reactions were observed at the injection site following immunization. An ongoing study is evaluating the immune and biological responses to this vaccine preparation in domestic cats, including fertility trials. Results from the preliminary study suggest that this vaccine might be a promising tool for feral cat population control. Moreover, if approved for commercialization, the vaccine containing the non-purified recombinant protein would be able to be marketed at a more reasonable price compared with the purified protein.

14.7 Future research directions

In conclusion, the field of immunocontraception/sterilization offers vast opportunities for research and applications in human and animal sciences. Currently, few veterinary AF vaccines are commercially available (Table 14.1). One of the major obstacles to approving novel AF vaccines for marketing is low efficacy. Due to individual variations in immune response to AF vaccines, the percentage of treated animals that have satisfactory biological response to immunization is often below the minimum required by regulatory agencies. Improving immunogenicity without increasing toxicity still is one of the biggest challenges in vaccine development for fertility control. Future directions for AF vaccines will likely rely on genetic engineering to maximize immunogenicity of antigens while eliminating the need for strong, potentially toxic adjuvants.

Acknowledgments

The author is grateful to Russell Dudley and Dr. Jerry Reeves and Dr. Monica Stoops for the valuable comments on the manuscript.

Table 14.1 Past and present market-driven veterinary vaccines against reproductive antigens (adapted from Meeusen et al. 2007).

Brand name (target antigen)	Intended use	Vaccine preparation		Protocol	Market[2]	Current status
		Antigen	Adjuvant			
SpayVac® (ZP)	♀ deer, population control	porcine ZP	Adjuvac™	single dose	U.S. wildlife agencies	To be approved by the FDA
Vaxstrate™ (LHRH)	♀ cattle, immuno-sterilization.	LHRH-OVA	oil-based	two doses	Australia	No longer available
Improvac™ (LHRH)	♂ pigs, boar taint control	LHRH[1]	water-soluble	two doses	Oceania, Latin America,[3] S. Africa, S. Korea	Available
Equity™ (LHRH)	♀ horses, estrous behavior control	LHRH[1]	Quil-A-based	two doses	Australia, New Zealand	Available
GonaCon™; GonaCon-B™ (LHRH)	deer, population control	LHRH-KLH; LHRH-blue protein	AdjuVac™	single dose	United States	To be approved by the FDA
Canine Gonadotropin Releasing Factor immunother. (LHRH)	♂ dogs, treatment of benign prostatic hyperplasia	LHRH-DT	Quil-A-based	variable	United States	Available

[1]No information provided by manufacturer on the LHRH molecule or carrier protein.
[2]Distributors, comments, and references:
SpayVac®: nonprofitable organization SpayVac(for Wildlife Inc., U.S.A. This vaccine was shown to reduce fertility in white-tailed deer (Locke et al. 2007).
Vaxstrate: Websters Animal Health, Australia. This was the first commercially available LHRH vaccine. It was released in the Australian market in the 1980s but was discontinued in 1996 due to low efficacy, high incidence of abscesses, and the need of two immunizations, which was impractical for local husbandry practices. Scientific literature on this vaccine is scarce (Hoskinson et al. 1990).
Improvac™: Pfizer Animal Health, [3]Brazil, Costa Rica, Guatemala, Mexico. Improvac™ efficiently eliminated boar taint when given at 8 and 4 weeks before slaughter (23–26 weeks of age). Improvac-immunized boars grew faster than intact boars during the 4-week period following booster immunization, possibly due to reduced sexual and aggressive behaviors. Compared with barrows, immunized boars had superior feed conversion and leaner carcasses (Dunshea et al. 2001).
Equity™: Pfizer Animal Health, Australia. The antigen is a synthetic LHRH molecule conjugated to a carrier protein. Estrous behavior was suppressed for at least 3 months in mares receiving two immunizations with Equity™ (Elhay et al. 2007).
GonaCon™, GonaCon-B™: National Wildlife Research Center, U.S.A. GonaCon™ contains LHRH conjugated to KLH, while GonaCon-B™ contains LHRH-blue protein. A study compared the two formulations in female white-tailed deer; both reduced fertility but GonaCon-B™ had longer lasting contraceptive effects (Miller et al. 2008).
Canine Gonadotropin Releasing Factor immunotherapeutic: Pfizer Animal Health, U.S.A. This vaccine is used to suppress androgen production as part of a treatment of intact male dogs diagnosed with benign prostatic hyperplasia (USDA).
AdjuVac™: adjuvant developed by the National Wildlife Research Center, U.S.A. Contains killed *Mycobacterium avium* (Miller et al. 2008).

References

Asquith, K.L., Kitchener, A.L., and Kay, D.J. 2006. Immunisation of the male tammar wallaby (*Macropus eugenii*) with spermatozoa elicits epididymal antigen-specific antibody secretion and compromised fertilisation rate. *Journal of Reproductive Immunology* **69**(2): 127–147.

Ballas, Z.K., Rasmussen, W.L., and Krieg, A.M. 1996. Induction of NK activity in murine and human cells by CpG motifs in oligodeoxynucleotides and bacterial DNA. *The Journal of Immunology* **157**(5): 1840–1845.

Bandivdekar, A.H., Koide, S.S., and Sheth, A.R. 1991. Antifertility effects of human sperm antigen in female rats. *Contraception* **44**(5): 559–569.

Bégin, S., Bérubé, B., Boué, F., and Sullivan, R. 1995. Comparative immunoreactivity of mouse and hamster sperm proteins recognized by an anti-P26h hamster sperm protein. *Molecular Reproduction and Development* **41**(2): 249–256.

Billiau, A. and Matthys, P. 2001. Modes of action of Freund's adjuvants in experimental models of autoimmune diseases. *Journal of Leukocyte Biology* **70**(6): 849–860.

Bird, A.P. 1986. CpG-rich islands and the function of DNA methylation. *Nature* **321**(6067): 209–213.

Bird, A.P. 1987. CpG islands as gene markers in the vertebrate nucleus. *Trends in Genetics* **3**(12): 342–347.

Bonneau, M. and Enright, W.J. 1995. Immunocastration in cattle and pigs. *Livestock Production Science* **42**(2–3): 193–200.

Broderson, J.R. 1989. A retrospective review of lesions associated with the use of Freund's adjuvant. *Laboratory Animal Science* **39**(5): 400–405.

Clarke, I.J., Brown, B.W., Tran, V.V., Scott, C.J., Fry, R., Millar, R.P., and Rao, A. 1998. Neonatal immunization against gonadotropin-releasing hormone (GnRH) results in diminished GnRH secretion in adulthood. *Endocrinology* **139**(4): 2007–2014.

Conforti, V.A., de Avila, D.M., Cummings, N.S., Wells, K.J., Ülker, H., and Reeves, J.J. 2007. The effectiveness of a CpG motif-based adjuvant (CpG ODN 2006) for LHRH immunization. *Vaccine* **25**(35): 6537–6543.

Conforti, V.A., de Avila, D.M., Cummings, N.S., Zanella, R., Wells, K.J., Ülker, H., and Reeves, J.J. 2008. CpG motif-based adjuvant as a replacement for Freund's Complete Adjuvant in a recombinant LHRH vaccine. *Vaccine* **26**(7): 907–913.

Dunshea, F.R., Colantoni, C., Howard, K., McCauley, I., Jackson, P., Long, K.A., Lopaticki, S., Nugent, E.A., Simons, J.A., Walker, J., and Hennessy, D.P. 2001. Vaccination of boars with a GnRH vaccine (Improvac) eliminates boar taint and increases growth performance. *Journal of Animal Science* **79**(10): 2524–2535.

Elhay, M., Newbold, A., Britton, A., Turley, P., Dowsett, K., and Walker, J. 2007. Suppression of behavioural and physiological oestrus in the mare by vaccination against GnRH. *Australian Veterinary Journal* **85**(1–2): 39–45.

Ensrud, K.M. and Hamilton, D.W. 1991. Use of neonatal tolerization and chemical immunosuppression for the production of monoclonal antibodies to maturation-specific sperm surface molecules. *Journal of Andrology* **12**(5): 305–314.

Fayrer-Hosken, R.A., Grobler, D., van Altena, J.J., Bertschinger, H.J., and Kirkpatrick, J.F. 2000. Immunocontraception of African elephants. *Nature* **407**(6801): 149.

Freund, J., Casals, J., and Hismer, E.P. 1937. Sensitization and antibody formation after injection of tubercle bacilli and paraffin oil. *Proceedings of the Society for Experimental Biology and Medicine* **37**: 509.

Gendimenico, G.J. and Mezick, J.A. 1995. Effects of topical inflammatory agents on Freund's adjuvant-induced skin lesions in rats. *Inflammation Research* **44**(1): 16–20.

Gorman, S.P., Levy, J.K., Hampton, A.L., Collante, W.R., Harris, A.L., and Brown, R.G. 2002. Evaluation of a porcine zona pellucida vaccine for the immunocontraception of domestic kittens (*Felis catus*). *Theriogenology* **58**(1): 135–149.

Haak, T., Delverdier, M., Amardeilh, M.F., Oswald, I.P., and Toutain, P.L. 1996. Pathologic study of an experimental canine arthritis induced with complete

Freund's adjuvant. *Clinical and Experimental Rheumathology* **14**(6): 633–641.

Harrenstien, L.A., Munson, L., Chassy, L.M., Liu, I.K., and Kirkpatrick, J.F. 2004. Effects of porcine zona pellucida immunocontraceptives in zoo felids. *Journal of Zoo and Wildlife Medicine* **35**(3): 271–279.

Hartmann, G., Weeratna, R.D., Ballas, Z.K., Payette, P., Blackwell, S., Suparto, I., Rasmussen, W.L., Waldschmidt, M., Sajuthi, D., Purcell, R.H., Davis, H.L., and Krieg, A.M. 2000. Delineation of a CpG phosphorothioate oligodeoxynucleotide for activating primate immune responses in vitro and in vivo. *The Journal of Immunology* **164**(3): 1617–1624.

Hernandez, J.A., Zanella, E.L., Bogden, R., de Avila, D.M., Gaskins, C.T., and Reeves, J.J. 2005. Reproductive characteristics of grass-fed, luteinizing hormone-releasing hormone-immunocastrated *Bos indicus* bulls. *Journal of Animal Science* **83**(12): 2901–2907.

Hill, R.E., de Avila, D.M., Bertrand, K.P., Greenberg, N.M., and Reeves, J.J. 2003. Immunization against luteinizing hormone-releasing hormone fusion proteins does not decrease prostate cancer in the transgenic adenocarcinoma mouse prostate model. *Experimental Biology and Medicine* **228**(7): 818–822.

Holmdahl, R. and Kvick, C. 1992. Vaccination and genetic experiments demonstrate that adjuvant-oil-induced arthritis and homologous type II collagen-induced arthritis in the same rat strain are different diseases. *Clinical and Experimental Immunology* **88**(1): 96–100.

Hoskinson, R.M., Rigby, R.D., Mattner, P.E., Huynh, V.L., D'Occhio, M., Neish, A., Trigg, T.E., et al. 1990. Vaxstrate: An anti-reproductive vaccine for cattle. *Australian Journal of Biotechnology* **4**(3): 166–170, 176.

Joshi, S.A., Ranpura, S.A., Khan, S.A., and Khole, V.V. 2003a. Monoclonal antibodies to epididymis-specific proteins using mice rendered immune tolerant to testicular proteins. *Journal of Andrology* **24**(4): 524–533.

Joshi, S.A., Shaikh, S., Ranpura, S., and Khole, V.V. 2003b. Postnatal development and testosterone dependence of a rat epididymal protein identified by neonatal tolerization. *Reproduction* **125**(4): 495–507.

Khan, M.A.H., Ferro, V.A., Koyama, S., Kinugasa, Y., Song, M., Ogita, K., Tsutsui, T., Murata, Y., and Kimura, T. 2007a. Immunisation of male mice with a plasmid DNA vaccine encoding gonadotrophin releasing hormone (GnRH-I) and T-helper epitopes suppresses fertility *in vivo*. *Vaccine* **25**(18): 3544–3553.

Khan, M.A.H., Prevost, M., Waterston, M.M., Harvey, M.J.A., and Ferro, V.A. 2007b. Effect of immunisation against gonadotrophin releasing hormone isoforms (mammalian GnRH-I, chicken GnRH-II and lamprey GnRH-III) on murine spermatogenesis. *Vaccine* **25**(11): 2051–2063.

Khobarekar, B.G., Vernekar, V., Raghavan, V., Kamada, M., Maegawa, M., and Bandivdekar, A.H. 2008. Evaluation of the potential of synthetic peptides of 80 kDa human sperm antigen (80 kDaHSA) for the development of contraceptive vaccine for male. *Vaccine* **26**(29–30): 3711–3718.

Kirkpatrick, J.F., Liu, I.K.M., and Turner, J.W. Jr. 1990. Remotely-delivered immunocontraception in feral horses. *Wildlife Society Bulletin* **18**(3): 326–330.

Kirkpatrick, J.F., Liu, I.K.M., Turner, J.W. Jr, and Bernoco, M. 1991. Antigen recognition in feral mares previously immunized with porcine zona pellucida. *Journal of Reproduction and Fertility. Supplement* **44**: 321–325.

Kirkpatrick, J.F., Turner, J.W. Jr, Liu, I.K.M., and Fayrer-Hosken, R. 1996. Applications of pig zona pellucida immunocontraception to wildlife fertility control. *Journal of Reproduction and Fertility. Supplement* **50**: 183–189.

Kitchener, A.L., Edds, L.M., Molinia, F.C., and Kay, D.J. 2002. Porcine zonae pellucidae immunization of tammar wallabies (*Macropus eugenii*): Fertility and immune responses. *Reproduction, Fertility and Development* **14**(4): 215–223.

Krieg, A.M. 1996. An innate immune defense mechanism based on the recognition of CpG motifs in microbial DNA. *The Journal of Laboratory and Clinical Medicine* **128**(2): 128–133.

Krieg, A.M., Yi, A.-K., and Hartmann, G. 1999. Mechanisms and therapeutic applications of immune stimulatory CpG DNA. *Pharmacology and Therapeutics* **84**(2): 113–120.

Krieg, A.M., Yi, A.-K., Matson, S., Waldschmidt, T.J., Bishop, G.A., Teasdale, R., Koretzky, G.A., and Klinman, D.M. 1995. CpG motifs in bacterial DNA trigger direct B-cell activation. *Nature* **374**(6522): 546–549.

Lane, V.M., Liu, I.K., Casey, K., van Leeuwen, E.M., Flanagan, D.R., Murata, K., and Munro, C. 2007. Inoculation of female American black bears (*Ursus americanus*) with partially purified porcine zona pellucidae limits cub production. *Reproduction, Fertility and Development* **19**(5): 617–625.

Levy, J.K., Mansour, M., Crawford, P., Cynda, P., Bill, B., and Robert, G. 2005. Survey of zona pellucida antigens for immunocontraception of cats. *Theriogenology* **63**(5): 1334–1341.

Liu, I.K., Bernoco, M., and Feldman, M. 1989. Contraception in mares heteroimmunized with pig zonae pellucidae. *Journal of Reproduction and Fertility* **85**(1): 19–29.

Locke, S.L., Cook, M.W., Harveson, L.A., Davis, D.S., Lopez, R.R., Silvy, N.J., and Fraker, M.A. 2007. Effectiveness of Spayvac for reducing white-tailed deer fertility. *Journal of Wildlife Diseases* **43**(4): 726–730.

Lyda, R.O., Hall, J., Ron, K., and Jay, F. 2005. A comparison of Freund's complete and Freund's modified adjuvants used with a contraceptive vaccine in wild horses (*Equus caballus*). *Journal of Zoo and Wildlife Medicine* **36**(4): 610–616.

Mackenzie, S.M., McLaughlin, E.A., Perkins, H.D., French, N., Sutherland, T., Jackson, R.J., Inglis, B., Müller, W.J., van Leeuwen, B.H., Robinson, A.J., and Kerr, P.J. 2006. Immunocontraceptive effects on female rabbits infected with recombinant myxoma virus expressing rabbit ZP2 or ZP3. *Biology of Reproduction* **74**(3): 511–521.

McAllister, T.A., Stanford, K., Ayroud, M., Bray, T.M., and Yost, G.S. 2002. Management practices to control acute interstitial pneumonia in feedlot heifers. *Final Report*, Alberta Beef Industry Development, 59.

McShea, W.J., Monfort, S.L., Hakim, S., Kirkpatrick, J., Liu, I., Turner, J., Chassy, L., and Munson, L. 1997. The effect of immunocontraception on the behavior and reproduction of white-tailed deer. *The Journal of Wildlife Management* **61**(2): 560–569.

Meeusen, E.N., Walker, J., Peters, A., Pastoret, P.-P., and Jungersen, G. 2007. Current status of veterinary vaccines. *Clinical Microbiology Reviews* **20**(3): 489–510.

Miller, D.W., Fraser, H.M., and Brooks, A.N. 1998. Suppression of fetal gonadotrophin concentrations by maternal passive immunization to GnRH in sheep. *Journal of Reproduction and Fertility* **113**(1): 69–73.

Miller, L.A., Gionfriddo, J.P., Fagerstone, K.A., Rhyan, J.C., and Killian, G.J. 2008. The single-shot GnRH immunocontraceptive vaccine (GonaCon) in white-tailed deer: Comparison of several GnRH preparations. *American Journal of Reproductive Immunology* **60**(3): 214–223.

Molenaar, G.J., Lugard-Kok, C., Meloen, R.H., Oonk, R.B., de Koning, J., and Wensing, C.J. 1993. Lesions in the hypothalamus after active immunisation against GnRH in the pig. *Journal of Neuroimmunology* **48**(1): 1–11.

Naz, R.K. 2006. Effect of sperm DNA vaccine on fertility of female mice. *Molecular Reproduction and Development* **73**(7): 918–928.

O'Rand, M.G., Widgren, E.E., Sivashanmugam, P., Richardson, R.T., Hall, S.H., French, F.S., VandeVoort, C.A., Ramachandra, S.G., Ramesh, V., and Rao, A.J. 2004. Reversible immunocontraception in male monkeys immunized with eppin. *Science* **306**: 1189–1190.

Pearson, C.M. 1956. Development of arthritis, periarthritis and periostitis in rats given adjuvants. *Proceedings of the Society for Experimental Biology and Medicine* **91**: 95–101.

Raffel, S. 1948. The components of the tubercle bacillus responsible for the delayed type of "infectious" allergy. *Journal of Infectious Diseases* **82**: 267–293.

Seleem, M.N., Ali, M., Boyle, S.M., and Sriranganathan, N. 2008. Vectors for enhanced gene expression and protein purification in Salmonella. *Gene* **421**(1–2): 95–98.

Simms, M.S., Scholfield, D.P., Jacobs, E., Michaeli, D., Broome, P., Humphreys, J.E., and Bishop, M.C. 2000. Anti-GnRH antibodies can induce castrate levels of testosterone in patients with advanced prostate cancer. *British Journal of Cancer* **83**(4): 443–446.

Siskind, G.W. and Benacerraf, B. 1969. Cell selection by antigen in the immune response. *Advances in Immunology* **10**: 1–50.

Skinner, S.M., Mills, T., Kirchick, H.J., and Dunbar, B.S. 1984. Immunization with zona pellucida proteins results in abnormal ovarian follicular differentiation and inhibition of gonadotropin-induced steroid secretion. *Endocrinology* **115**(6): 2418–2432.

Sosa, J.M., Zhang, Y., de Avila, D.M., Bertrand, K.P., and Reeves, J.J. 2000. Technical note: Recombinant LHRH fusion protein suppresses estrus in heifers. *Journal of Animal Science* **78**(5): 1310–1312.

Stevens, J.D., Sosa, J.M., de Avila, D.M., Oatley, J.M., Bertrand, K.P., Gaskins, C.T., and Reeves, J.J. 2005. Luteinizing hormone-releasing hormone fusion protein vaccines block estrous cycle activity in beef heifers. *Journal of Animal Science* **83**(1): 152–159.

Stills, H.F. Jr. 2005. Adjuvants and antibody production: Dispelling the myths associated with Freund's Complete and other adjuvants. *Institute for Laboratory Animal Research Journal* **46**: 280–293.

Stoops, M.A., Liu, I.K.M., Shideler, S.E., Lasley, B.L., Fayrer–Hosken, R.A., Bernirschke, K., Murata, K., van Leeuwen, E.M.G., and Anderson, G.B. 2006. Effect of porcine zonae pellucidae immunisation on ovarian follicular development and endocrine function in domestic ewes (*Ovis aries*). *Reproduction, Fertility and Development* **18**(6): 667–676.

Tast, A., Love, R.J., Clarke, I.J., and Evans, G. 2000. Effects of active and passive gonadotrophin-releasing hormone immunization on recognition and establishment of pregnancy in pigs. *Reproduction, Fertility and Development* **12**(5–6): 277–282.

Turkstra, J.A. 2005. Active immunisation against gonadotropin-releasing hormone, an effective tool to block the fertility axis in mammals. Doctoral dissertation, Utrecht University, The Netherlands.

Turkstra, J.A., Schaaper, W.M.M., Oonkb, H.B., and Meloen, R.H. 2005. GnRH tandem peptides for inducing an immunogenic response to GnRH-I without cross-reactivity to other GnRH isoforms. *Vaccine* **23**(41): 4915–4920.

Turner, J.W. Jr., Kirkpatrick, J.F., and Liu, I.K.M. 1996. Effectiveness, reversibility, and serum antibody titers associated with immunocontraception in captive white-tailed deer. *The Journal of Wildlife Management* **60**(1): 45–51.

Turner, J.W. Jr., Liu, I.K.M., and Kirkpatrick, J.F. 1992. Remotely delivered immunocontraception in captive white-tailed deer. *The Journal of Wildlife Management* **56**(1): 154–157.

Ülker, H., Yilmaz, A., Karakus, F., Yörük, M., Budag, C., de Avila, D.M., and Reeves, J.J. 2005. The effects of immunization against LHRH using recombinant LHRH fusion protein, ovalbumin-LHRH-7, on development, histologic and ultrasonographic appearance of ram lamb testis. *Paper read at 8th International Symposium "Modern Trends in Livestock Production"*, Belgrade, Serbia, October 5–8.

Wakle, M.S., Joshi, S.A., and Khole, V.V. 2005. Monoclonal antibody from vasectomized mouse identifies a conserved testis-specific antigen TSA70. *Journal of Andrology* **26**(6): 761–771.

Zhang, Y., Rozell, T.G., de Avila, D.M., Bertrand, K.P., and Reeves, J.J. 1999. Development of recombinant ovalbumin-luteinizing hormone releasing hormone as a potential sterilization vaccine. *Vaccine* **17**(17): 2185–2191.

15

Proteomics of Male Seminal Plasma

Vera Jonakova, Jiri Jonak, and Marie Ticha

15.1 Introduction

Mammalian fertilization is a unique event in which morphologically disparate gametes recognize each other, bind and fuse. This event includes highly regulated biochemical interactions between molecules located on the surface of both gametes as well as substances present in the natural environment of gametes both in the male and the female reproductive organs. The following phases of the reproduction process can be distinguished: binding of seminal plasma proteins to the sperm surface during ejaculation, interaction of sperm surface proteins with oviductal epithelial cells, sperm capacitation, gamete recognition, primary and secondary binding of the sperm to the zona pellucida (ZP), acrosome reaction of sperm, penetration of the sperm through the ZP of the ovum, and the fusion of the sperm and the egg (reviewed in Evans and Kopf 1998; Töpfer-Petersen et al. 1998, 2000, 2005, 2008; Visconti et al. 1998; Töpfer-Petersen 1999; Wassarman 1999; Jansen et al. 2001; Suarez 2001; Wassarman et al. 2001, 2005; Jonakova and Ticha 2004; Calvete and Sanz 2007; Jonakova et al. 2007; Manjunath et al. 2007; Tanphaichitr et al. 2007; Vadnais et al. 2007). Participation of the seminal plasma proteins of domestic animals in the individual steps of the reproduction process is reviewed in this chapter.

Mammalian seminal plasma is a complex mixture of secretions mainly produced by the testis, the epididymis, and the male accessory sex glands (seminal vesicles, ampulla, prostate, bulbourethral glands) and contains a variety of different substances (amino acid, lipids, fatty acids, saccharides, ions, peptides, and proteins). The complex content of the seminal plasma is designed to assure the successful fertilization of the oocyte by one of the spermatozoa present in the ejaculum. Proteins represent an important part of the high-molecular-weight substances in the seminal plasma (Yanagimachi 1994; Henault and Killian 1996).

As far as the seminal plasma of domestic animals is concerned, bull and boar proteins belong to the most studied ones. A number of papers concerning studies on the characterization of these protein molecules, as well as their properties and function, decrease in the following order: bovine (bull) ≫ porcine (boar) ≫ equine (stallion) > goat (buck) > ovine (ram) > poultry.

15.2 Proteins of seminal plasma

15.2.1 Structure and properties

The seminal plasma proteins of domestic animals discussed in this review can be approximately separated into three groups according to their structural characteristics:

1. Spermadhesins. They belong to a novel group of animal lectins. They form a subgroup of a superfamily of proteins with a single CUB domain (named after the proteins in which it was first identified: C, complement subcomponents C1r/C1s; U—uegf, urchin epidermal growth factor; B—Bmp1, bone morphogenetic protein) that has been found in a variety of developmentally regulated processes (Bork and Beckmann 1993).
2. Proteins containing Fibronectin type II (Fn-2) domains. They belong to a large family of cell and matrix adhesion proteins, which include seminal plasma proteins, fibronectins, and large cell surface receptors (Potts and Campbell 1994). A list of well-characterized proteins from various species of domestic animals (spermadhesins and Fn-2) is given in Table 15.1.
3. Different proteins exhibiting enzymatic, inhibitory, and other activities. Proteins belonging to the first two groups represent the major protein constituents of the mammalian seminal fluid. However, the relative abundance of these proteins varies in different species: in bull seminal plasma (BSP), the BSP proteins containing Fn-2 domains comprise about 65% of the total protein; homologous proteins in stallion and boar seminal plasma represent only 20% and 1.1% of the total protein, respectively (Manjunath et al. 2007). Proteins of CUB and Fn-2 structural groups are present in seminal plasma both as non-modified polypeptide chains and as differently glycosylated isoforms as described for boar spermadhesins by Calvete et al. (1993) and Solis et al. (1997) and for BSP proteins from bull seminal plasma by (Manjunath and Sairam 1987).

The third group of proteins detected in seminal plasma need not be directly

Table 15.1 Proteins of seminal plasma of various species containing fibronectin type II domains and homologous to boar spermadhesins.[1]

Species	Fibronectin type II domain proteins	Spermadhesin proteins
Boar	DQH (pB1)	AWN family AQN family PSP-I/II
Bull	BSP-A1/A2 (PDC-109) BSP-A3 BSP-30kDa	aSFP Z13
Stallion	HSP1 HSP2 HSP-12	HSP-7
Ram	RSP-15 kDa RSP-16 kDa RSP-22 kDa RSP-24 kDa	15 kDa protein
Buck	GSP-15 kDa GSP-20 kDa GSP-22 kDa	BSFP

[1]Data from the tables in Manjunath et al. (2007); references to individual proteins are given in Chapter 5 and Tables 15.2–15.7. DQH, AQN, AWN—boar seminal plasma proteins (designations are in accordance with their N-terminal amino acid sequence). PSP, porcine seminal plasma; BSP, bovine (bull) seminal plasma, BSP-A1/A2 equals PDC-109; HSP, horse seminal plasma; RSP, ram seminal plasma; GSP, goat seminal plasma; BSFP, buck seminal fluid protein.

involved in the reproduction process. They may have a protective role (as, e.g., antioxidant enzymes [Marti et al. 2007; Jelezarsky et al. 2008]), participate in a modulation of activity of seminal plasma proteins in male and female reproductive tract (Meyer et al. 1997; Soubeyrand et al. 1997; Cibulkova et al. 2007), or participate in the inhibition of enzymes affecting sperm function (Jonakova et al. 1992; Soubeyrand and Manjunath 1997; Jelinkova et al. 2003). The role of several other proteins present in seminal plasma *in vivo*, for example, lactoferrin, β-microseminoprotein, RNAase dimer, is still not clear.

The binding properties of homologous proteins in different species do not always need to be similar; acidic Seminal Fluid Protein (aSFP) from bull seminal plasma could serve as an example. This protein displays about 50% amino acid sequence identity with boar spermadhesins. Nevertheless, it possesses neither carbohydrate nor ZP-binding activity (Calvete and Sanz 2007). The representation of spermadhesins and Fn-2 proteins in seminal plasma of various investigated animals appears to significantly differ. Human seminal plasma was found to be very low in spermadhesin-like proteins (Kraus et al. 2001, 2005).

It was shown that under physiological conditions the seminal plasma proteins of boar, bull, and stallion form variable aggregates (homo- and hetero-oligomers) differing in relative molecular mass, number of individual spermadhesins and fibronectin type II domain proteins, and in interaction properties (Calvete et al. 1995a, 1997; Gasset et al. 1997; Solís et al. 1998; Jonakova et al. 2000; Manaskova et al. 2000, 2003; Jelinkova et al. 2004a).

The aggregated forms of boar seminal plasma proteins AQN, AWN, PSP, and DQH make the protein coverage of the sperm surface (Jonakova et al. 2000; Manaskova et al. 2002, 2003). The protein coating layers of sperm that are formed during ejaculation are subject to remodeling in the female reproductive tract.

The aggregation state of the seminal plasma proteins could be modulated by solute components, phosphorylcholine or heparin, or by substances of the native environment of gametes as demonstrated for PDC-109, the major protein of bull seminal plasma (Gasset et al. 1997; Liberda et al. 2001, 2002a; Talevi and Gualtieri 2001; Jelinkova et al. 2004a).

On the other hand, it was reported that the rates of heparin or phosphorylcholine binding to bull major seminal plasma proteins (Jelinkova et al. 2004a) and heparin binding to seminal plasma proteins of stallion (HSP—horse seminal plasma) proteins (Calvete et al. 1995a) are significantly affected by the extent and character of the seminal plasma protein oligomerization. These association/dissociation processes thus result in changes in their interaction properties as described, for example, in the case of PSP dimer from boar seminal plasma (Calvete et al. 1995d; Campanero-Rhodes et al. 2006), or BSP proteins (Jelinkova et al. 2004a).

Proteins in a polydisperse form were described to be present in bull seminal plasma (Manjunath and Sairam 1987; Calvete et al. 1999). Changes in the polydispersity of bull seminal proteins in the fertilization and the mechanism of sperm capacitation by PDC-109 has been proposed by Calvete and Sanz (2007) based on detailed structural studies of PDC-109 complexes with phosphorylcholine (Wah et al. 2002).

Out of all domestic animals, a detailed proteomic analysis of only bull seminal plasma has been published. The proteomic approach involving 2-D and 1-D electrophoretic separation and mass spectroscopy analysis revealed the presence of about 250 protein spots, out of which 99 were identified (Kelly et al. 2006). A similar approach was used to compare the protein composition of the accessory gland fluid from individual Holstein bulls (Moura et al. 2006a,b).

15.2.2 Localization and expression

Spermadhesins AWN, AQN1, AQN3, PSP-I, and PSP-II cDNAs were amplified from total RNA of porcine seminal vesicles (Ekhlasi-Hundrieser et al. 2002). This revealed, in the case of AWN, a 459 nucleotide open reading frame, comprising a signal peptide (amino acids 1–20) and 133 amino acid residues polypeptide, followed by 232 nucleotides of the 3′-untranslated region. AWN cDNA derived amino acid sequence differed in two positions, Tyr92 and Glu/Gln98, from that obtained by direct sequencing of AWN1 protein by Sanz et al. (1992b) who reported Arg and His respectively, at these positions. The amino acid sequence deduced from the cDNA fragment encoding AQN1 protein showed the complete amino acid sequence identity to that determined from the mature protein (Sanz et al. 1992a), whereas the AQN3 cDNA deduced amino acid sequence differed in position Thr78 and Glu95 from that obtained by protein sequencing (Gly78 and Asp95). The unidentified residue at position 85 was shown to be a serine residue. The presence of the 11-mer peptide, LNLXCGKEYV/LE, found in all mature porcine spermadhesins at positions 49–59, was also confirmed by cDNA sequencing. Besides seminal vesicles, AWN transcripts were also detected in extracts from prostate and caudal epididymis. No PCR products could be generated from RNA extracts of testis, rete testis, caput epididymis, corpus epididymis, and bulbourethral glands (or liver). AQN1 and AQN3 specific transcripts were also found in seminal vesicles, prostate, and cauda epididymis.

The mRNA transcripts of the DQH gene were possible to detect and clone from boar seminal vesicles (Plucienniczak et al. 1999), but not from other reproductive organs, such as testis, epididymis, or prostate (Manaskova et al. 2007). The DQH cDNA derived amino acid sequence showed complete identity with the covalent structure of the boar DQH sperm surface protein, which was determined by Edman degradation, MALDI-MS, and post-source decay (PSD; Bezouska et al. 1999). The DQH protein consists of the N-terminal O-glycosylated peptide followed by two fibronectin type II repeats. This approach also allowed detection of O-glycosidically linked carbohydrates attached to Thr 10 of the isolated N-terminal glycopeptide.

The cDNA sequences of spermadhesins PSP-I, PSP-II obtained from the total RNA of porcine seminal vesicles were originally described by Kwok et al. (1993). Amplified PSP-I and PSP-II cDNA products could be also generated from total RNA of rete testis, caudal epididymis, seminal vesicles, and prostate. A low expression of PSP-I mRNA was further detected in the testis, corpus, and caput epididymis.

The porcine spermadhesin genes were located on pig chromosome 14q28–q29. The pig contains five closely linked spermadhesin genes, whereas only two spermadhesin genes are present in the cattle genome (Haase et al. 2005).

Using a monoclonal avian antibody directed against purified porcine AWN and a rabbit polyclonal antibody generated against porcine AQN1, homologs of both spermadhesins were detected in extracts of seminal vesicles and prostate of the boar by Western blot analysis (Jonakova et al. 1998; Ekhlasi-Hundrieser et al. 2002). In stallion, in contrast, the seminal plasma protein HSP-7, a homolog of the boar AWN spermadhesin, was found to be secreted in the cauda epididymis and its localization on the ejaculated spermatozoa was shown to be on their equatorial segment (Reinert et al. 1997). Further studies revealed expression of both PSP proteins in boar testis, caput, and corpus epididymis, and in bulbourethral glands (García et al. 2008).

Indirect immunofluorescence on tissue sections from boar proved the presence of AQN and AWN spermadhesins in the lumen of epididymis, seminal vesicles, and prostate (Veselsky et al. 1992, 1999), whereas signals from PSP-I and PSP-II spermadhesins were detected in the secretory tissues of corpus epididymis, seminal vesicles, prostate, and Cowper's glands but not in testes (Manaskova and Jonakova 2008). Both PSP proteins were also detected in extracts from boar epididymis, seminal vesicles, prostate, and Cowper's glands (Ekhlasi-Hundrieser et al. 2002; Manaskova et al. 2002; García et al. 2008; Manaskova and Jonakova 2008). The AQN and AWN antibodies interacted with the acrosomal region of both epididymal and ejaculated boar spermatozoa (Veselsky et al. 1999). An interaction of PSP-I and PSP-II antibodies was observed with the acrosomal head region and the mid-piece of the epididymal spermatozoa but only with the acrosomal head region of the ejaculated sperm (Manaskova and Jonakova 2008). Besides, the PSP-II antibody stained the principal piece of the flagellum of the ejaculated spermatozoa (Manaskova and Jonakova 2008).

In boar, the AQN, AWN, and PSP spermadhesins and the DQH sperm surface protein were also detected on the surface of epididymal spermatozoa (Jonakova et al. 1998; Manaskova et al. 2007; Manaskova and Jonakova 2008), ß-microseminoprotein (ß-MSP) was isolated from the seminal plasma and immunodetected in the prostate extract (Jeng et al. 2001; Manaskova et al. 2002). Similarly, in humans, the homologous protein (PSP94) was described as a prostate secretory protein (Ohkubo et al. 1995).

In bull, expression products of the PDC-109 gene, both the PDC-109 mRNA and the PDC-109 protein, the major seminal vesicle secretory protein, were detected in and isolated from extracts of seminal vesicles (Scheit 1990). This finding confirmed and extended previous results showing immunoreactivity of the seminal vesicle epithelium and of the neck region and mid-piece of testicular and ejaculated spermatozoa with antiserum against PDC-109 (Aumüller et al. 1988; Scheit et al. 1988). Interestingly, neither the epididymal epithelium nor the seminal vesicle tissue of the calf gave any reaction to the PDC-109 antibody (Scheit et al. 1988). Finally, Wempe et al. (1992) reported preparation of the aSFP cDNA from extracts of bull seminal vesicle tissue. Expression and localization of acrosin inhibitor in boar reproductive tract is described in Davidova et al. (2009).

15.3 Function of seminal plasma proteins

Proteins of seminal plasma participate in almost all phases of the reproduction process; not only do they affect the properties and thus the behavior of sperm in both the male and the female reproductive tracts, but they also modulate a natural environment in which individual steps of the complex process proceed. Major seminal plasma proteins are mostly multifunctional substances; their function is not limited to one phase of the reproduction process only; it is frequently more complex and not yet fully understood.

Detailed studies mostly performed with bull and boar seminal plasma proteins showed that these proteins are involved in or at least might participate in the following steps of the reproduction process: remodeling of sperm surface, establishment of the oviductal reservoir, modulation of capacitation, gamete interaction, sperm membrane protection, sperm destruction, and modulation of sperm motility. They can also influence the antimicrobial activity of the seminal fluid and function as enzyme inhibitors (reviewed in, e.g., Töpfer-Petersen et al. 1998; Jansen et al. 2001; Jonakova and Ticha 2004; Calvete and Sanz 2007; Manjunath et al. 2007).

A correlation of structure and function of ungulate seminal plasma proteins was recently discussed by Calvete and Sanz. They showed that the sperm membrane remodeling events occurring in the female genital tract are essentially conserved although different seminal plasma proteins participate in these steps in different species (Calvete and Sanz 2007).

15.3.1 Establishment of the oviductal reservoir

In many mammals when sperm cells reach the uterotubal junction–isthmus of the oviductal tract, its epithelium cells trap the spermatozoa to form a sperm reservoir. The main function of the reservoir is to maintain a given population of spermatozoa viable for an extensive period of time, until ovulation, and to prevent polyspermy (Suarez 1998, 2001, 2002, 2007, 2008).

The functional sperm reservoir ensures that suitable numbers of viable and potentially fertile spermatozoa are available for fertilization at the ampullary isthmic junction. *In vivo*, the most viable spermatozoa in the preovulatory sperm reservoir are uncapacitated. Capacitation rates significantly increase after ovulation. Bicarbonate appears to be a common primary effector of the membrane destabilizing change that encompasses the first stages of capacitation. Sperm activation can be delayed or even reversed by co-incubation with membrane proteins of the tubal lining, isthmic fluid, or specific tubal glycosaminoglycans, such as hyaluronan (Rodriguez-Martinez 2007). Hyaluronan, on the other hand, increased capacitation in the post-ovulation period (Tienthai et al. 2004).

Formation of the sperm oviductal reservoir belongs to one of the saccharide-mediated events of the fertilization process and is probably species-specific. The following molecules are involved in the attachment of sperm to the oviduct and in the sperm release to meet an oocyte: (1) glycosylated components of the oviductal epithelium, (2) constituents of oviductal fluid, and (3) proteins localized on the sperm surface.

In cow, the oviductal reservoir is formed by the binding of sperm to L-fucose-containing glycoconjugates on the surface of oviductal epithelium cells (Lefebvre et al. 1995, 1997; Revah et al. 2000; Suarez 2001; Suarez and Ignotz 2001). The L-fucose-binding molecule that promotes bull sperm attachment to the oviductal epithelium was identified as PDC-109 (BSP-A1/A2), a protein secreted by seminal vesicles and associated with the sperm plasma membrane upon ejaculation (Gwathmey et al. 2001, 2003). Two other proteins of bovine seminal plasma (BSP-30-kDa and BSP-A3) enhance the sperm binding to oviductal cells (Gwathmey et al. 2006). Binding of epididymal bull sperm to the epithelium is low (Gwathmey et al. 2003). L-Fucose-binding molecules are lost during capacitation and, at the same time, D-mannose-binding sites are uncovered for the interaction with the ovum (Revah et al. 2000; Ignotz et al. 2001).

Annexins isolated from the apical plasma membrane of bovine oviductal epithelium were suggested to be candidates for bull sperm receptors in the sperm oviductal reservoir formation (Ignotz et al. 2007).

Contrary to the bovine model, the formation of the porcine oviductal sperm reservoir was shown to comprise a participation of D-mannosyl residues (Green et al. 2001); a detailed study showed a high affinity of sperm to oligomannosyl residues (Wagner et al. 2002). As epididymal spermatozoa showed significantly lower capability to bind to oviductal epithelium than ejaculated sperm, participation of the components of seminal plasma especially of sperm coating proteins in this binding process was suggested (Petrunkina et al. 2001). The presence of highly mannosylated structures on porcine oviductal epithelium that could be recognized by boar AQN1 spermadhesin has been demonstrated (Ekhlasi-Hundrieser et al. 2005a, 2008) as well as the affinity of boar spermadhesins to yeast mannan (Jelinkova et al. 2004b). Heparin-binding proteins of boar seminal plasma and especially AQN1 protein displayed the strongest interaction with the oviductal epithelium that was inhibited by yeast mannan (Liberda et al. 2006). On the other hand a glycoprotein (SPG) isolated from porcine oviductal cells containing Gal-ß1-3-GalNAc disaccharide chain was also described to bind to boar sperm (Marini and Cabada 2003; Teijeiro et al. 2007).

Contrary to the data concerning the participation of seminal plasma proteins in the process of sperm reservoir formation, much less information is available about their fate in the course of sperm release from the oviductal epithelium and capacitation (Rodríguez-Martínez et al. 2005; Suarez 2007). Capacitated bull sperm showed a reduced binding of sperm to the oviductal epithelium, as well as to the saccharide ligands (Revah et al. 2000; Ignotz et al. 2001). The loss of binding affinity could be explained by a release of sperm coating proteins during heparin-induced capacitation (Gwathmey et al. 2003). The role of the constituents of oviductal fluid both of the protein and the glycosaminoglycan (e.g., hyaluronic acid) nature on these processes cannot be neglected (Liberda et al. 2006; Rodriquez-Martinez 2007).

15.3.2 *Modulation of capacitation*

Sperm capacitation is a gradual multistep event of the reproduction process occurring in the female reproductive tract (Yanagimachi 1994; Rodríguez-Martínez et al. 2005). It involves a release of sperm from the sperm reservoir (Fazeli et al. 1999), removal of decapacitation substances, mainly of adsorbed epidydimal and seminal plasma proteins, from the sperm surface, reorganization of sperm membrane as a result of the promotion of membrane lipid disorder with consequent protein relocation, and so on (Yanagimachi 1994; Tienthai et al. 2004).

Participation of seminal plasma proteins in sperm capacitation has been studied in detail only with major seminal proteins and in bull. BSP proteins (BSP-A1/A2 [PDC-109], BSP-A3, and BSP-30K), which are secreted by seminal vesicles, are adsorbed to the sperm surface upon ejaculation (Manjunath et al. 1994a). These proteins are specifically bound to phosphorylcholine containing phospholipids present in the sperm membrane (Desnoyers and Manjunath 1992). In addition to the phosphorylcholine-binding activity, BSP proteins interact with high-density lipoprotein (HDL; Manjunath et al. 1989; Thérien et al. 1997, 2001) and glycosaminoglycans (GAGs, e.g., heparin; Thérien et al. 2005). Both types of these substances (HDL and GAGs) are physiological inducers of sperm capacitation and are present in oviductal and follicular fluids. BSP proteins potentiate sperm capacitation induced by either HDL or heparin (Thérien et al. 1997, 2001). In addition, BSP binding to sperm induces cholesterol and choline phospholipid efflux from sperm (Thérien et al. 1998, 1999) and thus modulates the capacitation of bull sperm cells (Manjunath and Therien et al. 2002; Tannert et al. 2007a,b).

Based on the results of these studies, two types of mechanism of the participation of BSP proteins in sperm capacitation mediated either by HDL or GAGs were proposed (Manjunath et al. 2007).

Only a limited amount of information is available on the role of seminal plasma proteins from other species. Boar seminal plasma contains low amounts of a homologous protein of the BSP family, named pB1 (DQH). This protein, as well as BSP-A1/A2 proteins from bull seminal plasma, potentiated boar epididymal sperm capacitation (Lusignan et al. 2007).

The heterodimer of boar spermadhesins PSP-I/PSP-II (present in post-sperm-rich fraction of boar ejaculate) acts as leukocyte chemoattractant both *in vitro* and *in vivo*, contributing to the phagocytosis of those spermatozoa not reaching the sperm reservoir (Rodríguez-Martínez et al. 2005).

15.3.3 *Gamete interaction*

Sperm-ovum interaction occurs in two sequential steps. It starts with the primary binding of acrosome intact sperm to the ZP, which is followed by the secondary binding of acrosome-reacted sperm to the ZP (Bleil et al. 1988). Primary mammalian sperm binding to the ovulated egg is not strictly species-specific, but the glycoprotein envelope of the ovum is supposed to represent a significant barrier to many, if not most, heterospecific interactions *in vitro* (Wassarman 1990; Yanagimachi 1994; Wassarman et al. 2005). These restrictions are attributed to the presence of receptors for spermatozoa on the ovum surface. The ability of sperm to interact with the egg surface can be detected by using solubilized ZP(Gwatkin 1977). The majority of experimental data using solubilized ZP or its components implicate their O-linked and N-linked oligosaccharide chains in the primary sperm binding. It has been estimated that about 75–80% of sperm–ZP binding is of the lectin-like nature, while the remaining ones are based on the protein-protein interactions (Litscher et al. 1995; Clark and Dell 2006).

The mammalian glycoprotein egg envelope has been the most extensively studied in the case of mouse (Wassarman et al. 2005) and much less attention has been paid to other species. The murine ZP3 glycoprotein is supposed to be the primary sperm receptor (Wassarman 1999; Tanphaichitr et al. 2007); in the pig: a heterodimer complex of ZPB and ZPC glycoprotein is suggested to participate in the primary binding (Yurewicz et al. 1998).

While there are only a few glycoprotein components of ZP that were shown to be involved in the primary gamete interaction, a large number of protein molecules were described to possess the ZP-binding ability (Tanphaichitr et al. 2007). This group of substances involves mostly sperm membrane-bound proteins, and in the case of the porcine model, proteins from seminal plasma. The boar spermadhesins were shown to be tightly bound to the sperm membrane of *in vitro* capacitated spermatozoa, and not removed by the capacitation process (Sanz et al. 1993; Dostalova et al. 1994; Calvete et al. 1995b). Sperm-egg binding test and other experimental data demonstrated that intact proteins on the sperm surface (e.g., AQN1, AWN1, DQH) are required for the primary binding of the sperm with the ZP of the ovum (Veselsky et al. 1992, 1999; Dostalova et al. 1995; Ensslin et al. 1995; Calvete et al. 1996a; Rodríguez-Martinez et al. 1998; Manaskova et al. 2000, 2007; Caballero et al. 2005).

There exists a list of many other candidates of sperm components involved in the gamete primary binding and this step of the fertilization process remains unresolved.

15.3.4 *Seminal plasma proteins as enzyme inhibitors*

Seminal plasma also contains proteins that regulate the activity of enzymes occurring in the ejaculate. Proteinase inhibitors are present in all tissues and body fluids. They interfere with the activity of the proteinases and thus maintain the homeostasis. In the male reproductive tract, proteolytic enzymes occur in the sperm acrosome, in the epididymal fluid, and in the seminal plasma. The main role of proteinase inhibitors is the inactivation of prematurely released acrosin from occasionally damaged spermatozoa, and thus protecting the male and the female genital tract against proteolytic degradation. The presence of several proteinase inhibitors in the seminal plasma of different species has been described. Serine proteinase inhibitors, of Kazal type, belong to the most studied inhibitors from the seminal plasma of domestic animals, such as boar (Fritz et al. 1976; Jonakova and Cechova 1985; Jonakova et al. 1991b, 1992; Jelinkova et al. 2003) or bull (Cechova and Jonakova 1981; Meloun et al. 1983, 1985). The acrosin inhibitor isolated from boar seminal plasma (SPAI) (Fritz et al. 1976) is structuraly related to the sperm-associated acrosin inhibitor (SAAI) isolated from boar spermatozoa and sequenced (Jonakova et al. 1991b, 1992). The protein sequence of SAAI was confirmed by the sequencing of its cDNA (Kwok et al. 1994). Proteinase inhibitors may also protect spermatozoa from a proteolytic damage. Spermadhesins (AQN, AWN) attached to the sperm head at ejaculation are acceptor molecules for SAAI. The attachment of the inhibitor to surface molecules of the sperm can stabilize its binding site for the ZP of the oocyte (Sanz et al. 1992c). Formation of a complex between SAAI and AQN1 spermadhesin was proven by gel chromatography (Jelinkova et al. 2003), by Western blotting (in this case also with AWN spermadhesin), and by the detection of SAAI-AQN1 complex on the surface of boar capacitated spermatozoa (Sanz et al. 1992c). It is thought that SAAI protects the ZP binding sites of spermadhesins on the sperm surface against proteolytic degradation from the moment of ejaculation until the sperm-egg encounter (Sanz et al. 1992c; Jonakova et al. 1995).

Besides proteinase inhibitors, the presence of an inhibitor of another enzyme was shown. Major BSP proteins from bovine seminal plasma inhibited the activity of bovine seminal phospholipase A_2 (PLA_2) that has been shown to be a platelet-activating factor acetylhydrolase (PAF-AH). BSP proteins modulate PLA_2 activity and therefore, phospholipid metabolism. They may act as spermatozoa-stabilizing agents by preventing premature lipolysis of the sperm surface (Manjunath et al. 1994b; Soubeyrand et al. 1997; Soubeyrand and Manjunath 1997; Soubeyrand et al. 1998).

15.4 *In vitro* effects of seminal plasma proteins

Seminal plasma contains various components, including those of a protein nature, that are beneficial and/or detrimental not only to the sperm function but also to sperm storage *in vitro*.

In vitro handling of spermatozoa in preparation for artificial insemination, involving processes such as dilution, cooling, freezing, re-warming and sperm sexing by flow cytometric sorting, may modify the proteins bound to the sperm surface, and thus the sperm membranes may be destabilized (Maxwell et al. 2007; de Graaf et al. 2008).

Mammalian sperm preservation in extenders containing egg yolk and milk has been used for a long time. In the case of bull sperm, the mechanism of the protective action of these substances was investigated. Upon binding to sperm surface the phospholipid-binding proteins present in bull seminal plasma (BSP proteins) induce cholesterol and phospholipid removal from sperm membrane and its destabilization in the course of storage. Sequestration of BSP proteins by their interaction with low-density lipoproteins (LDL) of egg yolk was suggested as a major mechanism of sperm protection by LDL (Bergeron and Manjunath 2006; Manjunath et al. 2007). The mechanism of sperm protection by skimmed milk was suggested to involve BSP protein-casein micelle interactions (Bergeron et al. 2006, 2007).

In the case of boar seminal plasma proteins, the heparin-binding proteins (HBP) exert opposite effects on viability, motility, and mitochondrial activity of highly diluted spermatozoa compared with PSP-I/PSP-II spermadhesins. The addition of the HBP had a detrimental effect on these parameters, whereas PSP-I/PSP-II heterodimer contributed to maintaining the functionality of the highly diluted boar spermatozoa (Centurion et al. 2003) and improved the *in vivo* fertilizing ability of sex-sorted boar spermatozoa (García et al. 2006, 2007). On the other hand, the PSP-I/PSP-II effect was found to be deleterious when the frozen-thawed spermatozoa activity to penetrate oocytes was checked (Caballero et al. 2004, 2008).

15.5 Properties of major proteins of seminal plasma of domestic animals

Proteins of seminal plasma isolated from bull, boar, stallion, ram, buck, and poultry are listed according to their origin in Tables 15.2–15.7. The Tables also summarize their relative molecular mass and basic characteristics. As previously mentioned, literature data on boar, bull, and stallion seminal plasma proteins predominate. The seminal plasma protein primary structures were determined for BSP proteins and spermadhesins from bull, spermadhesins and DQH protein from boar, and HSP proteins and spermadhesin from stallion (Tables 15.2–15.4); in some cases the structure of saccharide chains was also determined (e.g., O-linked saccharide chains of PDC-109 from bull [Calvete et al. 1994c; Gerwig et al. 1996], N-linked oligosaccharides of PSP-I/PSP-II spermadhesins from boar seminal plasma [Nimtz et al. 1999]).

Investigation of the crystal structure of heterodimer PSP-I/PSP-II from boar seminal plasma (Romero et al. 1997; Varela et al. 1997) and PDC-109 (Wah et al. 2002), aSFP (Romão et al. 1997; Romero et al. 1997), and ribonuclease (Mazzarella et al. 1993; Vitagliano et al. 1998) from bull seminal plasma was a subject of other studies.

Binding properties of individual proteins are summarized in Tables 15.2–15.3. The following interactions of seminal proteins that participate in the formation of sperm coating layers belong to the most studied ones:

Table 15.2 Major proteins isolated from boar seminal plasma.

Protein type	Protein	Relative molecular mass	Binding properties, inhibition	References
Spermadhesin	PSP-I, PSP-II	14,000 16,000	Hep^-	Parry et al. 1992; Rutherfurd et al. 1992; Calvete et al. 1993, 1995d; Solis et al. 1997, 1998
	AQN1 AQN3	13,000 12,000	Hep^+; saccharide-, ZP-binding	Jonakova et al. 1991a, 1998; Sanz et al. 1991,1992a; Calvete et al. 1993,1996a; Jelinkova et al. 2004b
	AWN family	14,000–16,000	Hep^+; saccharide-, ZP-binding	Sanz et al. 1992b; Calvete et al. 1994a; Jonakova et al. 1998; Jelinkova et al. 2004b
Fibronectin type II domains (Fn-2 type protein)	DQH (pB1) (pAIF)	13,000	Hep^+; P-choline-, mannan-, ZP-binding	Sanz et al. 1993; Calvete et al. 1997; Jonakova et al. 1998; Bezouska et al. 1999; Liberda et al. 2002a; Jelinkova et al. 2004b; Manaskova et al. 2007
Proteinase inhibitors	SPAI SAAI	12,000 8,000	Acrosin Acrosin, trypsin	Fritz et al. 1976; Jonakova et al. 1991b, 1992; Jelinkova et al. 2003; Davidova et al. 2009
Others	ß-MSP	10,000		Fernlund et al. 1994; Manaskova et al. 2002; Wang et al. 2005
	Lactoferrin	70,000		Roberts and Boursnell 1975

ZP, zona pellucida; P-choline, phosphorylcholine; Hep^+, heparin-binding protein; Hep^-, non-heparin-binding protein; ß-MSP, ß-microseminoprotein; SAAI, sperm-associated acrosin inhibitor; SPAI, seminal plasma acrosin inhibitor.

Table 15.3 Major proteins isolated from bull seminal plasma.

Protein type	Protein	Relative molecular mass	Binding properties, inhibition	References
Fibronectin type II domains (Fn-2 type protein)	BSP-A1/A2 PDC-109	13,000	Hep^+; gelatin-, P-choline, mannan-binding	Esch et al. 1983; Manjunath and Sairam 1987; Manjunath and Thérien 2002; Liberda et al. 2002b
	BSP-A3		Hep^+; gelatin-, P-choline-, mannan-binding	Manjunath and Sairam 1987; Manjunath and Thérien 2002; Seidah et al. 1987
	BSP-30	26,000	Hep^+; gelatin-, P-choline-, mannan-binding	Calvete et al. 1996b,c; Liberda et al. 2002b
Spermadhesin	aSFP	14,000		Einspanier et al. 1991, 1994
	Z13		Hep^-	Tedeschi et al. 2000
Proteinase inhibitors	BUSI I	9,000	Acrosin, trypsin, elastase,cathepsin G,	Cechova and Jonakova 1981; Meloun et al. 1983, 1985
	BUSI II	6,000	Acrosin, trypsin	
Others	PAF-AH	60,000		Soubeyrand et al. 1997; Soubeyrand and Manjunath 1997; Soubeyrand et al. 1998
	RNAase dimer	29,000	Mannan-binding	Di Donato and D'Alessio 1981; D'Alessio et al. 1991; Calvete et al. 1996b; Liberda et al. 2002b

Hep^+, heparin-binding protein; Hep^-, non-heparin-binding protein; PAF-AH, platelet-activating factor acetylhydrolase with phospholipase A2 (PLA_2) activity (Soubeyrand et al. 1998); ZP, zona pellucida; P-choline, phosphorylcholine; BUSI I, bull seminal plasma inhibitor I; BUSI II, bull seminal plasma inhibitor II; BSP, bovine seminal plasma proteins; aSFP, acidic seminal fluid protein.

Table 15.4 Major proteins isolated from stallion seminal plasma.

Protein type	Protein	Relative molecular mass	Properties	References
Fibronectin type II domains (Fn-2 type protein)	HSP-1	14,000	Hep^{+}; gelatin-, P-choline-binding	Calvete et al. 1994b, 1995a,c; Ekhlasi-Hundrieser et al. 2005b
	HSP-2	15,000	Hep^{+}; gelatin-, P-choline-binding	Calvete et al. 1994b; Ekhlasi-Hundrieser et al. 2005b
	HSP-12 (EQ-12)		P-choline-binding	Calvete et al. 1994b; Greube et al. 2004
Spermadhesin	HSP-7	15,000	Saccharide-, ZP-binding	Reinert et al. 1996
Others	CRISP proteins (HSP-3)			Magdaleno et al. 1997; Schambony et al. 1998
	HPK	29,000		Carvalho et al. 2002
	Lactoferrin			Inagaki et al. 2002
	D-gal-binding protein	25,000	D-galactose-, sperm-binding	Sabeur and Ball 2007

Hep^{+}, heparin-binding protein; Hep^{-}, non-heparin-binding protein; ZP, glycoproteins of zona pellucida; P-choline, phosphorylcholine; HPK, horse prostate kallikrein; HSP, horse seminal plasma protein.

Table 15.5 Characterized proteins from ram seminal plasma.

Protein type	Protein	Properties	References
Fibronectin type II domains (Fn-2 type protein)	RSP-15 kDa	Gelatin-binding	Bergeron et al. 2005
	RSP-16 kDa	Gelatin-binding	
	RSP-22 kDa	Gelatin-binding, Hep^{+}	
		Gelatin-binding, Hep^{+}	
	RSP-24 kDa		
	P14		Barrios et al. 2005; Cardozo et al. 2008
Spermadhesin	15.5 kDa protein[1]	Hep^{+}	Bergeron et al. 2005
Others	Phospholipase A2 (PLA2)		Upreti et al. 1999
	P20		Barrios et al. 2005; Cardozo et al. 2008

[1] Major ram seminal plasma protein.
Hep^{+}, heparin-binding protein; RSP, ram seminal plasma protein.

Table 15.6 Characterized proteins from buck seminal plasma.

Protein type	Protein	Relative molecular mass	Binding properties	References
Fibronectin type II domains (Fn-2 type protein)	GSP-14	14,000	Gelatin-binding, Hep^{-}	Villemure et al. 2003
	GSP-15	15,000	Gelatin-binding, Hep^{-}	
	GSP-20	20,000	Gelatin-binding, Hep^{+}	
	GSP-22	22,000	Gelatin-binding, Hep^{+}	
Spermadhesin	BSFP	12,500	Hep^{-}	Teixeira et al. 2002; 2006
Others	Phospholipase A			Sias et al. 2005

BSFP, buck seminal fluid protein; Hep^{+}, heparin-binding protein; Hep^{-}, non-heparin-binding protein; GSP-14, GSP-15, GSP-20, GSP-22, goat seminal plasma proteins.

Table 15.7 Characterized proteins from seminal plasma of poultry.

	Protein	Relative molecular mass	Properties	References
Chicken, roaster (*Gallus domesticus*)	Proteinase inhibitor	6,000	Acrosin inhibitor	Lessley and Brown 1978
	Acid phosphatase		Crystallization	Dumitru and Dinischiotu 1994
	SMIF	78,000	Antibacterial activity	Mohan et al. 1995
	UPSEBP		Fragment of prosaposin	Hammerstedt et al. 2001
Turkey (*Meleagris gallopavo*)	TSPE	28,000 –30,000+ 38,000 –44,000	Not identical with acrosin	Thurston et al. 1993

SMIF, Universal primary sperm-egg binding protein; UPSEBP, Sperm motility inhibiting factor; TSPE, Turkey seminal plasma protease.

- interaction with different types of glycoconjugates,
- interaction with membrane phospholipids,
- interactions between proteins.

Saccharide-based interactions of seminal plasma proteins play an important role in the interaction of sperm with glycoconjugates present in the female reproductive tract; binding sperm to the oviductal epithelium or primary sperm binding to ZP belong to the most investigated ones. The phosphorylcholine-binding activity of seminal plasma proteins is responsible for their adsorption to the sperm membrane and participation in sperm membrane modulation during capacitation. Interactions between proteins participate in the arrangement and remodeling of sperm-coating layers and modulate binding properties or other activities of protein monomer forms (compared above).

Properties of the proteins isolated from ram seminal plasma are summarized in Table 15.5. Similarly as in the case of buck seminal plasma (Table 15.6), their differential binding affinities to heparin and gelatin were used for their separation (Bergeron et al. 2005). The ram proteins belong to the Fn-2 type and to the spermadhesin family. The protein profile of ram seminal plasma was investigated using 2D PAGE. More than 20 spots were detected, out of these 3-5 interacted with antibodies against BSP-A1/A2 proteins (from bull seminal plasma; Jobim 2005). The same electrophoretic method was used to assess monthly variations in ram seminal plasma proteins (Cardozo et al. 2006).

Properties of the buck seminal plasma proteins are summarized in Table 15.6. The protein composition of the buck seminal plasma seems to be similar to that of bull and stallion. It contains proteins with Fn-2 domain (GSP; Villemure et al. 2003) that are characterized by their gelatin-binding ability, and it also contains a protein of spermadhesin family (BSFP;Teixeira et al. 2002, 2006). Buck gene encoding BSFP protein was characterized and its expression along the genital tract was investigated (Melo et al. 2008, 2009).

Only a limited amount of information in the literature is available on the characterization of proteins obtained from poultry seminal plasma. These studies mostly concern chicken and turkey seminal plasma proteins. A list of proteins isolated from those sources and at least partially characterized is presented in Table 15.7. Similarly as in the case of mammalian proteins, the removal of surface-associated proteins from chicken sperm affected the sperm function *in vivo*, especially migration in the female reproductive tract (Thurston et al. 1993; Steele et al. 1996).

15.6 Future research directions

Mammalian seminal plasma is a very complex fluid containing both the low-molecular and the high-molecular components, and in this chapter, attention was paid to the protein constituents. The physiological functions of a large number of seminal plasma proteins have not yet been fully elucidated, despite the fact that some proteins have been intensively studied and well characterized. Future research in this field will probably be directed to detailed proteomic studies of mammalian seminal plasma proteins including low-abundant protein components coupled with plasma proteins gene expression profiling and regulation, localization, and functional studies. This approach can contribute to better knowledge of changes in protein structure and protein modifications, which may alter their properties and help explain some steps in fertilization physiology and pathology.

Proteomics also provides a tool for understanding the interactions of seminal plasma proteins with spermatozoa, with other components of seminal plasma, as well as with substances present in the natural environment of gametes both in the male and the female reproductive organs. The functional proteomics will probably contribute to better characterization of seminal plasma protein function in the reproductive process. The development of mass spectrometric (MS) techniques now allows investigation of very complex protein mixtures. Seminal plasma has not yet received much attention from this point of view.

Better understanding of a function of seminal plasma proteins will provide a sophisticated support in our attempts to reduce infertility and improve fertility in breeding populations of agriculturally important animals, as well as in human population. Proteomic studies on seminal plasma proteins can thus also contribute to an assessment of animal and human fertility by monitoring changes in their reproductive tracts and to the improvement of the conditions of mammalian sperm preservation.

References

Aumüller, G., Vesper, M., Seitz, J., Kemme, M., and Scheit, K.H. 1988. Binding of a major secretory protein from bull seminal vesicles to bovine spermatozoa. *Cell Tissue Research* **252**(2): 377–384.

Barrios, B., Fernández-Juan, M., Muiño-Blanco, T., and Cebrián-Pérez, J.A. 2005. Immunocytochemical localization and biochemical characterization of two seminal plasma proteins that protect ram spermatozoa against cold shock. *Journal of Andrology* **26**(4): 539–549.

Bergeron, A., Villemure, M., Lazure, C., and Manjunath, P. 2005. Isolation and characterization of the major proteins of ram seminal plasma. *Molecular Reproduction and Development* **71**(4): 461–470.

Bergeron, A. and Manjunath, P. 2006. New insights towards understanding the mechanisms of sperm protection by egg yolk and milk. *Molecular Reproduction and Development* **73**(10): 1338–1344.

Bergeron, A., Brindle, Y., Blondin, P., and Manjunath, P. 2007. Milk caseins decrease the binding of the major bovine seminal plasma proteins to sperm and prevent lipid loss from the sperm membrane during sperm storage. *Biology of Reproduction* **77**(1): 120–126.

Bezouska, K., Sklenar, J., Novak, P., Halada, P., Havlicek, V., Kraus, M., Ticha, M., and Jonakova, V. 1999. Determination of the complete covalent structure of the major glycoform of DQH sperm surface protein, a novel trypsin-resistant boar seminal plasma O-glycoprotein related to pB1 protein. *Protein Science* **8**(7): 1551–1556.

Bleil, J.D., Greve, J.M., and Wassarman, P.M. 1988. Identification of a secondary sperm receptor in the mouse egg zona pellucida: Role in maintenance of binding of acrosome-reacted sperm to eggs. *Developmental Biology* **128**(2): 376–385.

Bork, P. and Beckmann, G. 1993. The CUB Domain. A widespread module in developmentally regulated proteins. *Journal of Molecular Biology* **231**(2): 539–545.

Caballero, I., Vazquez, J.M., Gil, M.A., Calvete, J.J., Roca, J., Sanz, L., Parrilla, I., Garcia, E.M., Rodriguez-Martinez, H., and Martinez, E.A. 2004. Does seminal plasma PSP-I/PSP-II spermadhesin modulate the ability of boar spermatozoa to penetrate homologous oocytes in vitro? *Journal of Andrology* **25**(6): 1004–1012.

Caballero, I., Vázquez, J.M., Rodríguez-Martínez, H., Gill, M.A., Calvete, J.J., Sanz, L., Garcia, E.M., Roca, J., and Martínez, E.A. 2005. Influence of seminal plasma PSP-I/PSP-II spermadhesin on pig gamete interaction. *Zygote* **13**(1): 11–16.

Caballero, I., Vazquez, J.M., García, E.M., Parrilla, I., Roca, J., Calvete, J.J., Sanz, L., and Martínez, E.A. 2008. Major proteins of boar seminal plasma as a tool for biotechnological preservation of spermatozoa. *Theriogenology* **70**(8): 1352–1355.

Calvete, J.J., Solís, D., Sanz, L., Díaz-Mauriño, T., Schäfer, W., Mann, K., and Töpfer-Petersen, E. 1993. Characterization of two glycosylated boar spermadhesins. *European Journal of Biochemistry* **218**(2): 719–725.

Calvete, J.J., Solís, D., Sanz, L., Díaz-Mauriño, T., and Töpfer-Petersen, E. 1994a. Glycosylated boar spermadhesin AWN-1 isoforms. Biological origin, structural characterization by lectin mapping, localization of O-glycosylation sites, and effect of glycosylation on ligand binding. *Biological Chemistry Hoppe-Seyler* **375**(10): 667–673.

Calvete, J.J., Nessau, S., Mann, K., Sanz, L., Sieme, H., Klug, E., and Töpfer-Petersen, E. 1994b. Isolation and biochemical-characterization of stallion seminal plasma proteins. *Reproduction in Domestic Animals* **29**(6): 411–426.

Calvete, J.J., Raida, M., Sanz, L., Wempe, F., Scheit, K.H., Romero, A., and Töpfer-Petersen, E. 1994c. Localization and structural characterization of an oligosaccharide O-linked to bovine PDC-109. Quantitation of the glycoprotein in seminal plasma and on the surface of ejaculated and capacitated spermatozoa. *FEBS Letters* **350**(2–3): 203–6.

Calvete, J.J., Reinert, M., Sanz, L., and Töpfer-Petersen, E. 1995a. Effect of glycosylation on the heparin-binding capability of boar and stallion seminal plasma proteins. *Journal of Chromatography A* **711**(1): 167–173.

Calvete, J.J., Sanz, L., Dostalova, Z., and Töpfer-Petersen, E. 1995b. Sparmadhesins: Sperm-coating proteins involved in capacitation and zona pellucida binding. *Fertilität* **11**: 35–40.

Calvete, J.J., Mann, K., Schäfer, W., Sanz, L., Reinert, M., Nessau, S., Raida, M., and Töpfer-Petersen, E. 1995c. Amino acid sequence of HSP-1, a major protein of stallion seminal plasma: Effect of glycosylation on its heparin-and gelatin-binding capabilities. *Biochemical Journal* **310**(Pt 2): 615–622.

Calvete, J.J., Mann, K., Schäfer, W., Raida, M., Sanz, L., and Töpfer-Petersen, E. 1995d. Boar spermadhesin PSP-II: Location of posttranslational modifications, heterodimer formation with PSP-I glycoforms and effect of dimerization on the ligand-binding capabilities of the subunits. *FEBS Letters* **365**(2–3): 179–182.

Calvete, J.J., Carrera, E., Sanz, L., and Töpfer-Petersen, E. 1996a. Boar spermadhesins AQN-1 and AQN-3: Oligosaccharide and zona pellucida binding characteristics. *Biological Chemistry Hoppe-Seyler* **377**(7–8): 521–527.

Calvete, J.J., Varela, P.F., Sanz, L., Romero, A., Mann, K., and Töpfer-Petersen, E. 1996b. A procedure for the large-scale isolation of major bovine seminal plasma proteins. *Protein Expression and Purification* **8**(1): 48–56.

Calvete, J.J., Mann, K., Sanz, L., Raida, M., and Töpfer-Petersen, E. 1996c. The primary structure of BSP-30K, a major lipid-, gelatin-, and heparin-binding glycoprotein of bovine seminal plasma. *FEBS Letters* **399**(1–2): 147–152.

Calvete, J.J., Raida, M., Gentzel, M., Urbanke, C., Sanz, L., and Töpfer-Petersen, E. 1997. Isolation and characterization of heparin- and phosphorylcholine-binding proteins of boar and stallion seminal plasma. Primary structure of porcine pB1. *FEBS Letters* **407**(2): 201–206.

Calvete, J.J., Campanero-Rhodes, M.A., Raida, M., and Sanz, L. 1999. Characterisation of the conformational and quaternary structure-dependent heparin-binding region of bovine seminal plasma protein PDC-109. *FEBS Letters* **444**(2–3): 260–264.

Calvete, J.J., and Sanz, L. 2007. Insights into structure-function correlations of ungulate seminal plasma proteins. *Society of Reproduction and Fertility Supplement* **65**: 201–215.

Campanero-Rhodes, M.A., Menéndez, M., Saiz, J.L., Sanz, L., Calvete, J.J., and Solís, D. 2006. Zinc ions induce the unfolding and self-association of boar spermadhesin PSP-I, a protein with a single CUB domain architecture, and promote its binding to heparin. *Biochemistry* **45**(27): 8227–8235.

Cardozo, J.A., Fernandez-Juan, M., Forcada, F., Abecia, A., Muino-Blanco, T., and Cebrian-Perez, J.A. 2006. Monthly variations in ovine seminal plasma proteins analyzed by two-dimensional polyacrylamide gel electrophoresis. *Theriogenology* **66**(4): 841–850.

Cardozo, J.A., Fernández-Juan, M., Cebrián-Pérez, J.A., and Muiño-Blanco, T. 2008. Identification of RSVP14 and RSVP20 components by two-dimensional electrophoresis and Western-blotting. *Reproduction in Domestic Animals* **43**(1): 15–21.

Carvalho, A.L., Sanz, L., Barettino, D., Romero, A., Calvete, J.J., and Romão, M.J. 2002. Crystal structure of a prostate kallikrein isolated from stallion seminal plasma: A homologue of human PSA. *Journal of Molecular Biology* **322**(2): 325–337.

Cechova, D., and Jonakova, V. 1981. Bull seminal plasma proteinase inhibitors. *Methods of Enzymology* **80**: 792–803.

Centurion, F., Vazquez, J.M., Calvete, J.J., Roca, J., Sanz, L., Parrilla, I., Garcia, E.M., and Martinez, E.A. 2003. Influence of porcine spermadhesins on the susceptibility of boar spermatozoa to high dilution. *Biology of Reproduction* **69**(2): 640–646.

Cibulkova, E., Manaskova, P., Jonakova, V., and Ticha, M. 2007. Preliminary characterization of multiple hyaluronidase forms in boar reproductive tract. *Theriogenology* **68**(7): 1047–1054.

Clark, G.F. and Dell, A. 2006. Molecular models for murine sperm-egg binding. *Journal of Biological Chemistry* **281**(20): 13853–13856.

D'Alessio, G., Di Donato, A., Parente, A., and Piccoli, R. 1991. Seminal RNase: A unique member of the ribonuclease superfamily. *Trends in Biochemical Sciences* **16**(3): 104–106.

Davidova, N., Jonakova, V., and Manaskova-Postlerova, P. 2009. Expression and localization of acrosin inhibitor in boar reproductive. *Cell Tissue Research* **338**(2): 303–311.

de Graaf, S.P., Leahy, T., Marti, J., Evans, G., and Maxwell, W.M. 2008. Application of seminal plasma in sex-sorting and sperm cryopreservation. *Theriogenology* **70**(8): 1360–1363.

Desnoyers, L. and Manjunath, P. 1992. Major proteins of bovine seminal plasma exhibit novel interactions with phospholipid. *Journal of Biological Chemistry* **267**(14): 10149–10155.

Di Donato, A. and D'Alessio, G. 1981. Heterogeneity of bovine seminal ribonuclease. *Biochemistry* **20**(25): 7232–7237.

Dostalova, Z., Calvete, J.J., Sanz, L., and Töpfer-Petersen, E. 1994. Quantitation of boar spermadhesins in accessory sex gland fluids and on the surface of epididymal, ejaculated and capacitated spermatozoa. *Biochimica et Biophysica Acta* **1200**(1): 48–54.

Dostalova, Z., Calvete, J.J., Sanz, L., and Töpfer-Petersen, E. 1995. Boar spermadhesin AWN-1. Oligosaccharide and zona pellucida binding characteristics. *European Journal of Biochemistry* **230**(1): 329–336.

Dumitru, I.F. and Dinischiotu, A. 1994. Cock seminal plasma acid phosphatase: Active site directed inactivation, crystallization and in vitro denaturation-renaturation studies. *International Journal of Biochemistry* **26**(4): 497–503.

Einspanier, R., Einspanier, A., Wempe, F., and Scheit, K.H. 1991. Characterization of a new bioactive protein from bovine seminal fluid. *Biochemical Biophysical Research Communications* **179**(2): 1006–1010.

Einspanier, R., Krause, I., Calvete, J.J., Töpfer-Petersen, E., Klostermeyer, H., and Karg, H. 1994. Bovine seminal plasma aSFP: Localization of disulfide bridges and detection of three different isoelectric forms. *FEBS Letters* **344**(1): 61–64.

Ekhlasi-Hundrieser, M., Sinowatz, F., deWilke, I.G., Waberski, D., and Töpfer-Petersen, E. 2002. Expression of spermadhesin genes in porcine male and female reproductive tracts. *Molecular Reproduction and Development* **61**(1): 32–41.

Ekhlasi-Hundrieser, M., Gohr, K., Wagner, A., Tsolova, M., Petrunkina, A., and Töpfer-Petersen, E. 2005a. Spermadhesin AQN1 is a candidate receptor molecule involved in the formation of the oviductal sperm reservoir in the pig. *Biology of Reproduction* **73**(3): 536–545.

Ekhlasi-Hundrieser, M., Schäfer, B., Kirchhoff, C., Hess, O., Bellair, S., Müller, P., and Töpfer-Petersen, E. 2005b. Structural and molecular characterization of equine sperm-binding fibronectin-II module proteins. *Molecular Reproduction and Development* **70**(1): 45–57.

Ekhlasi-Hundrieser, M., Calvete, J.J., von Rad, B., Hettel, C., Nimtz, M., and Töpfer-Petersen, E. 2008. Point mutations abolishing the mannose-binding capability of boar spermadhesin AQN-1. *Biochimica et Biophysica Acta* **1784**(5): 856–862.

Ensslin, M., Calvete, J.J., Thole, H.H., Sierralta, W.D., Adermann, K., Sanz, L., and Töpfer-Petersen, E. 1995. Identification by affinity chromatography of boar sperm membrane-associated proteins bound to immobilized porcine zona pellucida. Mapping of the phosphorylethanolamine-binding region of spermadhesin AWN. *Biological Chemistry Hoppe-Seyler* **376**(12): 733–738.

Esch, F.S., Ling, N.C., Böhlen, P., Ying, S.Y., and Guillemin, R. 1983. Primary structure of PDC-109, a major protein constituent of bovine seminal plasma. *Biochemical Biophysical Research Communications* **113**(3): 861–867.

Evans, J. and Kopf, G. 1998. Molecular mechanisms of sperm-egg interactions and egg activation. *Andrologia* **30**(4–5): 297–307.

Fazeli, A., Duncan, A.E., Watson, P.F., and Holt, W.V. 1999. Sperm-oviduct interaction: Induction of capacitation and preferential binding of uncapacitated spermatozoa to oviductal epithelial cells in porcine species. *Biology of Reproduction* **60**(4): 879–886.

Fernlund, P., Granberg, L.B., and Roepstorff, P. 1994. Amino acid sequence of beta-microseminoprotein from porcine seminal plasma. *Archives of Biochemistry and Biophysics* **309**(1): 70–76.

Fritz, H., Tschesche, H., and Fink, E. 1976. Proteinase inhibitors from boar seminal plasma. *Methods of Enzymology* **45**: 834–847.

Garcia, E.M., Vázquez, J.M., Calvete, J.J., Sanz, L., Caballero, I., Parrilla, I., Gil, M.A., Roca, J., and Martinez, E.A. 2006. Dissecting the protective effect of the seminal plasma spermadhesin PSP-I/PSP-II on boar sperm functionality. *Journal of Andrology* **27**(3): 434–443.

García, E.M., Vázquez, J.M., Parrilla, I., Calvete, J.J., Sanz, L., Caballero, I., Roca, J., Vazquez, J.L., and Martínez, E.A. 2007. Improving the fertilizing ability of sex sorted boar spermatozoa. *Theriogenology* **68**(5): 771–778.

García, E.M., Vázquez, J.M., Parrilla, I., Ortega, M.D., Calvete, J.J., Sanz, L., Martínez, E.A., Roca, J. and Rodríguez-Martínez, H. 2008. Localization and expression of spermadhesin PSP-I/PSP-II subunits in the reproductive organs of the boar. *International Journal of Andrology* **31**(4): 408–417.

Gasset, M., Saiz, J.L., Laynez, J., Sanz, L., Gentzel, M., Töpper-Petersen, E., and Calvete, J.J. 1997. Conformational features and thermal stability of bovine seminal plasma protein PDC-109 oligomers and phosphorylcholine-bound complexes. *European Journal of Biochemistry* **250**(3): 735–744.

Gerwig, G.L., Calvete, J.J., Töpfer-Petersen, E., and Vliegenthart, J.F. 1996. The structure of the O-linked carbohydrate chain of bovine seminal plasma protein PDC-109 revised by H-NMR spectroscopy A correction. *FEBS Letters* **387**(1): 99–100.

Green, C.E., Bredl, J., Holt. W.V., Watson, P.F., and Fazeli, A. 2001. Carbohydrate mediation of boar sperm binding to oviductal epithelial cells in vitro. *Reproduction* **122**(2): 305–315.

Greube, A., Müller, K., Töpfer-Petersen, E., Herrmann, A., and Müller, P. 2004. Interaction of fibronectin type II proteins with membranes: The stallion seminal plasma protein SP-1/2. *Biochemistry* **43**(2): 464–472.

Gwathmey, T.M., Ignotz, G.G., and Suarez, S.S. 2001. PDC-109 mediates the binding of bovine sperm to oviductal epithelium. *Biology of Reproduction* **64**: 112–112.

Gwathmey, T.M., Ignotz, G.G., and Suarez, S.S. 2003. PDC-109 (BSP-A1/A2) promotes bull sperm binding to oviductal epithelium in vitro and may be involved in forming the oviductal sperm reservoir. *Biology of Reproduction* **69**(3): 809–815.

Gwathmey, T.M., Ignotz, G.G., Müller, J.L., Manjunath, P., and Suarez, S.S. 2006.

Bovine seminal plasma proteins PDC-109, BSP-A3, and BSP-30-kDa share functional roles in storing sperm in the oviduct. *Biology of Reproduction* **75**(4): 501–507.

Gwatkin, R.B.L. 1977. *Fertilization Mechanism in Man and Mammals*. New York: Plenum Press, p. 161.

Haase, B., Schlötterer, C.H., Ekhlasi-Hundrieser, M., Kuiper, H., Distl, O., Töpfer-Petersen, E., and Leeb, T. 2005. Evolution of the spermadhesin gene family. *Gene* **352**: 20–29.

Hammerstedt, R.H., Cramer, P.G., Barbato, G.F., Amann, R.P., O'Brien, J.S., and Griswold, M.D. 2001. A fragment of prosaposin (SGP-1) from rooster sperm promotes sperm-egg binding and improves fertility in chickens. *Journal of Andrology* **22**(3): 361–375.

Henault, M.A. and Killian, G.J. 1996. Effect of homologous and heterologous seminal plasma on the fertilizing ability of ejaculated bull spermatozoa assessed by penetration of zona-free bovine oocytes. *Journal of Reproduction and Fertility* **108**(2): 199–204.

Ignotz, G.G., Lo, M.C., Perez, C.L., Gwathmey, T.M., and Suarez, S.S. 2001. Identification of a fucose-binding protein from bull sperm and seminal plasma that participates in forming the oviductal sperm reservoir. *Biology of Reproduction* **64**(6): 329.

Ignotz, G.G., Cho, M.Y., and Suarez, S.S. 2007. Annexins are candidate oviductal receptors for bovine sperm surface proteins and thus may serve to hold bovine sperm in the oviductal reservoir. *Biology of Reproduction* **77**(6): 906–913.

Inagaki, M., Kikuchi, M., Orino, K., Ohnami, Y., and Watanabe, K. 2002. Purification and quantification of lactoferrin in equine seminal plasma. *Journal of Veterinary Medical Science* **64**(1): 75–77.

Jansen, S., Ekhlasi-Hundrieser, M., and Töpfer-Petersen, E. 2001. Sperm adhesion molecules: Structure and function. *Cells Tissues Organs* **168**(1–2): 82–92.

Jelezarsky, L., Vaisberg, C.H., Chaushev, T., and Sapundjiev, E. 2008. Localization and characterization of glutathione peroxidase (GPx) in boar accessory sex glands, seminal plasma, and spermatozoa and activity of GPx in boar semen. *Theriogenology* **69**(2): 139–145.

Jelinkova, P., Manaskova, P., Ticha, M., and Jonakova, V. 2003. Proteinase inhibitors in aggregated forms of boar seminal plasma proteins. *International Journal of Biological Macromolecules* **32**(3–5): 99–107.

Jelinkova, P., Ryslava, H., Liberda, J., Jonakova, V., and Ticha, M. 2004a. Aggregated forms of bull seminal plasma proteins and their heparin-binding activity. *Collection of Czechoslovak Chemical Communications* **69**(3): 616–630.

Jelinkova, P., Liberda, J., Manaskova, P., Ryslava, H., Jonakova, V., and Ticha, M. 2004b. Mannan-binding proteins from boar seminal plasma. *Journal of Reproductive Immunology* **62**(1–2): 167–182.

Jeng, H., Chu, H.-H., Cheng, W., Chang, W.-C., and Su, S.-J. 2001. Secretory origin and temporal appearance of the porcine ß-microseminoprotein (sperm motility inhibitor) in the boar reproductive system. *Molecular Reproduction and Development* **58**(1): 63–68.

Jobim, M.I.M., Oberst, E.R., Salbego, C.G., Wald, V.B., Horn, A.P., and Mattos, R.C. 2005. BSP A1/A2-like proteins in ram seminal plasma. *Theriogenology* **63**(7): 2053–2062.

Jonakova, V. and Cechova, D. 1985. Demonstration of an anionic acrosin inhibitor in spermatozoa epididymal fluid and seminal plasma of the boar. *Andrologia* **17**(5): 466–471.

Jonakova, V., Sanz, L., Calvete, J.J., Henschen, A., Cechova, D., and Töpfer-Petersen, E. 1991a. Isolation and biochemical characterization of a zona pellucida-binding glycoprotein of boar spermatozoa. *FEBS Letters* **280**(1): 183–186.

Jonakova, V., Cechova, D., Töpfer-Petersen, E., Calvete, J.J., and Veselsky, L. 1991b. Variability of acrosin inhibitors in boar reproductive tract. *Biomedica Biochimica Acta* **50**(4–6): 691–695.

Jonakova, V., Calvete, J.J., Mann, K., Schäfer, W., Schmid, E.R., and Töpfer-Petersen, E. 1992. The complete primary structure of three isoforms of a boar sperm-associated acrosin inhibitor. *FEBS Letters* **297**(1–2): 147–150.

Jonakova V., Ticha, M., Kraus, M., and Cechova, D. 1995. Multifunctional sperm protein in gametic interaction. *Fertilität* **11**: 115–118.

Jonakova, V., Kraus, M., Veselsky, L., Cechova, D., Bezouska, K., and Ticha, M. 1998. Spermadhesins of the AQN and AWN families, DQH sperm surface protein and HNK protein in the heparin-binding fraction of boar seminal plasma. *Journal of Reproduction and Fertility* **114**(1): 25–34.

Jonakova, V., Manaskova, P., Kraus, M., Liberda, J., and Ticha, M. 2000. Sperm surface proteins in mammalian fertilization. *Molecular Reproduction and Development* **56** (**2** Suppl): 275–277.

Jonakova, V. and Ticha, M. 2004. Boar seminal plasma proteins and their binding properties. A review. *Collection of Czechoslovak Chemical Communications* **69**(3): 461–475.

Jonakova, V., Manaskova, P., and Ticha, M. 2007. Separation, characterization and identification of boar seminal plasma proteins. *Journal of Chromatography B, Analytical Technologies in the Biomedical and Life Sciences* **849**(1–2): 307–314.

Kelly, V.C., Kuy, S., Palmer, D.J., Xu, Z., Davis, S.R., and Cooper, G.J. 2006. Characterization of bovine seminal plasma by proteomics. *Proteomics* **6**(21): 5826–5833.

Kraus, M., Ticha, M., and Jonakova, V. 2001. Heparin-binding proteins of human seminal plasma homologous with boar spermadhesin. *Journal of Reproductive Immunology* **51**(2): 131–144.

Kraus, M., Ticha, M., Zelezna, B., Peknicova, J., and Jonakova, V. 2005. Characterization of human seminal plasma proteins homologous to boar AQN spermadhesins. *Journal of Reproductive Immunology* **65**(1): 33–46.

Kwok, S.C., Yang, D., Dai, G., Soares, M.J., Cheng, S., and McMurtry, J.P. 1993. Molecular cloning and sequence analysis of two porcine seminal proteins. PSP-I and PSP-II, new members of the spermadhesin family. *DNA and Cell Biology* **12**: 605–610.

Kwok, S.C.M., Dai, G., and McMurtry, J.P. 1994. Molecular cloning and sequence analysis of the cDNA encoding porcine acrosin inhibitor. *DNA and Cell Biology* **13**(4): 389–394.

Lefebvre, R., Chenoweth, P.J., Drost, M., LeClear, C.T., MacCubbin, M., Dutton, J.T., and Suarez, S.S. 1995. Characterization of the oviductal sperm reservoir in cattle. *Biology of Reproduction* **53**(5): 1066–1074.

Lefebvre, R., Lo, M.C., and Suarez, S.S. 1997. Bovine sperm binding to oviductal epithelium involves fucose recognition. *Biology of Reproduction* **56**(5): 1198–1204.

Lessley, B.A. and Brown, K.I. 1978. Purification and properties of a proteinase inhibitor from chicken seminal plasma. *Biology of Reproduction* **19**(1): 223–234.

Liberda, J., Kraus, M., Ryslava, H., Vlasakova, M., Jonakova, V., and Ticha, M. 2001. D-fructose-binding proteins in bull seminal plasma: Isolation and characterization. *Folia Biologica (Praha)* **47**(4): 113–119.

Liberda, J., Manaskova, P., Svestak, M., Jonakova, V., and Ticha, M. 2002a. Immobilization of L-glyceryl phosphorylcholine: Isolation of phosphorylcholine proteins from seminal plasma. *Journal of Chromatography B, Analytical Technologies in the Biomedical and Life Sciences* **770**: 101–110.

Liberda, J., Ryslava, H., Jelinkova, P., Jonakova, V., and Ticha, M. 2002b. Affinity chromatography of bull seminal proteins on mannan-Sepharose. *Journal of Chromatography B, Analytical Technologies in the Biomedical and Life Sciences* **780**(2): 231–239.

Liberda, J., Manaskova, P., Prelovska, L., Ticha, M., and Jonakova, V. 2006. Saccharide-mediated interactions of boar sperm surface proteins with components of the porcine oviduct. *Journal of Reproductive Immunology* **71**(2): 112–125.

Litscher, E.S., Juntunen, K., Seppo, A., Penttilä, L., Niemelä, R., Renkonen, O., and Wassarman, P.M. 1995. Oligosaccharide constructs with defined structures that inhibit binding of mouse sperm to unfertilized eggs in vitro. *Biochemistry* **34**(14): 4662–4669.

Lusignan, M.F., Bergeron, A., Crête, M.H., Lazure, C., and Manjunath, P. 2007. Induction of epididymal boar sperm capacitation by pB1 and BSP-A1/-A2 proteins, members of the BSP protein family. *Biology of Reproduction* **76**(3): 424–432.

Magdaleno, L., Gasset, M., Varea, J., Schambony, A.M., Urbanke, C., Raida, M., Töpfer-Petersen, E., and Calvete, J.J. 1997. Biochemical and conformational characterisation of HSP-3, a stallion seminal plasma protein of the cysteine-rich secretory protein (CRISP) family. *FEBS Letters* **420**(2–3): 179–185.

Manaskova, P., Liberda, J., Ticha, M., and Jonakova, V. 2000. Aggregated and monomeric forms of proteins in boar seminal plasma: Characterization and binding properties. *Folia Biologica (Praha)* **46**(4): 143–151.

Manaskova, P., Liberda, J., Ticha, M., and Jonakova, V. 2002. Isolation of non-heparin-binding and heparin-binding proteins of boar prostate. *Journal of Chromatography B, Analytical Technologies in the Biomedical and Life Sciences* **770**(1–2): 137–143.

Manaskova, P., Balinova, P., Kraus, M., Ticha, M., and Jonakova, V. 2003. Mutual interactions of boar seminal plasma proteins studied by immunological and chromatographic methods. *American Journal of Reproductive Immunology* **50**(5): 399–410.

Manaskova, P., Peknicova, J., Elzeinova, F., Ticha, M., and Jonakova, V. 2007. Origin, localization and binding abilities of boar DQH sperm surface protein tested by specific monoclonal antibodies. *Journal of Reproductive Immunology* **74**(1–2): 103–113.

Manaskova, P. and Jonakova, V. 2008. Localization of porcine seminal plasma (PSP) proteins in the boar reproductive tract and spermatozoa. *Journal of Reproductive Immunology* **78**(1): 40–48.

Manjunath, P. and Sairam, M.R. 1987. Purification and biochemical characterization of three major acidic proteins (BSP-A1, BSP-A2 and BSP-A3) from bovine seminal plasma. *Biochemical Journal* **241**(3): 685–692.

Manjunath, P., Marcel, Y.L., Uma, J., Seidah, N.G., Chrétien, M., and Chapdelaine, A.

1989. Apolipoprotein A-I binds to a family of bovine seminal plasma proteins. *Journal of Biological Chemistry* **264**(28): 16853–16857.

Manjunath, P., Chandonnet, L., Leblond, E., and Desnoyers, L. 1994a. Major proteins of bovine seminal vesicles bind to spermatozoa. *Biology of Reproduction* **50**(1): 27–37.

Manjunath, P., Soubeyrand, S., Chandonnet, L., and Roberts, K.D. 1994b. Major proteins of bovine seminal plasma inhibit phospholipase A2. *Biochemical Journal* **303**(Pt 1): 121–128.

Manjunath, P. and Thérien, I. 2002. Role of seminal plasma phospholipid-binding proteins in sperm membrane lipid modification that occurs during capacitation. *Journal of Reproductive Immunology* **53**(1–2): 109–119.

Manjunath, P., Bergeron, A., Lefebvre, J., and Fan, J. 2007. Seminal plasma proteins: Functions and interaction with protective agents during semen preservation. *Society of Reproduction and Fertility Supplement* **65**: 217–228.

Marini, P.E. and Cabada, M.O. 2003. One step purification and biochemical characterization of a spermatozoa-binding protein from porcine oviductal epithelial cells. *Molecular Reproduction and Development* **66**(4): 383–390.

Marti, E., Mara, L., Marti, J.I., Muiño-Blanco, T., and Cebrián-Pérez, J.A. 2007. Seasonal variations in antioxidant enzyme activity in ram seminal plasma. *Theriogenology* **67**(9): 1446–1454.

Maxwell, W.M., de Graaf, S.P., Ghaoui, R.-H., and Evans, G. 2007. Seminal plasma effects on sperm handling and female fertility. *Society of Reproduction and Fertility Supplement* **64**: 13–38.

Mazzarella, L., Capasso, S., Demasi, D., Di Lorenzo, G., Mattia, C.A., and Zagari, A. 1993. Bovine seminal ribonuclease: Structure at 1.9 A resolution. *Acta Crystallographica. Section D, Biological Crystallography* **49**(Pt 4): 389–402.

Melo, L.M., Teixeira, D.I.A., Havt, A., Da Cunha, R.M.S., Martins, D.B.G., Castelletti, C.H.M., De Souza, P.R.E., De Lima, J.L., Freitas, V.J.D., Cavada, B.S., and Radis-Baptista, G. 2008. Buck (Capra hircus) genes encode new members of the spermadhesin family. *Molecular Reproduction and Development* **75**(1): 8–16.

Melo, L.M., Nascimento, A.S., Silveira, F.G., Cunha, R.M., Tavares, N.A., Teixeira, D.I., Lima-Filho, J.L., Freitas, V.J., Cavada, B.S., and Rádis-Baptista, G. 2009. Quantitative expression analysis of Bodhesin genes in the buck (Capra hircus) reproductive tract by real-time polymerase chain reaction (qRT-PCR). *Animal Reproduction Science* **110**(3–4): 245–255.

Meloun, B., Cechova, D., and Jonakova, V. 1983. Homologies in the structures of bull seminal plasma acrosin inhibitors and comparison with other homologous proteinase inhibitors of the Kazal type. *Hoppe-Seyler's Zeitschrift für physiologische Chemie* **364**(12): 1665–1670.

Meloun, B., Jonakova, V., and Henschen, A. 1985. Acidic acrosin inhibitors from bull seminal plasma. Structural differences. *Biological Chemistry Hoppe-Seyler* **366**(12): 1155–1160.

Meyer, M.F., Kreil, G., and Aschauer, H. 1997. The soluble hyaluronidase from bull testes is a fragment of the membrane-bound PH-20 enzyme. *FEBS Letters* **413**(2): 385–388.

Mohan, J., Saini, M., and Joshi, P. 1995. Isolation of a spermatozoa motility inhibiting factor from chicken seminal plasma with antibacterial property. *Biochimica et Biophysica Acta-General Subjects* **1245**(3): 407–413.

Moura, A.A., Chapman, D.A., Koc, H., and Killian, G.J. 2006a. Proteins of the cauda epididymal fluid associated with fertility of mature dairy bulls. *Journal of Andrology* **27**(4): 534–541.

Moura, A.A., Koc, H., Chapman, D.A., and Killian, G.J. 2006b. Identification of proteins in the accessory sex gland fluid associated with fertility indexes of dairy bulls: A proteomic approach. *Journal of Andrology* **27**(2): 201–211.

Nimtz, M., Grabenhorst, E., Conradt, H.S., Sanz, L., and Calvete, J.J. 1999. Structural characterization of the oligosaccharide chains of native and crystallized boar seminal plasma spermadhesin PSP-I and PSP-II glycoforms. *European Journal of Biochemistry* **265**(2): 703–718.

Ohkubo, I., Tada, T., Ochiai, Y., Ueyama, H., Eimoto, T., and Sasaki, M. 1995. Human seminal plasma ß-microseminoprotein: Its purification, characterization, and immunohistochemical localization. *International Journal of Biochemistry and Cell Biology* **27**(6): 603–611.

Parry, R.V., Baker, P.J., and Jones, R. 1992. Characterization of low Mr zona pellucida binding proteins from boar spermatozoa and seminal plasma. *Molecular Reproduction and Development* **33**(1): 108–115.

Petrunkina, A.M., Gehlhaar, R., Drommer, W., Waberski, D., and Töpfer-Petersen, E. 2001. Selective sperm binding to pig oviductal epithelium *in vitro*. *Reproduction* **121**(6): 889–896.

Plucienniczak, G., Jagiello, A., Plucienniczak, A., Holody, D., and Strzezek, J. 1999. Cloning of complementary DNA encoding the pB1 component of the 54-kilodalton glycoprotein of boar seminal plasma. *Molecular Reproduction and Development* **52**(2): 303–309.

Potts, J.R. and Campbell, I.D. 1994. Fibronectin structure and assembly. *Current Opinion in Cell Biology* **6**(5): 648–655.

Reinert, M., Calvete, J.J., Sanz, L., Mann, K., and Töpfer-Petersen, E. 1996. Primary structure of stallion seminal plasma protein HSP-7, a zona pellucida-binding protein of the spermadhesin family. *European Journal of Biochemistry* **242**(3): 636–640.

Reinert, M., Calvete, J.J., Sanz, L., and Töpfer-Petersen, E. 1997. Immunohistochemical localization in the stallion genital tract, and topography on spermatozoa of seminal plasma protein SSP-7, a member of the spermadhesin protein family. *Andrologia* **29**(4): 179–186.

Revah, I., Gadella, B.M., Flesch, F.M., Colenbrander, B., and Suárez, S.S. 2000. Physiological state of bull sperm affects fucose- and mannose-binding properties. *Biology of Reproduction* **62**(4): 1010–1015.

Roberts, T.K. and Boursnell, J.C. 1975. The isolation and characterization of lactoferrin from sow milk and boar seminal plasma. *Journal of Reproduction and Fertility* **42**(3): 579–582.

Rodríguez-Martinez, H., Iborra, A., Martínez, P., and Calvete, J.J. 1998. Immunoelectronmicroscopic imaging of spermadhesin AWN epitopes on boar spermatozoa bound *in vivo* to the zona pellucida. *Reproduction, Fertility, and Development* **10**(6): 491–497.

Rodríguez-Martínez, H., Saravia, F., Wallgren, M., Tienthai, P., Johannisson, A., Vázquez, J.M., Martínez, E., Roca, J., Sanz, L., and Calvete, J.J. 2005. Boar spermatozoa in the oviduct. *Theriogenology* **63**(2): 514–535.

Rodriguez-Martinez, H. 2007. Role of the oviduct in sperm capacitation.

Theriogenology **68**(Supplement 1): S138–46.

Romão, M.J., Kölln, I., Dias, J.M., Carvalho, A.L., Romero, A., Varela, P.F., Sanz, L., Töpfer-Petersen, E., and Calvete, J.J. 1997. Crystal structure of acidic seminal fluid protein (aSFP) at 1.9 A resolution: A bovine polypeptide of the spermadhesin family. *Journal of Molecular Biology* **274**(4): 650–660.

Romero, A., Romão, M.J., Varela, P.F., Kölln, I., Dias, J.M., Carvalho, A.L., Sanz, L., Töpfer-Petersen, E., and Calvete, J.J. 1997. The crystal structures of two spermadhesins reveal the CUB domain fold. *Nature Structural Biology* **4**(10): 783–788.

Rutherfurd, K.J., Swiderek, K.M., Green, C.B., Chen, S., Shively, J.E., and Kwok, S.C. 1992. Purification and characterization of PSP-I and PSP-II, two major proteins from porcine seminal plasma. *Archives of Biochemistry and Biophysics* **295**(2): 352–359.

Sabeur, K. and Ball, B.A. 2007. Characterization of galactose-binding proteins in equine testis and spermatozoa. *Animal Reproduction Science* **101**(1–2): 74–84.

Sanz, L., Calvete, J.J., Mann, K., Schäfer, W., Schmid, E.R., and Töpfer-Petersen, E. 1991. The amino acid sequence of AQN-3, a carbohydrate-binding protein isolated from boar sperm. Location of disulphide bridges. *FEBS Letters* **291**(1): 33–36.

Sanz L., Calvete, J.J., Mann, K., Schäfer, W., Schmid, E.R., and Töpfer-Petersen, E. 1992a. The complete primary structure of the boar spermadhesin AQN-1, a carbohydrate-binding protein involved in fertilization. *European Journal of Biochemistry* **205**(2): 645–652.

Sanz, L., Calvete, J.J., Mann, K., Schäfer, W., Schmid, E.R., Amselgruber, W., Sinowatz, F., Ehrhard, M., and Töpfer-Petersen, E. 1992b. The complete primary structure of the spermadhesin AWN, a zona pellucida-binding protein isolated from boar spermatozoa. *FEBS Letters* **300**(3): 213–218.

Sanz, L., Calvete, J.J., Jonakova, V., and Töpfer-Petersen, E. 1992c. Boar spermadhesins AQN-1 and AWN are sperm-associated acrosin inhibitor acceptor protein. *FEBS Letters* **300**(1): 63–66.

Sanz, L., Calvete, J.J., Mann, K., Gabius, H.J., and Töpfer-Petersen, E. 1993. Isolation and biochemical characterization of heparin-binding proteins from boar seminal plasma: A dual role for spermadhesins in fertilization. *Molecular Reproduction and Development* **35**(1): 37–43.

Schambony, A., Gentzel, M., Wolfes, H., Raida, M., Neumann, U., and Töpfer-Petersen, E. 1998. Equine CRISP-3: Primary structure and expression in the male genital tract. *Biochimica et Biophysica Acta* **1387**(1–2): 206–216.

Scheit, K.H., Kemme, M., Aumüller, G., Seitz, J., Hagendorff, G., and Zimmer, M. 1988. The major protein of bull seminal plasma biosynthesis and biological function. *Bioscience Reports* **8**(6): 589–608.

Scheit, K.H. 1990. Gene expression in bovine seminal vesicles. *Andrologia* **22**(Supplement 1): 74–82.

Seidah, N.G., Manjunath, P., Rochemont, J., Sairam, M.R., and Chrétien, M. 1987. Complete amino acid sequence of BSP-A3 from bovine seminal plasma. Homology to PDC-109 and to the collagen-binding domain of fibronectin. *Biochemical Journal* **243**(1): 195–203.

Sias, B., Ferrato, F., Pellicer-Rubio, M.T., Forgerit, Y., Guillouet, P., Leboeuf, B., and Carrière, F. 2005. Cloning and seasonal secretion of the pancreatic lipase-related protein 2 present in goat seminal plasma. *Biochimica et Biophysica Acta* **1686**(3): 169–180.

Solís, D., Calvete, J.J., Sanz, L., Hettel, C., Raida, M., Diaz-Mauriño, T., and Töpfer-Petersen, E. 1997. Fractionation and characterization of boar seminal plasma spermadhesion PSP-II glycoforms reveal the presence of uncommon N-acetylgalactosamine-containing N-linked oligosaccharides. *Glycoconjugate Journal* **14**(2): 275–280.

Solís, D., Romero, A., Jiménez, M., Díaz-Mauriño, T., and Calvete, J.J. 1998. Binding of mannose-6-phosphate and heparin by boar seminal plasma PSP-II, a member of the spermadhesin protein family. *FEBS Letters* **431**(2): 273–278.

Soubeyrand, S., Khadir, A., Brindle, Y., and Manjunath, P. 1997. Purification of a novel phospholipase A2 from bovine seminal plasma. *Journal of Biological Chemistry* **272**(1): 222–227.

Soubeyrand, S. and Manjunath, P. 1997. Novel seminal phospholipase A2 is inhibited by the major proteins of bovine seminal plasma. *Biochimica et Biophysica Acta* **1341**(2): 183–188.

Soubeyrand, S., Lazure, C., and Manjunath, P. 1998. Phospholipase A2 from bovine seminal plasma is a platelet-activating factor acetylhydrolase. *Biochemical Journal* **329**(Pt 1): 41–47.

Steele, M.G. and Wishart, G.J. 1996. The effect of removing surface-associated proteins from viable chicken spermatozoa on sperm function *in vivo* and *in vitro*. *Animal Reproduction Science* **45**(1–2): 139–147.

Suarez, S.S. 1998. The oviductal sperm reservoir in mammals: Mechanisms of formation. *Biology of Reproduction* **58**(5): 1105–1107.

Suarez, S.S. 2001. Carbohydrate-mediated formation of the oviductal sperm reservoir in mammals. *Cells Tissues Organs* **168**(1–2): 105–112.

Suarez, S.S. and Ignotz, G.G. 2001. Fucosylated glycoproteins from oviductal epithelium bind PDC-109 and may be involved in creating the reservoir of sperm in the bovine oviduct. *Biology of Reproduction* **64**: 329.

Suarez, S.S. 2002. Formation of a reservoir of sperm in the oviduct. *Reproduction in Domestic Animals* **37**(3): 140–143.

Suarez, S.S. 2007. Interactions of spermatozoa with the female reproductive tract: Inspiration for assisted reproduction. *Reproduction Fertility and Development* **19**(1): 103–110.

Suarez, S.S. 2008. Regulation of sperm storage and movement in the mammalian oviduct. *International Journal of Develpomental Biology* **52**(5–6): 455–462.

Talevi, R. and Gualtieri, R. 2001. Sulfated glycoconjugates are powerful modulators of bovine sperm adhesion and release from the oviductal epithelium in vitro. *Biology of Reproduction* **64**(2): 491–498.

Tannert, A., Kurz, A., Erlemann, K.R., Müller, K., Herrmann, A., Schiller, J., Töpfer-Petersen, E., Manjunath, P., and Müller, P. 2007a. The bovine seminal plasma protein PDC-109 extracts phosphorylcholine-containing lipids from the outer membrane leaflet. *European Biophysical Journal* **36**(4–5): 461–475.

Tannert, A., Töpfer-Petersen, E., Herrmann, A., Müller, K., and Müller, P. 2007b. The lipid composition modulates the influence of the bovine seminal plasma protein PDC-109 on membrane stability. *Biochemistry* **46**(41): 11621–11629.

Tanphaichitr, N., Carmona, E., Bou Khalil, M., Xu, H., Berger, T., and Gerton, G.L. 2007. New insights into sperm-zona pellucida interaction: Involvement of sperm lipid rafts. *Frontiers in Bioscience* **12**: 1748–1766.

Tedeschi, G., Oungre, E., Mortarino, M., Negri, A., Maffeo, G., and Ronchi, S. 2000. Purification and primary structure of a new bovine spermadhesin. *European Journal of Biochemistry* **267**(20): 6175–6179.

Teijeiro, J.M., Cabada, M.O., and Marini, P.E. 2007. Sperm binding glycoprotein (SBG) produces calcium and bicarbonate dependent alteration of acrosome morphology and protein tyrosine phosphorylation on boar sperm. *Journal of Cell Biochemistry* **103**(5): 1413–1423.

Teixeira, D.I., Cavada, B.S., Sampaio, A.H., Havt, A., Bloch, C. Jr., Prates, M.V., Moreno, F.B., Santos, E.A., Gadelha, C.A., Gadelha, T.S., Crisóstomo, F.S., and Freitas, V.J. 2002. Isolation and partial characterisation of a protein from buck seminal plasma (Capra hircus), homologous to spermadhesins. *Protein and Peptide Letters* **9**(4): 331–335.

Teixeira, D.I., Melo, L.M., Gadelha, C.A., Cunha, R.M., Bloch, C. Jr., Rádis-Baptista, G., Cavada, B.S., and Freitas, V.J. 2006. Ion-exchange chromatography used to isolate a spermadhesin-related protein from domestic goat (Capra hircus) seminal plasma. *Genetics and Molecular Research* **5**(1): 79–87.

Thérien, I., Soubeyrand, S., and Manjunath, P. 1997. Major proteins of bovine seminal plasma modulate sperm capacitation by high-density lipoprotein. *Biology of Reproduction* **57**(5): 1080–1088.

Thérien, I., Moreau, R., and Manjunath, P. 1998. Major proteins of bovine seminal plasma and high-density lipoprotein induce cholesterol efflux from epididymal sperm. *Biology of Reproduction* **59**(4): 768–776.

Thérien, I., Moreau, R., and Manjunath, P. 1999. Bovine seminal plasma phospholipid-binding proteins stimulate phospholipid efflux from epididymal sperm. *Biology of Reproduction* **61**(3): 590–598.

Thérien, I., Bousquet, D., and Manjunath, P. 2001. Effect of seminal phospholipid-binding proteins and follicular fluid on bovine sperm capacitation. *Biology of Reproduction* **65**(1): 41–51.

Thérien, I., Bergeron, A., Bousquet, D., and Manjunath, P. 2005. Isolation and characterization of glycosaminoglycans from bovine follicular fluid and their effect on sperm capacitation. *Molecular Reproduction and Development* **71**(1): 97–106.

Thurston, R.J., Korn, N., Froman, D.P., and Bodine, A.B. 1993. Proteolytic-enzymes in seminal plasma of domestic turkey (Meleagris-gallopavo). *Biology of Reproduction* **48**(2): 393–402.

Tienthai, P., Johannisson, A., and Rodriguez-Martinez, H. 2004. Sperm capacitation in the porcine oviduct. *Animal Reproduction Science* **80**(1–2): 131–146.

Töpfer-Petersen, E., Romero, A., Varela, P.F., Ekhlasi-Hundrieser, M., Dostalova, Z., Sanz, L., and Calvete, J.J. 1998. Spermadhesins: A new protein family. Facts, hypotheses and perspectives. *Andrologia* **30**(4–5): 217–224.

Töpfer-Petersen, E. 1999. Molecules on the sperm's route to fertilization. *Journal of Experimental Zoology* **285**(3): 259–266.

Töpfer-Petersen, E., Petrounkina, A.M., and Ekhlasi-Hundrieser, M. 2000. Oocyte-sperm interactions. *Animal Reproduction Science* **60**(S1): 653–662.

Töpfer-Petersen, E., Ekhlasi-Hundrieser, M., Kirchhoff, C., Leeb, T., and Sieme, H. 2005. The role of stallion seminal proteins in fertilisation. *Animal Reproduction Science* **89**(1–4): 159–170.

Töpfer-Petersen, E., Ekhlasi-Hundrieser, M., and Tsolova, M. 2008. Glycobiology of fertilization in the pig. *International*

Journal of Develpomental Biology **52**(5–6): 717–736.

Upreti, G.C., Hall, E.L., Koppens, D., Oliver, J.E., and Vishwanath, R. 1999. Studies on the measurement of phospholipase A2 (PLA2) and PLA2 inhibitor activities in ram semen. *Animal Reproduction Science* **56**(2): 107–121.

Vadnais, M.L., Galantino-Homer, H.L., and Althouse, G.C. 2007. Current concepts of molecular events during bovine and porcine spermatozoa capacitation. *Archives of Andrology* **53**(3): 109–123.

Varela, P.F., Romero, A., Sanz, L., Romão, M.J., Töpfer-Petersen, E., and Calvete, J.J. 1997. The 2.4 A resolution crystal structure of boar seminal plasma PSP-I/PSP-II: A zona pellucida-binding glycoprotein heterodimer of the spermadhesin family built by a CUB domain architecture. *Journal of Molecular Biology* **274**(4): 635–649.

Veselsky, L., Jonakova, V., Sanz, M.L., Töpfer-Petersen, E., and Cechová, D. 1992. Binding of a 15kDa glycoprotein from spermatozoa of boars to surface of zona pellucida and cumulus oophorus cells. *Journal of Reproduction and Fertility* **96**(2): 593–602.

Veselsky, L., Peknicova, J., Cechova, D., Kraus, M., Geussova, G., and Jonakova, V. 1999. Characterization of boar spermadhesins by monoclonal and polyclonal antibodies and their role in binding to oocytes. *American Journal of Reproductive Immunology* **42**(3): 187–197.

Villemure, M., Lazure, C., and Manjunath, P. 2003. Isolation and characterization of gelatin-binding proteins from goat seminal plasma. *Reproductive Biology and Endocrinology* **1/39**: 1–10.

Visconti, P.E., Galantino-Homer, H., Moore, G.D., Bailey, J.L., Ning, X., Fornes, M., and Kopf, G.S. 1998. The molecular basis of sperm capacitation. *Journal of Andrology* **19**(2): 242–248.

Vitagliano, L., Adinolfi, S., Riccio, A., Sica, F., Zagari, A., and Mazzarella, L. 1998. Binding of a substrate analog to a domain swapping protein: X-ray structure of the complex of bovine seminal ribonuclease with uridylyl(2′,5′)adenosine. *Protein Science* **7**(8): 1691–1699.

Wagner, A., Ekhlasi-Hundrieser, M., Hettel, C.H., Petrunkina, A.M., Waberski, D., Nimtz, M., and Töpfer-Petersen, E. 2002. Carbohydrate-based interactions of oviductal sperm reservoir formation-studies in the pig. *Molecular Reproduction and Development* **61**: 249–257.

Wah, D.A., Fernández-Tornero, C., Sanz, L., Romero, A., and Calvete, J.J. 2002. Sperm coating mechanism from the 1.8 A crystal structure of PDC-109-phosphorylcholine complex. *Structure* **10**(4): 505–514.

Wang, I., Lou, Y.C., Wu, K.P., Wu, S.H., Chang, W.C., and Chen, C. 2005. Novel solution structure of porcine beta-microseminoprotein. *Journal of Molecular Biology* **346**(4): 1071–1082.

Wassarman, P.M. 1990. Profile of a mammalian sperm receptor. *Development* **108**(1): 1–17.

Wassarman, P.M. 1999. Mammalian fertilization: Molecular aspects of gamete adhesion, exocytosis, and function. *Cell* **96**(2): 175–183.

Wassarman, P.M., Jovine, L., and Litscher, E.S. 2001. A profile of fertilization in mammals. *Nature Cell Biology* **3**(2): E59–64.

Wassarman, P.M., Jovine, L., Qi, H., Williams, Z., Darie, C., and Litscher, E.S. 2005. Recent aspects of mammalian fertilization research. *Molecular and Cellular endocrinology* **234**(1–2): 95–103.

Wempe, F., Eimspanier, R., and Scheit, K.H. 1992. Characterization by cDNA cloning of the mRNA of a new growth factor from bovine seminal plasma: Acidic seminal fluid protein. *Biochemical Biophysical Research Communications* **183**(1): 232–237.

Yanagimachi, R. 1994. Mammalian fertilization. In: Knobil, E. and Neill, J.D. (eds.), *The Physiology of Reproduction*. New York: Raven Press, pp. 189–318.

Yurewicz, E.C., Sacco, A.G., Gupta, S.K., Xu, N., and Gage, D.A. 1998. Hetero-oligomerization-dependent binding of pig oocyte zona pellucida glycoproteins ZPB and ZPC to boar sperm membrane vesicles. *Journal of Biological Chemistry* **273**(13): 7488–7494.

16

Evolutionary Genomics of Sex Determination in Domestic Animals

Eric Pailhoux and Corinne Cotinot

16.1 Introduction

In vertebrates, sex is set up at fertilization and depends on the sex chromosome received from the heterogametic parent (XY/XX system when male is heterogametic; ZZ/ZW when female is heterogametic). Even if zygotes are genetically different, no sex difference has been clearly evidenced before gonad differentiation occurs. Thus, individuals of both sexes seem morphologically identical during early development. The first sign of sexual dimorphism appears when the undifferentiated gonad engages its differentiation onto a testis or an ovary, following a sex-determining signal. In heterogametic vertebrates, this genetic signal is located on sex chromosomes. Although dependent on the considered species, this signal could be more or less influenced by environmental factors such as temperature, steroid hormones, or population constitution (Figure 16.1).

Gonad differentiation depending on the temperature of egg incubation, called TSD for Temperature-dependent Sex Determination, has been intensely investigated and described in different reptiles (reviewed in Pieau and Dorizzi 2004). In these TSD species, the egg incubation temperature seems directly linked to steroid hormone production and more precisely to the enzyme P_{450} aromatase (*CYP19* gene). This gene is directly responsible for male to female steroidogenesis reversal by converting androgens into estrogens (reviewed in Conley and Hinshelwood 2001).

Steroid hormone treatments have been shown to be critical in sex differentiation of numerous vertebrates. With the exception of placental mammals, an estrogen or an anti-aromatase treatment could reverse the genetically or temperature-dependent predetermined sex of the gonad (Scheib 1983; Elbrecht and Smith 1992; Guiguen et al. 1999; Krisfalusi and Cloud 1999; Pieau et al. 1999; Coveney et al. 2001).

Population constitution and/or social factors influencing sex determination have been described in some fish hermaphroditic

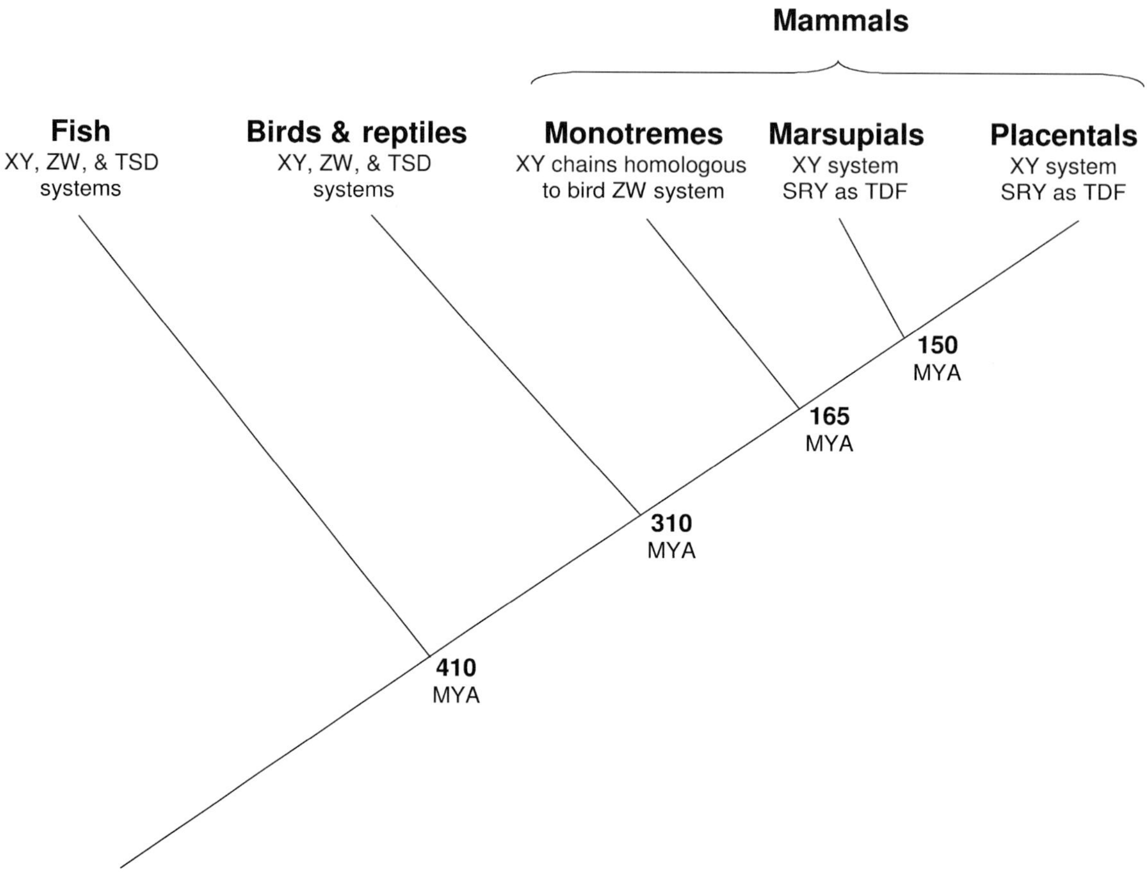

Figure 16.1 Phylogeny displaying the different sex chromosome systems of major vertebrate groups (TSD, Temperature Sex-Dependent). Divergence times are derived from Veyrunes et al. (2008).

species (Fishelson 1970). As a general feature, a perturbation of social interactions in a given subpopulation results in a complete sex reversal of one or several individuals (reviewed in Baroiller et al. 1999).

Following the determining switch, early differentiating gonads will secrete sexual hormones controlling the development of different sexual features of the species. This hormonal control has been clearly demonstrated for some mammalian traits such as the genital tract development (Jost 1947); sex-specific brain features (Arnold and Gorski 1984; Gorski 1984; McEwen 1992); sex-specific liver metabolisms (Gustafsson et al. 1983; Roy and Chatterjee 1983; Robins 2005); or secondary sexual characteristics such as horn development in different species (Toledano-Díaz et al. 2007) or feathers in some birds (Wilson et al. 1987). Sex-specific features closely or distantly linked to the initial sex-determining signal are numerous and diverse, even in the way human males and females chew gum (Gerstner and Parekh 1997). Intrigued or not by the abundant diversity of sexual dimorphisms, humans have developed many ways of understanding sex differences that originate from sex determination.

Most data on sex determination and gonad differentiation have been obtained on humans and mice. In this review, we will present the main commonalities in mammalian sex differentiation, and then species-specific features will be discussed for domestic mammals (mainly pig, sheep, and goat) and briefly for nonmammal vertebrates (mainly chicken).

16.2 State of knowledge of sex differentiation

16.2.1 Mammalian sex determination: The key role of SRY

Our present knowledge of mammalian sex determination is based on studies performed over 50 years ago, on Klinefelter and Turner syndromes in humans and mice that revealed the dominant Y chromosome factor in male differentiation. Individuals with Turner's syndrome are XO and are phenotypically females, whereas individuals with Klinefelter's syndrome are XXY and phenotypically males (Ford et al. 1959; Jacobs and Strong 1959). This identified the Y chromosome as the factor that engenders maleness and generated a long quest for identifying the testis-determining factor (TDF). In 1990, studies in human XX male patients led to the discovery of *SRY* (sex-determining region of the Y; *Sry* in mice) as the primary testis-determining factor (Sinclair et al. 1990).

Primary sex determination in mammals appears to be focused on the cell-fate decision that occurs in the supporting cell lineage precursor when the cells chose to differentiate into Sertoli or follicular (granulosa) cells. In mice, *Sry* is transiently expressed for a few hours in Sertoli cell precursors in the XY gonads between embryonic days 10.5 and 12.5 (Koopman et al. 1990; Hacker et al. 1995; Albrecht and Eicher 2001). Both the spatial and temporal regulations of *Sry* levels are critical to correctly direct the testis differentiation (Salas-Cortes et al. 1999; Bullejos and Koopman 2001; Sekido et al. 2004). Low or delayed expressions lead to the development of an ovotestis, a gonad containing a mixture of male and female tissues (Bullejos and Koopman 2005; Taketo et al. 2005).

SRY contains a high-mobility group (HMG)-box DNA-binding domain characteristic of the *SOX* gene family of transcription factors. Consistent with the importance of the DNA-binding function of *SRY*, most sex-reversed mutations occur within the HMG domain (Harley et al. 1992; Mitchell and Harley 2002). Furthermore, this domain is the only conserved feature across mammalian *SRY*. Regions outside this domain have evolved rapidly and present a large variability across species (Whitfield et al. 1993).

Since its discovery, a variety of mechanisms has been proposed by which *SRY* might initiate the testis differentiation from early bipotential gonads: (1) as a repressor of a repressor of male development (McElreavey et al. 1993), (2) through effects on local chromatin structure (Pontiggia et al. 1994), (3) through a role in mRNA splicing (Ohe et al. 2002), and (4) as a transcriptional activator of one or more critical male-specific targets, through partner proteins (Dubin and Ostrer 1994; Poulat et al. 1997; Thevenet et al. 2005). Recent works have shown that *Sry* binds to multiple *Sox9* elements located within a gonad-specific enhancer in mice, supporting a model for the positive regulation of *Sox9* expression in the male mouse gonad (Sekido and Lovell-Badge 2008).

16.2.2 The cascade after the switch

The first gene known to be expressed downstream of *SRY* is *SOX9*, a closely related family member. *Sox9* is expressed in various tissues during embryogenesis (Ng et al. 1997) and in Sertoli cells of male gonads (Morais da Silva et al. 1996). In contrast to *Sry*, *Sox9* is well conserved among mammals and also in vertebrates that have another sex chromosome system such as birds, reptiles, and fishes.

Mutations or deletions of the *Sox9* gene lead to male-to-female sex reversal (Wagner et al. 1994; Foster 1996; Chaboissier et al. 2004; Smyk et al. 2007) whereas duplication or overexpression of *Sox9* is responsible for female-to-male sex reversal (Huang et al. 1999; Bishop et al. 2000; Vidal et al. 2001). These studies indicate that *Sox9* can replace *Sry*, even leading to fertile males when expressed at sufficient levels in XY *Sry* null embryos (Qin and Bishop 2005). In mice, the expression of *Sry* is turned off just after that *Sox9* reached a critical threshold.

Several extracellular signaling pathways have been involved in recruiting the cells of the gonad to the testis pathway; among these are prostaglandin D2(*PGD2*) and Fibroblast growth factor 9 (*FGF9*). Both induce *Sox9* expression in XX cells *in vitro* and promote Sertoli cell differentiation (Malki et al. 2005, 2007; Wilhelm et al. 2005, 2007). In the absence of *Fgf9* in KO mice, *Sry* is expressed normally and *Sox9* expression is initiated but rapidly silenced. Subsequently the cells of the XY *Fgf9*$^{-/-}$ gonads express genes characteristic of the female pathway (Colvin et al. 2001). These data indicate that *Fgf9* is necessary for the maintenance of *Sox9* expression and promotes testis differentiation *in vivo* (Schmahl et al. 2004; Kim et al. 2006). *Sox9* initiates *Fgf9* transcription, and *Fgf9* maintains *Sox9* expression and induces nuclear localization of *Fgfr2* in Sertoli cell precursors. Moreover, it has been shown that *Sox9* is required for *Fgfr2* nuclear localization and, conversely, that *Fgfr2* is important for the maintenance of *Sox9* expression. These results suggest that *Fgfr2* and *Sox9* regulate each other through a *Sox9–Fgf9–Fgfr2* positive signaling loop (Kim et al. 2007; Bagheri-Fam et al. 2008).

The ability of extracellular signals to recruit cells to the testis developmental pathway has previously been hypothesized by XX-XY chimera experiments. These studies have shown that in XX↔XY chimerical mouse embryos, the ratio of XX to XY cells is ~50:50 in all tissues, the only exception being Sertoli cells. These cells were found to be more than 90% XY, indicating that the differentiation of Sertoli cells needs the presence of the Y chromosome (Palmer and Burgoyne 1991). However, these experiments also imply that *Sry* is not essential for the differentiation of all Sertoli cells. XX cells were recruited to differentiate into Sertoli cells contributing almost one-tenth of the total number of Sertoli cells. In addition to the cell-autonomous mechanism, at least one noncell-autonomous mechanism exists to ensure the differentiation of a sufficient number of Sertoli cells, above the estimated threshold of 20% to guarantee testis differentiation (Burgoyne et al. 1988; Patek et al. 1991).

Proliferation is also a critical event for testis development. The use of proliferation inhibitors or the disruption of *Fgf9* pathway leads to male-to-female sex reversal (Schmahl et al. 2000, 2004; Kim et al. 2006). It has also been shown that the insulin receptor tyrosine kinase family, comprising *Ir*, *Igf1r*, and *Irr*, is required for the appearance of male gonads and thus for male sexual differentiation. XY mice that are mutant for all three receptors develop ovaries and show a female phenotype. Reduced expression of both *Sry* and the early testis-specific marker *Sox9* indicated that the insulin signaling pathway is required for male sex determination (Nef et al. 2003).

Based on the established role of the platelet-derived growth factor (*PDGF*) family of ligands and receptors in cell migration, proliferation, and differentiation in various organ systems, the role of *PDGF* in testis organogenesis has been investigated. *Pdgfr-*$\alpha^{-/-}$ XY gonads displayed disruptions in the organization of the vasculature and in the partitioning of interstitial and testis cord compartments. Closer examination revealed severe reductions in characteristic XY proliferation, mesonephric cell migration, and fetal Leydig cell differentiation. This work identified *PDGF* signaling through the alpha receptor as an important event downstream of *Sry* in testis organogenesis and Leydig cell differentiation (Brennan et al. 2003).

Once the fate of supporting cell precursor is determined by *Sry*, feedback loops reinforcing the male pathway are initiated. *Sox9* and *Fgf9* upregulate each other and generate the first cellular events toward Sertoli cell fate. In addition, extracellular signals work to recruit other cells in the male pathway. Defects in these signaling loops could explain disorders in sexual development such as ovotestis formation and ambiguous genitalia.

The critical event in testis organogenesis is the specification of somatic cell lineages including Sertoli cells, peritubular myoid cells, and Leydig cells. Specification of these lineages is crucial for the establishment of testis morphology and the production of hormones. Autonomous expression of *Sry* in somatic cells and production of extracellular factors in the XY gonad lead to differentiation of Sertoli cells. Differentiating gonadal cells induce migration of cells from the mesonephros into the gonad. The migrating cells contribute to precursors of the peritubular myoid and vascular cell lineages (Martineau et al. 1997; Capel et al. 1999; Tilmann and Capel 1999). Differentiation of peritubular myoid cells and the consequent formation of testis cords are regulated by Desert hedgehog (*Dhh*), a signaling protein produced by Sertoli cells (Clark et al. 2000; Pierucci-Alves et al. 2001). It has been shown that Leydig cells, the male steroidogenic cells, differentiate under the action of *Dhh* and platelet-derived growth factor A (*PdgfA*), two factors produced by Sertoli cells (Yao et al. 2002; Brennan et al. 2003). Consequently, Sertoli cells appear as the conductor of testis differentiation.

All vertebrate males have testes that are similar in anatomy. Despite a variety of sex chromosome systems, a large number of genes acting in the differentiation of testes and male genitalia are conserved in vertebrates. The primary switch controlling sex determination is highly divergent across species, but it seems that the pathways downstream of the switch call on the same factors. Nevertheless the combination of these factors or their spatiotemporal expression can vary between species.

16.2.3 The ovarian pathway

In contrast to the male, the molecular bases of mammalian female sex determination are poorly understood. Indeed female sexual development was considered for a long time as a passive process due to the fact that female external genitalia can be established in the absence of a gonad whereas two active factors (testosterone and *AMH*) are needed to promote male sexual development (Jost 1947; Josso et al. 1993). In the last decade, three factors have been isolated having essential roles in ovary determination: *WNT4*, *FOXL2*, and *RSPO1*.

Wnt4 is expressed in the bi-potential gonad of both sexes and is then downregulated in the testis and upregulated in the ovary at 11.5 dpc (Vainio et al. 1999; Yao et al. 2004b). Several *Wnt* receptors are expressed in somatic cells of the gonads such as *Fzd6*. They might mediate autocrine/paracrine signaling for *Wnt4* in cells that are engaged in sex determination. The inactivation of *Wnt4* leads to incomplete sex reversal with early production of testosterone and male internal genitalia in XX mice (Vainio et al. 1999; Chassot et al. 2008). Germ cells started oogenesis before degenerating, and somatic supporting cells acquired partial testis-like features lately (Vainio et al. 1999; Yao et al. 2004b)s. In the absence of *Wnt4*, *Fgf9* and *Sox9* expression are transiently upregulated in the XX gonad. This suggests that *Wnt4* normally represses the male pathway in female gonad and that additional factors are needed to reinforce the female pathway.

Based on genetic analysis of natural mutations in goats (detailed below) or XX male human patients, two other genes have been proposed as candidate female sex-determining genes: *FoxL2* and *RSPO1* (Pailhoux et al. 2001a; Parma et al. 2006). *FoxL2* expression is uniquely female, undetectable in XY gonads of all tested species (Cocquet et al. 2002; Pisarska et al. 2004; Uda et al. 2004). In mice, it is activated at 12.5 dpc in the fetal ovary and increases in level steadily until early postnatal life, with a maximum of expression in supporting cells of primordial follicles. *FOXL2* has also been identified as the gene mutated in human patients with a syndromic form of premature ovarian failure called BPES (Crisponi et al. 2001). Experimental ablation of *Foxl2* in mice gives premature ovarian failure with only partial secondary sex reversal (Schmidt et al. 2004; Uda et al. 2004; Ottolenghi et al. 2005). In addition, forced expression of *Foxl2* impairs testis tubule differentiation in XY transgenic mice (Ottolenghi et al. 2007). This result is consistent with an anti-testis role of *Foxl2*.

Although *Wnt4* and *Foxl2* are independently expressed, they show complementary phenotypes in ovary morphogenesis, with *Wnt4* being required in early stroma differentiation and oocyte survival and *FoxL2* being involved in supporting cell lineage differentiation. The *Wnt4*$^{-/-}$/*Foxl2*$^{-/-}$ double knockout ovaries produce testis-like tubules and spermatogonia (Ottolenghi et al. 2007). This demonstrates that female sex-determining genes are required to suppress an alternative male fate in the ovary.

Recently, mutations in the *R-Spondin 1* (*RSPO1*) gene have been identified in human XX patients with testis development (Parma et al. 2006). This is the first human mutation that results in complete female-to-male sex reversal. *RSPO1* has also been shown to activate the canonical *β-catenin* signaling pathway, which raises the possibility that *Wnt4* and *Rspo1* act cooperatively to block the male pathway in XX gonads (Kim et al. 2006, 2008; Chassot et al. 2008). *Rspo1* knockout mice show masculinized gonads. Molecular analyses demonstrate an absence of female-specific activation of *Wnt4* and as a consequence XY-like vascularization and steroidogenesis. Moreover, germ cells of XX *Rspo1*$^{-/-}$ knockout embryos show changes in cellular adhesions and a failure to enter XX specific meiosis (Chassot et al. 2008). Sex cords develop around birth, when *Sox9* becomes strongly activated. These experiments demonstrate a balance between *Sox9* and *β-catenin* activation to determine the fate of the gonad, with *Rspo1* acting as a crucial regulator of the canonical *β-catenin* signaling required for female development.

Parallel to gene inactivation, multiple types of differential transcriptome analyses have been used to identify genes involved in testicular and ovarian differentiation, including cDNA microarray (Grimmond et al. 2000; Boyer et al. 2004; Nef et al. 2005; Small et al. 2005; Olesen et al. 2007), differential display (Nordqvist 1995; Nordqvist and Töhönen 1997; Töhönen et al. 1998), and representational difference analysis (Perera et al. 2001; Adams and McLaren 2002). A large amount of expressional data has been obtained; however, assigning roles for genes in particular morphological pathways has been a rate-limiting difficulty. The quantity of the data obtained likely reflects the complexity of the rapid and overlapping changes that occur during testis determination and differentiation. It also shows clearly that initiation and maintenance of ovarian pathway involves the active regulation of many genes and is not a passive/default developmental process.

16.2.4 Sexual dimorphism of germ cells

The developmental fate of primordial germ cells in the mammalian gonad depends on their environment. In the XY gonad, *Sry* induces a cascade of molecular and cellular events leading to the organization of testis cords. Germ cells are sequestered inside testis cords by 12.5 dpc where they arrest in mitosis. In contrast to male gonad, germ cells are crucial for the formation and maintenance of ovarian structures. In the absence of germ cells, ovarian follicles do not assemble, and when germ cells are lost, ovarian follicles rapidly degenerate (McLaren 1988). By 13.5 dpc, germ cells in the XX gonad enter meiosis and they arrest in prophase I by birth (McLaren 1988). The timing of germ cell entry into meiosis appears to be based on an intrinsic clock. By generating XX/XY recombinant aggregates in culture, it has been shown that the physical presence of germ cells inhibits initiation of the testis pathway. The developmental stage when germ cells from XX gonads inhibit the male pathway is temporally correlated with the time needed for germ cells to spontaneously enter meiosis (Yao et al. 2003). Thus, it has been proposed that once germ cells commit to meiosis, they reinforce ovarian fate by antagonizing the testis pathway.

The development of germ cells in concordance with sexual fate of the somatic cells in the gonad seems important not only for fertility but also for contributing to the fate of the gonad and for participating in the maintenance of a testis or an ovary.

16.2.5 The critical balance

A model (Figure 16.2) has been proposed to explain the *SRY*-negative XX female-to-male sex reversal existence in which *SRY* could repress a repressor of male development, called "Z." Based on this model, mutations in "Z" could lead to derepression of the male pathway in XX gonads (McElreavey et al. 1993). The sum of the current studies suggests that multiple redundant anti-testis activities ("Z factors") are deployed in fetal ovaries. It seems likely that activation of the *Wnt4/Rspo1* and *FoxL2* pathways antagonizes the establishment of *Sox9* in supporting cell precursors. This finding leads to a new model of sex determination in which the fate of somatic cells in the gonad depends on the predominance of *Sox9* versus *Wnt4/Rspo1* and *FoxL2* downstream signals (Kim et al. 2006). In mammals, *SRY* normally acts as the testis determinant by promoting

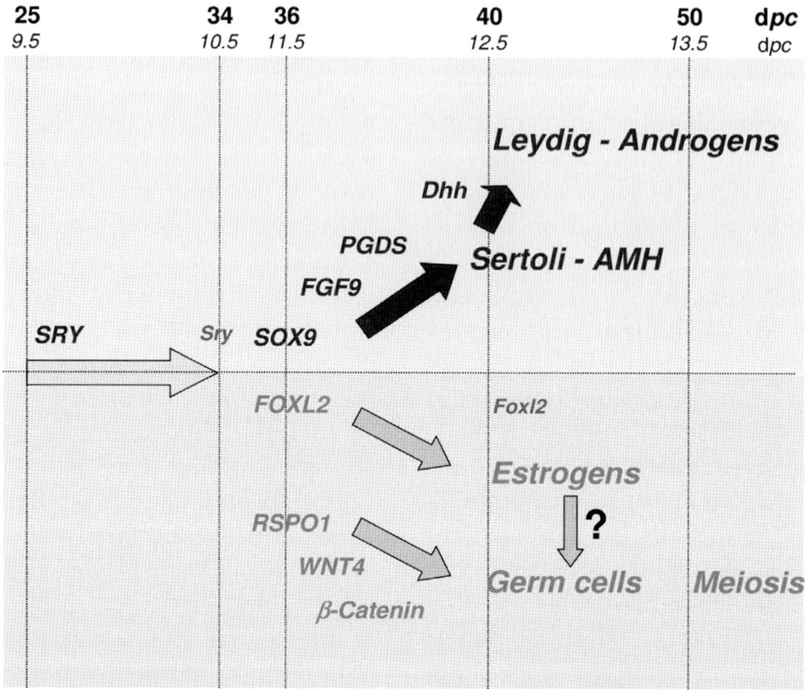

Figure 16.2 Schematic representation of genes involved in gonad differentiation in mammals. The upper numbers indicate the developmental stages in goat (bold) and mouse (italic). *Sry* and *Foxl2* in lower cases present when these genes are expressed in mouse, time points that seem to be delayed compared with the goat ones.

SOX9 expression. However, this outcome can also be promoted by loss of *RSPO1* or *FOXL2*. Within the bi-potential gonad, the somatic cells seem to be highly plastic and can differentiate as cells of the ovary or cells of the testis during the window of fetal sex determination. There is also evidence that XX cells can trans-differentiate to male cell fate under certain conditions in adult life. Perhaps the critical balance between these signaling pathways helps to explain the underlying "bi-potential" property of the gonadal cells and suggests a molecular mechanism by which this balance is tipped in one or another direction to regulate the fate of gonadal cells (DiNapoli and Capel 2008).

16.3 Sex differentiation in domestic mammals

Since the discovery of *SRY* in 1990, much progress has been made in the understanding of sex differentiation in mammals, especially by studying numerous naturally occurring sex-reversed mutants (mainly in human) or produced by specific gene targeting (in mouse). According to the current knowledge on sex differentiation in different vertebrate species, some mechanisms appear to be well conserved and some others more variable between species. In this section we will highlight, using the goat as a model, two features appearing variable between mouse and other mammals: (1) the *SRY* gene and (2) the ovarian differentiating pathway. Two additional examples will also be presented in pig and sheep in order to illustrate differences in sexual differentiation between mammals.

16.3.1 *SRY conservation across species*

The *SRY* gene has been equated as the testis-determinant mainly in human and mouse by mutation analysis and mouse transgenesis (reviewed in Polanco and Koopman 2007). Furthermore, *SRY* orthologs, located on the Y chromosome, have been identified for numerous mammalian species except in monotremes (Wallis et al. 2007) and in some rare cases of rodents such as voles (Just et al. 1995) and Japanese rats (Soullier et al. 1998; Sutou et al. 2001). *SRY* studies in different species have pointed out an unexpected feature of poor conservation of this master gene at the structural (Tucker and Lundrigan 1993; Whitfield et al. 1993; Payen and Cotinot 1994) and expressional levels. In contrast to mice, *SRY* gonadal expression persists many days after Sertoli cells differentiation in pig (Daneau et al. 1996; Parma et al. 1999), sheep (Payen et al. 1996), dog (Meyers-Wallen 2003), human (Salas-Cortés et al. 1999; Hanley et al. 2000), goat (Pannetier et al. 2006a), and tamar (Harry et al. 1995). In tamar, *SRY* expression was detected in different non-gonadic tissues (Harry et al. 1995) and in goat, *SRY* was found expressed from as early as the first sign of genital ridges formation to adulthood (Pannetier et al. 2006a). According to these results it appears that depending on the considered species, *SRY* expression is more or less focused on the crucial period preceding Sertoli cell differentiation with a strict regulation in mouse compared with other species. It could directly reflect species-specific *SRY* regulation or *SRY* implication in processes other than testis determination in non-rodent species. To partially answer this, we showed that the goat *SRY* gene is able to induce testis differentiation in XX mouse despite the poor conservation of SRY between both species and despite the goat-type expression profile of the transgene (Pannetier et al. 2006a).

Thus, *SRY* must be expressed at the beginning of XY gonad differentiation in order to determine the testis fate of the gonad. However, as this gene is present only in XY individuals, one can imagine that its strict spatiotemporal regulation would not be absolutely required, should its "mis-expression" (non-gonadic or gonadic after sex differentiation) have no detrimental effects. Accordingly, goat *SRY* was found expressed in non-gonadic tissues and in gonads of all stages in transgenic XX and XY mice and no additional phenotype was observed aside from the XX sex reversal one (Pannetier et al. 2006a).

16.3.2 *The goat as model for early ovarian differentiation*

Based on previous isolation of key factors for testis differentiation by linkage analyses and positional cloning (Gessler et al. 1990; Sinclair et al. 1990; Pelletier et al. 1991; Foster et al. 1994; Wagner et al. 1994) and according to the Z hypothesis (McElreavey et al. 1993), we have attempted to isolate key ovarian differentiating factors by studying XX sex-reversal pathologies. Apart from human, such pathologies have been described in at least four mammalian domestic species: dog, horse, pig, and goat (Pailhoux et al. 1994; Meyers-Wallen et al. 1999; Buoen et al. 2000; Pailhoux et al. 2005). In goat, as the polled trait was shown to be associated with XX sex reversal (Asdell 1944), we used families based on heterozygous polled males in order to localize and clone the Polled Intersex Syndrome (PIS) mutation (Vaiman et al. 1996; Schibler et al. 2000; Pailhoux et al. 2001a). This PIS mutation, responsible for both traits, polled (dominant) and XX sex reversal (recessive), was shown to be an 11.7-kb deletion located on goat chromosome 1 (Pailhoux et al. 2001a). This 11.7-kb DNA fragment encompassed no gene or part of gene but exerted transcriptional regulatory effects on at least three genes located in the vicinity, PIS regulated transcript number 1 (*PISRT1*), Promoter *FOXL2* inverse complementary (*PFOXic*), and Forkhead box L2 (*FOXL2*). Among these genes, only *FOXL2* appears to be a classical one encoding a transcription factor. Indeed, *PISRT1* encodes a poly-adenylated mono-exonic 1.5-kb transcript devoid of open-reading frame and *PFOXic*, a putative but nonconserved protein. All these three genes are transcriptionally controlled by the PIS region. Their expression depends on the PIS genotype and on the considered tissue. In the female gonads, the three genes are expressed from the beginning of ovarian formation (34 days post-coïtum [dpc] in goat) until adulthood in normal PIS$^{+/+}$ and PIS$^{+/-}$ animals, but their expression is loss in PIS$^{-/-}$ XX gonads. In the horn buds of both sexes, these three genes are not expressed under a PIS$^{+/+}$ wild-type genotype, but their expression is turn on in PIS$^{+/-}$ and PIS$^{-/-}$ horn buds.

Since the characterization of the mutation in 2001, studies have been developed in order (1) to understand the role of each PIS-regulated genes and (2) to decipher the molecular mechanism involved in the long-range regulation of these genes by the 11.7-kb PIS region (*FOXL2* and *PFOXic* lie at more than 300 kb apart from the PIS region). As results have mainly been accumulated on the role of each PIS-regulated gene, the second point on long-range regulation will not be treated here.

16.3.3 FOXL2 seems to be the major PIS-regulated gene

Following the discovery of *FOXL2* as the gene responsible for Blepharophimosis Ptosis Epicanthus inversus Syndrome (BPES, MIM #110100; Crisponi et al. 2001) and its potential involvement in XX sex reversal in PIS$^{-/-}$ goats, *Foxl2* invalidation has been performed in mouse (Schmidt et al. 2004; Uda et al. 2004). In contrast to goat, XX *Foxl2*$^{-/-}$ mice developed premature ovarian failure (POF) with a blockage in the first steps of follicle formation, as observed in XX BPES type 1 patients heterozygous for *FOXL2* mutations (De Baere et al. 2003). In these XX *Foxl2*$^{-/-}$ mice, the sole sign of XX sex reversal appears only 2 days after birth and consists of an overexpression of *Sox9* in the somatic granulosa cells (Ottolenghi et al. 2005). One of our major goals has thus been to understand the origin of the phenotype discrepancy between XX *Foxl2*$^{-/-}$ mice with POF and XX PIS$^{-/-}$ goats with sex reversal. A first hypothesis could be that other PIS-regulated genes could sustain the sex-reversal phenotype besides *FOXL2*. Alternatively, the phenotype discrepancy could result from species-specific differences in the role of *FOXL2*. Following the first hypothesis and based on the spatio-temporal expression profile of *PISRT1* that was shown to be decoupled from that of *FOXL2* on the earliest stages of ovarian development and after birth (Pailhoux et al. 2001a), *PISRT1* was first considered as a potential anti-testis Z gene. However, *PISRT1* expression ectopically restored in XX PIS$^{-/-}$ goat gonads had no effect on the sex-reversal phenotype (Boulanger et al. 2008). According to this result and to the fact that *PFOXic* was shown to be involved in

FOXL2 local regulation via a bidirectional promoter (Pannetier et al. 2005), *FOXL2* remains the sole "acting" gene of the PIS locus and might be responsible for both PIS-associated phenotypes. *FOXL2* invalidation in goat is currently in progress in order to highlight its species-specific function.

16.3.4 *FOXL2 as a female steroidogenic factor*

One important feature of FOXL2 is its ability to increase *CYP19* gene expression at the transcriptional level (Pannetier et al. 2006b). It was demonstrated by our team in goat following the observation that *CYP19* gene expression was drastically decreased in early developing XX PIS$^{-/-}$ gonads, as a consequence of the PIS-regulated genes extinction (Pailhoux et al. 2002). Thereafter, an impressive work demonstrated the crucial role of FOXL2 in the control of female steroidogenesis orientation in the nonmammalian fish species tilapia (Wang et al. 2007). In this study, *FOXL2* was shown to act with *Ad4BP/SF-1* on different promoters of key steroidogenic gene (including *CYP19*) in order to increase estrogen production. This close relation of *FOXL2* with estrogen production was also evidenced in other nonmammalian species such as birds, reptiles, and other fishes, suggesting an ancient and conserved mechanism (Baron et al. 2004; Govoroun et al. 2004; Hudson et al. 2005; Liu et al. 2007b; Nakamoto et al. 2007; Rhen et al. 2007).

The following observations on the phenotype discrepancy observed between *Foxl2*$^{-/-}$ mice and PIS$^{-/-}$ goat may help in its understanding: (1) In contrast to goat, mouse fetal ovaries are not steroidogenically active before meiosis; (2) *Foxl2/FOXL2* expression starts around 12.5 dpc (1 day before germ cell meiosis) in mouse and at 34 dpc (~20 days before meiosis) in goat. Consequently, in goat there is a period of around 20 days before germ cell meiosis during which ovarian-specific genes such as *FOXL2* and *CYP19* are turned on and consequently estrogen production begins (Mauléon et al. 1977; Pannetier et al. 2006b). This period seems to have no equivalent in mouse ovarian development and this difference could account for the phenotype discrepancy observed.

16.3.5 *Early ovarian organization in goat*

Following the recent discovery of a new gene *RSPO1* involved in human XX sex reversal associated with palmoplantar hyperkeratosis (PPK; Parma et al. 2006), the four R-spondin genes have been studied in goat (Kocer et al. 2008). From this study it appears that *FOXL2* and *RSPO1* are expressed by two different somatic cell types in early developing ovaries and that *FOXL2* extinction in XX PIS$^{-/-}$ gonads does not primarily affect *RSPO1* expression (Kocer et al. 2008). Indeed, at 40 dpc, 5 days after *FOXL2* extinction, *RSPO1* remains expressed in XX PIS$^{-/-}$ gonads even if these gonads begin to express *SOX9* and show clear histological signs of masculinization (Pailhoux et al. 2002; Kocer et al. 2008). Conclusively, the two unique genes, *RSPO1* and *FOXL2*, characterized today for their involvement in mammalian XX sex reversal, act on two different ovarian pathways. Efforts must now be developed in order to decipher putative cross talking between these two pathways.

According to immuno-histological studies, it seems clear now that ovarian differentiation in goat begins at the same time as testis differentiation (34–36 dpc). The first sign of ovarian differentiation consists of germ cell location in the cortical area just under the coelomic epithelium. By contrast, germ cells occupy all the medullar region of the XY

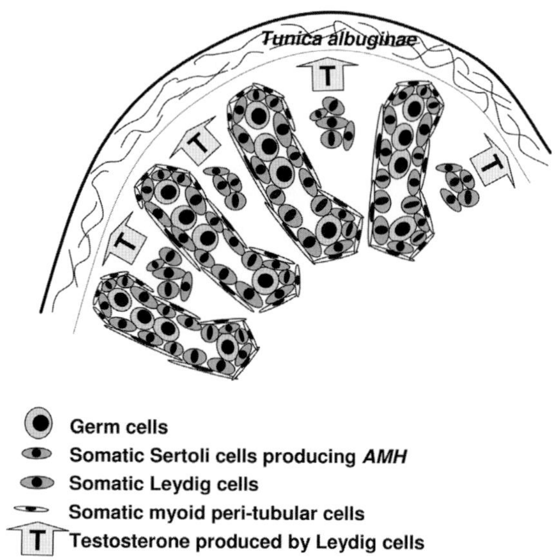

Figure 16.3 Schematic representation of a goat testis at developmental stages 40–45 dpc. Endothelial and fibroblastic cells are not represented.

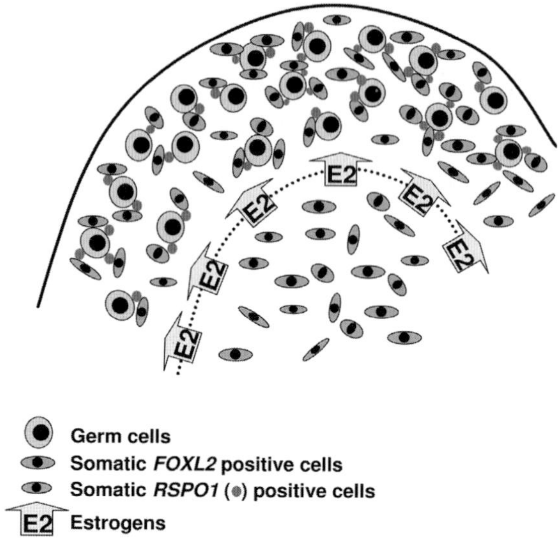

Figure 16.4 Schematic representation of a goat ovary at developmental stages 40–45 dpc. Endothelial and fibroblastic cells are not represented.

testes (Figure 16.3). Moreover, at these early stages of ovarian development, two somatic cell types exist (Figure 16.4). One type expresses *RSPO1* and *WNT4* and is mainly located in the cortical area of the gonad in close relation with the germ cells. The other one expresses *FOXL2*, produces estrogens, and is mainly localized in the medulla part of the ovary (Pannetier et al. 2006b; Kocer et al. 2008). Under this scheme it is interesting to notice that in XX PIS mutant animals the affected somatic cells are those expressing *FOXL2*, consequently those producing steroids. In XX PIS$^{-/-}$ gonads, these steroidogenically active cells will trans-differentiate into Sertoli-like cells that express *SOX9* and *AMH* but are devoided of steroidogenic activity.

16.3.6 Conservation of goat ovarian differentiation features

By contrast to mouse, early developing goat ovaries show two main features: (1) early gonadal regionalization—following their migration in the female gonads, the germ cells localize and stay under the coelomic epithelium—and (2) early steroidogenesis—estrogens are produced by the medulla part of the female gonads before meiosis. Interestingly, both features have been described in different nonmammalian vertebrates and seem to be part of conserved mechanisms of gonad development. Under this scheme, the mouse species might have delayed (regionalization) or switched off (early estrogen production) these features.

Early ovarian regionalization in cortex containing germ cells and medulla has been described in reptiles (Pieau et al. 1999, and references therein) and chicken (Smith and Sinclair 2004, and references therein). In the red-eared slider turtle *Trachemys scripta*, before gonadal differentiation, the undifferentiated gonads of both sexes contain primitive sex cords in their medulla part and the germ cells are located outside these cords just under the coelomic epithelium. At later stages, following gonadal differentiation, germ cells enter the sex cords in the medulla of male gonads and stay outside the sex cords in the cortex of female gonads. Concomitantly, sex cords increase in size and pursue their development in testes but regress in ovaries (Pieau et al. 1999; Yao et al. 2004a).

In chicken, early regionalization has been described since the undifferentiated stages (Stahl and Carlon 1973). Then, gonad differentiation depends upon which component, the cortex or the medulla, develops and maintains the germ cells (Smith and Sinclair 2004). Chicken ovary development seems to be closely similar to that of goat ovaries. In chicken the medulla part of the left gonad expresses *FOXL2*, *CYP19* and produces estrogens; the cortical area contains the germ cells that are in close relation with somatic cells expressing *RSPO1* and *WNT4* (Nakabayashi et al. 1998; Smith et al. 2008). Also interesting in this species is the asymmetric ovarian development: the left ovary develops but the right regresses following an absence of cortical development. Importantly, it has been shown that even if estrogens are produced by both left and right ovaries, the estrogen receptor is expressed unilaterally in the cortical area of the left gonad (Nakabayashi et al. 1998). According to this observation, the role of estrogens in cortical cell proliferation, including the germinal lineage, appears likely. This early ovarian estrogen production might represent the endocrine link between both ovarian somatic cell types, those expressing *FOXL2/CYP19* in the medulla and those expressing *RSPO1/WNT4* in the cortex.

16.3.7 Perspectives on sex differentiation in goat

One of our future prospects on the goat species will be to precisely determine the developmental stages at which the germ cells show a differential localization between sexes (cortical in female, medullar in male). If this event occurs as expected during the undifferentiated period, it will be of great interest to decipher the cellular and molecular mechanisms sustaining this sex-dimorphic feature. Moreover, the role of estrogens produced before germ cell meiosis should also be investigated.

More generally, different pieces of evidence seem to indicate that the major differences between domestic mammals and the mouse model lie on the germinal lineage development, especially before meiosis. Future studies will be developed in order to determine the role of germ cells in sex differentiation of domestic mammals, especially small ruminants such as sheep and goat.

16.3.8 The pig species as a counterexample

Early developing ovaries of pigs also show compartmentalization with a cortical area containing germ cells and a medulla part (Pelliniemi 1975, 1985). The difference in this species is the fact that the medulla part does not express *CYP19* and consequently cannot produce estrogens (Parma et al. 1999; Pailhoux et al. 2001b). Furthermore, in XY pigs, testes expressed *CYP19* since the beginning of Sertoli cell differentiation and trace amounts of estrone has been detected as early as when testosterone secretion starts (Raeside et al. 1993; Parma et al. 1999). According to the results in pig and mouse, it seems clear that some eutherian mammals could develop ovaries without estrogen production before germ cell meiosis. Consequently for sex differentiation mechanisms, some species-specific differences exist and mechanisms appearing highly conserved should be reconsidered in a given species of interest in order to avoid misunderstanding.

16.3.9 Mono-ovulatory/poly-ovulatory folliculogenesis

Folliculogenesis is the development of the follicle from the primordial stage through a series of morphologically defined stages: primary, secondary, antral, and subsequently culminating in the Graafian or preovulatory mature follicle. The development of ovarian follicle has been differentiated as a two-phase process: the initial recruitment of the follicle from the primordial pool to pre-antral follicles and the cyclic recruitment of the growing follicles, involving gonadotropin-dependent stages of rapid growth from pre-antral to mature Graafian follicles.

On the basis of gene-targeting studies, several factors have been shown to play an important role in the transition from resting primordial follicles to the growing phase (Kuroda et al. 1988; Huang et al. 1993; Carabatsos et al. 1998; Elvin et al. 1999). Among those are *GDF9* and *BMP-15*, two oocyte-secreted factors. Both *BMP-15* and *GDF9* are known to be important determinants of ovulation quota and litter size, whereas homozygous mutations lead to infertility with an arrest at the primary stage of folliculogenesis (Galloway et al. 2000; McNatty et al. 2003; Hanrahan et al. 2004).

The first indication that *BMP-15* may have distinct functions in mono-ovulatory versus poly-ovulatory species came with the development of the *Bmp-15* null mouse. Unlike the *Gdf9* null mice, mice lacking *Bmp-15* exhibit normal folliculogenesis but are sub-fertile due to defects in ovulation and early embryonic development (Yan et al. 2001). In mono-ovulatory sheep and human species, bioactive mature BMP-15 must be present at the primary follicle stage for folliculogenesis to proceed normally because mutations in the *BMP-15* gene in ewes and women cause the arrest of primary follicle growth (Galloway et al. 2000; Otsuka et al. 2000; McNatty et al. 2003; Hanrahan et al. 2004). Whereas ewes homozygous for *BMP-15* mutations are infertile, heterozygous ewes exhibit an increased ovulation quota (Galloway et al. 2000; McNatty et al. 2003; Hanrahan et al. 2004). It has been hypothesized that the poly-ovulatory nature of mice might be associated with the lack (or very low level) of functional Bmp-15 mature protein during folliculogenesis (Moore et al. 2004). However, mice overexpressing *Bmp-15* suffer from an early onset of acyclicity. This indicates that the lack of *Bmp-15* in wild type mice during early folliculogenesis is important in restraining follicle development to prevent a premature decline in the ovarian follicle pool (McMahon et al. 2008).

This example of phenotype difference in mice and ewes lacking *BMP-15* illustrates multiple differences existing in gonadal development within mammals. It thus reinforces the idea that knockout studies in mice could not reflect all human mutation phenotypes and that domestic animals might represent alternative pertinent models for understanding ovarian function.

16.4 Sex determination in nonmammal domestic species

In nonmammal vertebrate and in monotremes, *SRY* orthologs have not been detected and sex-determining signals seem to be different from that of therian mammals and seem to be variable between species. By contrast, sex-differentiating key genes such as *SOX9* for the male pathway and *FOXL2* for the female one were found highly conserved in all studied vertebrates, from fishes, batrachians, reptiles, and birds to mammals (Kent et al. 1996; Morais da Silva et al. 1996; Western et al. 1999; Choudhary et al. 2000; Takase et al. 2000; Vaillant et al. 2001; Valleley et al. 2001; Pask et al. 2002; Loffler et al. 2003; Zhou et al. 2003; Baron et al. 2004; Govoroun et al. 2004; Baron et al. 2005; Rodríguez-Marí et al. 2005; Nakamoto et al. 2006; Takada et al. 2006; Liu et al. 2007a; Rhen et al. 2007; Wotton et al. 2007; Alam et al. 2008; Ijiri et al. 2008). Moreover, for *SOX9*, a *Drosophila* equivalent (Sox100B) has been shown to be involved in the determination of the male-specific somatic gonadal precursors, supporting a throughout conservation of this gene for testis differentiation (DeFalco et al. 2003).

Conclusively, upregulation of *SOX9* and downregulation of female genes such as *FOXL2* and *RSPO1* are well-conserved prerequisites for testis differentiation. By contrast, the ways by which each species controls these genes seem more species-specific (Wilkins 1995).

In chicken, sex is genetically determined but, on the opposite, with therian mammals, female is the heterogametic sex (ZW) while the male is homogametic (ZZ). The avian sex chromosome (ZW) has been shown to be completely different from the mammalian XY system. Indeed, both systems evolve from different ancestral autosomes and do not have any gene in common (Graves and Shetty 2001). Interestingly is the situation in monotremes with the platypus *Ornithorynchus anatinus* as example. It possesses a complex male heterogametic system with five male-specific Y chromosomes and five X chromosomes; the female harbors two copies of the five X (Rens et al. 2004; McMillan et al. 2007). It has recently been shown that in contrast with some previous reports, platypus sex chromosomes share no homology with the ancestral therian X chromosome but display strong homology with the bird ZZ/ZW system (Veyrunes et al. 2008).

In chicken, there is presently no clear evidence in favor of a male Z dosage effect or a female W dominant effect as a primary sex-determining signal. Studies of ZZW triploid animals suggest that the W chromosome carries a female determinant that can be antagonized by the dosage of a Z-linked male-determinant (Smith and Sinclair 2004, and references therein). Importantly, the Z-linked gene *DMRT1* supports the Z-dosage model carrying a testis-determinant gene. Indeed, orthologs of *DMRT1* have been shown to be involved in different aspects of sexual differentiation, from invertebrates (*Drosophila* and *Caenorhabditis*) to human (Raymond et al. 1998). In fishes, reptiles, birds, and mammals, *DMRT1* appears to be involved in testes development, at different levels depending on the considered species (reviewed in Ferguson-Smith 2007). In the medaka fish *Oryzias latipes*, a *DMRT1* ortholog called *DMRT1bY/DMY* has been described on the male Y chromosome and is considered as the testis-determining gene in this species (Matsuda et al. 2002; Nanda et al. 2002). Indeed *DMRT1bY* is expressed exclusively in testis and mutations of this gene cause sex reversal (Kondo et al. 2006).

In addition to the Z-dosage hypothesis with *DMRT1* as a strong putative male-determining gene in chicken, some W female-specific genes strengthen the female W-dominant model. The small W chromosome (like the Y chromosome in mammals) has degenerated during evolution, suggesting that it carried a female sex-determining gene (reviewed in Stiglec et al. 2007). The chicken W chromosome encompasses few genes (~30–40) and many of them have homologs on the Z chromosome (reviewed in Smith 2007). Indeed two genes appear to be strictly specific to the W-chromosome (*FET1* and *2d-2D9*) but different evidence, such as their expression profiles, argue against the role of these genes as being ovarian-determining factors (Smith 2007, and references therein). Presently, the best female-determining candidate gene in chicken remains *HINTW* (also called *ASW* or *WPKCI*). This gene is located on the W chromosome and has a homolog *HINTZ* on the Z chromosome (Hori et al. 2000). *HINT* genes encode a family of nucleotide hydroxylase enzymes that need a histidine triad (HIT motif) to be functional. Interestingly, *HINTZ* has a HIT motif but *HINTW* does not. Moreover, as these *HINT* genes act as dimers, it has been postulated that *HINTW* may act as a dominant negative in avian sex determination by forming heterodimers with *HINTZ* (Hori et al. 2000; Moriyama et al. 2006).

In conclusion on sex determination in chicken, the actual view supposes that *DMRT1* acts as a testis-determining gene, having a dose-dependent effect because it is located on the Z chromosome, being inhibited in female by the presence of a single locus together with a W-located female determinant that could be *HINTW* or another yet uncharacterized W-linked gene.

16.5 Future research directions

Although sex is used as the mode of reproduction in vertebrates, sex-determining signals appear highly variable among the different phyla. Actually, two genes have been equated as sex-determining genes, *SRY* in mammals and *DMRT1bY* in medaka. Fishes represent an inexhaustible resource of mechanisms involved in sex determination. As an example, the sex-determining gene of the platyfish *Xiphophorus maculatus* has been narrowed to a short male-specific region on the Y chromosome. Interestingly, this region encompasses no *DMRT1*-related gene and is evolutionarily different from the sex-determining region of the medaka, and from mammalian and avian sex chromosomes (reviewed in Schultheis et al. 2006 and in Volff et al. 2007).

Even in the closely related group of mammals, some species-specific differences exist, not in the sex-determining mechanism itself, but in some aspects of gonad differentiation (e.g., estrogen production in bovidae compared with that in pig or rodents). Consequently, intending to control sex in a specific species requires a good knowledge of the molecular mechanisms involved in gonadal differentiation in the species of interest.

Presently in domestic mammals, production of only male progeny will be possible simply by adding an *SRY* transgene on the X chromosome of an XY animal. This could be interesting in some breed-specific bovine lines for beef production. Such XY founder animals will, however, present some disadvantages: (1) 50% of the offspring will be sterile, being XX-SRY+ males; (2) such an XY founder cannot reproduce naturally (only via the animal cloning technology), and (3) this founder and 50% of his progeny will fall into the highly controversial Genetically Modified Organism (GMO) category.

Consequently, today, the best way to control sex in a given species will be to create a heterogametic founder carrying an inducible killer transgene on one of his sexual chromosomes. The idea is to specifically eliminate spermatozoa of a given sex constitution (X or Y population depending on which sex chromosome the transgene is located) by activating the transgene in collected semen samples. Such a founder animal will circumvent all three disadvantages linked to an X(SRY+)Y founder and will allow also the production of only-male or only-female fertile and non-GMO progeny. Such strategy could be an alternative to sperm separation on a cell sorter that is used today in different species and that will be surely extended in the future (reviewed in Cran 2007).

Acknowledgments

We thank Jean-Luc Vilotte for the critical reading of this manuscript.

References

Adams, I. and McLaren, A. 2002. Sexually dimorphic development of mouse primordial germ cells: Switching from oogenesis to spermatogenesis. *Development* **129**: 1155–1164.

Alam, M., Kobayashi, Y., Horiguchi, R., Hirai, T., and Nakamura, M. 2008. Molecular cloning and quantitative expression of sexually dimorphic markers Dmrt1 and Foxl2 during female-to-male sex change in Epinephelus merra. *General and Compartive Endocrinology* **157**: 75–85.

Albrecht, K. and Eicher, E. 2001. Evidence that Sry is expressed in pre-Sertoli cells and Sertoli and granulosa cells have a common precursor. *Developmental Biology* **240**: 92–107.

Arnold, A. and Gorski, R. 1984. Gonadal steroid induction of structural sex differences in the central nervous system. *Annual Review of Neuroscience* **7**: 413–442.

Asdell, S. 1944. The genetic sex of intersexual goats and a probable linkage with the gene for hornlessness. *Science* **99**: 124.

Bagheri-Fam, S., Sim, H., Bernard, P., Jayakody, I., Taketo, M., Scherer, G., and Harley, V. 2008. Loss of Fgfr2 leads to partial XY sex reversal. *Developmental Biology* **314**: 71–83.

Baroiller, J.-F., Guiguen, Y., and Fostier, A. 1999. Endocrine and environmental aspects of sex differentiation in fish. *Cellular and Molecular Life Sciences* **55**: 910–931.

Baron, D., Cocquet, J., Xia, X., Fellous, M., Guiguen, Y., and Veitia, R. 2004. An evolutionary and functional analysis of FoxL2 in rainbow trout gonad differentiation. *Journal of Molecular Endocrinology* **33**: 705–715.

Baron, D., Houlgatte, R., Fostier, A., and Guiguen, Y. 2005. Large-scale temporal gene expression profiling during gonadal differentiation and early gametogenesis in rainbow trout. *Biology of Reproduction* **73**: 959–966.

Bishop, C., Whitworth, D., Qin, Y., Agoulnik, A., Agoulnik, I., Harrison, W., Behringer, R., and Overbeek, P. 2000. A transgenic insertion upstream of sox9 is associated with dominant XX sex reversal in the mouse. *Nature Genetics* **26**: 490–494.

Boulanger, L., Kocer, A., Daniel, N., Pannetier, M., Chesné, P., Heyman, Y., Renault, L., Mandon-Pépin, B., Chavatte-Palmer, P., Vignon, X., Vilotte, J.-L., Cotinot, C., Renard, J.-P., and Pailhoux, E. 2008. Attempt to rescue sex-reversal by transgenic expression of the PISRT1 gene in XX $PIS^{-/-}$ goats. *Sex Devevelopment* **2**: 142–151.

Boyer, A., Lussier, J., Sinclair, A., McClive, P., and Silversides, D. 2004. Pre-sertoli specific gene expression profiling reveals differential expression of Ppt1 and Brd3 genes within the mouse genital ridge at the time of sex determination. *Biology of Reproduction* **71**: 820–827.

Brennan, J., Tilmann, C., and Capel, B. 2003. Pdgfr-alpha mediates testis cord organization and fetal Leydig cell development in the XY gonad. *Genes Development* **17**: 800–810.

Bullejos, M. and Koopman, P. 2001. Spatially dynamic expression of Sry in mouse genital ridges. *Developmental Dynamics* **221**: 201–205.

Bullejos, M. and Koopman, P. 2005. Delayed Sry and Sox9 expression in developing mouse gonads underlies B6-Y(DOM) sex reversal. *Developmental Biology* **278**: 473–481.

Buoen, L., Zhang, T., Weber, A., and Ruth, G. 2000. SRY-negative, XX intersex horses: The need for pedigree studies to examine the mode of inheritance of the condition. *Equine Veterinary Journal* **32**: 78–81.

Burgoyne, P., Buehr, M., Koopman, P., Rossant, J., and McLaren, A. 1988. Cell-autonomous action of the testis-determining gene: Sertoli cells are exclusively XY in XX—XY chimaeric mouse testes. *Development* **102**: 443–450.

Capel, B., Albrecht, K., Washburn, L., and Eicher, E. 1999. Migration of mesonephric cells into the mammalian gonad depends on Sry. *Mechanisms of Development* **84**: 127–131.

Carabatsos, M., Elvin, J., Matzuk, M., and Albertini, D. 1998. Characterization of oocyte and follicle development in growth differentiation factor-9-deficient mice. *Developmental Biology* **204**: 373–384.

Chaboissier, M., Kobayashi, A., Vidal, V., Lützkendorf, S., van de Kant, H., Wegner, M., de Rooij, D., Behringer, R., and Schedl, A. 2004. Functional analysis of Sox8 and Sox9 during sex determination in the mouse. *Development* **131**: 1891–1901.

Chassot, A., Ranc, F., Gregoire, E., Roepers-Gajadien, H., Taketo, M., Camerino, G., de Rooij, D., Schedl, A., and Chaboissier, M. 2008. Activation of beta-catenin signaling by Rspo1 controls differentiation of the mammalian ovary. *Human Molecular Genetics* **17**: 1264–1277.

Choudhary, B., Ganesh, S., and Raman, R. 2000. Evolutionary conservation of the gene Cvsox9 in the lizard, Calotes versicolor, and its expression during gonadal differentiation. *Development Genes and Evolution* **210**: 250–257.

Clark, A., Garland, K., and Russell, L. 2000. Desert hedgehog (Dhh) gene is required in the mouse testis for formation of adult-type Leydig cells and normal development of peritubular cells and seminiferous tubules. *Biology of Reproduction* **63**: 1825–1838.

Cocquet, J., Pailhoux, E., Jaubert, F., Servel, N., Xia, X., Pannetier, M., De Baere, E. et al. 2002. Evolution and expression of FOXL2. *Journal of Medical Genetics* **39**: 916–921.

Colvin, J., Green, R., Schmahl, J., Capel, B., and Ornitz, D. 2001. Male-to-female sex reversal in mice lacking fibroblast growth factor 9. *Cell* **104**: 875–889.

Conley, A. and Hinshelwood, M. 2001. Mammalian aromatases. *Reproduction* **121**: 685–695.

Coveney, D., Shaw, G., and Renfree, M. 2001. Estrogen-induced gonadal sex reversal in the tammar wallaby. *Biology of Reproduction* **65**: 613–621.

Cran, D. 2007. XY sperm separation and use in artificial insemination and other ARTs. *Society of Reproduction and Fertility Supplement* **65**: 475–491.

Crisponi, L., Deiana, M., Loi, A., Chiappe, F., Uda, M., Amati, P., Bisceglia, L. et al. 2001. The putative forkhead transcription factor FOXL2 is mutated in blepharophimosis/ptosis/epicanthus inversus syndrome. *Nature Genetics* **27**: 159–166.

Daneau, I., Ethier, J., Lussier, J., and Silversides, D. 1996. Porcine SRY gene locus and genital ridge expression. *Biology of Reproduction* **55**: 47–53.

De Baere, E., Beysen, D., Oley, C., Lorenz, B., Cocquet, J., De Sutter, P., Devriendt, K. et al. 2003. FOXL2 and BPES: Mutational hotspots, phenotypic variability, and revision of the genotype-phenotype correlation. *American Journal of Human Genetics* **72**: 478–487.

DeFalco, T., Verney, G., Jenkins, A., McCaffery, J., Russell, S., and Van Doren, M. 2003. Sex-specific apoptosis regulates sexual dimorphism in the Drosophila embryonic gonad. *Developmental Cell* **5**: 205–216.

DiNapoli, L. and Capel, B. 2008. SRY and the standoff in sex determination. *Molecular Endocrinology* **22**: 1–9.

Dubin, R. and Ostrer, H. 1994. Sry is a transcriptional activator. *Molecular Endocrinology* **8**: 1182–1192.

Elbrecht, A. and Smith, R. 1992. Aromatase enzyme activity and sex determination in chickens. *Science* **255**: 467–470.

Elvin, J., Yan, C., Wang, P., Nishimori, K. and Matzuk, M. 1999. Molecular characterization of the follicle defects in the growth differentiation factor 9-deficient ovary. *Molecular Endocrinology* **13**: 1018–1034.

Ferguson-Smith, M. 2007. The evolution of sex chromosomes and sex determination in vertebrates and the key role of DMRT1. *Sex Dev* **1**: 2–11.

Fishelson, L. 1970. Protogynous sex reversal in the fish Anthias squamipinnis (Teleostei, Anthiidae) regulated by the presence or absence of a male fish. *Nature* **227**: 90–91.

Ford, C., Jones, K., Miller, O., Mittwoch, U., Penrose, L., Ridler, M., and Shapiro, A. 1959. The chromosomes in a patient showing both mongolism and the Klinefelter syndrome. *Lancet* **1**: 709–710.

Foster, J. 1996. Mutations in SOX9 cause both autosomal sex reversal and campomelic dysplasia. *Acta Paediatr Japonica* **38**: 405–411.

Foster, J., Dominguez-Steglich, M., Guioli, S., Kowk, G., Weller, P., Stevanović, M., Weissenbach, J., Mansour, S., Young, I., and Goodfellow, P. 1994. Campomelic dysplasia and autosomal sex reversal caused by mutations in an SRY-related gene. *Nature* **372**: 525–530.

Galloway, S., McNatty, K., Cambridge, L., Laitinen, M., Juengel, J., Jokiranta, T., McLaren, R. et al. 2000. Mutations in an oocyte-derived growth factor gene (BMP15) cause increased ovulation rate and infertility in a dosage-sensitive manner. *Nature Genetics* **25**: 279–283.

Gerstner, G. and Parekh, V. 1997. Evidence of sex-specific differences in masticatory jaw movement patterns. *Journal of Dental Research* **76**: 796–806.

Gessler, M., Poustka, A., Cavenee, W., Neve, R., Orkin, S., and Bruns, G. 1990. Homozygous deletion in Wilms tumours of a zinc-finger gene identified by chromosome jumping. *Nature* **343**: 774–778.

Gorski, R. 1984. Critical role for the medial preoptic area in the sexual differentiation of the brain. *Progress and Brain Research* **61**: 129–146.

Govoroun, M., Pannetier, M., Pailhoux, E., Cocquet, J., Brillard, J., Couty, I., Batellier, F., and Cotinot, C. 2004. Isolation of chicken homolog of the FOXL2 gene and comparison of its expression patterns with those of aromatase during ovarian development. *Developmental Dynamics* **231**: 859–870.

Graves, M.J. and Shetty, S. 2001. Sex from W to Z: Evolution of vertebrate sex chromosomes and sex determining genes. *Journal of Experimental Zoology* **290**: 449–462.

Grimmond, S., Van Hateren, N., Siggers, P., Arkell, R., Larder, R., Soares, M., de Fatima Bonaldo, M. et al. 2000. Sexually dimorphic expression of protease nexin-1 and vanin-1 in the developing mouse gonad prior to overt differentiation suggests a role in mammalian sexual development. *Human Molecular Genetics* **9**: 1553–1560.

Guiguen, Y., Baroiller, J., Ricordel, M., Iseki, K., Mcmeel, O., Martin, S., and Fostier, A. 1999. Involvement of estrogens in the process of sex differentiation in two fish species: The rainbow trout (Oncorhynchus mykiss) and a tilapia (Oreochromis niloticus). *Molecular Reproduction and Development* **54**: 154–162.

Gustafsson, J., Mode, A., Norstedt, G., and Skett, P. 1983. Sex steroid induced changes in hepatic enzymes. *Annual Review of Physiology* **45**: 51–60.

Hacker, A., Capel, B., Goodfellow, P., and Lovell-Badge, R. 1995. Expression of Sry, the mouse sex determining gene. *Development* **121**: 1603–1614.

Hanley, N., Hagan, D., Clement-Jones, M., Ball, S., Strachan, T., Salas-Cortés, L., McElreavey, K. et al. 2000. SRY, SOX9, and DAX1 expression patterns during human sex determination and gonadal development. *Mechanisms of Development* **91**: 403–407.

Hanrahan, J., Gregan, S., Mulsant, P., Mullen, M., Davis, G., Powell, R., and Galloway, S. 2004. Mutations in the genes for oocyte-derived growth factors GDF9 and BMP15 are associated with both increased ovulation rate and sterility in Cambridge and Belclare sheep (Ovis aries). *Biology of Reproduction* **70**: 900–909.

Harley, V., Jackson, D., Hextall, P., Hawkins, J., Berkovitz, G., Sockanathan, S., Lovell-Badge, R., and Goodfellow, P. 1992. DNA binding activity of recombinant SRY from normal males and XY females. *Science* **255**: 453–456.

Harry, J., Koopman, P., Brennan, F., Graves, J., and Renfree, M. 1995. Widespread expression of the testis-determining gene SRY in a marsupial. *Nature Genetics* **11**: 347–349.

Hori, T., Asakawa, S., Itoh, Y., Shimizu, N., and Mizuno, S. 2000. Wpkci, encoding an altered form of PKCI, is conserved widely on the avian W chromosome and expressed in early female embryos: Implication of its role in female sex determination. *Molecular Biology of the Cell* **11**: 3645–3660.

Huang, B., Wang, S., Ning, Y., Lamb, A., and Bartley, J. 1999. Autosomal XX sex reversal caused by duplication of SOX9. *American Journal of Medical Genetics* **87**: 349–353.

Huang, E., Manova, K., Packer, A., Sanchez, S., Bachvarova, R., and Besmer, P. 1993. The murine steel panda mutation affects kit ligand expression and growth of early ovarian follicles. *Developmental Biology* **157**: 100–109.

Hudson, Q., Smith, C., and Sinclair, A. 2005. Aromatase inhibition reduces expression of FOXL2 in the embryonic chicken ovary. *Developmental Dynamics* **233**: 1052–1055.

Ijiri, S., Kaneko, H., Kobayashi, T., Wang, D., Sakai, F., Paul-Prasanth, B., Nakamura, M., and Nagahama, Y. 2008. Sexual dimorphic expression of genes in gonads during early differentiation of a teleost fish, the Nile tilapia Oreochromis niloticus. *Biology of Reproduction* **78**: 333–341.

Jacobs, P. and Strong, J. 1959. A case of human intersexuality having a possible XXY sex-determining mechanism. *Nature* **183**: 302–303.

Josso, N., Lamarre, I., Picard, J., Berta, P., Davies, N., Morichon, N., Peschanski, M., and Jeny, R. 1993. Anti-müllerian hormone in early human development. *Early Human Development* **33**: 91–99.

Jost, A. 1947. Recherches sur la différenciation sexuelle de l'embryon de lapin. *Archives d'Anatomie Microscopique et de Morphologie Expérimentale* **36**: 271–315.

Just, W., Rau, W., Vogel, W., Akhverdian, M., Fredga, K., Graves, J., and Lyapunova, E. 1995. Absence of Sry in species of the vole Ellobius. *Nature Genetics* **11**: 117–118.

Kent, J., Wheatley, S., Andrews, J., Sinclair, A., and Koopman, P. 1996. A male-specific role for SOX9 in vertebrate sex determination. *Development* **122**: 2813–2822.

Kim, K., Wagle, M., Tran, K., Zhan, X., Dixon, M., Liu, S., Gros, D. et al. 2008. R-spondin family members regulate the wnt pathway by a common mechanism. *Molecular Biology of the Cell* **19**: 2588–2596.

Kim, Y., Bingham, N., Sekido, R., Parker, K., Lovell-Badge, R., and Capel, B. 2007. Fibroblast growth factor receptor 2 regulates proliferation and Sertoli differentiation during male sex determination. *Proceedings of the National Academy of Sciences of the United States of America* **104**: 16558–16563.

Kim, Y., Kobayashi, A., Sekido, R., DiNapoli, L., Brennan, J., Chaboissier, M., Poulat, F., Behringer, R., Lovell-Badge, R., and Capel, B. 2006. Fgf9 and Wnt4 act as antagonistic signals to regulate mammalian sex determination. *PLoS Biology* **4**: e187.

Kocer, A., Pinheiro, I., Pannetier, M., Renault, L., Parma, P., Radi, O., Kim, K., Camerino, G., and Pailhoux, E. 2008. R-spondin1 and FOXL2 act into two distinct cellular types during goat ovarian differentiation. *BMC Developmental Biology* **8**: 36.

Kondo, M., Hornung, U., Nanda, I., Imai, S., Sasaki, T., Shimizu, A., Asakawa, S. et al. 2006. Genomic organization of the sex-determining and adjacent regions of the sex chromosomes of medaka. *Genome Research* **16**: 815–826.

Koopman, P., Münsterberg, A., Capel, B., Vivian, N., and Lovell-Badge, R. 1990. Expression of a candidate sex-determining gene during mouse testis differentiation. *Nature* **348**: 450–452.

Krisfalusi, M. and Cloud, J. 1999. Gonadal sex reversal in triploid rainbow trout (Oncorhynchus mykiss). *Journal of Experimental Zoology* **284**: 466–472.

Kuroda, H., Terada, N., Nakayama, H., Matsumoto, K., and Kitamura, Y. 1988. Infertility due to growth arrest of ovarian follicles in Sl/Slt mice. *Developmental Biology* **126**: 71–79.

Liu, J., Liu, S., Tao, M., Li, W., and Liu, Y. 2007a. Isolation and expression analysis of testicular type Sox9b in allotetraploid fish. *Marine Biotechnology (New York)* **9**: 329–334.

Liu, Z., Wu, F., Jiao, B., Zhang, X., Hu, C., Huang, B., Zhou, L. et al. 2007b. Molecular cloning of doublesex and mab-3-related transcription factor 1, forkhead transcription factor gene 2, and two types of cytochrome P450 aromatase in Southern catfish and their possible roles in sex differentiation. *The Journal of Endocrinology* **194**: 223–241.

Loffler, K., Zarkower, D., and Koopman, P. 2003. Etiology of ovarian failure in blepharophimosis ptosis epicanthus inversus syndrome: FOXL2 is a conserved, early-acting gene in vertebrate ovarian development. *Endocrinology* **144**: 3237–3243.

Malki, S., Bibeau, F., Notarnicola, C., Roques, S., Berta, P., Poulat, F., and Boizet-Bonhoure, B. 2007. Expression and biological role of the prostaglandin D synthase/SOX9 pathway in human ovarian cancer cells. *Cancer Letters* **255**: 182–193.

Malki, S., Nef, S., Notarnicola, C., Thevenet, L., Gasca, S., Méjean, C., Berta, P., Poulat, F., and Boizet-Bonhoure, B. 2005. Prostaglandin D2 induces nuclear import of the sex-determining factor SOX9 via its cAMP-PKA phosphorylation. *The EMBO Journal* **24**: 1798–1809.

Martineau, J., Nordqvist, K., Tilmann, C., Lovell-Badge, R., and Capel, B. 1997. Male-specific cell migration into the developing gonad. *Current Biology* **7**: 958–968.

Matsuda, M., Nagahama, Y., Shinomiya, A., Sato, T., Matsuda, C., Kobayashi, T., Morrey, C. et al. 2002. DMY is a Y-specific DM-domain gene required for male development in the medaka fish. *Nature* **417**: 559–563.

Mauléon, P., Bézard, J., and Terqui, M. 1977. Very early and transient 17β-estradiol secretion by fetal sheep ovary. *In vito* study. *Annales de Biologie Animale, Biochimie, Biophysique* **17**: 399–401.

McElreavey, K., Vilain, E., Abbas, N., Herskowitz, I., and Fellous, M. 1993. A regulatory cascade hypothesis for mammalian sex determination: SRY represses a negative regulator of male development. *Proceedings of the National Academy of Sciences of the United States of America* **90**: 3368–3372.

McEwen, B. 1992. Steroid hormones: Effect on brain development and function. *Hormone Research* **37**(Supplement 3): 1–10.

McLaren, A. 1988. Somatic and germ-cell sex in mammals. *Philosophical Transctions of the Royal Society of London. Series B, Biological Sciences* **322**: 3–9.

McMahon, H., Hashimoto, O., Mellon, P., and Shimasaki, S. 2008. Oocyte-specific overexpression of mouse bone morphogenetic protein-15 leads to accelerated folliculogenesis and an early onset of acyclicity in transgenic mice. *Endocrinology* **149**: 2807–2815.

McMillan, D., Miethke, P., Alsop, A., Rens, W., O'Brien, P., Trifonov, V., Veyrunes, F. et al. 2007. Characterizing the chromosomes of the platypus (Ornithorhynchus anatinus). *Chromosome Research* **15**: 961–974.

McNatty, K., Juengel, J., Wilson, T., Galloway, S., Davis, G., Hudson, N., Moeller, C. et al. 2003. Oocyte-derived growth factors and ovulation rate in sheep. *Reproduction Supplement* **61**: 339–351.

Meyers-Wallen, V. 2003. Sry and Sox9 expression during canine gonadal sex determination assayed by quantitative reverse transcription-polymerase chain reaction. *Molecular Reproduction and Development* **65**: 373–381.

Meyers-Wallen, V., Schlafer, D., Barr, I., Lovell-Badge, R., and Keyzner, A. 1999. Sry-negative XX sex reversal in purebred dogs. *Molecular Reproduction and Development* **53**: 266–273.

Mitchell, C. and Harley, V. 2002. Biochemical defects in eight SRY missense mutations causing XY gonadal dysgenesis. *Molecular Genetics and Metabolism* **77**: 217-225.

Moore, R., Erickson, G., and Shimasaki, S. 2004. Are BMP-15 and GDF-9 primary determinants of ovulation quota in mammals? *Trends in Endocrinology and Metabolism* **15**: 356–361.

Morais da Silva, S., Hacker, A., Harley, V., Goodfellow, P., Swain, A., and Lovell-Badge, R. 1996. Sox9 expression during gonadal development implies a conserved role for the gene in testis differentiation in mammals and birds. *Nature Genetics* **14**: 62–68.

Moriyama, S., Ogihara, J., Kato, J., Hori, T., and Mizuno, S. 2006. PKCI-W forms a heterodimer with PKCI-Z and inhibits the biological activities of PKCI-Z in vitro, supporting the predicted role of PKCI-W in sex determination in birds. *Journal of Biochemistry* **139**: 91–97.

Nakabayashi, O., Kikuchi, H., Kikuchi, T., and Mizuno, S. 1998. Differential expression of genes for aromatase and estrogen receptor during the gonadal development in chicken embryos. *Journal of Molecular Endocrinology* **20**: 193–202.

Nakamoto, M., Matsuda, M., Wang, D., Nagahama, Y., and Shibata, N. 2006. Molecular cloning and analysis of gonadal expression of Foxl2 in the medaka, Oryzias latipes. *Biochemical and Biophysical Research Communications* **344**: 353–361.

Nakamoto, M., Wang, D., Suzuki, A., Matsuda, M., Nagahama, Y., and Shibata, N. 2007. Dax1 suppresses P450arom expression in medaka ovarian follicles. *Molecular Reproduction and Development* **74**: 1239–1246.

Nanda, I., Kondo, M., Hornung, U., Asakawa, S., Winkler, C., Shimizu, A., Shan, Z. et al. 2002. A duplicated copy of DMRT1 in the sex-determining region of the Y chromosome of the medaka, Oryzias latipes. *Proceedings of the National Academy of Sciences of the United States of America* **99**: 11778–11783.

Nef, S., Schaad, O., Stallings, N., Cederroth, C., Pitetti, J., Schaer, G., Malki, S. et al. 2005. Gene expression during sex determination reveals a robust female genetic program at the onset of ovarian development. *Developmental Biology* **287**: 361–377.

Nef, S., Verma-Kurvari, S., Merenmies, J., Vassalli, J., Efstratiadis, A., Accili, D., and Parada, L. 2003. Testis determination requires insulin receptor family function in mice. *Nature* **426**: 291–295.

Ng, L., Wheatley, S., Muscat, G., Conway-Campbell, J., Bowles, J., Wright, E., Bell, D., Tam, P., Cheah, K., and Koopman, P. 1997. SOX9 binds DNA, activates transcription, and coexpresses with type II collagen during chondrogenesis in the mouse. *Development Biology* **183**: 108–121.

Nordqvist, K. 1995. Sex differentiation—Gonadogenesis and novel genes. *The International Journal of Developmental Biology* **39**: 727–736.

Nordqvist, K. and Töhönen, V. 1997. An mRNA differential display strategy for cloning genes expressed during mouse gonad development. *The International Journal of Developmental Biology* **41**: 627–638.

Ohe, K., Lalli, E., and Sassone-Corsi, P. 2002. A direct role of SRY and SOX proteins in pre-mRNA splicing. *Proceedings of the National Academy of Sciences of the United States of America* **99**: 1146–1151.

Olesen, C., Nyeng, P., Kalisz, M., Jensen, T., Møller, M., Tommerup, N., and Byskov, A. 2007. Global gene expression analysis in fetal mouse ovaries with and without meiosis and comparison of selected genes with meiosis in the testis. *Cell and Tissue Research* **328**: 207–221.

Otsuka, F., Yao, Z., Lee, T., Yamamoto, S., Erickson, G., and Shimasaki, S. 2000. Bone morphogenetic protein-15. Identification of target cells and biological functions. *The Journal of Biological Chemistry* **275**: 39523–39528.

Ottolenghi, C., Omari, S., Garcia-Ortiz, J., Uda, M., Crisponi, L., Forabosco, A., Pilia, G., and Schlessinger, D. 2005. Foxl2 is required for commitment to ovary differentiation. *Human Molecular Genetics* **14**: 2053–2062.

Ottolenghi, C., Pelosi, E., Tran, J., Colombino, M., Douglass, E., Nedorezov, T., Cao, A., Forabosco, A., and Schlessinger, D. 2007. Loss of Wnt4 and Foxl2 leads to female-to-male sex reversal extending to germ cells. *Human Molecular Genetics* **16**: 2795–2804.

Pailhoux, E., Vigier, B., Chaffaux, S., Servel, N., Taourit, S., Furet, J., Fellous, M. et al. 2001a. A 11.7-kb deletion triggers intersexuality and polledness in goats. *Nature Genetics* **29**: 453–458.

Pailhoux, E., Mandon-Pepin, B., and Cotinot, C. 2001b. Mammalian gonadal differentiation: The pig model. *Reprod Suppl* **58**: 65–80.

Pailhoux, E., Popescu, P., Parma, P., Boscher, J., Legault, C., Molteni, L., Fellous, M., and Cotinot, C. 1994. Genetic analysis of 38XX males with genital ambiguities and true hermaphrodites in pigs. *Animal Genetics* **25**: 299–305.

Pailhoux, E., Vigier, B., Schibler, L., Cribiu, E., Cotinot, C., and Vaiman, D. 2005. Positional cloning of the PIS mutation in goats and its impact on understanding mammalian sex-differentiation. *Genetics, Selection, Evolution* **37**(Supplement 1): S55–64.

Pailhoux, E., Vigier, B., Vaiman, D., Servel, N., Chaffaux, S., Cribiu, E., and Cotinot, C. 2002. Ontogenesis of female-to-male sex-reversal in XX polled goats. *Developmental Dynamics* **224**: 39–50.

Palmer, S. and Burgoyne, P. 1991. In situ analysis of fetal, prepuberal and adult XX—XY chimaeric mouse testes: Sertoli cells are predominantly, but not exclusively, XY. *Development* **112**: 265–268.

Pannetier, M., Tilly, G., Kocer, A., Hudrisier, M., Renault, L., Chesnais, N., Costa, J. et al. 2006a. Goat SRY induces testis development in XX transgenic mice. *FEBS Letters* **580**: 3715–3720.

Pannetier, M., Fabre, S., Batista, F., Kocer, A., Renault, L., Jolivet, G., Mandon-Pépin, B., Cotinot, C., Veitia, R., and Pailhoux, E. 2006b. FOXL2 activates P450 aromatase gene transcription: Towards a better characterization of the early steps of mammalian ovarian development. *Journal of Molecular Endocrinology* **36**: 399–413.

Pannetier, M., Renault, L., Jolivet, G., Cotinot, C., and Pailhoux, E. 2005. Ovarian-specific expression of a new gene regulated by the goat PIS region and transcribed by a FOXL2 bidirectional promoter. *Genomics* **85**: 715–726.

Parma, P., Pailhoux, E., and Cotinot, C. 1999. Reverse transcription-polymerase chain reaction analysis of genes involved in gonadal differentiation in pigs. *Biology of Reproduction* **61**: 741–748.

Parma, P., Radi, O., Vidal, V., Chaboissier, M., Dellambra, E., Valentini, S., Guerra, L., Schedl, A., and Camerino, G. 2006. R-spondin1 is essential in sex determination, skin differentiation and malignancy. *Nature Genetics* **38**: 1304–1309.

Pask, A., Harry, J., Graves, J., O'Neill, R., Layfield, S., Shaw, G., and Renfree, M. 2002. SOX9 has both conserved and novel roles in marsupial sexual differentiation. *Genesis* **33**: 131–139.

Patek, C., Kerr, J., Gosden, R., Jones, K., Hardy, K., Muggleton-Harris, A., Handyside, A., Whittingham, D., and Hooper, M. 1991. Sex chimaerism, fertility and sex determination in the mouse. *Development* **113**: 311–325.

Payen, E. and Cotinot, C. 1994. Sequence evolution of SRY gene within Bovidae family. *Mammalian Genome* **5**: 723–725.

Payen, E., Pailhoux, E., Abou Merhi, R., Gianquinto, L., Kirszenbaum, M., Locatelli, A., and Cotinot, C. 1996. Characterization of ovine SRY transcript and developmental expression of genes involved in sexual differentiation. *The International Journal of Developmental Biology* **40**: 567–575.

Pelletier, J., Bruening, W., Li, F., Haber, D., Glaser, T., and Housman, D. 1991. WT1 mutations contribute to abnormal genital system development and hereditary Wilms' tumour. *Nature* **353**: 431–434.

Pelliniemi, L. 1975. Ultrastructure of gonadal ridge in male and female pig embryos. *Anatomy and Embryology* **147**: 20–34.

Pelliniemi, L. 1985. Sexual differentiation of the pig gonad. *Archives d'Anatomie Microscopique et de Morphologie Expérimentale* **74**: 76–80.

Perera, E., Martin, H., Seeherunvong, T., Kos, L., Hughes, I., Hawkins, J., and Berkovitz, G. 2001. Tescalcin, a novel gene encoding a putative EF-hand Ca(2+)-binding protein, Col9a3, and renin are expressed in the mouse testis during the early stages of gonadal differentiation. *Endocrinology* **142**: 455–463.

Pieau, C. and Dorizzi, M. 2004. Oestrogens and temperature-dependent sex determination in reptiles: all is in the gonads. *The Journal of Endocrinology* **181**: 367–377.

Pieau, C., Dorizzi, M., and Richard-Mercier, N. 1999. Temperature-dependent sex determination and gonadal differentiation in reptiles. *Cellular and Molecular Life Sciences* **55**: 887–900.

Pierucci-Alves, F., Clark, A., and Russell, L. 2001. A developmental study of the Desert hedgehog-null mouse testis. *Biology of Reproduction* **65**: 1392–1402.

Pisarska, M., Bae, J., Klein, C., and Hsueh, A. 2004. Forkhead l2 is expressed in the ovary and represses the promoter activity of the steroidogenic acute regulatory gene. *Endocrinology* **145**: 3424–3433.

Polanco, J. and Koopman, P. 2007. Sry and the hesitant beginnings of male development. *Developmental Biology* **302**: 13–24.

Pontiggia, A., Rimini, R., Harley, V., Goodfellow, P., Lovell-Badge, R., and Bianchi, M. 1994. Sex-reversing mutations affect the architecture of SRY-DNA complexes. *The EMBO Journal* **13**: 6115–6124.

Poulat, F., Barbara, P., Desclozeaux, M., Soullier, S., Moniot, B., Bonneaud, N., Boizet, B., and Berta, P. 1997. The human testis determining factor SRY binds a nuclear factor containing PDZ protein interaction domains. *The Journal of Biological Chemistry* **272**: 7167–7172.

Qin, Y. and Bishop, C. 2005. Sox9 is sufficient for functional testis development producing fertile male mice in the absence of Sry. *Human Molecular Genetics* **14**: 1221–1229.

Raeside, J., Wilkinson, C., and Farkas, G. 1993. Ontogenesis of estrogen secretion by porcine fetal testes. *Acta Endocrinologica (Copenhagen)* **128**: 549–554.

Raymond, C., Shamu, C., Shen, M., Seifert, K., Hirsch, B., Hodgkin, J., and Zarkower, D. 1998. Evidence for evolutionary conservation of sex-determining genes. *Nature* **391**: 691–695.

Rens, W., Grützner, F., O'brien, P., Fairclough, H., Graves, J., and Ferguson-Smith, M. 2004. Resolution and evolution of the duck-billed platypus karyotype with an X1Y1X2Y2X3Y3X4Y4X5Y5 male sex chromosome constitution. *Proceedings of the National Academy of Sciences of the United States of America* **101**: 16257–16261.

Rhen, T., Metzger, K., Schroeder, A., and Woodward, R. 2007. Expression of putative sex-determining genes during the thermosensitive period of gonad development in the snapping turtle, Chelydra serpentina. *Sex Dev* **1**: 255–270.

Robins, D. 2005. Androgen receptor and molecular mechanisms of male-specific gene expression. *Novartis Foundation Symposium* **268**: 42–52; discussion 53-46, 96–49.

Rodríguez-Marí, A., Yan, Y., Bremiller, R., Wilson, C., Cañestro, C., and Postlethwait, J. 2005. Characterization and expression pattern of zebrafish Anti-Müllerian hormone (Amh) relative to sox9a, sox9b, and cyp19a1a, during gonad development. *Gene Expression Patterns* **5**: 655–667.

Roy, A. and Chatterjee, B. 1983. Sexual dimorphism in the liver. *Annual Review of Physiology* **45**: 37–50.

Salas-Cortés, L., Jaubert, F., Barbaux, S., Nessmann, C., Bono, M., Fellous, M., McElreavey, K., and Rosemblatt, M. 1999. The human SRY protein is present in fetal and adult Sertoli cells and germ cells. *The International Journal of Developmental Biology* **43**: 135–140.

Scheib, D. 1983. Effects and role of estrogens in avian gonadal differentiation. *Differentiation* **23**(Supplement): S87–92.

Schibler, L., Cribiu, E., Oustry-Vaiman, A., Furet, J., and Vaiman, D. 2000. Fine mapping suggests that the goat Polled Intersex Syndrome and the human Blepharophimosis Ptosis Epicanthus Syndrome map to a 100-kb homologous region. *Genome Research* **10**: 311–318.

Schmahl, J., Eicher, E., Washburn, L., and Capel, B. 2000. Sry induces cell proliferation in the mouse gonad. *Development* **127**: 65–73.

Schmahl, J., Kim, Y., Colvin, J., Ornitz, D., and Capel, B. 2004. Fgf9 induces proliferation and nuclear localization of FGFR2 in Sertoli precursors during male sex determination. *Development* **131**: 3627–3636.

Schmidt, D., Ovitt, C., Anlag, K., Fehsenfeld, S., Gredsted, L., Treier, A., and Treier, M. 2004. The murine winged-helix transcription factor Foxl2 is required for granulosa cell differentiation and ovary maintenance. *Development* **131**: 933–942.

Schultheis, C., Zhou, Q., Froschauer, A., Nanda, I., Selz, Y., Schmidt, C., Matschl, S. et al. 2006. Molecular analysis of the sex-determining region of the platyfish Xiphophorus maculatus. *Zebrafish* **3**: 299–309.

Sekido, R., Bar, I., Narváez, V., Penny, G., and Lovell-Badge, R. 2004. SOX9 is up-regulated by the transient expression of SRY specifically in Sertoli cell precursors. *Developmental Biology* **274**: 271–279.

Sekido, R. and Lovell-Badge, R. 2008. Sex determination involves synergistic action of SRY and SF1 on a specific Sox9 enhancer. *Nature* **453**: 930–934.

Sinclair, A., Berta, P., Palmer, M., Hawkins, J., Griffiths, B., Smith, M., Foster, J., Frischauf, A., Lovell-Badge, R., and Goodfellow, P. 1990. A gene from the human sex-determining region encodes a protein with homology to a conserved DNA-binding motif. *Nature* **346**: 240–244.

Small, C., Shima, J., Uzumcu, M., Skinner, M., and Griswold, M. 2005. Profiling gene expression during the differentiation and development of the murine embryonic gonad. *Biology of Reproduction* **72**: 492–501.

Smith, C. 2007. Sex determination in birds: HINTs from the W sex chromosome? *Sex Dev* **1**: 279–285.

Smith, C., Shoemaker, C., Roeszler, K., Queen, J., Crews, W., and Sinclair A. 2008. Cloning and expression of *R-Spondin1* in different vertebrates suggests a conserved role in ovarian development. *BMC Developmental Biology* **8**: 72.

Smith, C. and Sinclair, A. 2004. Sex determination: Insights from the chicken. *Bioessays* **26**: 120–132.

Smyk, M., Obersztyn, E., Nowakowska, B., Bocian, E., Cheung, S., Mazurczak, T., and Stankiewicz, P. 2007. Recurrent SOX9 deletion campomelic dysplasia due to somatic mosaicism in the father. *American Journal of Medical Genetics. Part A* **143A**: 866–870.

Soullier, S., Hanni, C., Catzeflis, F., Berta, P., and Laudet, V. 1998. Male sex determination in the spiny rat Tokudaia osimensis (Rodentia: Muridae) is not Sry dependent. *Mammalian Genome* **9**: 590–592.

Stahl, A. and Carlon, N. 1973. Morphogenesis of the sex cords and the significance of the medullary zone of the gonad in the chick embryo. *Acta Anatomica (Basel)* **85**: 248–274.

Stiglec, R., Ezaz, T., and Graves, J. 2007. Reassignment of chicken W chromosome sequences to the Z chromosome by fluorescence in situ hybridization (FISH). *Cytogenetic and Genome Research* **116**: 132–134.

Sutou, S., Mitsui, Y., and Tsuchiya, K. 2001. Sex determination without the Y chromosome in two Japanese rodents Tokudaia osimensis osimensis and Tokudaia osimensis spp. *Mammalian Genome* **12**: 17–21.

Takada, S., Ota, J., Kansaku, N., Yamashita, H., Izumi, T., Ishikawa, M., Wada, T. et al. 2006. Nucleotide sequence and embryonic expression of quail and duck Sox9 genes. *General and Comparative Endocrinology* **145**: 208–213.

Takase, M., Noguchi, S., and Nakamura, M. 2000. Two Sox9 messenger RNA isoforms: Isolation of cDNAs and their expression during gonadal development in the frog Rana rugosa. *FEBS Letters* **466**: 249–254.

Taketo, T., Lee, C., Zhang, J., Li, Y., Lee, C., and Lau, Y. 2005. Expression of SRY proteins in both normal and sex-reversed XY fetal mouse gonads. *Developmental Dynamics* **233**: 612–622.

Thevenet, L., Albrecht, K., Malki, S., Berta, P., Boizet-Bonhoure, B., and Poulat, F. 2005. NHERF2/SIP-1 interacts with mouse SRY via a different mechanism than human SRY. *The Journal of Biological Chemistry* **280**: 38625–38630.

Tilmann, C. and Capel, B. 1999. Mesonephric cell migration induces testis cord formation and Sertoli cell differentiation in the mammalian gonad. *Development* **126**: 2883–2890.

Toledano-Díaz, A., Santiago-Moreno, J., Gómez-Brunet, A., Pulido-Pastor, A., and López-Sebastián, A. 2007. Horn growth related to testosterone secretion in two wild Mediterranean ruminant species: The Spanish ibex (Capra pyrenaica hispanica) and European mouflon (Ovis orientalis musimon). *Animal Reproduction Science* **102**: 300–307.

Tucker, P. and Lundrigan, B. 1993. Rapid evolution of the sex determining locus in Old World mice and rats. *Nature* **364**: 715–717.

Töhönen, V., Osterlund, C., and Nordqvist, K. 1998. Testatin: A cystatin-related gene expressed during early testis development. *Proceedings of the National*

Academy of Sciences of the United States of America **95**: 14208–14213.

Uda, M., Ottolenghi, C., Crisponi, L., Garcia, J., Deiana, M., Kimber, W., Forabosco, A., Cao, A., Schlessinger, D., and Pilia, G. 2004. Foxl2 disruption causes mouse ovarian failure by pervasive blockage of follicle development. *Human Molecular Genetics* **13**: 1171–1181.

Vaillant, S., Magre, S., Dorizzi, M., Pieau, C., and Richard-Mercier, N. 2001. Expression of AMH, SF1, and SOX9 in gonads of genetic female chickens during sex reversal induced by an aromatase inhibitor. *Development Dynamics* **222**: 228–237.

Vaiman, D., Koutita, O., Oustry, A., Elsen, J., Manfredi, E., Fellous, M., and Cribiu, E. 1996. Genetic mapping of the autosomal region involved in XX sex-reversal and horn development in goats. *Mammalian Genome* **7**: 133–137.

Vainio, S., Heikkilä, M., Kispert, A., Chin, N., and McMahon, A. 1999. Female development in mammals is regulated by Wnt-4 signalling. *Nature* **397**: 405–409.

Valleley, E., Cartwright, E., Croft, N., Markham, A., and Coletta, P. 2001. Characterisation and expression of Sox9 in the Leopard gecko, Eublepharis macularius. *Journal of Experimental Zoology* **291**: 85–91.

Veyrunes, F., Waters, P., Miethke, P., Rens, W., McMillan, D., Alsop, A., Grützner, F. et al. 2008. Bird-like sex chromosomes of platypus imply recent origin of mammal sex chromosomes. *Genome Research* **18**: 965–973.

Vidal, V., Chaboissier, M., de Rooij, D., and Schedl, A. 2001. Sox9 induces testis development in XX transgenic mice. *Nature Genetics* **28**: 216–217.

Volff, J., Nanda, I., Schmid, M., and Schartl, M. 2007. Governing sex determination in fish: Regulatory putsches and ephemeral dictators. *Sex Dev* **1**: 85–99.

Wagner, T., Wirth, J., Meyer, J., Zabel, B., Held, M., Zimmer, J., Pasantes, J., Bricarelli, F., Keutel, J., and Hustert, E. 1994. Autosomal sex reversal and campomelic dysplasia are caused by mutations in and around the SRY-related gene SOX9. *Cell* **79**: 1111–1120.

Wallis, M., Waters, P., Delbridge, M., Kirby, P., Pask, A., Grützner, F., Rens, W., Ferguson-Smith, M., and Graves, J. 2007. Sex determination in platypus and echidna: autosomal location of SOX3 confirms the absence of SRY from monotremes. *Chromosome Research* **15**: 949–959.

Wang, D., Kobayashi, T., Zhou, L., Paul-Prasanth, B., Ijiri, S., Sakai, F., Okubo, K., Morohashi, K., and Nagahama, Y. 2007. Foxl2 up-regulates aromatase gene transcription in a female-specific manner by binding to the promoter as well as interacting with ad4 binding protein/steroidogenic factor 1. *Molecular Endocrinology* **21**: 712–725.

Western, P., Harry, J., Graves, J., and Sinclair, A. 1999. Temperature-dependent sex determination: Upregulation of SOX9 expression after commitment to male development. *Developmental Dynamics* **214**: 171–177.

Whitfield, L., Lovell-Badge, R., and Goodfellow, P. 1993. Rapid sequence evolution of the mammalian sex-determining gene SRY. *Nature* **364**: 713–715.

Wilhelm, D., Hiramatsu, R., Mizusaki, H., Widjaja, L., Combes, A., Kanai, Y., and Koopman, P. 2007. SOX9 regulates prostaglandin D synthase gene transcription in vivo to ensure testis development. *The Journal of Biological Chemistry* **282**: 10553–10560.

Wilhelm, D., Martinson, F., Bradford, S., Wilson, M., Combes, A., Beverdam, A.,

Bowles, J., Mizusaki, H., and Koopman, P. 2005. Sertoli cell differentiation is induced both cell-autonomously and through prostaglandin signaling during mammalian sex determination. *Developmental Biology* **287**: 111–124.

Wilkins, A. 1995. Moving up the hierarchy: A hypothesis on the evolution of a genetic sex determination pathway. *Bioessays* **17**: 71–77.

Wilson, J., Leshin, M., and George, F. 1987. The Sebright bantam chicken and the genetic control of extraglandular aromatase. *Endocrine Reviews* **8**: 363–376.

Wotton, K., French, K., and Shimeld, S. 2007. The developmental expression of foxl2 in the dogfish Scyliorhinus canicula. *Gene Expression Patterns* 7: 793–797.

Yan, C., Wang, P., DeMayo, J., DeMayo, F., Elvin, J., Carino, C., Prasad, S. et al. 2001. Synergistic roles of bone morphogenetic protein 15 and growth differentiation factor 9 in ovarian function. *Molecular Endocrinology* **15**: 854–866.

Yao, H., DiNapoli, L., and Capel, B. 2003. Meiotic germ cells antagonize mesonephric cell migration and testis cord formation in mouse gonads. *Development* **130**: 5895–5902.

Yao, H., DiNapoli, L., and Capel, B. 2004a. Cellular mechanisms of sex determination in the red-eared slider turtle, Trachemys scripta. *Mechanisms of Development* **121**: 1393–1401.

Yao, H., Matzuk, M., Jorgez, C., Menke, D., Page, D., Swain, A., and Capel, B. 2004b. Follistatin operates downstream of Wnt4 in mammalian ovary organogenesis. *Developmental Dynamics* **230**: 210–215.

Yao, H., Whoriskey, W., and Capel, B. 2002. Desert Hedgehog/Patched 1 signaling specifies fetal Leydig cell fate in testis organogenesis. *Genes Development* **16**: 1433–1440.

Zhou, R., Liu, L., Guo, Y., Yu, H., Cheng, H., Huang, X., Tiersch, T., and Berta, P. 2003. Similar gene structure of two Sox9a genes and their expression patterns during gonadal differentiation in a teleost fish, rice field eel (Monopterus albus). *Molecular Reproduction and Development* **66**: 211–217.

17

Toxicogenomics of Reproductive Endocrine Disruption

Ulf Magnusson and Lennart Dencker

17.1 Introduction

In the past few decades public awareness and scientific knowledge on how xenobiotics exert hormone-like activity in humans and animals has expanded rapidly. The driving force behind this expansion has been the public and scientific concern that this phenomenon—called endocrine disruption—is a severe threat to human and wildlife health. Attention has been paid to environmental contaminants such as industrial and consumer chemicals as well as pesticides. However, this hormone-like activity may also be exerted by natural compounds such as plant phytoestrogens. Overall, most scientific reports within this field deal with the disruption of the reproductive endocrine system.

The mechanism of action for these endocrine-disrupting chemicals was initially regarded to be restricted to activation via certain hormone receptors. However, it has now become increasingly clear that there are other mechanisms of action that make the picture more complex. Another challenging aspect of the understanding of the endocrine disruption is that several of the chemicals of concern show non-monotic dose–response curves in experimental settings. Furthermore, in real life humans or animals are rarely exposed to one chemical at a time, but, rather, to a mixture of chemicals that may act antagonistically, additively, or synergistically. Moreover, the endocrine-disrupting chemicals generally act at relatively low concentrations, which do not cause any overt effects on the individual at the time of exposure. Usually the disruption affects the individual at certain vulnerable periods during development, and the effects become obvious first at adulthood. This also contributes to making the study of endocrine disruption very complex.

Hence, there is a need for powerful and precise research tools to dissect and understand the world of endocrine disruption. The disciplines of toxicogenomics and ecotoxicogenomics—that is, toxicology at the molecular level—may, by their precise readouts and generality across species, yield valuable contributions.

17.2 Reproductive endocrine disruption

17.2.1 The concept

Endocrine disruption, especially of the reproductive endocrine system, is recognized as an important environmental concern. Public research funding directed to the field has been substantial; the involvement of non-governmental organizations and industries has been considerable; and the scientific society has, in the last decade, generated much new knowledge related to this phenomenon.

The concept and concern of endocrine disruption emerged in the early 1990s (Colborn and Clement 1992), and data from observations in wildlife, humans, laboratory, and domestic animals have increased ever since. There are some slightly different opinions on how to define an endocrine-disrupting chemical. One of the most widely agreed upon is that put forward by the International Program on Chemical Safety in 2002: "An endocrine disruptor is an exogenous substance or mixture that alters function(s) of the endocrine system and consequently causes adverse health effects in an intact organism, or its progeny, or (sub)populations" (Damstra et al 1992). Notably, in domestic animals endocrine disruption of the reproductive system has long been recognized in the form of phytoestrogens showing adverse effects on grazing ruminants' fertility.

17.2.2 Observations in wildlife

The data from wildlife are most often field data with weak to moderate associations between cause and effects. Sometimes the causality has been confirmed by experiments on wildlife in captivity under controlled conditions. In large part these observations are from species living in a highly contaminated aquatic environment or from species high up in the food chain. One classic example from the former category is how the biocide tributylin causes imposex in female prosobranch gastropods. Imposex is an imposition of male reproductive organs onto female snails that can impair their reproductive ability. Imposex has been documented in some 150 species of these marine snails worldwide (Horiguchi 2006). The use of tributylin in anti-fouling paints of ships therefore became restricted in many countries in the 1990s. This restriction has been followed by the recovery of several of the gastropod populations. Another well-recognized association between endocrine-disrupting chemicals and reproductive disorders in wildlife is that between the industrial chemical polychlorinated biphenyl (PCB) and uterine occlusion in seals in the Baltic Sea (Helle et al. 1976). However, in this case the mechanism of action is more poorly understood than imposex in marine snails.

17.2.3 Indications in humans

As for wildlife, the data on endocrine disruption in humans result from studies where the association between exposure to a certain chemical and effect is often rather weak. However, the magnitude of the problem of estrogenic effects was greatly spurred by an early disaster caused by administration of the synthetic estrogen diethylstilbestrol (previously an important drug in veterinary medicine) during human pregnancy; this led to major malformations and dysfunction of the reproductive organs in the offspring (for review, see Giusti et al. 1995). Lately, a particularly interesting epidemiological study is one in which the exposure of the mother to plastic softeners during pregnancy was inversely related to the anogenital distance in newborn boys (Swan et al. 2005). A shortened anogenital distance in male laboratory rodents is regarded as a sign of feminization achieved during development. The most convincing data on endocrine disruption in humans, though, are from the field of occupational medicine or from cases of accidental exposure such as the case of the diethylstilbestrol mentioned above. Other such examples are mothers who had PCB-contaminated rice oil during pregnancy and who gave birth to boys who in adulthood displayed abnormal sperm morphology and motility (Guo et al. 2000) and male workers in a chemical industry that manufactured a stilbene derivate who had lower serum testosterone levels and suffered from decreased libido and impotence (Grajewski et al. 1996; Whelan et al. 1996)

17.2.4 Experimental evidence from laboratory animals

In laboratory animals there are solid experimental data showing endocrine-disrupting features in several groups of chemicals. Mainly in these experimental settings the exposure is at a relatively low dose and does not cause any acute or general signs of intoxication. Typically the exposure has to occur during the pre- or early postnatal period in order to cause the most dramatic effects. Effects have been observed in both female and male animals, although the majority of studies concern the male reproductive system. This is likely due to the fact that several of the first reported endocrine-disrupting chemicals were estrogenic and were therefore expected to cause more harm to the male than to the female reproductive system. However, the concept of reproductive endocrine disruption has now been expanded to include compounds with androgenic and anti-androgenic effects. The disruptive effects have been observed on the morphology and function of the reproductive systems as well as on the reproductive behavior (for reviews, see Gray et al. 2004; Stokes 2004). Interestingly, several of the experimental exposure–effect studies in laboratory rodents do back up the associations between exposure and effects seen in the field and epidemiological studies. For instance, the so-called testicular dysgenesis syndrome in men, including cryptorchidism, hypospadiasis, testicular cancer, and poor semen quality, can be mimicked in rats by fetal exposure to plastic softeners (Hutchison et al. 2008).

17.2.5 Data from domestic animals

The data on endocrine disruption in domestic animals are sparse compared with those from wildlife, humans, and laboratory animals, and are mainly of two types: case reports and controlled studies where farm animals have been used as experimental animals.

As mentioned above, reproductive endocrine disruption has a long history in farm animals. The so-called sweet clover disease is one of the earliest reported cases of endocrine disruption affecting mammals. The disease is caused by the phytoestrogens genestein and formononetin present in high concentrations in clover (Cox 1978). Typical effects observed in sheep grazing such clover are prolapse of the uterus and reduced fertility attributable to embryonic death. Another well-known case of reproductive endocrine disruption in livestock is when pigs are affected by the phytoestrogen zearalenon (reviewed by Diekman and Green 1992). This phytoestrogen is an acid lactone compound produced by the fungi *Fusarium*. Prepubertal gilts seem particularly sensitive and may show severe clinical signs of hyperestrogenecity such as vaginal or rectal prolapses. Sexually mature sows that have eaten feedstuff contaminated with the fungus have shown serious reproductive disorders such as abortion, fetal mummification, stillbirths, and abnormal return to estrus intervals.

Besides these observations related to phytoestrogens there are very few reports on accidental exposure of farm animals to environmental chemicals associated with disruption of the reproductive endocrinology. One possible example of such exposure and effect association is the case of dairy heifers that were drinking surface water in direct contact with sewerage overflow and that showed an increased age at first calving (Meijer et al. 1999). However, neither an analysis of chemicals nor any endocrinological measurements were reported in that study.

In addition to these observations from the field there is a set of controlled experimental studies where endocrine disruption has been investigated in various species of farm animals. There are at least three rationales for using farm animals as experimental animals (for review, see Magnusson 2005).

The first rationale is that by using farm animals one tests or challenges the generality of data generated in the classical laboratory mammalians (mainly rabbits, mice, and rats). The international testing strategy for chemicals is much regulated by use of the latter species. Confirmatory study in domestic animals is, however, highly relevant, given the diversity of the reproductive and metabolic systems throughout the animal kingdom, and may thus contribute to the risk assessment of chemicals for human and environmental health. A recent and remarkable finding that underpins the importance of confirmatory studies is that transgenerational fertility problems caused by the antiandrogenic compound vinclozolin after intraperitoneal injections to pregnant rats (Anway et al. 2005) could not be repeated by the same substance given orally to pregnant rats (although a different strain; Schneider et al. 2008).

The second rationale is that farm animals can in fact be better models for human than laboratory species when it comes to physiological aspects relevant for studies of endocrine disruption. For instance, many endocrine-disrupting chemicals reach the endocrine system through the oral route. Because the pig, for example, is an omnivorous and intermittent eater like humans, the porcine digestive system shows many similarities to the human digestive system (Moughan et al. 1994). It is therefore likely that pigs have an advantage over laboratory species as a model for human oral exposure to endocrine-disrupting chemicals. The kinetics of an endocrine-disrupting plastic softener following oral exposure in pigs is indeed more similar to that in primates than to that in rats (Ljungvall et al. 2004). In addition, farm animals do have longer embryofetal and prepubertal periods compared with laboratory species. Thus, they are, in this sense, more similar to humans. Since exposure to endocrine-disrupting chemicals is typically long-term and individuals are in general particularly sensitive to exposure during these two periods, farm animals also have an advantage as a model for humans in this respect.

The third rationale is that there might be methodological advantages to use farm animals as model species. Obviously, larger samples can be collected from farm animals

Table 17.1 Examples of studies where farm animals have been used as models for studying reproductive endocrine disruption.

Species	Exposure	Effect	Reference
Goat	Gestational and lactational; PCB 153	Reduced testosterone concentration and testis size and increased proportion damaged sperms	Oskam et al. 2005
Goat	Gestational and lactational; PCB 153	Lowered prepubertal luteinizing hormone concentration and delayed puberty in female goats	Lyche et al. 2004
Pig	Prepubertal; Di(2-ethylhexyl)phthalate	Increased testosterone concentration and Leydig cell number at adulthood	Ljungvall et al. 2005
Pig	Prepubertal; Di(2-ethylhexyl)phthalate	Precocious development of bulbourethral glands	Ljungvall et al. 2008
Sheep	Prepubertal; Bisphenol A	Suppressed luteinizing hormone pulse frequency	Evans et al. 2004
Sheep	Gestational; Octylphenol	Decreased testis size and Sertoli cell number at birth	Sweeney et al. 2000

than from smaller laboratory species, both *in vivo* and from euthanized animals. Also, repeated sampling can more easily be performed in farm animals. This holds true for instance for semen collection as well as blood sampling; the latter can, in cattle and pigs, be performed through venous catheters (Basu and Kindahl 1987; Rojkittikhun et al. 1991). Finally, since there are indications that sexually dimorphic behavior is sensitive to endocrine-disrupting chemicals (Palanza et al. 2002), the reproductive behavior is of interest for studies in this context. The mating behavior of farm animals is well described and is in some species relatively long and complex compared with laboratory species and is thus very suitable for study in the context of endocrine disruption, for instance, in the pig (Ljungvall et al. 2006) or the quail (Brunström et al. 2003).

So which species of farm animals have been used in experimental studies? What kinds of compounds have been investigated and which effects have been observed? In Table 17.1 samples of studies on endocrine disruption *in vivo* in farm animals is presented. It should be noted that there are also some interesting *in vitro* data on reproductive endocrine disruption in farm animals, for instance on bovine oocyte maturation and subsequent development (Pocar et al. 2001). Collectively the *in vivo* data in Table 17.1 show that studies in farm animals may contribute novel and sometimes opposing data compared with those generated in laboratory species.

17.3 Reproductive endocrine disruptors

17.3.1 Chemicals of concern

As indicated above there is a wide range of chemicals that may act as endocrine disruptors. The majority of these are man-made, but there are some disruptors that are naturally occurring in the ecosystem, such as the previously discussed phytoestrogens. In the so-called developed world, environmental pollution caused by chemicals hazardous to humans and the environment has changed over the last decades; thanks to cleaner industrial procedures, pollution from industries is very much reduced. However, there is a more recently discovered type of pollution, and that is the diffuse and constant exposure to consumer chemicals that are abundant in our environment. This is called background exposure. Despite this change in the major source of pollution, several of the hazardous industrial chemicals that have been banned for years are still an environmental concern due to their continuous presence in the environment. In addition, an emerging pollution concern that is particularly interesting in the context of endocrine disruption is pharmaceuticals. Created to be biologically potent, when used in the wrong context they are potentially harmful to wildlife as well as to humans.

Another challenge when trying to estimate risks with individual, potential endocrine disruptors in the environment is that they are mostly present with other chemicals (Kolpin et al 2002). The vast majority of toxicological studies on endocrine disruption have been on single chemicals. However, it is well established that several of the chemicals of concern may act as additives or even synergists on the endocrine system (Halldin et al. 2005). Obviously this makes the risk assessment in real life more complex. To establish a solid causality between exposure to a chemical and effects on the environment in the real-life situation is difficult. Causality has instead been confirmed in laboratory studies when associations between chemical exposure and effect on the reproductive system have been suspected in the environment. In Table 17.2 some examples of associations between exposure and effect in the real-life situation are presented. Very likely such a table may be expanded over the coming years due to the intensive research within this field.

17.3.2 Vulnerable windows and late effects

The effects of endocrine-disrupting compounds differ in a general pattern depending on when the organism or individual is exposed. Exposure during development generally causes irreversible organizational effects on organs or organ systems, whereas exposure of the adult generally causes activational effects that are reversible

Table 17.2 Examples of associations between chemicals in the environment and effects on the reproductive system in humans or animals.

Chemicals	Species affected	Reproductive effect	Reference
Industrial chemicals and pesticides			
PCB, DDE	White-tailed sea eagle	Reduced reproductive success	Helander et al. 1982
	American alligator	Reduced phallus size	Guillette et al.1999
Consumer's chemicals			
Phthalates	Human	Decreased anogenital distance in newborn boys	Swan et al. 2005
Pharmaceuticals			
Trenbolone	Fathead minnow	Lower testicular testosterone synthesis and smaller testis size	Orlando et al. 2004.
Ethinyl estradiol	White sucker	Female-biased sex ratio and increased intersex	Vajda et al. 2008
Phytoestrogens			
Genestein and formononetin	Sheep	Uterine prolapse, embryonic death	Cox 1978

(McLachlan 2001). Most of the attention to endocrine disruption has been given to the irreversible effects during development, since a lower dose of exposure is usually needed to produce an effect during this life stage compared with adulthood. In other words, the developing organism is in general more sensitive to endocrine disruption than the adult. Notably, also during development there is a variability in sensitivity; the so-called windows of vulnerability may vary by endocrine-disrupting compound, species, and organ or physiological system affected. In one elegant study showing the concept of windows of vulnerability, male rabbits were exposed to the plasticizer dibuthyl phthalate *in utero* during adolescence or post puberty (Higuchi et al. 2003). The most dramatic effects were seen in the group exposed *in utero* with decreased number of sperm ejaculated and reduced testis as adults. These windows may be very narrow, a matter of days, as shown by studies on the effect of estrogens on leopard frogs (Hogan et al. 2008).

Another timing aspect of endocrine disruption, particularly relevant to disruption of the reproductive system, are the so-called late effects. This means that there is a long time gap between exposure, typically during development, and overt effects on the reproductive system, typically at adulthood when the individual is beginning to be sexually functional. This is of course obvious for several reproductive endpoints that are very difficult to measure before puberty, such as number of sperm in an ejaculate or number of eggs at ovulation. However, in our own split-litter designed studies in pigs that were exposed to a plastic softener for some weeks soon after birth, we showed that at adulthood, that is, 5–6 months after the end of exposure, the plasma testosterone level was elevated and bulbouthretral gland size was larger in exposed pigs compared with controls. Interestingly, this effect was not seen in littermates given the same exposure in parallel when examined directly after exposure (Ljungvall et al. 2005, 2008).

While biologically intriguing, vulnerable windows and late effects present troublesome challenges both for regulatory toxicology when assessing the health risk for chemicals and for environmental monitoring and tracking of effects of chemical exposure in the ecosystem.

17.3.3 Different mechanisms of action

The chemicals that exert endocrine disruption are structurally very diverse with different physiochemical properties; one may therefore assume that they use various mechanisms of action for their disrupting effects. Classically endocrine-disrupting chemicals have been regarded, or defined, as chemicals that modulate the endocrine system by binding to hormone receptors and having an agonistic or antagonistic effect. Such a receptor-mediated mechanism has been elegantly shown for the drug diethylstilbestrol by using estrogen receptor knockout mice (Henley and Korach 2006). The downstream effect during the development of the fetal mice by the diethylstilbestrol disruption seems then to be a decrease in *Hox* and *Wnt* gene expression, critical for the development of the female genital tract.

Besides this orthodox receptor-mediated endocrine disruption it has become increasingly clear that there are additional mechanisms of action for endocrine-disrupting chemicals, such as hormone synthesis and clearance. One such proposed mechanism of particular interest for reproductive steroid hormone metabolism is the increased aromatase activity reported for atrazine in frogs that increases the conversion of androgens to estrogens (Hayes et al. 2003). Another example of modulating hormone metabolism is that exerted by methoxychlor, which activates two nuclear receptors: the human steroid and xenobiotic receptor/rodent preganane X receptor and the constitutive androstane receptor (reviewed by Tabb and Blumberg 2006). These receptors are highly expressed in the liver and mediate the induction of cytochrome P450 and conjugation enzymes, whereby they may severely affect steroid hormone metabolism.

Yet another non-receptor-mediated mechanism is modulation of the proteosome-mediated degradation of steroid receptors (Wijayaratne and McDonnell. 2001). For instance, it has been reported that Bisphenol A slows the degradation of the estrogen receptor alpha (Masuyama and Hiramatsu 2004). In addition, it has been put forward that endocrine-disrupting chemicals exert their effects by altering the levels of nuclear receptor coactivators (reviewed by Tabb and Blumberg 2006).

Possibly the most intriguing, and alarming, effects or mechanisms of action by endocrine-disrupting chemicals are the transgenerational effects on the reproductive system. The mechanisms for these effects are reported to be epigenetic, involving altered DNA methylation (reviewed by Anway and Skinner 2006). This is an emerging field for environmental research and regulatory bodies.

Finally, regarding the diversity of mechanisms of action for endocrine-disrupting chemicals: the very same chemical can show different effects in different species due to downstream dissimilarities even though the initial interaction is the same (Tabb et al. 2004). For the same reason a chemical may exert different effects in different tissues within the same individual (Lonard et al. 2004). This complexity calls for precise and powerful research tools such as those of toxicogenomics (and other "-omics") in molecular biology.

17.4 Toxicogenomics

17.4.1 Complexity of endocrine disruption

Endocrine-disrupting chemicals traditionally consider those that mimic or block transcriptional activation by endogenous hormones, and are not restricted to hormonal systems related to reproduction but include, for example, thyroid hormones. Working with endogenous substances or their synthetic analogs acting on receptors directly involved in transcriptional activation provides a "pure" study object when it comes to various aspects on toxicogenomics. If then cells are also studied *in vitro*, perhaps transfected by reporter genes directly coupled to the receptor complex, one gets a rather straightforward biological "answer."

When investigating the effects of, for example, environmental chemicals with significant but much lower affinity for a receptor under study—as compared with the endogenous ligand—one can demonstrate *in vitro* or *in vivo* if that affinity also translates into a biological effect. However, that particular chemical has to be administered at higher concentrations, or dose to an animal, making effects on other cellular systems likely to occur. If we consider the fate and biological activity of chemicals, be it receptor ligands or not, in an organism (experimental or domestic animals, etc.) the picture rapidly becomes more complicated. It is then reasonable to extend the definition of endocrine disruptors from receptor active substances only to include those exogenous, environmental molecules that affect, for example, the synthesis, secretion, transport, metabolism, protein binding, and catabolism of natural hormones in the body. On a cellular level, substances not only may act through receptors but may also interfere in various ways in the complex transcriptional regulatory machinery, including histone deacetylase (HDAC) inhibition and proteasomal degradation of receptor complexes (reviewed by Tabb and Blumberg 2006). Lately, the term "epigenetics" has come into fashion, being defined in this context as describing changes in gene expression that are more or less stable even between generations, without causing changes in the DNA sequence (reviewed by Szyf 2007). Epigenetics is discussed further below.

It has been shown that the potent environmental compound 2,3,7,8-tetrachlorodibenso-p-dioxin (TCDD) is an endocrine disruptor, although the mechanism of action has been obscure. Recently, it was found that the aryl hydrocarbon receptor nuclear translocator protein (ARNT), which is a necessary partner for the TCDD or aryl hydrocarbon receptor (AhR), acts as a coactivator for estrogen receptors. Reducing the levels of available ARNT by activating the AhR- or HIF (*hypoxia*-inducible factor)-pathways, or by targeted downregulation of ARNT by siRNA, decreased estrogen receptor (ER) transcriptional activity, suggesting that competition for ARNT may be at least partly responsible for the antiestrogenic effects of dioxins (Rüegg et al. 2007).

17.4.2 Global gene expression analysis and phenotypic anchoring

Toxicogenomics can be defined as an integration of toxicology with genomics, which in turn can be transcriptomics (gene expression), proteomics, metabolomics, peptidomics, etc. So far, there is more extensive information from transcriptomics experiments than from any of the other -omics techniques in toxicology (for review, see, e.g., Gant 2007; Gomase and Tagore 2008).

Changes in gene expression as measured by mRNA levels using polymerase chain reaction (PCR) as well as globally by using microarrays must, in order to give meaningful information, somehow be related to morphological and/or physiological changes in the cell/tissue. This is called "phenotypic anchoring." Typically, and of course based on temporal sequences, up- or downregulation of mRNA appears before observable cellular effects. This is important to consider in planning experiments. To get information as close as possible to the "source" of the change, that is, to more reliably obtain mechanistic information, measurement of global mRNA should be performed within hours after exposure. Measuring later may of course give valuable information, but the information achieved will be different, because one adds on secondary, tertiary, etc. inductions/repressions, while initial responses may be attenuated.

Interesting results have come from studies on temporal gene expression changes in the endometrium after estrogen exposure in rodents and women (reviewed by Groothuis et al. 2007). The endometrium may of course be an interesting tissue to study in domestic animals as well, considering its vital role in reproduction and perhaps as a sensitive target organ to endocrine disruptors.

The following are some important general considerations as derived from the review by Groothuis and collaborators.

1. There are dissimilarities in responses to estrogens between the mouse and human endometrium. This brings into question the likelihood of learning an easy lesson from these species and transferring it to

any domestic animal. Differences are both temporal and qualitative.

2. Comparison of gene array studies in rodents, both between experiments in the same lab and between labs, shows that rather few genes respond similarly, or in the same direction (up- or downregulated), despite attempts to standardize experimental conditions. Differences in platforms used play a role, and it has even been proposed that EE and E_2 may give different responses. As far as comparison between mouse and humans—sincc thcy respond very differently physiologically anyway—very few genes are regulated in the same way. What is common is that genes regulating the cell cycle are induced by estrogens in both species.
3. In the mouse, which is the easiest to study, there are interesting consecutive, temporal changes in gene expression, which can be related to the physiological state of the endometrium. In the first 4 h, there was an increased influx of fluid into the uterus, which may be due to the increased expression of vascular endothelial growth factor (VEGF) and thus increase in vascular permeability. Other vasoactive growth factors and vascular endothelial receptors were also upregulated. Following these changes, genes involved in transcriptional regulation (e.g., mRNA and protein synthesis) and signaling for growth and differentiation were seen to be upregulated. These gene expression patterns were observed in the absence of obvious histological changes. After that, genes involved in controlling chromosomal replication were upregulated as a "worm up" in the cell cycle, and at 24 h, a substantial increase in mitotic index could be observed. These genes being initially upregulated then fell back, often to below control levels. During this built-up phase of the endometrium, genes controlling proteins involved in anti-apoptosis were also upregulated, while pro-apoptotic genes were downregulated. One would expect a generality in these changes, but a comparison with other laboratories showed that, surprisingly, only some 14 genes were affected similarly.
4. As pointed out by Groothuis and collaborators in their review, it is well known that when studying whole tissues from *in vivo* studies, changes in gene expression may occur in a small population of cells, being of outmost physiological importance, while the responsible mRNAs and changes in their levels may not be detectable due to dilution in the global mRNA. In such a case laser capture microdissection can be applied to pick mRNA from specific regions of a tissue slice. Of course, one must know which cells to look for.

Yet another reason for phenotypic anchoring is that any perturbation such as exposure to chemicals is likely to change gene expression even in the absence of cellular changes; this may be considered a stress response or "background noise" and not related to obvious toxicity. Thus it is not always trivial to relate changes in the expression of a single gene or gene family to a biological effect.

17.4.3 Epigenetics

Alterations in the DNA-methylation (de- and remethylation) state of the embryo occur in a sophisticated way in waves, and very specifically in germ line cells. Methylation patterns can be affected by disease states, nutritional deficiencies, etc., during pregnancy in a way that alters the health of the offspring in adulthood. The same holds true for chemical exposure during pregnancy. As to endocrine disruptors, there is increasing evidence that the same mechanisms may be at play. For example, for diethylstilbestrol, it is likely that vaginal tumors occurring at adolescence and later in women exposed transplacentally as a result of maternal ingestion are caused by epigenetic changes. The other obvious examples are that of the anti-androgenic fungicide vinclozolin mentioned above and also methoxychlor, which has been used as a replacement for DDT as an insecticide. If endocrine disruptors are administered to pregnant rats during the period of sex determination, increased spermatogenetic cell apoptosis and decreased sperm number and mobility are observed in the next generation of males. Interestingly, this phenotype is transmitted transgenerationally through the male germ line. Not only were signs of decreased fertility observed, but also increased cancer rate, prostatic disease, and kidney and immune system problems.

At the molecular level it is likely that methylation of critical genes during embryonic gonadal sex determination can alter the male germ-line epigenetics, causing an epigenetic reprogramming that appears to be transmitted transgenerationally. At the transcriptome level, expression of 196 genes was found to be influenced (day 16 embryos), with the majority of genes being silenced. Interestingly, methyltransferases were affected in the F1 and F2 vinclozolin generation (at embryonic day 16) embryonic testis, being in line with an effect on DNA methylation (Anway et al. 2008). This complex area of research, especially on vinclozin and diethylstilbestrol, has been reviewed (Anway and Skinner 2006; Crews and McLachlan 2006). Epigenetic effects of chemicals certainly give new, and to some extent frightening, aspects on toxicology in general and on endocrine disruption in particular.

17.4.4 Advantage of domestic animals in toxicogenomics?

Artificial selection of domestic animals over thousands of years has created a great number of breeds (reviewed by Georges 2007). Lately, the whole genome has been mapped for the most common domestic animals. These late developments will give animal breeding a new jump in refinement. There is no doubt that the new genetic information and technologies as a spin-off effect will give us tools to refine our understanding of mechanisms of toxicology as well, including endocrine disruption. For example, can positional identification of genes underlying complex traits be used to study the influence of chemicals on characteristics such as behavior and sexual function? To what extent does toxicological influence depend on interaction with classical hormone receptors? To what extent does it involve the epigenetic machinery? Can information gathered by the different "-omics" techniques of transcriptomics, proteomics, and peptidomics be integrated into a more global understanding? It is difficult to foresee which direction research in endocrine disruption will take, but considering the available tools and the understanding of the biology of gene regulation, the only limitations to development will be our imagination. To get a broader view on all of these aspects of endocrine disruption, domestic animals have a place in research.

17.4.5 Experience on toxicogenomics in avians

It has been shown, for example, by Brunström et al. (2003), that male quail embryos exposed to estrogenic substance in the egg may have aberrant sexual behavior when adults. We thought that this would involve several regions of the brain and that aberrant morphology would be discovered close to the time of exposure to ethinyl estradiol. However, we found no change in gene expression related to this exposure, either in the quail or in the chicken. Interestingly, however, very stably expressed in the embryonic brain even from day 4 of incubation (the earliest time point studied) were a number of sex-specific, sex-chromosome linked genes (Scholz et al. 2006). Yet another finding in our work was the absence of evident dosage compensation of sex-linked genes (Ellegren et al. 2007).

As indicated in Section 17.4.2, point 4, it is possible that the effects of ethinyl estradiol are restricted to a few cells in a restricted area, in this case, for example, in the preoptic medial nucleus (POM) in the developing thalamus of the quail brain, which differs in size between males and females. To dissect this area in a similar way in treated and controlled embryos to get a fair comparison, however, is nearly impossible. Laser microdissection may then be an alternative not yet tested by our laboratory.

An alternative to transcriptomics to define estrogenic effects in the developing male quail brain was to study the neuropeptidome. Scholz in his doctoral thesis found that gonadotropin-inhibiting hormone related peptide 2 (GnIH-RP2), among hundreds of other peptides identified, was upregulated (Scholz 2008). This occurred in an embryonic period when GnIH-RP2 is known to commence its regulation of luteinizing hormone (LH) and testosterone levels in the quail. Several other neuropeptides showed a temporal change over the last period of *in ovo* development. This line of research is worth further development and refinement in the future.

17.5 Future research directions

The focus in research on endocrine disruption so far has been on chemicals interacting with hormone receptors, mainly due to the fact that handy and high-throughput cellular systems have been developed. We foresee that more *in vivo* studies will then be required and that domestic animals will be valuable in these confirmatory studies, especially to support or reject the generality of findings in the classical laboratory species. The current mapping of the genome of traditional domestic species will provide an excellent opportunity to combine research on endocrine disruption with toxicogenomics. In particular, phenotypic anchoring is an area where research on domestic animals can contribute, since reproductive functions including behavior are very well characterized in these species.

References

Anway, M.D., Cupp, A.S., Uzumcu, M., and Skinner, M.K. 2005. Epigenetic transgenerational actions of endocrine disruptors and male fertility. *Science* **308**(5727): 1466–1469.

Anway, M.D., Rekow, S.S., and Skinner, M.K. 2008 Transgenerational epigenetic programming of the embryon testis transcriptome. *Genomics* **91**: 30–40.

Anway, M.D. and Skinner, M.K. 2006. Epigenetic transgenerational actions of endocrine disruptors. *Endocrinology* **147**(Supplement 6): S43–S49.

Basu, S. and Kindahl, H. 1987. Development of a continuous blood collection technique and a detailed study of prostaglandin F2 alpha release during luteolysis and early pregnancy in heifers. *Zentralblat Veterinarmedicine A*. **34**(7): 487–500.

Brunström, B., Axelsson, J., and Halldin, K. 2003. Effects of endocrine modulators on sex differentiation in birds. *Ecotoxicology* **12**: 287–295.

Colborn, T. and Clement, C. 1992. *Chemically-Induced Alterations in Sexual and Functional Development: The Wildlife/Human Connection*. Princeton, NJ: Princeton Scientific Publishing Co., Inc.

Cox, R.I. 1978. Plant estrogens affecting livestock in Australia. In: Keeler, R.F., Van Kampen K.R., and James, L.F. (eds.), *Effects of Poisonous Plants of Livestock*. New York: Academic Press, p. 451.

Crews, D. and McLachlan, J.A. 2006. Epigenetics, evolution, endocrine disruption, health, and disease. *Endocrinology* **147**(Supplement 6): S4–S10.

Damstra, T., Barlow, S., Bergman, A., Kavlock, R., and Kraak, G. van der. 2002. Global assessment of the state-of-the-science of endocrine disruptors. International Programme on Chemical Safety, WHO/PCS/EDC/02.2 p. 1.

Diekman, M.A. and Green M.L. 1992. Mycotoxins and reproduction in domestic livestock. *Journal of Animal Science* **70**: 1615–1627.

Ellegren, H., Hultin-Rosenberg, L., Brunström, B., Dencker, L., Kultima, K., and Scholz, B. 2007. Faced with inequality: Chicken do not have a general dosage compensation of sex-linked genes. *BMC Biology* **5**: 40–51.

Evans, N.P., North, T., Dye, S., and Sweeney, T. 2004. Differential effects of the endocrine-disrupting compounds bisphenol-A and octylphenol on gonadotropin secretion, in prepubertal ewe lambs. *Domest Anim Endocrinol* **26**(1): 61–73.

Gant, T.W. 2007. Novel and future applications of microarrays in toxicological research. *Expert Opinion on Drug Metabolism and Toxicology* **3**: 599–608.

Georges, M. 2007. Mapping, fine mapping, and molecular dissection of quantitative trait Loci in domestic animals. *Annual Review of Genomics Human Genetics* **8**: 131–162.

Giusti, R.M., Iwamoto, K., and Hatch, E. E. 1995. Diethylstilbestrol revisited: A review of the long-term health effects. *Annales of internal medicine* **122**(10): 778–788.

Gomase, V.S. and Tagore, S. 2008. Toxicogenomics. *Current Drug Metabolism* **9**: 250–254.

Grajewski, B., Whelan, E.A., Schnorr, T.M., Mouradian, R., Alderfer, R., and Wild, D.K. 1996. Evaluation of reproductive function among men occupationally exposed to a stilbene derivative: I. Hormonal and physical status. *American Journal of Industrial Medicine* **29**(1): 49–57.

Gray, L.E. Jr., Wilson, V., Noriega, N., Lambright, C., Furr, J., Stoker, T.E., Laws, S.C., Goldman, J., Cooper, R.L., and Foster, P.M. 2004. Use of the laboratory rat as a model in endocrine disruptor screening and testing. *Institute of Laboratory Animal Resources Journal* **45**(4): 425–437.

Groothuis, P.G., Dassen, H.H., Romano, A., and Punyadeera, C. 2007. Estrogen and the endometrium: Lessons learned from gene expression profiling in rodents and human. *Human Reproduction Update* **13**: 405–417.

Guillette, L.J. Jr., Brock, J.W., Rooney, A.A., Woodward, A.R. 1999. Serum concentrations of various environmental contaminants and their relationship to sex steroid concentrations and phallus size in juvenile American alligators. *Archives of Environmental Contamination and Toxicology* **36**(4): 447–455.

Guo, Y.L., Hsu, P.C., Hsu, C.C., and Lamberst, G.H. 2000. Semen quality after prenatal exposure to polychlorinated biphenyls and dibenzofurans. *Lancet* **356**(9237): 1240–1241.

Halldin, K., Axelsson, J., and Brunström, B. 2005. Embryonic co-exposure to methoxychlor and Clophen A50 alters sexual behavior in adult male quail. *Archives in Toxicology* **79**(4):237–242.

Hayes, T., Haston, K., Tsui, M., Hoang, A., Haeffele, C., and Vonk, A. 2003. Atrazine-induced hermaphroditism at 0.1 ppb in American leopard frogs (*Rana pipiens*): Laboratory and field evidence. *Environmental Health Perspectives* **111**(4): 568–575.

Helander, B., Olsson, M., and Reutergårdh, L. 1982. Residue levels of organoclorine and mercury compounds in unhatched eggs in white-tailed sea eagles *(Haliaeetus albicilla)* in Sweden. *Holarctic Ecology* **5**: 349–366.

Helle, E., Olsson, M., and Jensen, S. 1976. PCB levels correlated with pathological changes in seal uteri. *Ambio* **5**: 261–263.

Henley, D.V. and Korach, K.S. 2006. Endocrine-disrupting chemicals use distinct mechanisms of action to modulate endocrine system function. *Endocrinology* **147**(Supplement 6): S25–S32

Higuchi, T.T., Palmer, J.S., Gray, L.E. Jr., and Veeramachaneni, D.N. 2003. Effects of dibutyl phthalate in male rabbits following in utero, adolescent, or postpubertal exposure. *Toxicol Sci* **72**(2): 301–313.

Hogan, N.S., Duarte, P., Wade, M.G., Lean, D.R., and Trudeau, V.L. 2008. Estrogenic exposure affects metamorphosis and alters sex ratios in the northern leopard frog (*Rana pipiens):* Identifying critically vulnerable periods of development. *General and Comparative Endocrinology* **156**(3): 515–523.

Horiguchi, T. 2006. Masculinization of female gastropod mollusks induced by organotin compounds, focusing on mechanism of actions of tributyltin and triphenyltin for development of imposex. *Environmental Science* **13**: 77–87.

Hutchison, G.R., Scott, H.M., Walker, M., McKinnell, C., Ferrara, D., Mahood, I.K., Sharpe, R.M. 2008. Sertoli cell development and function in an animal model of testicular dysgenesis syndrome. *Biology of Reproduction* **78**(2): 352–360.

Kolpin, D.W., Furlong, E.T., Meyer, M.T., Thurman, E.M., Zaugg, S.D., Barber, L.B., Buxton, H.T. 2002. Pharmaceuticals, hormones, and other organic wastewater contaminants in U.S. streams, 1999–2000: A national reconnaissance. *Environmental Science Technology* **36**(6): 1202–1211.

Ljungvall, K., Karlsson, P., Hultén, F., Madej, A., Norrgren, L., Einarsson, S., and Magnusson, U. 2005. Effects on the hypothalamic-pituitary-testis axis by Di(2-ethylhexyl)phthalate or oestradiol benzoate in the prepubertal boar. *Theriogenology* **64**(5): 1170–84.

Ljungvall, K., Spjuth, L., Hultén, F., Einarsson, S., Rodriguez-Martinez, H., Andersson, K., Magnusson, U. 2006. Early post-natal exposure to low dose oral di(2ethylhexyl) phthalate affects the peripheral LH-concentration in plasma, but does not affect mating behavior in the post-pubertal boar. *Reproductive Toxicology* **21**(2): 160–166.

Ljungvall, K., Tienport, B., David, F., Magnusson, U., and Törneke K. 2004. Kinetics of orally administered Di(2-ethylhexyl)phthalate and its metabolite Mono-ethylhexyl phthalate in male pigs. *Archives of Toxicology* **78**(7):384–389.

Ljungvall, K., Veeramachaneni, D.N., Hou, M., Hultén, F., and Magnusson, U. 2008. Morphology and morphometry of the reproductive organs in prepubertal and postpubertal male pigs exposed to di(2-ethylhexyl) phthalate before puberty: Precocious development of bulbourethral glands. *Theriogenology* **70**(6): 984–991.

Lonard, D.M., Tsai, S.Y., O'Malley, B.W. 2004. Selective estrogen receptor modulators 4-hydroxytamoxifen and raloxifene impact the stability and function of SRC-1 and SRC-3 coactivator proteins. *Molecular Cell Biology* **24**(1): 14–24.

Lyche, J.L., Oskam, I.C., Skaare, J.U., Reksen, O., Sweeney, T., Dahl, E., Farstad, W., and Ropstad, E. 2004. Effects of gestational and lactational exposure to low doses of PCBs 126 and 153 on anterior pituitary and gonadal hormones and on puberty in female goats. *Reprod Toxicol* (1): 87–95.

Magnusson, U. 2005. Can farm animals help to study endocrine disruption? *Domestic Animal Endocrinology* **29**(2): 430–435.

Masuyama, H. and Hiramatsu, Y. 2004. Involvement of suppressor for Gal 1 in the ubiquitin/proteasome-mediated degradation of estrogen receptors. *Journal of Biology and Chemistry* **279**(13): 12020–12026.

McLachlan, J.A. 2001. Environmental signaling: What embryos and evolution teach us about endocrine disrupting chemicals. *Endocrine Reviews* **22**(3): 319–341.

Meijer, G.A.L., de Bree, J.A., Wgenaar, J.A., and Spoelstra, S.F. 1999. Sewerage overflows put production and fertility of dairy cows at risk. *Journal of Environment Quality* **28**: 1381–1383.

Moughan, P.J., Cranwell, P.D., Darragh, A.J., and Rowan, A.M., 1994. The domestic pig as amodel animal for stuying digestion in humans. In: Souffrant, W.B. and Hagemester, H. (eds.), *Proceedings of the Sixth International Symposium on Digestive Physiology in Pigs*, Bad Doberan, Germany, October 4-6, pp. 389–396.

Orlando, E.F., Kolok, A.S., Binzcik, G.A., Gates, J.L., Horton, M.K., Lambright, C.S., Gray, L.E. Jr., Soto, A.M., and Guillette, L.J. Jr. 2004. Endocrine-disrupting effects of cattle feedlot effluent on an aquatic sentinel species, the fathead minnow. *Environment Health Perspectives* **112**(3):353–358.

Oskam, I.C., Lyche, J.L., Krogenaes, A., Thomassen, R., Skaare, J.U., Wiger, R., Dahl, E., Sweeney, T., Stien, A., Ropstad, E. 2005. Effects of long-term maternal exposure to low doses of PCB126 and PCB153 on the reproductive system and related hormones of young male goats. *Reproduction* **130**(5):731–742.

Palanza, P.L., Howdeshell, K.L., Parmigiani, S., and vom Saal, F.S. 2002. Exposure to a low dose of bisphenol A during fetal life or in adulthood alters maternal behavior in mice *Environmental Health Perspectives* **110**(Supplement 3): 415–422.

Pocar, P., Perazzoli, F., Luciano, A.M., and Gandolfi, F. 2001. In vitro reproductive toxicity of polychlorinated biphenyls: Effects on oocyte maturation and developmental competence in cattle. *Molecular Reproductive Development* **58**(4): 411–416.

Rojkittikhun, T., Einarsson, S., and Kindahl, H. 1991. A technique for continuously monitoring hormone levels in lactating sows and results obtained using it to study LH release: *Zentralblat Veterinarmedicine A* **38**(5): 344–349

Rüegg, J., Swedenborg, E., Wahlström, D., Escande, A., Balaguer, P., Pettersson, K., and Pongratz, I. 2007. The transcription factor aryl hydrocarbon receptor nuclear translocator functions as an estrogen receptor beta-selective coactivator, and its recruitment to alternative pathways mediates antiestrogenic effects of dioxin. *Molecular Endocrinology* **22**: 304–316.

Schneider, S., Kaufmann, W., Buesen, R., and van Ravenzwaay, B. 2008 Vinclozolin—the lack of a transgenerational effect after oral maternal exposure during organogenesis. *Reproductive Toxicology* **25**: 352–360.

Scholz, B. 2008. Genomic and peptidomic characterization of the developing avian brain. Doctoral dissertation, Uppsala University, Faculty of Pharmacy, Department of Pharmaceutical Biosciences, urn:nbn:se:uu:diva-8507)

Scholz, B., Kultima, K., Mattsson, A., Axelsson, J., Brunström, B., Halldin, K., Stigson, M., and Dencker, L. 2006. Sex-dependent gene expression in early brain development of chicken embryos. *BMC Neuroscience* 7: 12.

Stokes, W.S. 2004. Selecting appropriate animal models and experimental designs for endocrine disruptor research and testing studies. *Institute of Laboratory Animal Resources Journal* **45**(4): 387–393.

Swan, S.H., Main, K.M., Liu, F., Stewart S.L., Kruse, R.L., Calafat A.M., Mao C.S., Redmon J.B., Ternand C.L., Sullivan, S., and Teague, J.L. 2005. Decrease in anogenital distance among male infants with prenatal phthalate exposure. *Environmental Health Perspectives* **113**(8): 1056–1061

Sweeney, T., Nicol, L., Roche, J.F., and Brooks, A.N. 2000. Maternal exposure to octylphenol suppresses ovine fetal follicle-stimulating hormone secretion, testis size, and sertoli cell number. *Endocrinology* **141**(7): 2667–2673.

Szyf, M. 2007. The dynamic epigenome and its implications in toxicology. *Toxicological Science* **100**: 7–23.

Tabb, M.M. and Blumberg, B. 2006 New modes of action for endocrine-disrupting chemicals. *Molecular Endocrinology* **20**: 475–482.

Tabb, M.M., Kholodovych, V., Grün, F., Zhou, C., Welsh, W.J., and Blumberg, B. 2004. Highly chlorinated PCBs inhibit the human xenobiotic response mediated by the steroid and xenobiotic receptor (SXR). *Environment Health Perspectives* **112**(2): 163–169.

Vajda, A.M., Barber, L.B., Gray, J.L., Lopez, E.M., Woodling, J.D., and Norris, D.O. 2008. Reproductive disruption in fish downstream from an estrogenic wastewater effluent. *Environment Science Technology* **42**(9): 3407–3414

Whelan, E.A., Grajewski, B., Wild, D.K., Schnorr, T.M., and Alderfer, R. 1996. Evaluation of reproductive function among men occupationally exposed to a stilbene derivative: II. Perceived libido and potency. *American Journal of Industrial Medicine* **29**(1): 59–65.

Wijayaratne, A.L. and McDonnell, D.P., 2001. The human estrogen receptor-alpha is a ubiquitinated protein whose stability is affected differentially by agonists, antagonists, and selective estrogen receptor modulators. *Journal of Biology and Chemistry* **276**(38): 35684–35692.

18

Nutrigenomics for Improved Reproduction

John P. McNamara

18.1 Introduction

Reproduction is a function of nutritional state, and also a director of nutrient flux, and both have genetically inherited elements of control. These three systems function integrally in the animal, and as such there is no way that nutrition, reproduction, and genetics can be separated in research. There are several systematic, dynamic controls of nutrient flux involved in ovulation, gestation, and lactation. Glucose can alter the release of hormones from the hypothalamus to direct ovulation, and it can also direct the secretion of other hormones such as insulin or IGFI that affect metabolic activity in reproductive organs. Once an animal ovulates and fertilization occurs, additional interactive control systems are induced to help direct nutrients to the developing fetus(es) and then to the mammary gland. After thousands of years of human observations on the interactions of nutrition and fertility, and after two to three generations (80 to 100 years) of reductionist research, we can now integrate our detailed knowledge to describe the system as a whole. Using genetic and genomic approaches recognizes that nutrient use traits and reproductive traits are heritable and specific gene sequences associated with these traits can be identified. Introducing nutrigenetic and nutrigenomic approaches recognizes that nutrient status affects gene transcription in many organs, which in turn alters metabolic activity in reproductive organs and thus fertility, and gestational and lactational success. We now have technical tools (primarily transcription arrays) to help define specific mechanisms that connect nutritional fluxes with reproductive success. The adipose tissue plays a central role in reproductive success, not just as an energy storage and release organ, but perhaps also as a source of hormones and control factors of reproduction. In order to move forward both in research and application, we must use dynamic, integrated biomathematical modeling tools to help define those reproductive processes that respond to nutrient status and genetic selection and those changes in nutrient flux that respond to genetic selection and reproductive state.

Therefore, this chapter evaluates the effects of diet and nutrient management schemes on gene expression and thus addresses some of the nutrition limitations in reproductive performance.

18.2 Nutritional physiology of reproduction: A brief view

18.2.1 Nutrigenomics and nutrigenetics

Nutrigenomics is generally defined as the effect of dietary nutrients on gene transcription: "Nutrigenomics aims to determine the influence of common dietary ingredients on the genome, and attempts to relate different phenotypes to differences in the cellular and/or genetic response of the biological system" (Mutch et al. 2005). An example of this is studying the effect of changes in diet on gene transcription and metabolic flux in the adipose tissue during pregnancy and lactation and relating those changes to differences in reproductive processes. "Nutrigenetics, on the other hand, aims to understand how the genetic makeup of an individual coordinates their response to diet, and thus considers genetic polymorphisms" (Mutch et al. 2005). The practical application here is to identify the gene variants that relate to differential response to nutrients. Obviously there is tremendous overlap here and the two approaches can easily be related. Nutrigenomic studies might find similarities or differences in the nutrient effects on the transcriptome of phenotypically similar (or different!) animals, while nutrigenetics might find that animals with specific gene variants respond to a dietary change differently, whether in the transcriptome or post-transcriptional processes. A variation in a key controller like prolactin or IGF-I may be easy to explain (and perhaps manipulate), while changes in transcriptome expression present much more of a challenge to sort out changes in known key controlling genes from those of a more constitutive nature. For the purposes of this chapter, I am going to concentrate primarily on nutrigenomics.

18.2.2 Body fat and reproduction

With that quick introduction to the present and future, we need to visit the past to allow us to understand the true role of nutrigenomic work. The role of nutritional status in reproductive fertility was recognized early in human history. Ancient or historical texts, drawings, and writings speak to traditions and perceptions of body fatness, shape, and size in human fertility. Likewise, domesticated animals were fattened to become fertile and sleek fat cattle were desired for their fertility. Initial fertility and postpartum anestrous varies among species or even breeds, and can be attenuated or exacerbated by nutritional status; while a certain amount of body fat might correlate with improved fertility, too much may be detrimental. Today, we realize that in fact there is more to fertility than just fatness—some animals alter fertility after increases or decreases in body fatness (Wade and Schneider 1992). It is not simply the amount of body fat but flux of glucose or other nutrients such as vitamins and minerals that can alter fertility or gestational and lactational success (Wade and Schneider 1992; Wade and Jones 2004). There have been several excellent summaries and reviews on these complicated topics, and the new reader to this field is strongly encouraged to take the time to read them, as it is my experience that there is no real understanding or application of "genomics" outside of understanding the underlying nutritional and reproductive physiology, from the basic to applied in practice (Staples et al. 1990; Wade and Schneider 1992; Wade and Jones 2004; McNamara 2005; Roche 2006; Vinsky et al. 2006; Chagas et al. 2007, and many of the older papers listed in the bibliography and the bibliographies of these citations).

Early observations showed there was some connection with being "well-fed," if not fat, and fertility. Young females needed to mature to a certain range of body shape and fatness (species-dependent) before obvious outward signs of potential fertility, such as estrus behavior or receptivity, were observed. Even if those signs were observed, and acted upon by a male, actual fertilization and successful pregnancy, and lactation, also depended on some level of adequate nutrition. Human families and animal farmers alike acted on these cues and attempted to ensure adequate nutrient status prior to serious attempts at reproduction.

In the last 100 years or so, there have been many studies of biological connections between nutritional status and fertility. It was noted that gross or even moderate stunting of growth delayed sexual maturity in most females. Even when outward signs of fertility were seen, a number of situations in which actual fertility was delayed or reduced were noted, including insufficient total food, or the lack of certain food components. One eventual understanding was that some amount of adipose tissue was necessary for a successful reproductive cycle of ovulation, fertilization, implantation, pregnancy to term, *and not to be overlooked*, a successful lactation.

However, the story was not over. As noted above, there was a lot of variation in the amount of body fat, or body fat gain or loss and fertility; it was not a direct and complete connection. Now we understand that there is in fact a connection between a positive energy balance (energy in > energy out) and reproductive success was clear (Wade and Schneider 1992). We will cover this in more detail, but much of this work led to understanding the key role of glucose in reproduction. Yet other questions arose: "How can reproductive organs monitor and respond to the amount of body fat, or vice versa?" A corollary and related line of research asked the same question in relation to maintenance of body fat: "How does the body monitor and maintain a fairly constant body fat percentage, and what are the situations in which this system can fail and obesity (or extreme thinness) or reproductive problems ensue?"

Then, research in many species tested theories on the presence of signaling molecules or nervous activity to and from the adipose tissue. One major outcome of this research effort (truly spanning from early post-World War II until today) was that the amount of body fat alone did not account for even a small majority of the variation in fertility. Large variations in the amount of body fat, and rate of change of body fat, in reproductively successful females within and among species precluded body fat content as being the ultimate driving force (Butler and Smith 1989; Wade and Schneider 1992). Although there is a perception that body fat (or in dairy cows, body condition) directly relates to reproduction (Chagas et al. 2007), the preponderance of evidence is that although body fat is a key part of the system, it is the nutrient flux (energy balance or glucose supply) that is the mechanistic cause of changes in reproductive status and success (Wade and Jones 2004).

The focus on body fat and potential signaling pathways did in fact lead to important discoveries, one is that adipose tissue is an endocrine organ, secreting, among many other substances, insulin-like growth factor and the intake controlling hormone leptin.

These two molecules are critically important in tissue growth, ovarian metabolism, and food intake. In addition, glucose flux, even in ruminants, directly affects and is affected by the amount and activity of adipose tissue. The sum of all this work is, in my opinion, a great success story in nutritional and reproductive biology. The results of this research effort allow us to bring the tools of nutrigenomics to bear in a focused fashion on the role of genetics and gene expression in nutrient use and reproduction. The role of glucose interaction with the adipose tissue and reproduction may be expanded to changes in synthesis and secretion of IGFI, leptin, or other control molecules (e.g., cytokines; Zieba et al. 2008).

18.2.3 Metabolic flux and reproduction

One of the subsequent lines of research focused on the primary nutrient glucose. Many studies were done to ask the question "How does glucose status relate to reproductive success?" The majority of biological scientists now agree that a major driver of reproductive success is sufficient glucose flux in the body (Wade and Jones 2004; Chagas et al. 2007). Glucose flux into many cell types, including brain, adipose, liver, muscle, and ovary initiates many cascading signals that direct metabolic flux, including fat and protein synthesis. In addition to just the use of glucose for energy generation for anabolic reactions, included in these cascades in most instances are changes in gene expression. The multivariate role of glucose in regulation of metabolism extends from the short term (seconds, minutes)—enzyme activation, increases in ATP and NADPH concentrations—to longer term (days, weeks) changes in mRNA transcription and translation to make more or less of the enzymes that catalyze many synthetic reactions (Girard et al. 1997). Several of these effects relate directly or indirectly to reproductive processes.

Eventually the research led to the endocrine aspects of nutrient flux and reproduction. Nutritional scientists started to realize that "reproductive hormones" affected nutrient use, while reproductive scientists started to explore the effects of nutrients on reproductive processes. We cannot fully explain the connections between nutrient flux and reproduction without introducing the endocrine aspects. If one were to ask students of biology to list "reproductive hormones," estrogen, testosterone, and progesterone would probably be the first answers, also offered up would be luteinizing hormone (LH), follicle stimulating hormone, prolactin, placental lactogen, human chorionic gonadotropin, maybe oxytocin, and relaxin. But not many students (other than those with really good nutritional or reproduction specialists as teachers!) would also list insulin, somatotropin, insulin-like growth factor, thyroid hormone, and corticosteroids as involved in reproductive processes, yet, they are, both indirectly (regulating cell division and tissue growth) and directly (regulating glucose entry into the ovary, follicular growth and fetal and mammary gland differentiation, growth and metabolism). Recent studies also suggest roles for cytokines and inflammatory molecules (Trayhurn and Wood 2004; Loor et al. 2005, 2006; Chagas et al. 2007). These findings have come out of the integration of many different studies on many different aspects of nutrition and reproduction. Yet even though all these hormones have a role in various reproductive functions, the majority of them respond to, or (or also) affect, the major driving force behind truly successful reproduction: the glucose flux in the body.

18.2.4 Nutrigenomics for improved reproduction

To prove the point of the reality of the genetics if not the nutrigenomics of reproduction, recently, the Dairy Herd Improvement Association has added daughters' pregnancy rate, or days open, as a trait for bull selection, and other countries have done similar work (VanRaden et al. 2004; Harris 2005; Weigel 2006). The mass of empirical genetic data now show that in fact important traits of reproduction are heritable, as we have recognized for nutrient use for decades. We also know that the genome must be "properly fed" to fully express its potential. Much has been written in the last decade of the declining fertility of Holstein dairy cattle, primarily in the United States, with a myriad of suggested mechanisms, many of which actually have little data to back them up (Royal et al. 2002a, b; Chagas et al. 2007).

It is oft repeated that "increasing milk production decreases fertility" and many statistics are cited to "prove" fertility is lower in dairy cattle today. Yet on many herds and many hundreds of thousands of cows, simultaneously fast rates of milk secretion, feed intake, and good fertility (any way you measure it) occur all the time. Also, the recognition that the end result of successful rebreeding during lactation can be a selectable trait in the bull proofs "proves the point" that in fact, fertility is heritable, that it has many control factors involved, and that there is no direct no overriding reason why all "high-producing dairy cattle" should be sub-fertile.

The genetics of reproduction has not been ignored in the pork industry either. It has long been recognized and acted upon that sow traits, including reproductive processes such as ovulation, litter size, and return to estrus, have measurable heritabilities (Bergsma et al. 2007). Also, the interactions of nutrition and reproduction have been given strong attention, and the results are impressive (Bergsma et al. 2007, and many references therein) and the beginnings of a nutrigenomic awakening are present (Dawson 2006; Bonnet et al. 2008). An interesting anecdote applies here. In the ancient past (late 1960s) when this author was slaving away helping to raise pigs (and milk cows, you could do both at once then) in Bureau County, Illinois, he was oft teased by the wise old farmers that "you college types" could talk all you want about weaning nine pigs per litter, getting 2.3 litters per year, or you name it, but "it doesn't work out here in the real world." Well, funny enough, now the "real world" is showing us college types that in fact it does work and we better catch up with our research!

The long overdue "admission" that reproductive traits are heritable has already begun to improve reproduction in dairy cattle and pigs. If pregnancy rate or days open can be used as a selection trait, we should be able to describe the biochemical mechanisms that make it so. The accepted generality that "well fed" cattle are more fertile, coupled with the renewed focus on genetics of fertility, and the simultaneously fast rates of milk secretion and normal fertility support the concept that nutrigenomics and nutrigenetics are the "Venn Diagram" of nutrient use, genetic traits, and reproductive success.

18.3 Mechanistic connections between nutrient flux and reproductive processes

18.3.1 Integration of reproductive processes and nutrient flux

The earlier observations and research findings introduced above have given us

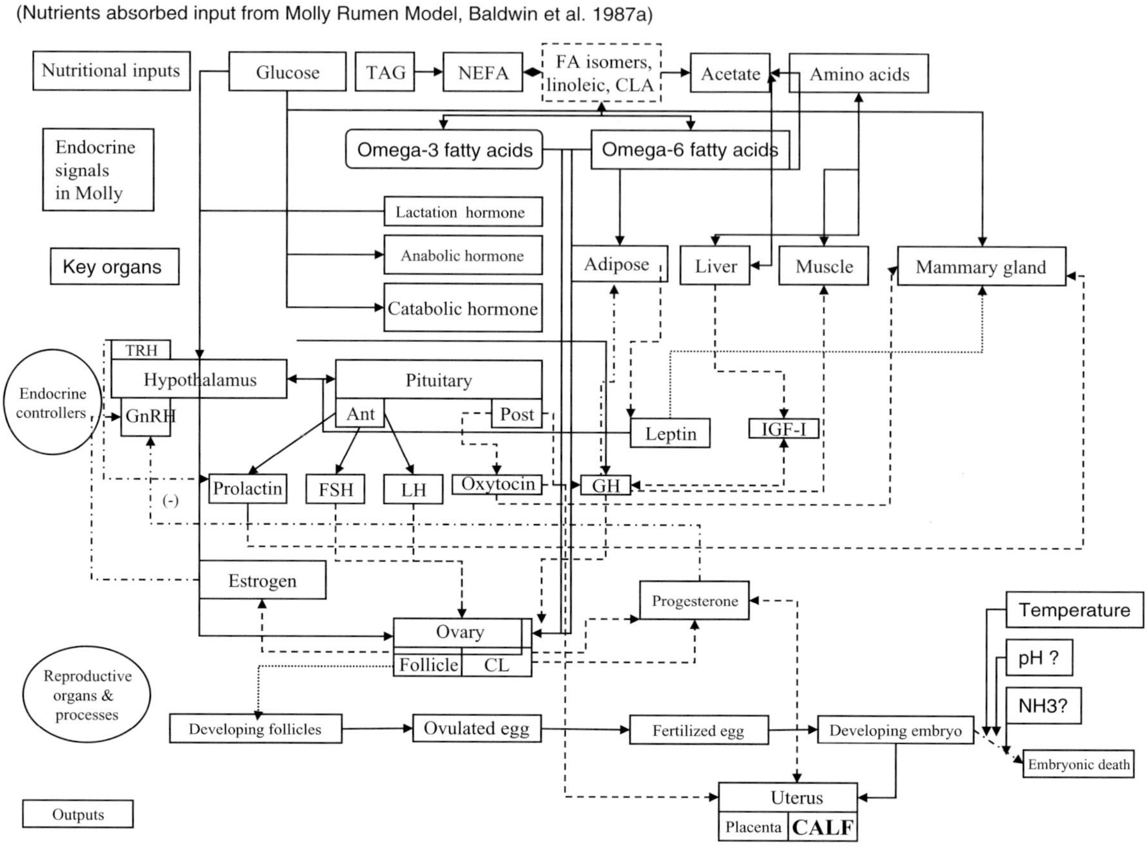

Figure 18.1 Schematic flow diagram of a model of nutrient flux and reproductive functions.

sufficient knowledge to seriously study the integrated functions of nutrient use, genetic expression, and reproductive processes. Figure 18.1 presents a simple flux diagram of a model of nutrient use and reproduction. An old, simpler version was published previously (McNamara 2005). Another more conceptual one can be found in the excellent review of Chagas et al. (2007). The flux diagram is basically species-independent, although there will be some variation among species in the mechanisms of control. It is aggregated at the nutrient flux level, not at specific biochemical reactions or gene transcription events in order to describe the basic processes in an animal that connect nutrient use and reproductive function. In addition, care is given to use the major driving factors as states and signals (hormones) as connections between states and fluxes. Lower levels of metabolic control (specific enzymes or gene transcription events) are in fact the mechanisms that define the "arrows" in the flux diagram.

In order to understand the context and details of such a model, we must revisit in brief research that allowed this flux diagram to be constructed. We can also point out components that have strong justification and validation, and others that are based on much less data. Then, we will revisit this model with an eye to how to move forward in an ordered nutrigenetic and nutrigenomic framework.

We can simplify the cycle of reproductive events to original sexual maturation, first ovulations and ability to conceive, to successful gestation, to the first lactation, and then, often, to renewed ovarian cyclicity and a second (and subsequent) gestation and lactation. In most species, certainly cattle and pigs, females must reach a certain physiological maturity before the hypothalamus, pituitary, and ovary can fully communicate and function to develop an oocyte capable of becoming fertilized (Senger 2004). The specifics of development of these reproductive organs are covered elsewhere in this book. The nutritional development of fertility is both direct and indirect. There is likely not any one nutrient that directs the first follicular waves, estrus behavior, and ovulation, but the end result of nutrient flux allowing development of mature organs (such as adipose tissue), and adequate glucose availability. The growth rate of the animals, a function of both genetics and nutrient supply, dictates that animals will arrive at a body composition and glucose flux state that supports the actions of the hypothalamus, pituitary, and ovary. Early research discovered that this initial fertility was much less a function of age than of physiological maturity (roughly monitored by body composition), and modern domestic breeds (cattle, pigs, poultry) certainly reach physiological maturity, and become pregnant, at much earlier ages than they did previously, based on our selection pressure on growth rate. Even when physiological maturity is reached, in general, the animal must still be "well-fed"—at maintenance or in positive energy balance, with sufficient circulating blood glucose to support follicular development and eventual LH release and ovulation.

18.3.2 The role of glucose

A key controller in the connection between the brain and the ovary for follicular development and ovulation is glucose. Glucose is known to have a direct effect on the hypothalamus that causes the release of GnRH, which in turn causes LH release from the pituitary (Wade and Jones 2004; Senger 2004). In addition, glucose elicits increases in circulating insulin and IGFI, which have positive effects on follicular growth. Although there appears to be a wide range of "effective" glucose flux rates or circulating concentrations to allow these effects, they are still critical. This is one reason that in most cases, fertility is not affected negatively until a serious deficit in glucose happens. In lactating sows and cattle, return to estrus after parturition is also closely connected with adequate glucose flux returning after the mammary gland starts to use large amounts of glucose and prior to sufficient increase in glucose. There are other aspects to the full return of fertility postpartum, but glucose is an important factor. The role of glucose in stimulating insulin and IGFI is likely also important in the return to ovulation of viable oocytes after parturition.

Thus it is not only nutrient flux effects on ovulation that are important, but on development of a viable oocyte and, perhaps, support of a uterine environment conducive to blastocyst development and implantation. Glucose likely plays a role via stimulation of insulin and IGFI, which helps to support anabolic metabolism and oocyte development.

18.3.3 The role of fatty acids

Certain classes of fatty acids, primarily the omega-3 series and omega-6 series and their metabolites, have also been identified as positive controllers (Ambrose et al. 2006; Bilby et al. 2006a,b). Some intriguing results have been reported in practical use of omega-3 and omega-6 fatty acids in improving fertility in lactating dairy cattle (Bilby et al. 2006a,b) empirically, yet molecular mechanisms here are not understood. Likely candidates include control of basic cell development and membrane function, and in reduction of inflammation or inflammatory molecules that may hinder oocyte development (Trayhurn and Wood 2004; Webb et al. 2004; Ambrose et al. 2006) There is ample evidence that specific fatty acids can alter gene expression in many tissues (Al-Hasani and Joost 2005, and many references therein). Thus the ground for finding specific nutrigenomic mechanisms for fatty acids and reproduction is quite fertile, so to speak. An obvious problem is that the massive complexity and thus large number of possible permutations will take the "will" for scientists to focus, together, on biomathematical solutions and complex models to move forward.

18.3.4 Early embryonic losses and nutritional status

Such challenges notwithstanding, as long as the mother is in a reasonably good range of nutrient supply, the embryo will develop normally. In dairy cattle, much work has been done to investigate early embryonic losses, usually categorized as animals first diagnosed pregnant (28 to 42 days post breeding) then showing open. The other losses from breeding to showing estrus again have myriad causes, including body temperature (Chagas et al. 2007) and uterine pH and ammonia concentrations (in turn likely a function of body temperature). Early research connected, I think mistakenly, early embryonic losses directly with excess protein leading to changes in uterine ammonia and pH, but follow-up research failed to show a strong connection. Lately, it has been suggested, based on larger empirical and mechanistic studies, that even moderate heat stress, leading to an increase in body temperature as little as 0.5 or 1°C can alter the uterine environment (pH and ammonia concentration) that may hinder embryonic development. Students of physics and chemistry will recognize the potential of the Arrhenius equation at work here: in general, for every 10°C decrease or increase in temperature, all chemical reaction rates are halved or doubled (McNaught and Wilkinson 1997). A 1°C difference could mean a 10 % change in metabolic buffering reactions, which could easily affect embryonic development. This is an exciting ongoing area of research, which promises to improve our understanding of nutrient use, the environment, and fertility. Because gene transcription events are functions of metabolic reaction rates, there is (an admittedly broad) potential involvement of nutrigenomic mechanisms.

In addition to the glucose, fatty acid ,and heat stress effects, there is potentially a role between protein nutrition, amino acid metabolism, genetics, and fertility. This is likely not a major function of dietary protein, but a subtle interaction between amino acids, gene transcription, and endocrine regulation. Genomic studies have suggested a connection between variants in the myostatin and calpastatin genes and fertility in the cow (Garcia et al. 2006; Mitchell et al. 2006; Chagas et al. 2007; Szyda and Komisarek 2007). The protein myostatin may in fact regulate glucose uptake in reproductive and other organs (Mitchell et al. 2006). These intriguing studies may provide initial evidence for a mechanistic link between protein metabolism, gene expression, and reproduction in a true nutrigenetics and nutrigenomics way.

18.4 History of integration of physiological state, nutrient flux, and reproduction

18.4.1 Hammond's seminal work

We cannot write a chapter on nutrition, genetics, and reproduction without a nod to what is likely the seminal work and first true study in this area by Sir John Hammond: "Physiological Factors Affecting Birth Weight" (Hammond 1944). In this artilce the concept of the partitioning of nutrients was first put forth. This concept captured the idea that each organ has a priority for nutrient use, with the brain having the highest priority, metabolic organs less, and muscle and adipose even less. However, once pregnancy or lactation occurred, these organs moved up to or close to the priority for the brain and in fact may alter the priority of other tissues. Dr. Hammond did a tremendous amount of research on genetics, nutrition, and reproduction before our knowledge allowed us to become much more specific, reductionist, and thus segregated as "geneticists," "nutritionists," and "reproductive physiologists." He was the first to pose the question (in print at least) of the connection between the genetics of the sire and dam, and thus the fetus, and the use of nutrients, for a specific reproductive outcome.

In an elegant experiment he bred large-breed horse sires to small-breed horse dams, and small-breed sires to large-breed dams, and proved the point (which any horse breeder knew but scientists had not figured out yet) that the fetus can direct nutrients to itself to meet its "pre-programmed" genetic pattern. Although that a large breed sire would produce a smaller offspring in a smaller dam, and vice versa was known, there was no understanding of why at the time. Why did not a large breed sire produce a huge foal in all dams? Of course, there is a range—breeding large sires to small dams will statistically result in larger foals, but not as large as they would be in larger dams. Dr. Hammond posited that there were some possible "special growth substances of maternal origin" that directed the "partitioning of nutrients" to the organs most important in that physiological state, and also that "…some limiting internal secretion or metabolic substance produced by the mother, as a controlling factor in foetal growth." However, he recognized that "It is possible that the limitation of the size of the crossbred foetus in the small mother is brought about by a higher rate of metabolism of the maternal tissues in the small breed than in the large breed, that is, by the greater competition of the maternal tissues."

The brain is always the most important (as the body will always sacrifice other organs when faced with nutrient deficits). During reproduction, however, the fetal-placental unit "takes charge" and directs nutrients to itself, even during periods of nutrient deficit. When lactation begins, the mammary gland does the same thing. The work of Dr. Hammond, lacking in chemical and biochemical technology, laid the conceptual framework for a large part of ensuing work in the nutrition, genetic, and reproductive biology of successful pregnancy and lactation.

18.4.2 Homeostasis

At that time the concept of homeostasis in physiological flux was well accepted. The interactions of glucose and insulin were known. But Dr. Hammond introduced the concept of a higher level, long-term regulation of metabolism, embodied in the term "partitioning of nutrients." Some factors altered the normal homeostatic fluxes such that, while insulin still stimulated glucose uptake in insulin-sensitive cells, the set point of control was tipped toward ensuring that the fetal-placental unit, or the mammary gland, was the highest priority tissues for glucose and other nutrient flux. Hormones of pregnancy were later identified as the factors controlling metabolism in maternal tissues, and mechanisms were later identified such as alterations in hormone receptor content, activity, and/or cellular responses.

18.4.3 Homeorhesis

Years later, based on a then large body of data on nutrition and endocrinology of pregnancy and lactation (summarized below), the concept of homeorhesis, the long-term alteration of metabolism in support of a dominant physiological state was developed (Bauman and Currie 1980). This concept embodies the importance of changing the partitioning of nutrients to support reproduction. Nutrigenetic and nutrigenomic mechanisms certainly play a role in this control.

18.5 Nutritional physiology of pregnancy and lactation

18.5.1 Pregnancy

Homeorhesis, nutrigenomics, and nutrigenetics are exemplified by metabolic control in pregnancy. The anabolism of pregnancy has been recognized for a long time: even on somewhat restricted intakes, pregnant animals will accrue more adipose tissue than similar nonpregnant ones. A major causative factor in this anabolism, which is usually strongest in early to mid-pregnancy is progesterone from the corpus luteum. Progesterone directs the ovary, uterus, and hypothalamus to cease follicular development and ovulation (Senger 2004). However, progesterone also directly affects metabolism in the liver, adipose tissue, and muscle, and promotes development of the mammary gland! Liver, adipose tissue, and muscle increase the synthesis of fat and protein to be used by the fetus, even at an unchanged feed or energy intake. To ensure support of fetal development and growth, however, progesterone also helps to increase food intake. The nutrigenomic mechanism is such that progesterone alters the transcription of anabolic enzymes to make fat or amino acids, and alters the gene expression of insulin or beta-adrenergic receptors or downstream signaling molecules. Then, when sympathetic nervous system release of norepinephrine or pancreatic release of insulin occur to alter glucose or fatty acid flux, the tissue responds, but at different rates, such that glucose is used for body fat synthesis or lipolysis is decreased more than normal to store body fat.

Then, in mid to later pregnancy, anabolism switches to catabolism to use the stored energy. The fetal-placental unit secretes a protein hormone that directs the liver, adipose tissue and muscle to supply an increased amount of fat and protein to be used by the fetus, even at an unchanged feed or energy intake. The hormone is named differently among species, but is basically the same protein: placental lactogen or chorionic gonadotropin. In some species (e.g., rabbits), the second cousin of this group, prolactin, serves the same role. These two hormones are part of three called the "placental lactogen gene family" and comprise a huge role in the reproductive success of many species (Harris et al. 2004). We will get to the third member later. In addition to supporting fetal development, these hormones also help control mammary development.

The nutrigenomic mechanism involved is that pregnancy hormones bind to receptors in the adipose tissue, for example, and alter the gene expression of insulin or beta-adrenergic receptors or downstream signaling molecules. Then, when sympathetic nervous system release of norepinephrine or pancreatic release of insulin occur to alter glucose or fatty acid flux, the tissue responds, but at a different rates, such that glucose use for body fat synthesis is decreased or lipolysis is increased more than normal to supply nutrients to the fetus(es) or mammary gland (McNamara 2005, 2006).

On the practical side, we use the knowledge of pregnancy anabolism in many agricultural arenas, and often "limit feed" pregnant animals, especially swine, during gestation to avoid the animal becoming too fat, which can lead to problems in lactation. This useful practice embodies the concept of partitioning of nutrients: we can limit feed a pregnant animal knowing that, within a range, the fetuses will "take care of themselves" and grow to a healthy weight, neither too small nor too large.

There is a large body of knowledge in many species on nutrient requirements for a successful pregnancy (including optimal fetal development), and also on the nutritional problems with pregnancy, including small or low birth weight, pregnancy diabetes, pregnancy toxemia, ketosis, and dystocia (Senger 2004, and other). Although beyond the scope of this chapter, the study of epigenetics, potential changes in the DNA of developing fetuses, is one nutrigenetic mechanism that is most exciting: Can specific nutrients or metabolites actually change the genomic or transcriptomic framework such that permanent alterations in fetal and later neonatal development ensue? I think the case for such a role in "extreme" situations such as small or large birth weight has been made. It is interesting to note that this exciting new area of research, with tremendous potential for improving human and livestock life and productivity, can be traced back directly to Dr. Hammond's original work.

18.5.2 Lactation

The same concepts apply to the final reproductive process in mammals—lactation. The pathway to a successful independent next generation must follow through the lactation for ultimate success. The hormones of pregnancy not only direct nutrients toward the fetus, but also begin the process of mammary development. Progesterone, estrogen, placental lactogen, prolactin, insulin, IGF1, and corticoids all play a role in mammary development. Significant amounts of nutrients are not usually needed in pregnancy for mammary development, but late in pregnancy and at lactation, the mammary gland goes "from 0 to 60" in a short period. Modern sows can make 700 g of lactose, 600 g of fat, and 450 g of protein a day. Work done many years ago demonstrated that the use of nutrients was not just a function of "the giant sucking sound" (also known as "metabolic pull" or more specifically "a demand function") from the mammary glands, but a coordinated effort of the hormones of pregnancy and lactation.

In late pregnancy, progesterone concentration declines and allows the lactogenic hormones to initiate the expression of several genes for catalysis of milk component synthesis. In addition, these same hormones, primarily progesterone, prolactin (placental lactogen in some species) and somatotropin, the final member of the placental lactogen gene family, coordinate metabolism in the adipose tissue and muscle to help direct glucose, fatty acids, and amino acids to the mammary gland for milk synthesis. The importance of the muscle cannot be overlooked, as it must supply many of the essential and nonessential amino acids for milk synthesis, and also glucogenic amino acids to the liver for conversion to glucose for lactose synthesis. We have tended to concentrate on body fat, for good reason, but a modern sow can lose 1 kg of muscle a day for several days during lactation. Also, as noted above, amino acid or protein metabolism may yet play a role in fertility at least in dairy cattle (Mitchell et al. 2006; Chagas et al. 2007).

We were able to conduct a series of studies during the 1980s and the 1990s that detailed several of the enzymatic and flux changes that occurred during late pregnancy and early lactation in dairy cattle and pigs (McNamara et al. 1985; McNamara 1998; McNamara and Boyd 1998; McNamara 2005, 2006, and several references therein), and laid a framework for the nutrigenomic work that was ongoing. In short, we investigated how animals of different genetic merit (in dairy cattle), or litter size (in pigs), fed different amounts of energy, expressed the receptors and enzymes of anabolism or catabolism. Expression and activity of enzymes and beta-adrenergic receptors in adipose tissue were increased from late pregnancy to early lactation (Parmley and McNamara 1996). Animals of greater genetic merit had a greater activity of the receptors and enzymes that controlled lipolysis, and had a greater fatty acid release, even at the same intakes. Animals limited in feed intake, however, reduced expression of lipogenic enzymes dramatically and had a lower response in lipolytic control. We demonstrated that there were clear differences in metabolic control that were either functions of the genetic merit of the animal, or functions of the diet, but that of course, were interconnected. Although a direct connection to fertility had yet to be made, these studies demonstrated that it was not just "body condition score" that related to lactational success.

Just last year, we were able to finally identify some of the genes involved in adipose tissue metabolic control in lactation (in dairy cattle) using RT-PCR and transcriptome arrays. We sampled several animals with a range of genetic merit for milk production and fed the same diets. We identified the beta-adrenergic receptors (all three subtypes), hormone sensitive lipase, and its cofactor perilipin as all increasing in transcription from pre-partum to postpartum, in the adipose tissue of dairy cattle (Figure 18.2). In addition, we extended and confirmed earlier findings that there is a reduction in expression of enzymes controlling and supporting lipogenesis (Figure 18.3).

As expected, however, there was great variation from animal to animal in both lipogenic and lipolytic control genes (Figure 18.4), suggesting that there is room for large individual genetic responses to diet and physiological state. The figure on the animal variation is presented to make a strong point here. Most nutritionists, physiologists, and reproduction specialists have been trained that to do an experiment, you need to reduce as much among animal variation, to get as homogenous a genetic pool as possible to increase the likelihood that you can demonstrate "statistical significance" in whatever hypothesis you are testing. In this way, many would look at these figures and "wave them away": "you have nothing here." This

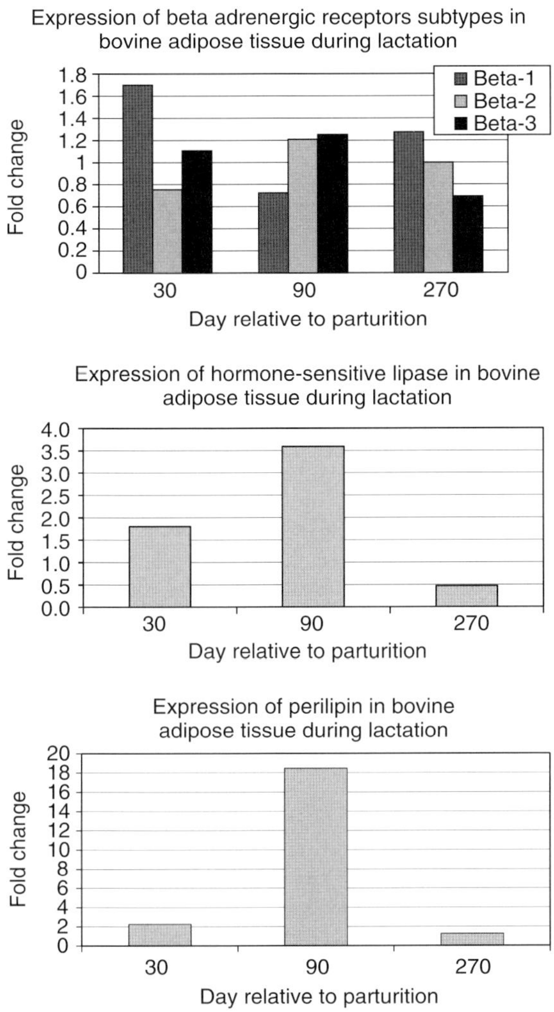

Figure 18.2 Expression of beta-adrenergic receptor subtypes, hormone sensitive lipase, and its cofactor perilipin in the adipose tissue of dairy cattle in lactation. Data are the fold change measured against the expression at 30 days pre-partum, measured by RT-PCR. Control of lipolysis in adipose tissue is a major contributing factor to reproductive success, including a successful lactation and fertility for rebreeding. Further mechanistic knowledge on control of adipose tissue metabolic control will help to define the specific roles in supporting reproduction.

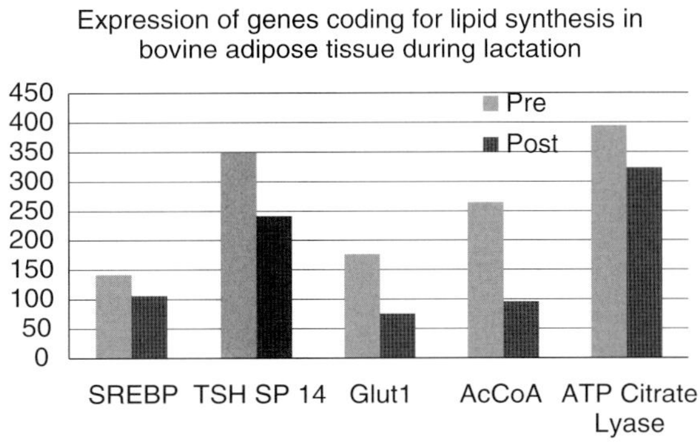

Figure 18.3 Expression of sterol regulatory element binding protein (SREBP), thyroid stimulating hormone spot 14, glucose transporter 1, acetyl CoA carboxylase, and ATP citrate lyase in adipose tissue of dairy cattle at 30 days pre-partum (pre) and 14 days postpartum (post). Data are in signal strength from the Affymetrix Bovine Gene Array, normalized to 125. Data show a consistent reduction in the family of control proteins that regulate carbohydrate conversion to fatty acids. Reductions in lipogenesis can affect overall reproductive efficiency, from a successful lactation to rebreeding in dairy cattle. Using gene array data and mathematical models, we can ask questions directly related to the causative relationships underlying fertility and lactation.

philosophy can be understood based on a lack of understanding of complex systems. In some cases you want to isolate the "nutrition" from the "genetics" to identify specific mechanisms. However, it is this author's opinion (and shared by some others), that in ignoring the genetic variation in response to nutrition, tremendous knowledge has been missed along the way, and it took the "genomic" work to bring scientists back together to understand the direct and undeniable connections between nutrition, reproduction, and genetics. The pace of advancement in appreciating and understanding complex biological systems is already increasing because of this.

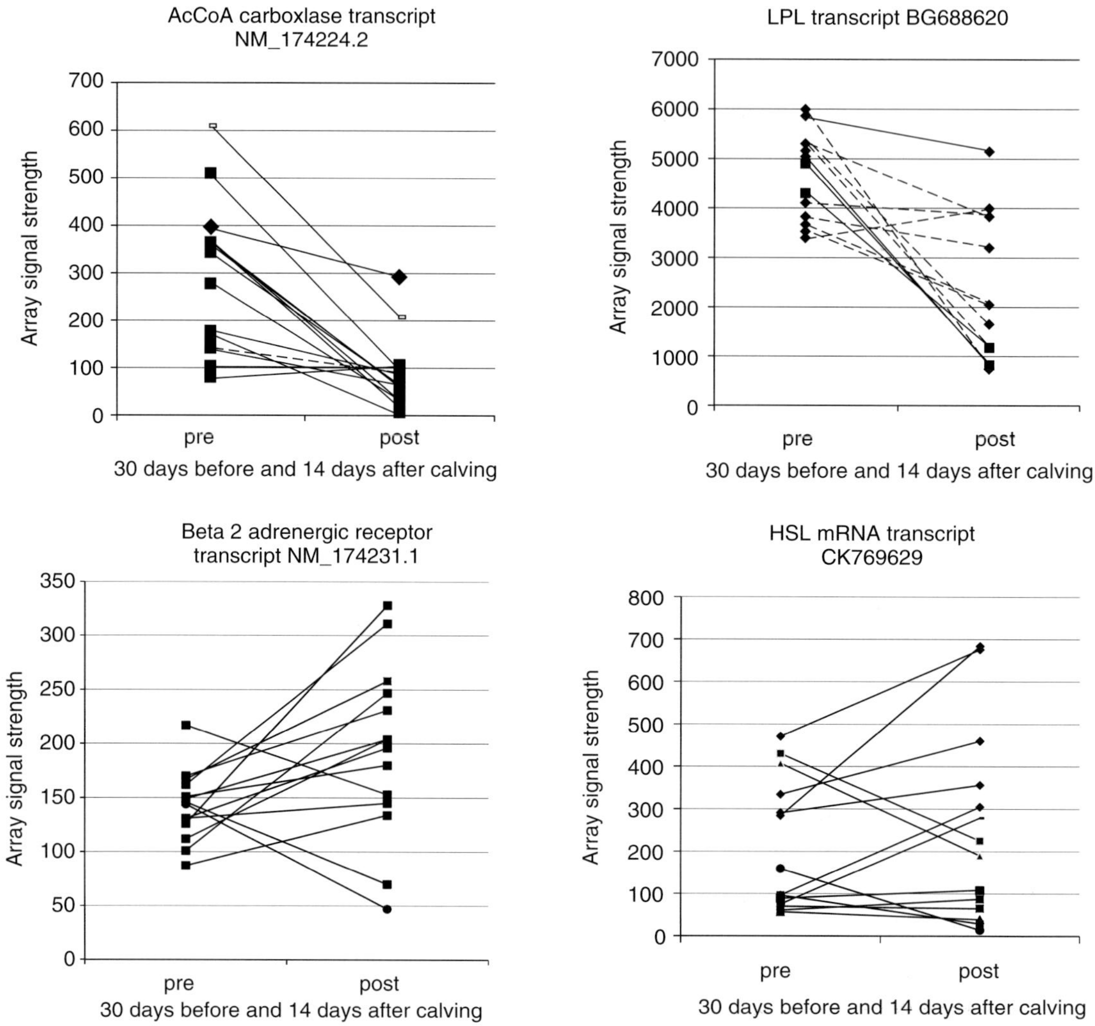

Figure 18.4 Expression of mRNA sequences for two major anabolic control enzymes and two lipolysis control proteins. Samples taken from 11 Holstein dairy cattle in same lactation, fed same diets. The question arises—What are the causes and consequences of animal variation in gene expression in adipose tissue to overall reproductive physiology?

We are presently conducting more specifically designed studies to determine the range in nutrigenomic response to diet in animals of varying genetic merit. Because the adipose tissue secretes several molecules that may affect ovarian function, there is potential for identifying some important control factors through this approach.

It is through the continuous loop of nutrient intake, hormonal response, gene expression, and nutrient partitioning to various organs of metabolism and reproduction that the nutrigenetics and nutrigenomics of reproduction occur. Now, with this brief recap of 60 years of directed effort, we can move forward.

18.6 Nutrigenetics and nutrigenomics approaches for improved fertility, pregnancy, and lactation

If one does a literature search (May 2008) on "nutrigenomics of reproduction" or various iterations, one does not find much. However, "nutrition and reproduction (or fertility)" yields a lifetime of reading material. The study of the role of gene expression in either nutrition or reproduction has been fairly extensive. It is up to interested scientists to now "make the connections", as we have done in Figure 18.1. The success of efforts to unravel the nutrigenetics and nutrigenomics of reproduction will rely on the construction, testing, and refinement of mechanistic, dynamic, biomathematical models of nutrient use and reproductive processes. The remainder of this chapter will present in brief two pertinent examples of such models, and some specific examples of transcriptomic work focusing on nutrient effects on gene transcription related to reproductive success, and how such knowledge can be integrated into mechanistic models.

The flood of information from the various genome works and the ability to generate large volumes of transcriptome data from animal studies have renewed calls for more integration of knowledge, including using biomathematical approaches. A model or a modeling approach to research may also be defined as an ordered way of describing knowledge of some real complex system. Such models have been useful in practical systems to describe, for example, drug metabolism, biochemical pathways, and nutrient requirements. A quantitative description of metabolic transactions is critical to enhance understanding and improvement of nutrient requirements, health, and longevity. Models of increasing complexity, ever grounded in validated research data, will continue to improve our quantitative understanding. It is this author's experience that information from genomic research can only be understood with the means of complex model systems, a philosophy shared by others (Dawson 2006).

18.6.1 The acceptance of integrative biology is critical

A major barrier to improvement of models remains lack of an accurate description of the phenotype of the animal being modeled, expressed as, for example, gene transcription control, enzyme activity, hormone and receptor kinetics, and intracellular signaling. If we are going to integrate nutrient status and reproductive physiology into research and practical systems, we need an ordered approach centered on well-constructed biomathematical, dynamic models of nutrient flux and reproduction.

An additional barrier continues to be the thought processes of scientists who are not trained in more complex regulation and theories and are uncomfortable with the ideas or skeptical of the value of integrative biology. The genome projects themselves are starting to change those attitudes, especially in younger scientists, because the central nature of gene transcription in metabolic regulation is better understood now than before, and because the sheer mass of information generated in genomic and transcriptomic work dictates mathematical methods and approaches to bring clarity from the data.

One underlying concept to such integrative work is that the amount and activity of all enzymes and hormones are genetically regulated, from immediate gene transcription and translation, to heritability of variations in hormone and enzyme synthesis and secretion. Some examples may be found in Girard et al. (1997) and Cornish-Bowden et al. (2007). However, some have small heritabilities, or are expressed constitutively, are members of redundant control systems, and are thus not relevant to metabolic control (Cornish-Bowden et al. 2007). To quote from a recent review on nutrition and fertility: "...reduced fertility is not caused simply by changes in management but also by changes in the genotype and underlying metabolic processes... . The demands impose by lactation intcract with the genetic makeup of the cow to have a major negative effect on the reproductive system..." (Chagas et al. 2007). That summarizes the genetic connection to "everything"; now the devil is in the details.

The statement "nutrigenomics for reproduction" explicitly recognizes (finally) that all three processes (genetics, nutrition, reproduction) are integrated without possibility of separation. That integration must be codified in a model of nutrient use and reproduction. The objective of a model dictates (or should) the model components. If an objective is to model metabolic flux in any one species of animal, allowing for description of variation among animals, then genetic control by definition must be included. There is not space to recapitulate all the proper and improper uses of models, or their importance to true understanding of complex systems. The reader is directed to some key references to pursue that further, but regardless of one's personal experience or opinion of the use of model systems in research, their importance and utility and effectiveness cannot be denied (Carson et al. 1981; McNamara et al. 1991; Pettigrew et al. 1992a,b; Baldwin 1995; NRC 2001; McNamara and Pettigrew, 2002a,b; Baldwin 2005; McNamara 2005, 2006; Cornish-Bowden et al. 2007).

18.6.2 A present basis for a nutrient-reproduction model

For most research models, the objective is to provide a framework to organize complex information to describe a system, set and test complex hypotheses, and evaluate usefulness of data for improving our quantitative understanding of complex systems. The objective of the model(s) described here are to develop dynamic, mechanistic models of digestion and metabolism (in cows or pigs) suitable for evaluation of hypotheses regarding underlying patterns of nutrient use and reproductive processes.

There already exist two solid and validated frameworks for models of nutrient flux that can provide a basis for a nutritional reproduction model in cattle and swine. The first is the 40 years of modeling work of Baldwin and his many colleagues (Baldwin et al. 1987a,b; Baldwin 1995), which led to tremendous improvement in understanding of the mechanistic connections between diet and animal performance. The model in question is titled "Molly" and the full history and detail can be found in the previous reference.

In 1968, Dr. Baldwin published an article titled "Estimation of Theoretical Calorific Relationships as a Teaching Technique: A Review." (Baldwin 1968). In it he described the aggregate biochemical pathways that in fact were the components of the net energy system of feeding cattle, a work that was just wrapping up after about 100 years of effort across the world by many scientists (Lofgreen and Garrett 1968; NRC 1968). This connection between the mechanisms of nutrient flux and practical, empirical cattle feeding led to 40 years of work on developing biomathematical models of nutrient use, and "spun off" many other related efforts.

Stemming from that work came the model of nutrient use in the sow, "Susie," developed by Pettigrew and colleagues (1992a, b) and since then developed and presently being extended to reproduction (McNamara 2005). This effort began more than 20 years ago, and in 1992, Jim Pettigrew and colleagues gave a start to the first model of nutrition and reproduction in pigs, and a direct quote from that article is in order (as I cannot say it any better!):

> The mechanisms connecting the diet to reproductive performance are presently unknown but may include variations in voluntary feed intake, digestion, absorption, metabolism of absorbed nutrients, and endocrine effects. Clear understanding and manipulation of this connection to optimize long-term sow herd performance requires ability to track, systematically and quantitatively, dietary effects through the various processes to reproductive performance. The project consists of the development of a mathematical model of one component of the connection, the metabolism of absorbed energy-containing nutrients, including amino acids, related to long-term feeding strategies in the lactation phase of the reproductive cycle of sows. (Pettigrew et al. 1992a)

These models describe pathway biochemistry, as aggregated pathways in a simple and scientifically correct fashion. There is not an attempt to model every reaction, but to model at the level of biological control most pertinent to the modeling objective. For a thorough discussion of the purposes and practices of metabolic models, see Baldwin (1995). We will focus on just two pathways in one tissue: lipogenesis and lipolysis in the adipose tissue. These are two critical pathways in fertility, as they are an important mechanism by which animals utilize excess glucose or respond to a deficit of glucose and direct fatty acids to reproductive tissues. The adipose tissue is chosen for its historic connection to fertility and for its more recent discovery as an endocrine organ (Mohamed-Ali et al. 1998). We are beginning to understand that some of the aspects of initial fertility (puberty) and successful pregnancy and lactation may in fact be the result of changes in endocrine activity of the adipose tissue. In addition, it is also recognized that many important reproductive traits in sows are heritable and responsive to nutrition (Quesnel et al. 2006; Schneider et al. 2006; Bergsma et al. 2007) A transcriptomic approach here can have great value in identifying the potential mechanisms involved and ruling out those that are not.

The genetic elements of any metabolic reaction can be incorporated into flux control models. Maximal rates and substrate sensitivities are genetically inherited and in some cases, have a measurable heritability. In an aggregate pathway, changes in substrate sensitivity can be measured (McNamara and Boyd 1998; McNamara 2005, 2006). We can also envision the V_{max} varying during the life cycle or by hormones related to environmental or physiological state.

Using lipogenesis as an example, kinetic flux can be described through the two Michaelis-Menten parameters of maximal velocity (V_{max}) and substrate sensitivity (K_m). The equation below is used in Pettigrew's model of metabolism in lactating sows to describe glucose conversion to body fat (Pettigrew et al. 1992a):

$$UglTs = vGlTs/(1+(MGlTs/cGl)),\ and \qquad (1)$$

$$MGlTs = MAGlTs*\left((cGlr/cGl)^{tAGlTs}\right); \qquad (2)$$

where MGlTs is the substrate sensitivity constant for glucose, and is controlled by the concentration of glucose, and by an "anabolic hormone," representing primarily insulin such that as glucose concentration rises (insulin increases); the sensitivity constant becomes smaller and reaction rate would increase. The representation of insulin (cGlr/cGl) is raised to a theta value that can alter the sensitivity of the reaction.

Let us explore the genetic elements in this equation. The V_{max} represents the total amount of catalytic activity available, in this case, in the sum of body adipose tissue (at other levels of aggregation, this may be in a specific organ, cell or single enzyme or receptor). This is controlled genetically, inhcrited from the parents. The V_{max} itself may be variable, decreased by periods of energy deficit that decreases the total mass of adipose tissue. The value of this parameter for a population of animals may be determined in studies combining direct measures of enzyme activity and measures of total body adipose protein content. The K variable, substrate sensitivity, is also inherited (but may not have a high heritability, as the catalytic sensitivity of this enzyme, as for most, is a function of the molecule, not concentration of the molecule).

The other important metabolic pathway in adipose tissue is the breakdown of triacylglycerol to fatty acids. This "fight or flight" syndrome has in fact generated so much interest over the years; the history of this quest is rich in itself, noted by many seminal research breakthroughs and leading to several Nobel prizes given for discoveries at many levels of metabolic control.

Triacylglycerol breakdown to free fatty acids is described as follows:

$$\begin{aligned} \mathrm{UtsFa} = {} & (\mathrm{vTsFa}/(1+(\mathrm{MTsFa}/\mathrm{Chl})))^{*} \\ & (\mathrm{Qts}^{**}0.67)^{*} \\ & (1-((\mathrm{QsTs}/\mathrm{Qts})^{**}\mathrm{thTsFa})) \end{aligned} \tag{3}$$

Such that the maximal velocity of lipolysis (vTsFa) is attenuated by a sensitivity constant, a hormone or hormones (Chl; lactation hormone) and is inhibited as body fat stores approach zero. The endocrine regulation of triacylglycerol lipolysis (UtsFa; Eq. 3) is recognized by introduction controlled by a maximal rate, as well as by "lactation hormone" (Chl),

The summary integration of metabolic flux is given here, using glucose as an example. If glucose flux (rate of entry, exit, or concentration) is important to reproductive organs (hypothalamus, ovary, uterus, mammary gland) then we must have a mathematical description of it to fully understand its mechanistic importance. It is glucose around which the major regulatory processes of the body have evolved. The regulatory mechanisms invoked as glucose availability changes have major effects on metabolic rates in other organs. Altering glucose use by food restriction, or by gene insertion for proteins such as the insulin-dependent glucose transporter result in changes in transcription of thousands of genes in mice (Fu et al. 2004).

Changes in glucose concentration or pool size (Gl) in the body are summed as:

$$\begin{aligned} DQGldt = {} & PaaGl + PabGl + PpaGl + PgyTsFa \\ & - UglTs - UgyFaTm - UgyFaTs \\ & - UglCd - UGlGc - UgyGlTs \\ & - UglLm - UglTm \end{aligned} \tag{4}$$

We sum the uptake of glucose, gluconeogenesis from amino acids (*PAaGl*), absorbed glucose (*PAbGl*), glycerol from lipolysis (*PGyTsFa*), and subtract the use of glucose for milk and body fat synthesis (*UGlTm, UGlTs*), use for glycerol in TAG (*UGyFaTm, UGyFaTs*), oxidation to carbon dioxide (*UGlCd*), glycogen (*UGlGc*), lactose (*Lm*). In this summative equation, all genetic effects are included in the equations describing each pathway as exemplified above.

The use of glucose has several dozens if not hundreds of possible control points throughout the body. Glucose use in the muscle *affects and is affected by every single other use of glucose* in the body. For example, the specific process in the muscle may in fact have a major effect on glucose dynamics, or might in fact be so overwhelmed or attenuated by other processes in other organs that the true physiological significance is minor. This becomes truly obvious only when we start to construct models that must make the connection. An example is that one animal or set of animals will have genetically controlled different maxima for gluconeogenesis from others, and this definitely will affect their glucose *and* amino acid use. In turn, this will affect nutrient and endocrine impacts on reproductive organs.

18.6.3 Integrating reproductive and nutritional functions

So now with a summary background, we can begin to construct equations that describe the fluxes represented in Figure 18.1, and thus the mechanistic connections between nutrient use and reproductive processes. Although the example given is for the cow, the same concepts apply to the sow. All variables are in mass, concentration, or rate of flux.

Fertilized Egg to Calf

Calf = Developing Embryo − Embryonic Death
Embryonic_Death_28 ∫ Σ [T°, pH, NH_3]
Embryonic_Death_45 ∫ Σ [T°, pH, NH_3]

These equations capture the following processes: a live calf is a function of a conception, minus the rate of embryonic or fetal death (here represented at 28 and 45 days post fertilization). Embryonic death is a function of uterine temperature, pH and NH_3 at day 28 through day 45 after conception.

Ovulated Egg to Fertilized Egg

Fertilized_Egg ∫ Σ [ovulated_egg, viable_sperm]
Max_Fert_Egg = 0.75 ovulated_egg

The equation here is based on data presented by J. Santos at the Dairy Cattle Reproductive Council Meeting in Denver, October 2006, in which he described that in fact, the rate of fertilization of an egg by the sperm, in dairy cattle, is approximately 75%.

Follicular Development and Dominant Follicle to Ovulated Egg

Ovulated_egg ∫ Σ [Dominant_follicle, Luteinizing Hormone].

Dominant follicle, second wave follicle, first wave follicle ∫ Σ [follicle stimulating hormone, progesterone, 1/estrogen, IGFI, insulin, glucose, 1/NEFA, growth hormone].

The equations here capture the knowledge that a dominant follicle, which will ovulate, is a function of three different waves in one cycle of 21 days. The first wave (recruitment) is a function, either directly or indirectly, of FSH, progesterone, insulin, glucose, and the reciprocal function of NEFA, estrogen and perhaps growth hormone. Here it is pertinent to state that after the original construction of potential equations, based on some knowledge, the next step is to actually find (or create through research) the data to set parameters for those equations. If no data are available, or research shows no such relation, then the equations are dropped from the model.

The role of the hormones of reproduction, in prose form as opposed to equations, include: progesterone is a function of the presence of corpus luteum, the placenta,

1/estrogen concentration, and of the rate of progesterone clearance in the liver. PGF2a concentration is a function of uterus PGF2a, and of luteal oxytoxin and perhaps of omega-6 fatty acid concentration. Luteinizing hormone is a function of the concentration of gonadotropin-releasing hormone (GnRH), low progesterone, and increased estrogen. Follicle-stimulating hormone is a function of the concentration of GnRH; of low progesterone and of the concentration of estrogen and inhibin. Gonadotropin-releasing hormone concentration is a function of the secretion of GnRH by the hypothalamus, which is a function of glucose concentration, the clearance of the GnRH by the liver, estrogen, low progesterone, and perhaps leptin concentration.

From these theoretical equations and functions, we see the connection of nutrient flux, primarily glucose, perhaps some specific fatty acids, perhaps NH_3 in the uterus (as a function of amino acid concentrations and also increased temperature) to reproductive physiology. The challenge, of course, is then to find sufficient data from the literature to set parameters for these equations. If none exists, specific experiments have to be designed to determine the parameters of the equation. If this is unsuccessful, then the scientists involved need to judge whether parameters cannot be obtained because the research tools are not there to measure them; there is some other reason for not being able to measure parameters (measuring the V_{max} of acetyl CoA Carboxylase in adipose tissue of a live sow is fairly difficult but can be done *in vitro*); or that in fact there is no mechanistic relation, and adjust the model accordingly. This is a process that scares many scientists, because it is much easier to say "glucose controls LH release" or "prostaglandin F2A causes regression of the corpus luteum" than to actually obtain data that allow a clear mechanistic, mathematical relationship to be established. But in the absence of the latter, no true understanding has been made. The experiment is not finished, even if we have constructed a large industry based on the empirical observations.

Empirical approaches have been excellent and have in fact made great strides in animal biology and production of food. But as biological scientists, if we cannot provide direct chemical, and thus mathematical, evidence, then we cannot truly move forward. The exactitude required in gene transcription control requires it. If as biochemists we are so strict on showing a direct molecular mechanism to "prove a hypothesis", why do we shy away from a mathematical one as required in physics or chemistry? In biology, as in physics and chemistry, they are one and the same.

In no case does an equation stay in a model unless there is a clear body of evidence to justify its inclusion. It is this author's opinion that if even just a few reproductive and nutritional scientists made a fair effort, we could have a working computer, dynamic, mechanistic model of nutrient use, and reproduction in cattle and pigs within 2 years.

18.6.4 One example of a transcriptomic approach to improve reproduction

Now, finally, we turn our attention to the title of the chapter. And with good reason, as stated earlier, we cannot invoke "nutrigenomics" of reproduction until we have laid the basis of nutrition, genetics, and reproduction. Here we will describe a recent experiment that we have conducted in dairy cattle, and refer to some other efforts that promise to be a starting point for integration of transcriptomic data into dynamic models of nutrient use and reproduction, in this case, of the dairy cow.

Adipose tissue has been discussed above. Several metabolic regulators and cytokines can be produced in and secreted from adipose tissue (Al-Hasani and Joost 2005). The objective of this study was to obtain a more in-depth understanding of the transcriptomic adaptations in adipose tissue of Holstein heifers from the transition from pregnancy to lactation, a key period in reproductive success—the establishment of lactation, and the "resetting" of the embryo and uterus for another ovulation and pregnancy.

We have conducted an initial analysis of the gene transcriptome in bovine adipose tissue during the transition from pregnancy to lactation (Sumner et al. 2009). We identified a set of heifers and first lactation animals that covered a range of genetic merits based on sires milk production transmission ability and the 305ME record of the first lactation animals. They were housed and fed similarly. We obtained adipose tissue by biopsy at 30 days pre-partum and 14 days postpartum and extracted the RNA. This was hybridized to the Affymetrix Genechip® Bovine Genome Array. Animals averaged 29.8 (SEM = 1.3 kg/d of milk for the first 60 DIM (range 18.6 to 44.8 kg/d). They lost 42.6 kg of BW (SEM 8.4, range +9.1 to −113.6) and 0.38 BCS units (SEM 0.10, range 0 to −1.0) from 0 to 14 DIM. This is a normal range for dairy cattle, housed and fed alike and gives a glimpse of the yet unknown effects of genetic variance in a similar population.

There are about 24,000 gene products on these chips, with an admittedly low level of confident annotation. Approximately 433 genes increased 100% or more, 3406 increased 25% to 100%; 1951 decreased 25% to 50%, 337 decreased 75% or more. Genes expressed in greatest amounts included collagen and ribosomal proteins, and fatty acid binding protein. Lipoprotein lipase was expressed at 4261 (SEM 509), the most highly expressed gene-regulating nutrient flux. Leptin receptor was expressed at 734 (50) pre-partum and was only 12% less at 14 DIM, thus leaving open the question if gene transcription is a mechanism for changes in leptin concentration in lactation (which is still an open question).

Genes involved in cell synthesis, transcriptional control, and inflammation increased fivefold or more, including beta-defensin, tenfold, cytokine inducible nuclear protein, eightfold, chromosomal reading frame 4, sixfold, sarcoplasmic Ca ATP-ase, fourfold, leucine-rich repeat-containing 2, 3.5-fold; voltage-dependent calcium channel subunit, 3.5-fold. Bos taurus uncoupling protein 3 increased threefold, indicating possible proton uncoupling in white adipose tissue. These data provide some initial insight into the global transcriptomic response of adipose tissue to lactation.

Anabolic pathway genes decreased ($P < 0.05$), including (mean (% change), (SEM)): SREBP, −25.1, (6.2); GLUT1, −57.3 (14.1); THRSP14, −30.8 (7.4); LPL, −48.4 (7.7), and AcCoA Carboxylase, −60.6 (13.0). The regression of transcript change on milk production was 0.18 for AcCoA carb and 0.26 for ATP-CL ($P < 0.05$). Lipolytic control elements increased, with much variation among animals, including Ca channel subunit 338% (203); B2AR 52.0 (8.8); PKC receptor 10.1 (2.6), and HSL mRNA 23.0 (17.9). The regression of transcript change on milk was 0.30 and 0.25 for B2AR and HSL mRNA.

Regressions among variables in a multivariate system are often misunderstood. Some scientists think only "high regressions" are important, while a geneticist can prove that an "R Squared" of 0.05 can mean millions of dollars when applied over many animals for several generations. The reality is, the regression is only what it is, and we can learn a fair bit from interpretation of linear, nonlinear, and multiple regressions.

We first need to understand the complexity of the animal, and obtaining a "high regression value" among variables at the organ and metabolic level is neither likely nor the objective. We need to have an ordered approach of both statistical and mechanistic, biomathematical research to identify the key components of a system. If in fact, 18% to 30% of the change in transcript amount for key metabolic control proteins can be related to milk production, that is a critical control point. If we categorize temporal phases of research of metabolic control into animal observations, empirical relationships, direct cause and effect studies, and building of biochemical and mathematical pathways, then we may say that we are presently in the next phase: "transcriptomic studies," both empirical and mechanistic. Upon reflection and interpretation, we can draw a direct line through every phase: "animals that make more milk at the same food intake lose more fat"; "increasing or decreasing feed intake at the same milk production alters body fat"; "animals with less body fat, or greater rates of body fat loss have decreased fertility"; "enzymes involved in fat synthesis and release vary with milk production and feed intake"; and "transcripts for enzymes involved in metabolic reactions relate to changes in milk production." With the iterative, supportive evidence at hand, we can, with some confidence, identify gene transcription of a few critical control proteins as a mechanism of the relationship of nutrient flux in the adipose tissue with reproductive processes (in this case lactation).

It remains to be mined from the data, if transcriptional regulation for proteins directly involved with ovarian function is altered in early lactation. Even if they are not, we will have a more complete picture of the adaptive mechanisms of nutrient flux to reproduction and vice versa. Changes in lipogenesis and lipolysis are functions of changes in gene transcripts, with lipogenesis more related to changes in flux not directly related to mammary function, and lipolysis more directly related. This result is completely consistent with all the animal feeding, metabolic flux, enzyme activity, and endocrine studies that have gone before, and has provided more knowledge of the mechanisms of the partitioning of nutrients to support reproduction.

Although direct studies of nutrigenomics of reproduction have been hard to find, we should note at least two other ongoing studies that relate directly to this. This is the effort of scientists at the University of Illinois to determine transcriptomic changes in the liver of pregnant and lactating cows as affected by lactation and plane of nutrition (Loor et al. 2005, 2006). There are also other transcriptomic studies in specific reproductive organs, but as yet a true nutrigenomic approach has not been reported (Rhoads et al. 2008a, b). It is only a matter of time and will that such an effort will expand.

18.7 Future research directions

We have a long way to go. We need a reinvigorated, multi-investigator, multidisciplinary integrated approach to solve the present and future problems of reproduction, and specific to the role of nutrigenetics and nutrigenomics for improved reproduction, this research effort will require construction and testing of mechanistic biomathematical models. Finally, we need to train students, scientists, and professionals in the importance of using integrative biology and biomathematical models to identify, solve, and prevent reproductive problems.

References

Al-Hasani, H. and Joost, H.-G. 2005. Nutrition/diet-induced changes in gene expression in white adipose tissue. *Best Practice & Research. Clinical Endocrinology & Metabolism* **19**: 589–603.

Ambrose, D.J., Kastelic, J.P., Corbett, R., Pitney, P.A., Petit, H.V., Small, J.A., and Zalkovic, P. 2006. Lower pregnancy losses in lactating dairy cows fed a diet enriched in alpha-linolenic acid. *Journal of Dairy Science* **89**: 3066–3074.

Baldwin, R.L. 1968. Estimation of theoretical calorific relationships as a teaching technique: A review. *Journal of Dairy Science* **51**: 104.

Baldwin, R.L. 1995. *Modelling Ruminant Digestion and Metabolism*. New York: Chapman & Hall, pp. 469–518.

Baldwin, R.L., France, J., and Gill, M. 1987a. Metabolism of the lactating cow. I. Animal elements of a mechanistic model. *Journal of Dairy Research* **54**: 74–105.

Baldwin, R.L., France, J., Beever, D.E., Gill, M., Thornley, J.H.M. 1987b. Metabolism of the lactating cow. III. Properties of mechanistic models suitable for evaluation of energetic relationships and factors involved in the partition of nutrients. *Journal of Dairy Research* **54**: 133–145.

Bauman, D.E. and Currie, W.B. 1980. Partitioning of nutrients during pregnancy and lactation: A review of mechanisms involving homeostasis and homeorhesis. *Journal of Dairy Science* **63**: 1514–1529.

Bergsma, R., Kanis, E., Verstegen, M.W.A., and Knol, E.F. 2007. Genetic parameters and predicted selection results for maternal traits related to lactation efficiency in sows. *Journal of Animal Science* **86**: 1067–1080.

Bilby, T.R., Block, J., do Amaral, B.C., Sa Filho, O., Silvestre, F.T., Hansen, P.J., Staples, C.R., and Thatcher, W.W. 2006a. Effects of dietary unsaturated fatty acids on oocyte quality and follicular development in lactating dairy cows in summer. *Journal of Dairy Science* **89**: 3891–3903.

Bilby, T.R., Jenkins, T., Staples, C.R., and Thatcher, W.W. 2006b. Pregnancy, bovine somatropin, and dietary n-3 fatty acids in lactating dairy cows: III. Fatty Acid Distribution. *Journal of Dairy Science* **89**: 3386–3399.

Bonnet A., Lê Cao K.A., Sancristobal M., Benne F., Robert-Granié C., Law-So G., Fabre S., Besse P., De Billy E., Quesnel H., Hatey F., Tosser-Klopp G. 2008. In vivo gene expression in granulosa cells during pig terminal follicular development. *Reproduction* **136**(2): 211–224.

Butler, W.R. and Smith, R.D. 1989. Interrelationships between energy balance and postpartum reproductive function in dairy cattle. *Journal of Dairy Science* **72**: 767–783.

Carson, E.R., Cobelli, C., and Finkelstein, L. 1981. Modeling and identification of metabolic systems. *American Journal of Physiology* **240**: R120–129.

Chagas, L.M., Bass, J.J., Blache, D., Burke, C.R., Kay, J.K., Lindsay, D.R., Lucy, M.C., Martin, G.B., Meier, S., Rhodes, F.M., Roche, J.R., Thatcher, W.W., and Webb, R. 2007. Invited review: New perspectives on the roles of nutrition and metabolic priorities in the subfertility of high-producing dairy cows. *Journal of Dairy Science* **90**: 4022–4032.

Cornish-Bowden, A., Cárdenas, M.L., Letelier, J.C., and Soto-Andrade, J. 2007. Beyond reductionism: Metabolic circularity as a guiding vision for a real biology of systems. *Proteomics* **6**: 839–845.

Dawson, K.A. 2006. Nutrigenomics: Feeding the genes for improved fertility. *Animal Reproduction Science* **96**: 312–322.

Fu, C., Xi, L., Wu, Y. ,McCarter, R., Richardson, A., Hickey, M., and Eun-Soo Han, E.-S. 2004. Hepatic genes altered in expression by food restriction are not influenced by the low plasma glucose level in young male GLUT4 transgenic mice. *Journal of Nutrition* **134**: 2965–2974.

Garcia, M.D., Michal, J.J., Gaskins, C.T., Reeves, J.J., Ott, T.L., Liu, Y., and Jiang, Z. 2006. Significant association of the calpastatin gene with fertility and longevity in dairy cattle. *Animal Genetics* **37**: 293–307.

Girard, J., Ferre, P., and Foufelle, F. 1997. Mechanism by which carbohydrates regulate expression of genes for glycolytic and lipogenic enzymes. *Annual Review of Nutrition* **17**: 325–352.

Hammond, J. 1944. Physiological factors affecting birth weight. *The Proceedings of the Nutrition Society* **3**: 8–13.

Harris, B.L. 2005. Multiple trait fertility model for the national genetic evaluation. Livestock Improvement Corporation. 1–24. www.aeu.org.nz/page.cfm?id=59&nid=35

Harris, J., Stanford, P.M., Oakes, S.R., and Ormanday, C.J. 2004. Prolactin and the prolactin receptor: New targets of an old hormone. *Annals of Medicine* **36**: 414–425.

Lofgreen, G.P. and Garrett, W.N. 1968. A system for expressing net energy requirements and feed values for growing and finishing beef cattle. *Journal of Animal Science* **27**: 793–806.

Loor, J.J., Dan, H.M., Evarts, R.E., Oliveira, R., Green, C.A., Jan Vick Gruretzky, N.A., Green, C.A., Rodriguez-Zas, S.L., Lewin, H.A., and Drackley, J.K. 2005. Temporal gene expression profiling of liver from periparturient dairy cows reveals complex adaptive mechanisms in hepatic function. *Physiological Genomics* **23**: 217–226.

Loor, J.J., Dann, H.M., Janovick Gruretzky, N.A., Everts, R.E., Oliveira, R., Green, C.A., Litherland, N.B., Rodriguez-Zas, S.L., Lewin, H.A., and Drackley, J.K. 2006. Plane of nutrition alters hepatic gene expression and function in dairy cows as assessed by longitudinal transcript and metabolic profiling. *Physiological Genomics* **27**: 29–41.

McNamara, J.P. 1998. Interaction of glucose and amino acid metabolism in lactating sows: Estimating internal parameters of a model of metabolism. In: McKracken, K. (ed.), *14th Symposium on Energy Metabolism of Farm Animals*, EAAP Publication No., pp. 32–35.

McNamara, J.P. 2005. Research, Improvement and application of mechanistic, biochemical, dynamic models: From genetics to kinetics. Chapter 6. In: *Mathematical Modelling in Nutrition and Toxicology. Proceedings of the 8th International Conference on Mathematical Modeling in Nutrition and Health Science*, Georgia Center for Continuing Education, pp. 87–110. Mathematical Biology Press.

McNamara, J.P. 2006. Modelling metabolism in the lactating sow: Mechanistic modelling at the metabolic level. In: *Pig and Poultry Modeling, Symposium*, South Africa, April, 2005, in press.

McNamara, J.P. and Boyd, D.E. 1998. Hormones as quantitative controllers. In: Kyriazakis, I. (ed.), *A Quantitative Biology of the Pig*. London: CAB International, pp. 199–225.

McNamara, J.P., Dehoff, M.H., Bazer, F.W., and Collier, R.J. 1985. Adipose tissue fatty acid metabolism changes during pregnancy in swine. *Journal of Animal Science* **61**: 410–415.

McNamara, J.P. and Pettigrew, J.E. 2002a. Protein and energy intake in lactating sows. 1: Effects on milk production and body composition. *Journal of Animal Science* **80**: 2442–2451.

McNamara, J.P. and Pettigrew, J.E. 2002b. Protein and energy intake in lactating sows. 2. Challenging parameters of a model of metabolism. *Journal of Animal Science* **80**: 2452–2460.

McNamara, J.P., Pettigrew, J.E., Baldwin, R.L., Walker, B., Close, W.H., and Oltjen, J.W. 1991. Information needed for mathematical modelling of energy use by animals. In: *Energy Metabolism in Farm Animals*, Gruppe Emahrung, Zurich Switzerland, EAAP Pub. No. 58, pp. 468–472.

McNaught, A.D. and Wilkinson, A. 1997. *Compendium of Chemical Terminology. The Gold Book, Second Edition*. Oxford: Blackwell Science.

Mitchell, M.D., Osepchook, C.C., Leung, K.C., McMahon, C.D., and Bass, J.J. 2006. Myostatin is a human placental product that regulates glucose uptake. *The Journal of Clinical Endocrinology and Metabolism* **91**: 1434–1437.

Mohamed-Ali, V., Pinkney, J.H., and Coppack, S.W. 1998. Adipose tissue as an endocrine and paracrine organ. *International Journal of Obesity* **22**: 1145–1158.

Mutch, D.J.M., Wahli, W., and Williamson, G. 2005. Nutrigenomics and nutrigenetics: the emerging faces of nutrition. *Faseb J* **19**: 1602–1616.

National Research Council. 1968. *A Net Energy System for Cattle*. Washington, DC: National Academies Press.

National Research Council. 2001. *Nutrient Requirements of Dairy Cattle*, 9th Edition. Washington, DC: National Academies Press.

Parmley, K.L.S. and McNamara, J.P. 1996. Lipid metabolism in adipose tissue of pigs fed varying amounts of energy. *The Journal of Nutrition* **126**: 1644–1656.

Pettigrew, J.E., Gill, M., France, J., and Close, W.H. 1992a. A mathematical integration of energy and amino acid metabolism. *Journal of Animal Science* **70**: 3742–3761.

Pettigrew, J.E., Gill, M., France, J., and Close, W.H. 1992b. Evaluation of a mathematical model of lactating sow metabolism. *Journal of Animal Science* **70**: 3762–3773.

Phillips, G.J., Citron, T.L., Sage, J.S., Cummins, K.A., Cecava, M.J., and McNamara, J.P. 2003. Adaptations in body muscle and fat in transition dairy cattle fed differing amounts of protein and methionine hydroxy analog. *Journal of Dairy Science* **86**: 3634–3647.

Quesnel, H., Etienne, M., and Pere, M.-C. 2006. Influence of litter size on metabolic status and reproductive axis in primiparous sows. *Journal of Animal Science* **85**: 118–128.

Rhoads, M.L., Meyer, J.P., Kolath, S.J., Lamberson, W.R., and Lucy, M.C. 2008a. Growth hormone receptor, insulin-like growth factor (IGF)-1, and IGF-binding protein-2 expression in the reproductive tissues of early postpartum dairy cows. *Journal of Dairy Science* **91**: 1802–1813.

Rhoads, M.L., Meyer, J.P., Lamberson, W.R., Keisler, D.H., and Lucy, M.C. 2008b. Uterine and hepatic gene expression in relation to days postpartum, estrus, and pregnancy in postpartum dairy cows. *Journal of Dairy Science* **91**: 140–150.

Roche, J.F. 2006. The effect of nutritional management of the dairy cow on reproductive efficiency. *Animal Reproduction Science* **96**: 282–296.

Royal, M. D., Flint, A. P. F. , and Woolliams, J. A. 2002a. Genetic and phenotypic relationships among endocrine and traditional fertility traits and production traits in Holstein-Friesian dairy cows. *Journal of Dairy Science* **85**: 958–967.

Royal, M. D. Pryce, J. E. Woolliams, J. A. and Flint, A. P. F. 2002b. The genetic relationship between commencement of luteal activity and calving interval, body condition score, production, and linear type traits in Holstein-Friesian dairy cattle. *Journal of Dairy Science* **85**: 3071–3080.

Schneider, J.D., Tokach, M., Dritz, S.S., Nelssen, J.L., DeRouchey, J.M., and Goodband, R.D. 2006. Effects of feeding schedule on body condition, aggressiveness and reproductive failure in group-housed sows. *Journal of Animal Science* **85**: 3462–3469.

Senger, P. L. 2004. *Pathways to Pregnancy and Parturition*. Pullman, WA: CCI Publishing.

Staples, C.R., Thatcher, W.W., and Clark, J.H. 1990. Relationship between ovarian activity and energy status during the early postpartum period of high producing dairy cows. *Journal Dairy Science* **73**: 938–947.

Sumner, J.M., Schachtschneider, C., Hutjens, A., Youngquist, A., Duncan, G., Rocco, S., Miller, J., Vierck, J.L. and McNamara, J.P. 2009. Regulation of dairy cattle adipose tissue metabolism by adrenergic control systems and gene transcription mechanisms dictating increased overall efficiency. *Proceedings of the International Society of Ruminant Digestion and Physiology*, 452–453.

Szyda, J. and Komisarek, J. 2007. Statistical modeling of candidate gene effects on milk production traits in dairy cattle. *Journal of Dairy Science* **90**: 2971–2979.

Trayhurn, P. and Wood, I.S. 2004. Adipokines: Inflammation and the pleiotropic role of white adipose tissue. *The British Journal of Nutrition* **92**: 347–355.

VanRaden, P.M., Sanders, A.H., Tooker, M.E., Miller, R.H., Norman, H.D., Kuhn, M.T., and Wiggans, G.R. 2004. Development of a national genetic evaluation for cow fertility. *Journal of Dairy Science* **87**: 2285–2292.

Vinsky, M.D., Novak, S., Dixon, W.T., Dyck, M.K., and Foxcroft, G.R. 2006. Nutritional restriction in lactating primiparous sows selectively affects female embryo survival and overall litter development. *Reproduction, Fertility, and Development* **18**: 347–355.

Wade, G.N. and Jones, J.E. 2004. Neuroendocrinology of nutritional infertility. *American Journal of Physiology. Regulatory, Integrative and Comparative Physiology* **287**: R1277–R1296.

Wade, G.N. and Schneider, J.E. 1992. Metabolic fuels and reproduction in female mammals. *Neuroscience and Biobehavioral Reviews* **16**: 235–272.

Webb, R., Garnsworthy, P.C., Gong, J.G., and Armstrong, D.G. 2004. Control of follicular growth: local interactions and nutritional influences. *Journal of Animal Science* **82** (E. Suppl.): E63–E74.

Weigel, K.A. 2006. Impact of genetic selection on female fertility. *Proceedings of the Dairy Cattle Reproductive Council*, DCRC, November 2006.

Zieba, D. A., Szczesna, M., Klocek-Gorck, B., and Williams, G. L. 2008. Leptin as a nutritional signal regulating appetite and reproductive processes in seasonally-breeding ruminants. *Journal of Physiology and Pharmacology* **59**, Suppl 9: 7–18.

Index

Page numbers in *italics* refer to Figures; those in **bold** to Tables.